Gums and Stabilisers for the Food Industry 15

Gums and Stabilisers for the Food Industry 15

Edited by

Peter A. Williams
Centre for Water Soluble Polymers, Glyndwr University, Wrexham, UK

Glyn O. Phillips
Phillips Hydrocolloids Research Ltd, London, UK

RSC Publishing

The proceedings of the 15th Gums and Stabilisers for the Food Industry Conference held on 22nd - 26th June 2009 at Glyndwr University, Wrexham, UK.

Special Publication No. 325

ISBN: 978-1-84755-199-3

A catalogue record for this book is available from the British Library

Published by The Royal Society of Chemistry,
Thomas Graham House, Science Park, Milton Road,
Cambridge CB4 0WF, UK

Registered Charity Number 207890

For further information see our web site at www.rsc.org

Preface

This volume provides the specialised selected papers from the 15th Gums and Stabilisers for the Food Industry Conference which was held in June 2009 at Glyndwr University, Wrexham, UK. It thus becomes the landmark for 30 years of technical and scientific development in the field of food hydrocolloids. The planning of the conference took place in a disastrous economic climate; even banks were failing and going bankrupt. A regular international conference in this field was cancelled during the year and there were many prophets of doom and gloom around. The only acknowledgment we made to the depressing industrial climate was to shorten the conference by one day, pack in the papers in parallel sessions, work later and continue into the afternoon when we would normally have enjoyed either golf, a trip around Wales or a shopping expedition into Chester. This was the advice we obtained from the industrial members of the Organising Committee, paralleling the greater demand being now made upon them by their masters. But I can tell you that it was a relief to see a full lecture theatre facing me when I opened the conference! This volume too shows that the high standard of previous years was more than maintained. So shall we say that all's well that ends well and we invite you now to get with this volume your biennial fix – the publishers, the Royal Society of Chemistry tell us that the demand for these volumes is continually increasing and that they have now become an updating reference to what is current and new in this area of food hydrocolloids.

The emphasis was placed on the technofunctionality and the biofunctionality of hydrocolloids, giving an appropriate emphasis to the manipulative skills of the food scientist and recognising the special part hydrocolloids can play in supporting human health. In this respect the continuation of the Pilnik Memorial Lecture again drew attention to the dominant role which pectin is having in supporting the natural health ingredient movement. As you will see from the first Chapter, the Pilnik Lecture delivered by Dr Ralet was a *tour de force*.

One important feature which deserves mention is the increasingly global nature of the conference now. It was a pleasure to welcome so many from Iran, Iraq, India, Malaysia, Thailand, Japan, Australia, New Zealand, Sudan, Switzerland, Singapore, China, including our old friends from USA, Canada, the countries of Europe and, of course, the home team from the UK. We find the same expansion into these countries from the papers submitted and published in the Food Hydrocolloids journal. Could it be that it is from the non-traditional countries that we will get our new products in the future? Iran, China and Malaysia are now certainly determined to develop their own resources, and Canada are leading the way in supporting efforts to add value to the traditional agricultural products.

The papers on structure, characterisation, rheology/texture and emulsification show the creative nature of the manipulation of existing industrial hydrocolloids. This volume contains papers on food applications which result partly from the New Product Forum which was a new feature of the conference. "Users" are notably reluctant to explain their subject but there are excellent examples in this volume. I was personally fascinated by the novel textures described by Christoph Schmitt in his paper on polysaccharide-protein complexes.

The health and medical areas too are now becoming prominent. Human digestion, satiety and even the modelling of processes occurring in the small intestine illustrate what an important contribution food hydrocolloids can make to controlling the scourge of obesity. Enjoy the volume.

Finally my thanks to Peter Williams for his stewardship of the conference and this volume, Haydn Hughes our "Mr Fix-it" and of course the constant support of the Conference Committee and our faithful Sponsors.

Glyn O. Phillips
Chairman, Gums and Stabilisers Conference Organising Committee

Acknowledgements

The Food Hydrocolloids Trust is indebted to the members of the conference organising committee

Dr P. Boulenguer, Cargill France SAS
Dr M. Capelle, Nestle, France
Professor E. Dickinson, University of Leeds, UK
Professor E. A. Foegeding, North Carolina State University, USA
Dr T. J. Foster, ***(Treasurer),*** University of Nottingham, UK
Mr H. Hughes *(Administrative Secretary),* Glyndwr University
Dr A. Imeson, FMC Corporation
Dr M. Marrs
Dr R.G. Morley, Delphi Consultant Services Inc., USA
Profesoor J.R. Mitchell, University of Nottingham
Professor E. R. Morris, University College Cork, Ireland
Professor V.J. Morris, Institute of Food Research, Norwich
Dr J.C.F. Murray
Professor K. Nishinari, Osaka City University
Professor G.O.Phillips *(Chairman*), Phillips Hydrocolloids Research Ltd
Dr K. Philp, CyberColloids Ltd, Ireland
Dr C. Rolin, CP Kelco, Denmark
Dr C. Schorsch, Danone, France
Dr C. Speirs, CCFRA, UK
Dr G. Sworn, Danisco, France
Dr A. Tziboula-Clarke, FMC Corporation
Professor A.G.J. Voragen, University of Wageningen, The Netherlands
Professor P.A. Williams *(Scientific Secretary*), Glyndwr University

and also acknowledges financial support from

Alfred L. Wolff
Biopolymer Solutions
Campden BRI
Cargill
C P Kelco
Cyber Colloids Ltd
Danisco
Elsevier
EPNOE
Kerry Food and Beverage
FMC
Mars
National Starch and Chemical
Pepsico
Phillips Hydrocolloids Research Centre
Unilever
Woodhead Publishing Ltd

Contents

Technofunctionality

Hydrogels for Medical Applications

Structure and Characterisation

PECTIC OLIGOSACCHARIDES

M.-C. Ralet

INRA, UR1268 Biopolymères Interactions Assemblages, rue de la Géraudière, B.P. 71627, F-44316 Nantes, France.

1 INTRODUCTION

Pectin is a natural constituent of all terrestrial plants. It is one of the major plant cell wall components and probably the most complex macromolecule in nature since it can be composed of 17 different monosaccharides, some of them being esterified by methyl, acetyl or feruloyl groups.[1] Degrading cell wall materials or extracted pectins by purified pectolytic enzymes showed that the different monosaccharides are not randomly distributed along the pectin macromolecule but are concentrated within different pectic structural domains. Over the years, many pectic structural domains have been identified and pectins are currently viewed as multiblock co-biopolymers containing: (i) homogalacturonan (HG), (ii) xylogalacturonan (XGA), (iii) rhamnogalacturonan I backbone (RG-I), encompassing arabinan, galactan and arabinogalactan I and II side-chains, and (iv) rhamnogalacturonan II (RG-II) (Figure 1).[2]

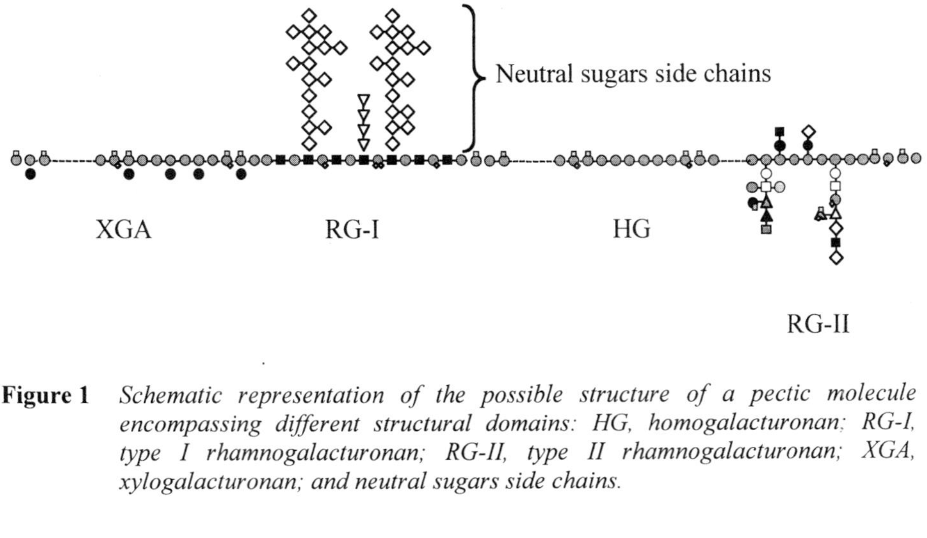

Figure 1 *Schematic representation of the possible structure of a pectic molecule encompassing different structural domains: HG, homogalacturonan; RG-I, type I rhamnogalacturonan; RG-II, type II rhamnogalacturonan; XGA, xylogalacturonan; and neutral sugars side chains.*

The proportion, fine structure and maybe length of each domain vary widely depending on the plant source and the extraction conditions applied. This impinges on the pectin physico-chemical and hydrodynamic properties and hence on the functionality in the specific cellular context as well as in the food application field. Although most plant tissues contain pectin, solely few plant sources are currently used for commercial pectin extraction and commercial pectin is produced almost exclusively from citrus peel or apple pomace.[3]

Since pectins are extremely complex multiblock biopolymers, analysis on extracted pectin whole macromolecules is not sufficient to give an insight into the pectin fine structure. To reveal its structural characteristics, pectin is commonly degraded into oligosaccharides by enzymatic means. After fractionation of the degradation products, different sets of pectic oligosaccharides, which are in the analytical range of several analytical techniques, can be obtained. These pectic oligosaccharides can be used for structural analysis, to search for biological activities, or to generate specific antibodies.

2 METHOD AND RESULTS

2.1 Fine structure of homogalacturonan domains: methylation

Homogalacturonan (HG) is the simplest and the most abundant structural domain. It consists of a linear chain of around 100 (1,4)-linked α-D-galacturonic acid (GalA) units.[4] GalA residues are partly methyl-esterified at C-6 and both the degree of methyl-esterification (DM, number of methylated GalA residues for 100 total GalA residues) and the distribution of methyl-esterified residues have a deep impact on pectin associative properties *via* calcium ions.[5] It has been reported that 7-20 contiguous free GalA residues are required for calcium bridging.[6,7] Pectins thereby exhibit stronger interaction with calcium when the DM is low and when methyl esters are blockwisely distributed.[8]

In the past ten years, a lot of progress has been made in the understanding of methyl esters distribution due to:

(i) the generation of adequate pectic substrates

(ii) the increased availability of pure pectolytic enzymes for deesterification or hydrolysis purposes

(iii) the development of techniques capable of separating methyl-esterified oligogalacturonides

(iv) the development of mass spectrometry for oligosaccharides structural characterization.

Series of model pectins were generated by separately treating the same mother pectin with base, fungal pectin-methyl-esterase (f-PME) and plant-pectin-methyl-esterase (p-PME).[9] Chemical deesterification and deesterification by f-PME produce a random deesterification pattern while the action of p-PME is believed to follow a single chain mechanism deesterifying pectin in a blockwise fashion (Figure 2).[8,10] Model pectins differing solely by their degree and pattern of methyl-esterification were thereby produced.

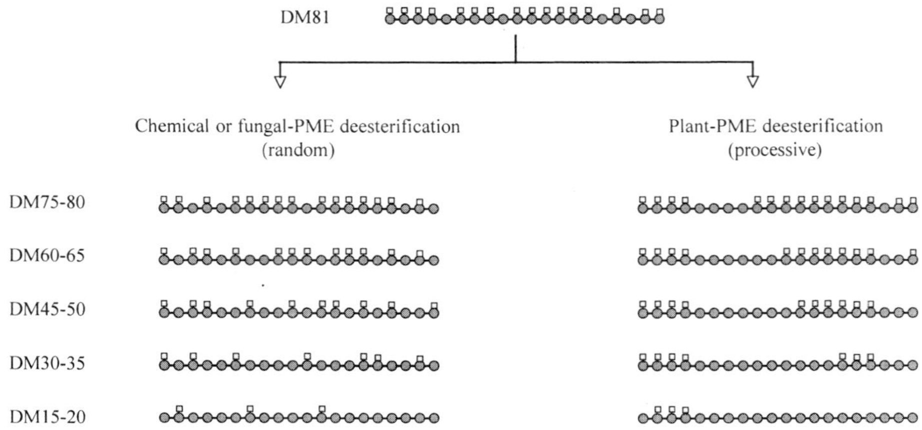

Figure 2 *Model pectins have been generated by chemical or enzymatic (fungal or plant pectin-methyl esterases, PME) deesterification of a very-highly methylated mother pectin. Chemical means and fungal PME lead to a random deesterification while plant-PMEs processive enzymes, generating stretches of demethylated GalA residues.*

One important breakthrough was the development of HPAEC at low pH followed by PAD-detection for the analysis of partly methyl-esterified GalA oligomers.[11] Different commercial pectins were hydrolyzed by a purified endo-polygalacturonase of *Kluyveromyces fragiles*. This enzyme degrades HG backbone only when more than 4 adjacent non-methylesterified GalA residues are present (Figure 3). Complex elution patterns were obtained by HPAEC analysis at pH 5 with a series of non-esterified GalA oligomers of dp 1 to 3 (very small amounts of dp4 were also detected for some samples) and various peaks attributed to partly methylated GalA oligomers on the basis of off-line Matrix-Assisted Laser Desorption/Ionisation Time-of-Flight Mass Spectrometry (MALDI-TOF-MS) analyses.[11] Quantification of non-esterified GalA oligomers by HPAEC at pH5 allowed the authors to introduce the concept of "degree of blockiness, DB" [11] and, more recently, the concept of "absolute degree of blockiness, DB_{abs}" [12], DB being the amount of unesterified monomers, dimers and trimers liberated by the enzyme related to the total amount of non-esterified GalA units in the pectin and DB_{abs} the amount of unesterified monomers, dimers and trimers liberated by the enzyme related to the total amount of GalA units (including methylesterified ones) in the pectin (Figure 3). Limberg *et al.* used a slightly different fingerprinting approach using endo- and exo-polygalacturonases from *Aspergillus niger*.[13] The combined use of these two enzymes allowed quantifying the amount of GalA units located in deesterified blocks (so-called blocksequence, BS, by the authors) by simply measuring the amount of GalA monomers released. Both approaches proved to be powerful in differentiating pectin samples of similar DM but different methyl-esterification patterns.

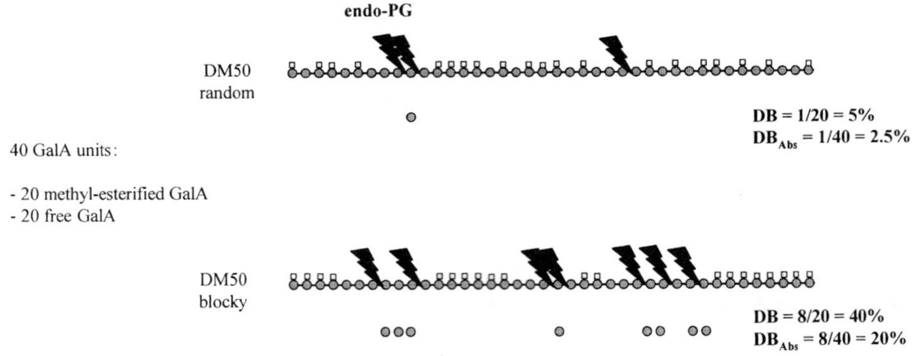

Figure 3 *Schematic representation of enzymatic digestion with endo-polygalacturonase of* Kluyveromyces fragiles *on random and blocky DM50 pectins. Description of the parameters degree of blockiness (DB) and absolute degree of blockiness (DB_{abs}). Adpated from Guillotin et al.(2005).*[12]

Partly methylated oligomers generated by enzymatic digestion of pectin can be efficiently separated by HPAEC at low pH [11,13] and by capillary electrophoresis.[14,15] The various methyl-esterified oligomers can be easily recognized on MALDI-TOF-MS spectra. The development of powerful analyzers with MS/MS or MS^n capability further allowed the application of mass spectrometry techniques to reveal the exact location of the methyl esters within the GalA oligomers.[16] Unfortunately, quantification of the partly methylated oligomers is not yet possible neither by MS, nor by HPAEC or CE due to the lack of adequate standards. Although this part of the information about methyl-ester distribution is still lacking, DB_{abs} [15] and BS (Figure 4) proved to be good indicators of pectin calcium sensitivity

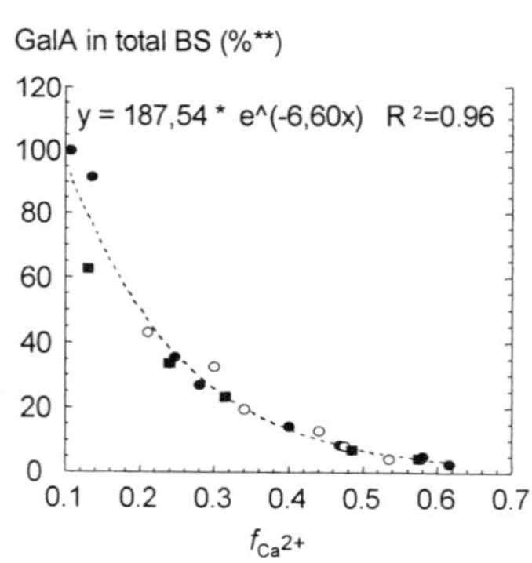

Figure 4 *GalA in blocksequence (values from Limberg et al., 2000)*[13] *as a function of calcium transport coefficient (values from Ralet et al., 2001)*[17] *for base-deesterified (black squares), plant-PME-deesterified (white circles) and f-PME-deesterified (black circles) pectin samples.*

2.2 Fine structure of homogalacturonan domains: acetylation

In some plant species such as sugar beet for example, GalA units in HG regions are partly acetyl-esterified at O-2 or O-3 on top of methyl-esterification at C-6.[5] Acetylation of HG domains strongly alter pectins associative properties *via* calcium ions.[18-22] On the other hand, acetylation has been related to the ability of sugar beet pectin to stabilize emulsions.[23,24] To get a better understanding of the relationship between the acetylation of pectin and their associative or emulsifying properties, more information about the distribution of acetyl groups onto HG domains is needed. As a consequence, sugar beet pectin was degraded by purified pectolytic enzymes and degradation products were separated by anion-exchange chromatography (Figure 5).[25,26]

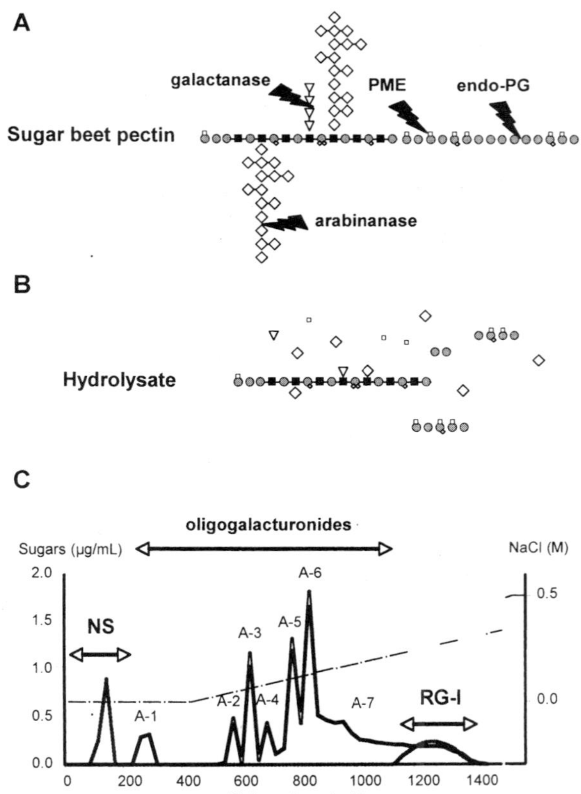

Figure 5 *A. Schematic representation of sugar beet pectin enzymatic digestion with pectolytic enzymes.*
B. Schematic representation of the hydrolysate.
C. Anion-exchange elution pattern (solid line, GalA; dotted line, neutral sugars). NS, neutral sugars; RG-I, rhamnogalacturonan I.

Several fractions corresponding to oligomers of GalA were recovered (A-1 to A-7) (Figure 5) and analysed for their dp, DM and DAc. The quantative recovery with respect to both GalA and acetyl groups was calculated. Next the different fractions were analysed by ESI-IT-MS^n. The electrospray ionisation followed by formation of fragment ions by collision-induced dissociation allows sequencing of oligosaccharides. Carbohydrates undergo two types of fragmentation, those related to cleavage of glycosidic bonds and those related to cross-ring cleavages. The carbohydrate fragmentation nomenclature was introduced by Domon and Costello.[27] The negative ion mode provides the simplest spectra with, for oligogalacturonates, two types of glycosidic cleavage ions arising from the C- and Z-series and a sole specific cross-ring cleavage ion arising from the A-series. The complementary glycosidic cleavage ions of the C- and Z-series that were observed allowed sensitive sequencing of the different oligogalacturonates (Figure 6).[16,25,26] Furthermore, the $^{0,2}A_n$ cross-ring cleavage ions proved to be highly diagnostic ions permitting precise assignment of acetyl groups to O-2 or O-3 of oligogalacturonates (Figure 6).[16,25,26,28]

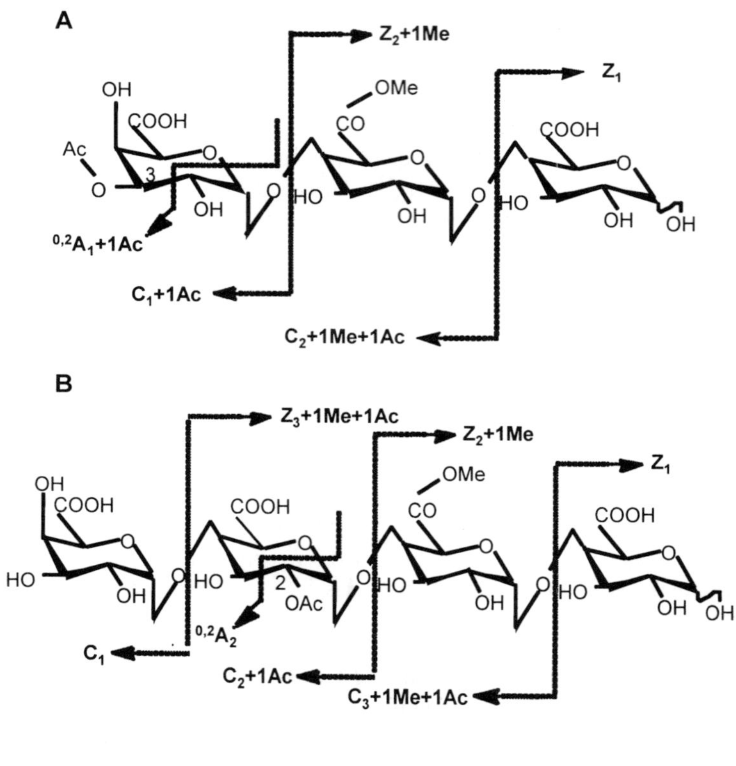

Figure 6 *Observed cleavages at different MS steps for two partly methylated and acetylated GalA oligomers. A, $GalA_3Me_1Ac_1$; B, $GalA_4Me_1Ac_1$. Ions from the C- and Z-series are glycosidic cleavage ions allowing sequencing. Ions from the A-series are cross-ring cleavage ions allowing acetyl group location to O-3 (A) or O-2 (B) of GalA units.*

The quantitative recovery with respect to both GalA and acetyl groups, the structural characterization by ESI-IT-MS of most of the oligogalacturonates and the known chain length of HG domains [4] permitted depiction as a quantitative representation of the different HG-derived oligomers.[25] So a list of oligogalacturonates sequences was obtained but the puzzle of in which order they were connected one to the other in the original sugar beet pectin had still to be solved. This was recently done by using different endo-polygalacturonases known to display variable tolerance towards methyl and acetyl groups.[26] A blockwise distribution of acetyl groups onto HG domains with, on average, four zones of 7-15 contiguous non-acetylated GalA units was evidenced, intermolecular discrepancies being however most probable.

2.3 Fine structure of rhamnogalacturonan I domains

Another major pectic domain is rhamnogalacturonan I (RG-I). The RG-I backbone consists of [2)-α-L-Rhap-(1,4)-α-D-GalpA-(1] repeats,[29-31] the rhamnosyl residues being substituted, mainly at O-4, by variable arabinose- and/or galactose-containing side-chains. RG-I domains are highly acetylated. Although RG-I is a major pectic structural domain, not much is known about its function in the molecular architecture of pectin. Acid-extracted pectins used as food ingredients are indeed very rich in HG domains, which determine to a large extent the physical properties of pectic macromolecules. However, in the plant cell wall context, the fine structure of "hairy regions" (encompassing RG-I domains and neutral sugars side chains) and how it evolves during growth and development is very important to understand how cell wall properties are regulated in different cell types or at different developmental stages. Modified pectic "hairy regions" (MHR) isolated from apple juice produced by the liquefaction process[32] were structurally characterised thanks to the isolation and purification of a rhamnogalacturonase, an enzyme able to cleave galactopyranosyluronic-rhamnopyranosyl linkages.[33,34] The occurrence of pectic "hairy regions" in various plant cell wall materials and their degradability by rhamnogalacturonase was then assessed.[35,36] Different sub-units were identified in MHR among which stubs of the backbone rich in long arabinan side chains that are not degraded by rhamnogalacturonase and subunits corresponding to the rhamnogalacturonase-degradable part of the molecule. NMR analyses allowed evidencing that rhamnogalacturonase oligomers consist of between 4 and 9 sugar units with a backbone of alternating Rha and GalA residues, partly substituted with galactose residues linked to C-4 of the Rha moiety.[36] The arrangement of the different subunits and particularly the distribution of the rhamnogalacturonase-degradable parts over the pectin molecules seems rather diverse depending on the plant source and the extraction conditions. More recently, selective chemical ß-eliminative degradation was successful in liberating RG-I fragments, which were previously not accessible.[37]

The complexity of pectin structure, and especially of RG-I domains, provides a multiplicity of structural epitopes and much effort has been devoted to the generation of diverse antibodies with the aim of identifying the different pectin structures in cell and development contexts.[38,39] Although antibodies against arabinan and (arabino)-galactan side-chains have been developed and are available,[40-43] well-characterised antibodies against the RG-I backbone have not been described yet. With the aim of producing antibodies against RG-I backbone, we recently took advantage of seed mucilage. Mucilage is released from myxospermous seeds upon imbibition, and, in *Arabidopsis thaliana*, consists of a water-soluble outer layer and an adherent inner layer, the water-soluble outer layer consisting of unbranched RG-I.[44] Water-soluble *Arabidopsis thaliana* seed mucilage was therefore degraded by a rhamnogalacturonase for the production of unbranched RG-I oligomers of various dp. Oligomers were separated by size-exclusion chromatography and fractions corresponding to dp 6-10 were pooled. Analysis by ESI-IT-MS^n showed that oligomers had an alternating sequence of Rha and GalA units and always encompassed a

GalA residue at the reducing end. These oligomers were used for neoglycoprotein preparation by reductive amination (Figure 7).

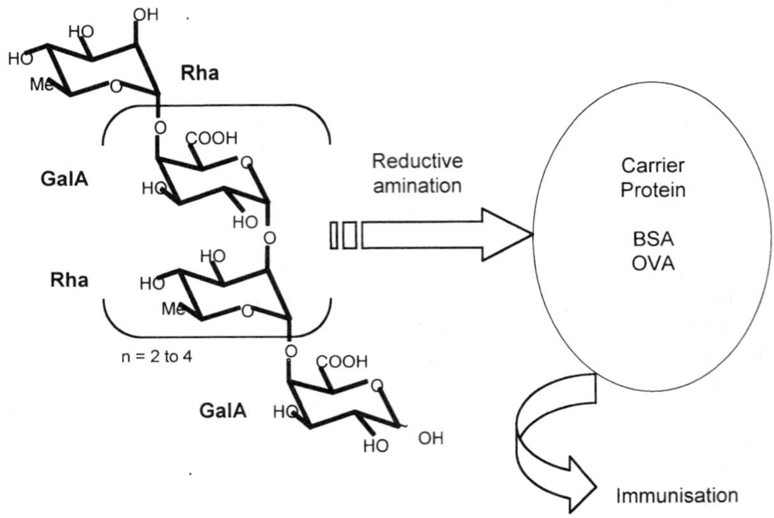

Figure 7 *Preparation of a neoglycoprotein immunogen, in which oligomers of RG-I are coupled to bovine serum albumin (BSA) or ovalbumin (OVA) through reductive amination.*

Several monoclonal antibodies were generated and characterised; they are currently used to detect RG-I in cell walls in a range of plant species at different developmental stages and will thereby help to understand the biosynthesis and the role of this peculiar pectic domain.

CONCLUSION

Structural analysis of pectin is a big challenge due to the complexity of this multi-block co-biopolymer. Revealing fine structure of cell-wall polysaccharides is crucial for understanding the fine tuning of polysaccharides in the cell wall context and how this fine tuning impacts on polymer interaction properties, cell wall architecture and hence cell wall mechanical properties.

References

1 B.L. Ridley, M.A. O'Neill and D. Mohnen, *Phytochemistry*, 2001, **57**, 929.
2 H.A. Schols and A.G.J. Voragen, in *Pectins and their Manipulation*, eds. G.B. Seymour and J.P. Knox, Blackwell Publishing-CRC Press, Oxford, 2002, ch. 1, p. 1.
3 C. Rolin, in *Pectins and their Manipulation*, eds. G.B. Seymour and J.P. Knox, Blackwell Publishing-CRC Press, Oxford, 2002, ch. 8, p. 222.

4 J.-F. Thibault, C.M.G.C. Renard, M.A.V. Axelos, P. Roger and M.-J. Crépeau, *Carbohydr. Res.*, 1993, **238**, 271.
5 A.G.J. Voragen, W. Pilnik, J.-F. Thibault, M.A.V. Axelos, C.M.G.C. Renard, in *Food Polysaccharides and their Applications*, ed. A.M. Stephen, Marcel Dekker, New-York, 1995, ch. 10, p. 287.
6 R. Kohn, *Pure Appl. Chem.*, **42**, 371.
7 D.A. Powell, E.R. Morris, M.J. Gidley and D.A. Rees, *J. Mol. Biol.*, 1982, **155**, 517.
8 J.-F. Thibault and M. Rinaudo, *Biopolymers*, 1985, **24**, 2131.
9 G. Limberg, R. Körner, H.C. Buchholt, T.M.I.E. Christensen, P. Roepstorff and J.D. Mikkelsen, *Carbohydr. Res.*, 2000, **327**, 293.
10 R. Kohn, O. Markovic and E. Machova, *Collect. Czech. Chem. Commun.*, 1983, **48**, 790.
11 P.J.H. Daas, K. Meyer-Hansen, H.A. Schols, G.A. De Ruiter and A.G.J. Voragen, *Carbohydr. Res.*, 1999, **318**, 135.
12 S.E. Guillotin, E.J. Bakx, P. Boulenguer, J. Mazoyer, H.A. Schols and A.G.J. Voragen, *Carbohydr. Polym.*, 2005, **60**, 391.
13 G. Limberg, R. Körner, H.C. Buchholt, T.M.I.E. Christensen, P. Roepstorff and J.D. Mikkelsen, *Carbohydr. Res.*, 2000, **327**, 321.
14 M.A.K. Williams, T.J. Foster and H.A. Schols, *J. Agric. Food Chem.*, 2003, **51**, 1777.
15 A. Ström, P. Ribelles, L. Lundin, I. Norton, E.R. Morris and M.A.K. Williams, *Biomacromolecules*, 2007, **8**, 2668.
16 M.-C. Ralet, P. Lerouge and B. Quéméner, *Carbohydr. Res.*, 2009, in press.
17 M.-C. Ralet, V. Dronnet, H.C. Buchholt and J.-F. Thibault, *Carbohydr. Res.*, 2001, **336**, 117.
18 E.L. Pippen, R.M. McCready and H.S. Owens, *J. Am. Chem. Soc.*, 1950, **72**, 813.
19 R. Kohn and I. Furda, *Collect. Czech. Chem. Commun.*, 1968, **33**, 2217.
20 R. Kohn and A. Malovikova, *Collect. Czech. Chem. Commun.*, 1978, **43**, 1709.
21 C.M.G.C. Renard and M. Jarvis, *Carbohydr. Polym.*, 1999, **39**, 209.
22 M.-C. Ralet, M.-J. Crépeau, H.C. Buchholt and J.-F. Thibault, *Biochem. Eng. J.*, 2003, **16**, 191.
23 J. Leroux, V. Langendorff, G. Schick, V. Vaishnav and J. Mazoyer, *Food Hydrocolloids*, 2003, **17**, 455.
24 C.K. Siew, P.A. Williams, S.W. Cui and Q. Wang, *J. Agric. Food Chem.*, 2008, **56**, 8111.
25 M.-C. Ralet, J.C. Cabrera, E. Bonnin, B. Quéméner, P. Hellìn and J.-F. Thibault, *Phytochemistry*, 2005, **66**, 1832.
26 M.-C. Ralet, M.-J. Crépeau and E. Bonnin, *Phytochemistry*, 2008, **69**, 1903.
27 B. Domon and C.E. Costello, *Glycoconjugate J.*, 1988, **5**, 397.
28 B. Quéméner, J.C. Cabrera Pino, M.-C. Ralet, E. Bonnin and J.-F. Thibault, *J. Mass Spectrom.,* 2003, **38**, 641.
29 M. McNeil, A.G. Darvill and P. Albersheim, *Plant Physiol.*, 1980, **66**, 1128.
30 M. McNeil, A.G. Darvill, S.C. Fry and P. Albersheim, *Ann. Rev. Biochem.*, 1984, **53**, 625.
31 P. Albersheim, A.G. Darvill, M.A. O'Neill, H.A. Schols and A.G.J. Voragen, in *Pectins and Pectinases*, eds. J. Visser and A.G.J. Voragen, A. G. J., Elsevier Science B.V., Amsterdam, 1996, p. 47.
32 H.A. Schols, M.A. Posthumus and A.G.J. Voragen, *Carbohydr. Res.*, 1990, **206**, 117.
33 H.A. Schols, C.C.J.M. Geraeds, M.J.F. Searle-van Leeuwen, F.J.M. Kormelink and A.G.J. Voragen, *Carbohydr. Res.*, 1990, **206**, 105.

34 I.J. Colquhoun, G.A. de Ruiter, H.A. Schols and A.G.J. Voragen, *Carbohydr. Res.*, 1990, **206**, 131.

35 H.A. Schols and A.G.J. Voragen, *Carbohydr. Res.*, 1994, **256**, 83.

36 H.A. Schols, A.G.J. Voragen and I.J. Colquhoun, *Carbohydr. Res.*, 1994, **256**, 97.

37 G.-J. Coenen, Ph.D. Thesis, University of Wageningen (NL), 2007. ISBN 978-90-8504-779-7.

38 W.G.T. Willats, L. McCartney, W. Mackie and J.P. Knox, *Plant Mol. Biol.*, 2001, **47**, 9.

39 J.P. Knox, *Curr. Opin. Plant Biol.*, 2008, **11**, 308.

40 L. Jones, G.B. Seymour and J.P. Knox, *Plant Physiol.*, 1997, **113**, 1405.

41 W.G.T. Willats, S.E. Marcus and J.P. Knox, Carbohydr. Res., 1998, 308, 149.

42 W.G.T. Willats, Steele-King C.G., S.E. Marcus and J.P. Knox, *Plant J.*, 1999, **20**, 619.

43 Y. Verhertbruggen, S.E. Marcus, A. Haeger, H.A. Schols, B.V. McCleary, L. McKee, H.J. Gilbert and J.P. Knox, *Plant J.*, 2009, **59**, 413.

44 A. Macquet, M.-C. Ralet, J. Kronenberger, A. Marion-Poll, H.M. North, *Plant Cell Physiol.*, 2007, **48**, 984.

TAILORING PECTIN WITH SPECIFIC SHAPE, COMPOSITION AND ESTERIFICATION PATTERN

C. Rolin[†][1], I.B. Chrestensen[1,2,3], K.M. Hansen[1], J. Staunstrup[1], S. Sørensen[1]

[1] CP Kelco ApS, Ved Banen 16, DK-4623 Lille Skensved, Denmark
[2] Novozymes A/S, Krogshøjvej 36, DK-2880 Bagsværd, Denmark)
[3] Plant Biochemistry Laboratory, Department of Plant Biology, Faculty of Life Sciences, University of Copenhagen, Thorvaldsensvej 40, DK-1871 Frederiksberg C, Denmark

ABSTRACT

Citrus fruit is industrially utilised for co-producing fruit juice, aromatic oils, and peel for pectin production. Citrus peel is today by far the most important source of commercial pectin. Amongst the citrus fruits, orange is the more abundantly available source, and it is today used for a considerable amount of pectin. Anyway, since production of pectin from orange peel involves significant technological challenges, lime and lemon have a longer tradition in the pectin industry, and together they still account for the largest production volume. Commercial pectin from lime and lemon peel is characteristic by having a very high proportion of polymeric material that in analysis upon exhaustive hydrolysis yields galacturonic acid. The molecules are on average elongated having large $[\eta]/M_w$-proportion in comparison with pectin from at least some other studied sources.

The traditional pectin manufacturing process comprises incubation of the materials at elevated temperature and low pH in order to increase the yield of pectin and also reduce the pectin degree of methyl esterification (DM) to the level that suits commercial uses. The de-esterification with acid is believed to remove esterified methyl groups randomly, so that the substitution pattern approaches statistical randomness. In published literature it is sometimes postulated that the above traits, high galacturonan content and elongated shape, are a consequence of the massive acid use. Both traits are highly desirable for the traditional technofunctional uses for pectin, viz. gels and protein stabilisation, but the putative beneficial physiological effects that have been published for diverse, mostly non-commercial, pectin preparations may perhaps depend on other characteristics. Thus, it is interesting to study pectin samples of higher non-galacturonan content, or different shape, or different pattern of methyl ester substitution.

The use of enzymes for de-esterifying pectin has been commercially implemented in the pectin industry, and it is nowadays possible to produce citrus pectin of low DM without exposing the materials to excessive acidity. It has thereby become possible to assess to how large extent the typical traits of present-day commercial pectin are predetermined by the protopectin of the raw material, versus the alternative, that they are a consequence of manufacturing conditions. The present paper addresses this question as well as the differences in DM substitution pattern between acid de-esterified and enzyme de-esterified pectin.

1. INTRODUCTION

Pectin is a generic term for a diversity of complex materials that can be isolated from plants. When characterising these materials in scientific studies they are typically broken down in some limited and controlled way, and the thus treated materials are fractionated and characterised individually. In this way it has been established that pectin molecules contain characteristically different regions, for example homogalacturonan (HG) and at least two kinds of rhamnogalacturonans, RG-I and RG-II, which, in turn, possess galacturonan-containing regions and other regions that sometimes are referred to as "the neutral sugars". The designation "neutral" is because these polysaccharides in contrast to the galacturonans do not possess (acidic) carboxyl groups.

The HG that can be isolated by carefully controlled hydrolysis of pectin from citrus peel is a linear polymer of anhydrogalacturonic acid with Degree of Polymerization (DP) in the order of magnitude ≈ 100 [1, 2] and relatively stiff[3], while the rhamnogalacturonans are branched and more flexible[3]. Some of the carboxyl groups of the HG are esterified with methanol; the percentage of methyl-esterified carboxyl groups per total pectin carboxyl groups is known as the Degree of Methyl esterification (DM). This esterification of the HG varies with the location of the pectin within the plant cell and with the age, stage of development, and function of the cell. The HG is highly esterified when it is inserted into the plant cell wall, and the esterification may then later be lowered by native plant pectin esterase in order to obtain diverse functionalities such as water-binding, cell to cell adhesion, and cell differentiation in general[4, 5]. The enzyme works in a processive fashion creating blocks of consecutive non-esterified anhydrogalacturonan repeating units within the HG. Pectin with DM ≈ 0, termed *pectate*, forms insoluble salts with Ca^{++}, and if esterase-created blocks that locally resemble pectate are present in a pectin sample, solutions of that sample will thicken or gel upon addition of calcium salts[6, 7]. The proportion of non-esterified pectin units that inhabit blocks with at least three other non-esterified units, expressed as percent of total non-esterified units, is known as the Degree of Blockiness (DB)[8].

It is still being discussed whether only two HG segments can attach to one RG domain, so the molecule essentially is one long molecule of HG segments with interposing RG-domains at which comparatively short side-chains of neutral sugars may be found (so-called "hairy regions"[9]). -Or if, alternatively, three or more HG segments can attach to the same RG-domain, so that the molecules in their entirety are more branched[10].

Further details about pectin architecture may be found in review articles[11, 12].

Pectin has been defined in context with its commercial uses as an ingredient in food and pharmaceutical products[13-16]. These definitions describe pectin as a substance that has been isolated from a botanical raw material as the soluble part that ensues from aqueous extraction, and which if subjected to complete hydrolysis yields a certain minimum percentage of galacturonic acid (%GA), for example 65% w/w relative to described analytical procedures[13]. The material that must exist in the intact plant tissue and that upon exposure to the conditions of the extraction process becomes dissolved pectin is sometimes referred to as *protopectin*[17]. This terminology will also be used in the present manuscript in order to distinguish between the soluble isolates that lend themselves to characterization versus the originating material with which the options for direct study are more limited.

Commercial food uses for pectins (we use the plural form pectins as short for different qualities of pectin) mostly are for providing the foods with a specific rheology, typically a gelled or thickened texture of jams, jellies, fruit flavoured chewy candies and the like. Pectin is also used for stabilising suspensions and emulsions, most importantly drinkable yoghurt (or, in general, low-pH protein-containing drinks in which pectin keeps insoluble protein uniformly suspended), and also flavour oil emulsions (pectin prevents the oil drops from turning into a macroscopic oil phase). All of these "technofunctional" uses for pectin can be described by the models and techniques of physical chemistry; however, the structure-function relationships have almost entirely been discussed with reference to the structure of the HG. Thus, it is described how the DM influences the setting temperature of gels[17], which are typically prepared as hot liquid preparations and filled into jars and then solidify as the jar cools down. Further, it is known how the blockiness determines the strength of calcium-mediated gels[7], the stability of yoghurt drinks[18], and the ability to thicken yoghurt-serum simulating liquids[19]. The thickening or gelling with Ca^{++} has been described as specific binding of the Ca^{++} between two or more helical pectate segments[20, 21], so-called egg-box gelation[22, 23], Ca^{++} being the "egg's" and the pectate segments being the "box". It is also described that the efficacy of pectin for gelling is positively correlated to its apparent average molecular size[24-26]. As for the rhamnogalacturonan parts with their side-chains it has been described that they may entangle in solution when exceeding a certain concentration[27, 28]. It has been speculated that the rhamnogalacturonan regimes may sterically act to interrupt junction zones between HG parts in different molecules, so that these junction zones become of limited size. However, except from the above-mentioned entanglement, there has been no direct demonstration of what the amount of neutral sugars and their composition mean for pectin technofunctionality.

The biological effects ("biofunctionalities") of pectin are a diverse group of phenomena that cannot be rationalised with one common explanation for them all. In some cases the scientific understanding is more in its infancy. However, in stark contrast to the technofunctionality, it is not always reasonable to assume that the effect of pectin depends solely upon the structure of the HG part. In particular the use of modified pectin for inhibiting cancer metastasis has been demonstrated to relate to the galactan content of the pectin[29]. Likewise, the immune-system-stimulating effect of certain pectins depends upon RG-I presence[30, 31].

The present manuscript describes opportunities for tailoring pectin in pilot plant scale and full commercial production scale. One element is the overall shape of pectins from various fruit, and the proportion of galacturonic acid versus neutral sugars in the same pectins. Another element is the DM and the DB of the HG part. Also we will address the discussion of overall pectin architecture based on our point of view that extraction conditions can modify protopectin in alternative ways and thereby release different pectins.

2. MATERIALS AND METHODS

2.1 Pectin samples (except for experiments with "protopectinase")

The pectin samples were extracted commercially, or in pilot scale under conditions emulating commercial conditions, as described in general reviews[32, 33]. The 236 samples (133 from lemon peel, 67 from lime peel, 24 from orange peel, and 12 from the residue of sugar beet after sugar extraction) were selected or produced during the years from 2002 through 2009 for various research purposes. With all citrus pectins the acidity, extraction temperature and extraction duration were chosen according to commercial practice so the ensuing pectin samples got a DM that is typical for acidified milk stabilizing grades, in most cases between 70 and 77. Based on the prior impact of the esterases that are native to the citrus fruit, commercial dried peel for pectin production may generally differ with respect to the DB that the extracted pectins will acquire; the pectins of the present study differed for this reason in spite of having rather similar DM.

Some of the pectin samples were further de-esterified by exposing the liquid pectin extract either to pectin methyl esterase from *Aspergillus* (PME-A), or pectin methyl esterase from papaya (PME-P). In one experimental series pectin was de-esterified without enzymes by incubating the acidified pectin solution at pH $\approx 1.00 \pm 0.15$, 40°C ± 5°C and aliquots of different DM were withdrawn after different durations.

2.2 "Protopectinase"-aided pectin extraction

For "protopectinase"[34] experiments dried citrus peel was initially suspended in acidified water at pH 3.5 and 80°C for the purpose of inactivating native enzymes. It was then diluted with cold water and cooled to 42°C and gently stirred at pH 3.5 for two hours with about 3 mg Rohament PL (AB Enzymes GmbH, Postfach 101239, 64212 Darmstadt, Germany) per g dried peel, or without enzyme (blind experiment). Finally the mixtures were heated to 75°C and filtered through diatomaceous earth and further worked up as per usual commercial practices[32, 33]. Note: Rohament PL is almost pure polygalacturonase (PG); the "L" signifies "liquid" (not "lyase"). For PG treatment after extraction, the filtered pectin extract was treated with about 13 mg Rohament PL per g pectin for two hours at pH 3.5, 42°C.

2.3 Measurements of %GA, DM and DB

%GA and DM were determined by titration of an acid/alcohol-purified sample and by titration of the corresponding saponified sample as described by FAO/JECFA[13]. The DB was determined as per Daas et al[8].

2.4 Measurement of M_w and[η]

Molecular parameters were determined by size exclusion chromatography followed by three detectors: refractive index, right angle light scattering, and viscosity. Equipment: Viscotek Tri-Sec instrument (Viscotek Corp., Houston, Texas) with 3 TSK Columns GMPWXL (Supelco). Elüent: 0.3 M Li-acetate buffer pH 4.8mM, temperature 40°C. The molecular parameters were deduced from the detector signals by the software supplied by Viscotek, assuming dn/dc for pectin = 0.145

3. RESULTS

The M_w and the [η] in the buffer solution used for chromatography are shown for all of the studied pectin samples in figure 1. The observation is that the sugar beet pectins and the orange pectins tend to form each their distinct flock of data-points, while the lemon and lime pectins occur together and have the comparatively lowest Mw/[η]-proportions. The same materials are shown in figures 2a and 2b (2b being an enlarged view of a part of 2a) with %GA as x-axis and the Mw/[η]-proportion as y-axis, and it may be observed that the materials form species-dependent flocks in the same way. Also shown in figures 1 and 2b are the effects of using polygalacturonase enzyme (PG) during extraction (shown with a dotted line connecting the two compared samples with/without PG treatment) and after extraction in the solution that ensues after removing insoluble materials by filtration (solid connecting line). In the experiments, the use of PG during extraction substantially increased the M_w/[η]-proportion and decreased the %GA, while the use of PG after extraction changed these parameters only to a much smaller extent.

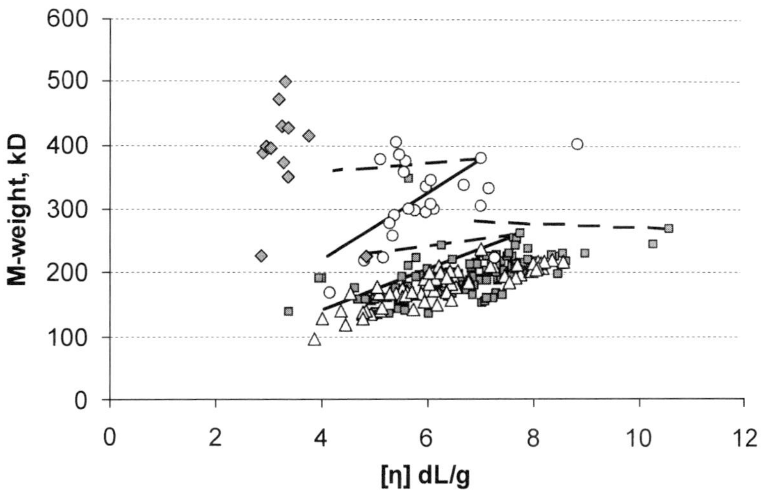

Figure 1. *M_w and [η] for diverse botanical sources. ■ = lemon, Δ = lime, ○ = orange, ♦ = beet. Dotted lines connect samples with PG during extraction (left) to their blind experiments without PG (right). Solid lines connect samples with/without PG used in the filtered liquid (see text)*

In order to describe the effect of acid use upon the content of neutral polysaccharides we compared acid de-esterification at pH ≈ 1 (dotted line) to Aspergillus enzyme de-esterification (PME-A, solid line) of the same parent pectin, see figure 3. Also shown in figure 3 are those other pectin samples from lime and lemon that had not been treated with

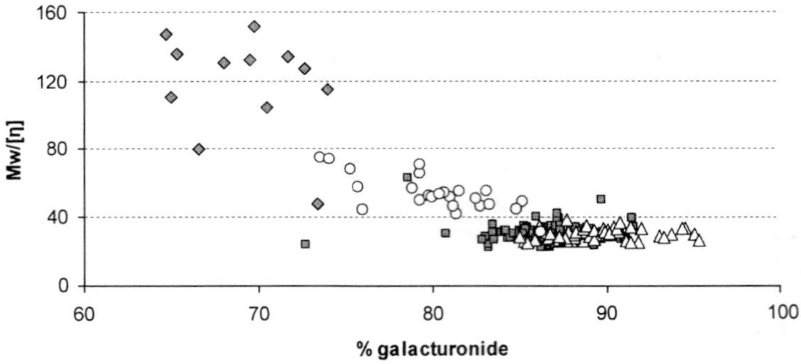

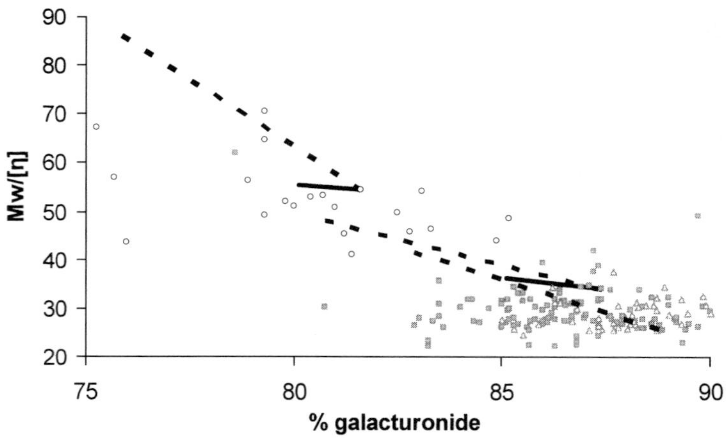

Figures 2a (above) and 2b *$M_w/[\eta]$ for diverse botanical sources. ■ = lemon, Δ = lime, ○ = orange, ♦ = beet. Dotted lines connect samples with PG during extraction (left) to their blind experiments without PG (right). Solid lines connect samples with/without PG used in the filtered liquid (see text)*

Aspergillus esterase (i.e., we have omitted sugar beet pectins from figure 3, and we have omitted all other PME-A treated citrus pectin samples than those represented by the solid line). The x-axis of figure 3 is the DM outside of blocks calculated as follows:

$$DM_{\text{outside blocks}} = 100\cdot DM/(DM+(1-DB/100)\cdot(100-DM))$$

With the deliberate omissions made in figure 3, we believe it is safe to use the DM outside of blocks as a measure of the severity of the total acid impact during processing (except with the PME-A de-esterified series marked up with a solid line): the further to the left on

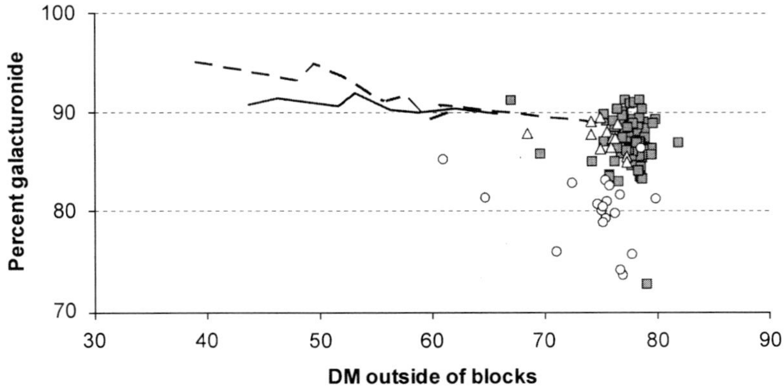

Figure 3. *Percent GA in function of DM by extraction. ■ = lemon, Δ = lime, ○ = orange, ♦ = beet. The dotted line connects samples de-esterified with acid, while the solid line connects samples de-esterified with PME-A.*

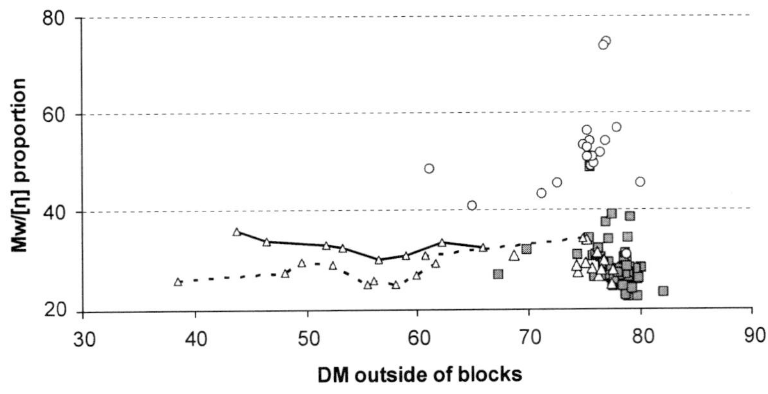

Figure 4. *Molecular shape in function of DM for diverse raw materials. ■ = lemon, Δ = lime, ○ = orange, ♦ = beet. The dotted line connects samples de-esterified with acid, while the solid line connects samples de-esterified with PME-A.*

the x-axis of figure 3, the more methyl ester has been removed by acid hydrolysis. The y-axis is the %GA; the closer to 100 (and higher up the y-axis), the less room for neutral sugar. When comparing the solid and the dotted lines we observe that acid use indeed does decrease the content of neutral sugars. However, the effect is moderate in comparison to the characteristic difference between typical orange versus typical lime and lemon. The same data is shown with the Mw/[η]-proportion as y-axis in figure 4 in which we again find a moderate tendency to more elongated molecules at high acid use.

Figure 5a shows the DB as y-axis against DM as x-axis for all of the studied pectin samples with the four raw material species each represented by their symbol. The DB tends to be low with all sugar beet pectin samples; this is perhaps an artefact because the HG of sugar beet pectin is known to possess considerable acetyl-esterification of O-2 and O-3 hydroxyl groups and this, in turn, may inhibit the PG-enzyme that is used as an analytical tool (segments are registered as blocks according to whether the enzyme can hydrolyse them). Other than that, figure 5a reveals no pronounced raw material related differences.

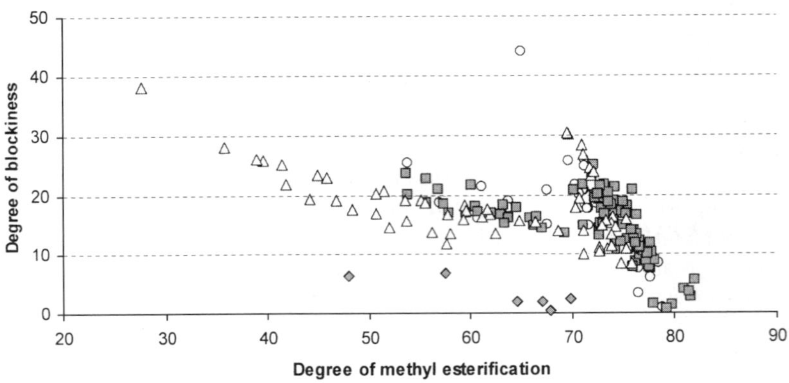

Figure 5a *DE and DB for diverse pectin samples of different botanical origin. ■ = lemon, Δ = lime, ○ = orange, ♦ = beet.*

Figures 5b, 5c, and 5d show the same data as 5a, but with various samples highlighted. In figure 5b series of samples from a common ancestor de-esterified to various DM with Aspergillus esterase (PME-A) have been highlighted. In figure 5c series of samples from a common ancestor de-esterified to various DM with plant pectin esterase (PME-P) have been highlighted, and in figure 5d acid de-esterification at pH ≈ 1 (dotted line) to Aspergillus enzyme de-esterification (PME-A, solid line) of the same parent pectin have been highlighted for comparison. The direct observation from diagrams 5b, 5c, and 5d is that de-esterification with plant pectin esterase creates much blockiness, while the same reduction in DM with acid or Aspergillus enzyme creates less blockiness.

4. DISCUSSION

As demonstrated by the results, by selection of appropriate raw materials and processing conditions pectin can be tailored for having M_w up to at least about 400 kD and [η] in the eluent for SEC analysis up to at least 10 dL/g. The proportions between the above figures

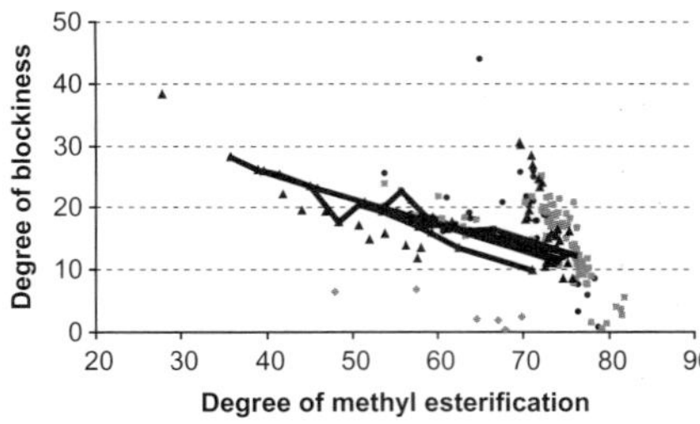

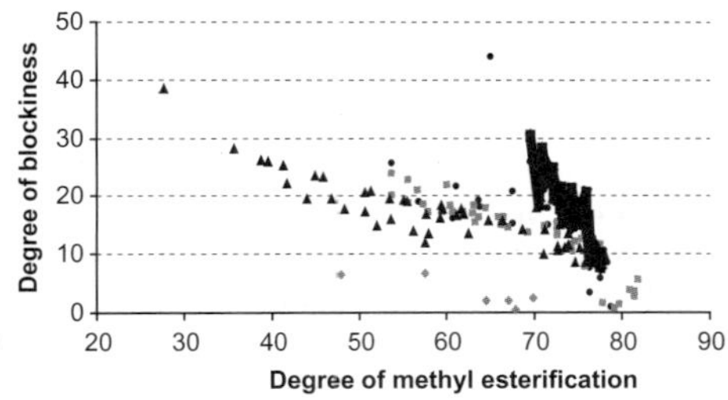

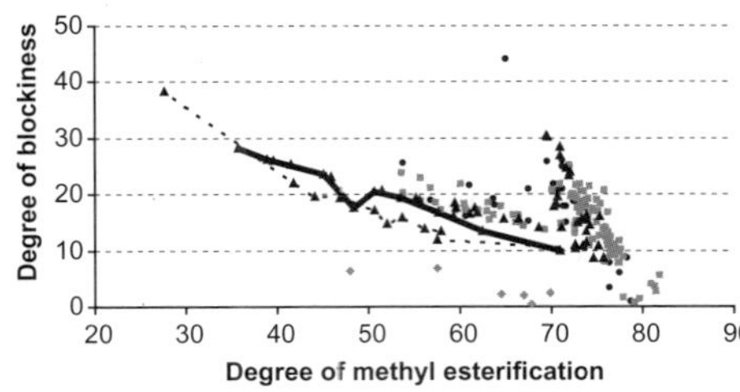

Figures 5b, 5c, and 5d. *Same as figure 5a but with specific de-esterifying treatments highlighted, PME-A (5b, top left), PME-P (5c top right), and PME-A (solid line) and acid (dotted line) compared (5d, bottom right)*

can vary from about 30 to about 140 kD•g/dL. The %GA by titration analysis can be varied from about 65 to about 95. Apparently, in our data, the $M_w/[\eta]$ proportion and the %GA are mutually correlated and correlated to the raw material species. At face value, sugar beet pectin seems to have high $M_w/[\eta]$ and the relatively lowest %GA, while lime and lemon pectins have the lowest $M_w/[\eta]$ and the highest %GA, and orange pectin is in between. The %GA can be slightly increased by exposing the pectin or the protopectin to acidity, but within the processing conditions that are commercially relevant for controlling the DM of pectin, the effect of the raw material species is more important.

The pectin DM can be varied from a maximum close to 80 (by our means of extracting pectin) and to any lower value. This can be accomplished with acid, or with enzymes in which in the latter case the de-esterification hardly changes the %GA or the molecular shape. The DB can be as low as about 5 at DM = 80 and can be changed by de-esterification so it increases steeply with the DM-change if plant pectin esterase is used, or it increases only moderately if microbial pectin esterase is used. These observations corroborate previous findings [6, 35].

In the experiments where polygalacturonase was used during the extraction process, the extracted pectin became richer in neutral polysaccharides and got a higher $M_w/[\eta]$ proportion, i.e. a more compact molecular shape. When, in contrast, polygalacturonase was used with a solution of already extracted pectin, then it reduced M_w as well as $[\eta]$ but without changing the $M_w/[\eta]$ proportion or the %GA very much.

The authors suggest interpreting from the above that the $M_w/[\eta]$ proportion and the %GA are characteristics of the raw material species when the pectin is extracted traditionally with warm dilute acid at process conditions within the ranges tested in our study. A similar difference between pectins from orange and lime was found with microwave-assisted extraction experiments at USDA [36, 37]. On the other hand, other values for the $M_w/[\eta]$

proportion and the %GA will result from using polygalacturonase for extracting the pectin. These interpretations are subject to two provisos. The observed difference between the citrus fruit species needs not, or not entirely, be a characteristic of the species *per se*, it is likely also influenced by the ripeness at which juice producers prefer to process the individual species and they thereby become available as raw materials for pectin production. As for the protopectinase (polygalacturonase) use the proviso obviously is the small number of experiments and lack of assessment of experimental variance.

Outside of the present study, in general, the structure of pectin obviously depends upon its botanical origin as well as the conditions during its isolation. The major part of the protopectin of the intact (unprocessed) raw materials does not spontaneously exude as dissolved pectin upon submerging these raw materials in water. For releasing pectin, *something* has to be broken, and we have no data, nor a clear definition, for determining whether this "something" is part of the protopectin. Alternatively, we could imagine, the "something" being a structure of something else than pectin, merely holding the pectin (which in such case would be identical to the protopectin) like a parcel holding its interior contents, or by entanglement like a padlock holding a chain. The progression of pectin extraction from lime and orange albedo with increasing time of microwave heating was described by Fishman, Chau, Hoagland and Hotchkiss[37] as the degradation of (super-molecular) network structures to clusters and branched structures and eventually to rod-like structures. Molecular structures as evidenced by size-exclusion chromatography with viscosity and light scattering detection went from high apparent M_w with high apparent $M_w/[\eta]$ proportion and low Mark-Houwink exponents to smaller but less compact molecules, i.e. lower M_w with lower $M_w/[\eta]$ proportion and higher Mark-Houwink exponent. The above authors interpreted this as the progressive dissociation of non-covalent aggregation, but (in the present author's interpretation) it still evidences irreversible molecular changes since the compact shapes were not observed to revert, and it could be interpreted as a conversion of protopectin to a different material, viz. pectin. Further, the data strongly suggests that the M_w inferred from light scattering detection may reflect the combined size of aggregated molecules rather than the size of covalently connected entities - a proviso of obvious relevance also for our present study. Other authors[38] found that a combined presence of a protease and a cellulase could facilitate the release of reasonably high-M_w soluble pectin from plant materials even though both enzyme preparations were having only minor side-activity towards pectin. Subject to the definition that protein and cellulose are not part of protopectin this could be taken as an argument for the opposite point of view, viz. that dissolved pectin can be structurally practically identical to what it was in the intact plant.

The present authors propose, which is clearly arguable in view of the latter cited investigation, that as long as we have no solid proof for a "parcel" or "padlock" picture, nor a definition of which plant-cell materials that are part of the pectin, then it may generally be more fruitful to think of pectin extraction as a modification of insoluble protopectin, rather than the mere liberation of putatively already existing pectin. Further, we propose that since the dominant purpose of commercial pectin production, and even the ultimate purpose of many investigations in academia, is technofunctional uses that ideally crave high $[\eta]$ of the pectin, then extraction conditions have typically been chosen so that they balance the desire of a reasonably high yield of isolated pectin with the conflicting desire of minimising molecular degradation. Investigators would likely choose such compromise based upon "common sense" even if they were not explicitly and consciously balancing yield versus decay. In view of typical average degrees of polymerisation for pectin in the order of magnitude 1000, the number of linkages broken by such a balanced extraction protocol is small compared to the size of the ensuing molecules as well as to the size of individual pectin regions. In turn, large remains of the regions that were snapped by the extraction must occur at terminal positions of the extracted molecules. Yet further, we propose that the few linkages that snap for releasing pectin are, depending upon the vulnerability of pectin structures to the conditions of the extraction process, not always of the same kind or placed in the same protopectin regions.

We suggest explaining the observations of USDA as well as our own present investigation, viz. that orange pectin seemingly has a more compact molecular structure than lemon and lime pectins, as differences in vulnerability of the HG towards the acidic extraction conditions. Pectin of low DM hydrolyses more readily under acidic conditions than pectin of high DM[39, 40]. Lime and lemon have inherently very high DM, and the HG region is therefore still undamaged when the extraction conditions have modified non-HG materials to such extent that pectin molecules can disentangle and exude. Orange pectin, in contrast, has owing to the native pectin esterase inherently lower DM, so there is slightly greater probability for snapping HG segments whereby the pectin molecule can be released before non-HG materials have been decayed as much as with lime/lemon extraction. Thereby, the molecules become of more compact average shape. If predominantly calcium-reactive esterase-created blocks snap by the pectin extraction, the remains of these blocks will occur at the ends of molecules. In turn, this may explain the otherwise mysterious finding that endo-polygalacturonase can abate orange pectin calcium-sensitivity considerably without causing a concomitant and devastating loss of gel functionality[41]. In the present investigation we have now further demonstrated this idea by replacing the acid effect during extraction with polygalacturonase, which just like the acid attacks the non-esterified HG segments, just in a more selective way and to a more pronounced extent, and we have observed a similar consequence for molecular shape. Further support for the point of view that the methyl-esterification and the substitution pattern are important for the vulnerability of HG during extraction was brought in an investigation of prospective commercial utilisation of pectin from *Aloe vera*[42]. Since the DM of Aloe pectin is intrinsically low, extraction at pH of 7.0-8.5 was preferred over the acidic conditions that are traditional with commercial pectin production.

5. CONCLUSION

We have demonstrated that pectin to large extent can be tailored for diverse DM, DB, and [η]. We have shown that the %GA and $M_w/[\eta]$ depend upon the raw material species, but within the citrus pectins we cannot infer from the present data to which extent this dependence reflects differences in the typical ripeness at harvesting rather than a true species characteristic. We speculate that pectin molecules obtain different shapes based upon the vulnerability of protopectin HG towards the extraction conditions, and we have demonstrated this by extracting with PG-enzyme as "protopectinase", but more data is needed for substantiating this speculation.

LITERATURE

1. Thibault, J.-F., C.M.G.C. Renard, M.A.V. Axelos, P. Roger & M.-J. Crepeau, *Carbohydr. Res.* **238** (1993) 271-280

2. Yapo, B.M., P. Lerouge, J.-F. Thibault & M.-C. Ralet, *Carbohydr. Polym.* **69** (2007) 426-435
3. Ralet, M.-C., M.-J. Crepeau, J. Lefebvre, G. Mouille, H. Hoefte & J.-F. Thibault, *Biomacromolecules* **9** (2008) 1454-1460
4. Mohnen, D., *Curr. Opin. Plant Biol.* **11** (2008) 266-277
5. Peaucelle, A., R. Louvet, J.N. Johansen, H. Hofte, P. Laufs, J. Pelloux & G. Mouille, *Curr. Biol.* **18** (2008) 1943-1948.
6. Kohn, R., I. Furda & Z. Kopec, Collect. Czech. Chem. Commun. **33** (1968) 264-269
7. Ström, A., P. Ribelles, L. Lundin, I. Norton, E. Morris & M.A.K. Williams, *Biomacromolecules* **8** (2007) 2668-2674
8. Daas, P.J.H., K. Meyer-Hansen, H.A. Schols, G.A. DeRuiter & A.G.J. Voragen, *Carbohydr. Res.* **318** (1999) 135-145
9. De Vries, J.A., F.M. Rombouts, A.G.J. Voragen & W. Pilnik, *Carbohydrd. Polym.* **2** (1982) 25-33
10. Vincken, J.-P., H.A. Schols, R.J.F.J. Oomen, M.C. McCann, P. Ulvskov, A.G.J. Vorangen & R.G.F. Visser, *Plant Physiol.* **132** (2003) 1781-1789.
11. Schols, H.A., G.-J. Coenen & A.G.J. Voragen, "Revealing pectin's structure" in H.A. Schols, R.G.F. Visser & A.G.J. Voragen (eds.) *Pectins and Pectinases* (book) pp. 19-34, Wageningen Academic Publishers, The Netherlands, **2009**.
12. Schols, H.A. & A.G.J. Voragen, "The chemical structure of pectin" pp. 1 - 29 in Seymour, G.B. & J.P. Knox (eds.) *Pectins and their Manipulation* (book), Blackwell Publishing, Oxford UK (2002)
13. *FAO JECFA Monographs* **4** (68th JECFA, 2007)
14. EU Directive of 27th August 2008 (2008/84/EC + updates)
15. Food Chemicals Codex 6th edition 2008 (incl. supplements)
16. United States Pharmacopeia USP 31 (incl. supplements)
17. Kertesz, Z.I., The Pectic Substances (book), Interscience Publishers, New York (1951)
18. Glahn, P.E. & C. Rolin, "Casein-pectin interactions in sour milk beverages" in *Food Ingredients Europe Conference Proceedings*, **1994**, Process Press Europe, Maarsen, The Netherlands.
19. Meyer-Hansen, K., H.L. Nielsen & C. Rolin, "Molecular shape and functionality of HM pectin" in H.A. Schols, R.G.F. Visser & A.G.J. Voragen (eds.) *Pectins and Pectinases* (book) pp. 57-70, Wageningen Academic Publishers, The Netherlands, **2009**.
20. Braccini, I. & S. Pérez, *Biomacromolecules* **2** (2001) 1089-1096
21. Fang, Y., S. Al-Assaf, G.O. Phillips, K. Nishinari, T. Funami & P.A. Williams, *Carbohydr. Polym.* **72** (2008) 334-341.
22. Morris, E.R., D.A. Rees & D. Thom, *Carbohydr. Res.* **66** (1978) 145-154.
23. Grant, G.T., E.R. Morris, D.A. Rees, P.J.C. Smith & D. Thom, *FEBS Lett.* **32** (1973) 195-198.
24. Christensen, P.E. *Food Res.* **19** (1954)163-172
25. Kim, W.J., V.N.M. Rao & C.J.B. Smit, *J. Food Sci.* **43** (1978) 572-575
26. Van Deventer-Schriemer, W.H. & W. Pilnik, *Acta Alimentaria* **16** (1987) 143-153.
27. Hwang, J. & J.L. Kokini, *Carbohydr. Polym.* **19** (1992) 41-50
28. Hwang, J.-K. & J.L. Kokini, *J. Korean Soc. Food Nutr.* **24** (1995) 389-395
29. Gunning, A.P., R.J.M. Bongaerts & V.J. Morris, *FASEB Journal* **23** (2009) 415-424.

30. Westereng, B., T.E. Michaelsen, A.B. Samuelsen & S.H. Knutsen, *Carbohydr. Polym.* **72** (2008) 32-42
31. Westereng, B., G.-J. Coenen, T.E. Michaelsen, A.G.J. Voragen, A.B. Samuelsen, H.A. Schols & S.H. Knutsen, *Molec. Nutr. Food Res.* **53** (2009) 780-789
32. Brejnholt, S., "Pectin" pp. 237-262 in A. Imeson (ed.) *Food Stabilisers, Thickeners and Gelling Agents* (book), Blackwell Publishing Ltd., Oxford, England (2009)
33. Rolin, C. "Commercial pectin preparations" pp. 222-241 in G.B. Seymour & J.P. Knox (eds.) Pectins and their Manipulation (book) Blacwell Publishing Ltd., Oxford, England (2002)
34. Sakai, T., T. Sakamoto, J. Hallaert & E.J. Vandamme, Advan. Appl. Microbiol. **39** (1993) 213-294
35. Ishii, S., K. Kiho, S. Sugiyama & H. Sugimoto, *J. Food Sci.* **44** (1979) 611-614
36. Fishman, M.L., H.K. Chau, D.R. Coffin & A.T. Hotchkiss, "A comparison of lime and orange pectin which were rapidly extracted from albedo" , pp. 107-122 in F. Voragen, H. Schols & R. Visser (eds.) Advances in Pectin and Pectinase Research (book), Kluwer Academic Publishers, The Netherlands, **2003**.
37. Fishman, M.L., H.K. Chau, P.D. Hoagland & A.T. Hotchkiss, *Food Hydrocol.* **20** (2006) 1170-1177
38. Zykwinska, A., M.-H. Boiffard, H. Kontkanen, J. Buchert, J.-F. Thibault & E. Bonnin, *J. Agric. Food Chem.* **56** (2008) 8926-8935.
39. Fraeye, I., A. De Roeck, T. Duvetter, I. Verlent, M. Hendrickx & A. Van Loey, *Food Chem.* **105** (2007) 555-563
40. Krall, S.M. & R.F. McFeeters, *J. Agric. Food Chem.* **46** (1998) 1311-1315
41. Rolin, C., "Calcium sensitivity of high ester citrus pectins" pp. 413-422 in G.O. Phillips, P.A. Williams & D.J. Wedlock (eds.) *Gums and Stabilisers for the Food Industry 7*, IRL Press, Oxford, England (1994)
42. Ni, Y., K.M. Yates & R. Zarzycki, *US patent* **5,929,051** (1999)

FLAXSEED GUM FROM FLAXSEED HULLS: EXTRACTION, FRACTIONATION, PHYSICOCHEMICAL AND FUNCTIONAL CHARACTERISATION

K.Y. Qian[1,2], S.W. Cui[2], Y. Wu[2] and H.D. Goff[1*]

[1]University of Guelph, Department of Food Science, Guelph, ON, Canada, N1G 2W1
[2]Agriculture and Agri-Food Canada, Guelph Food Research Centre, 93 Stone Rd. W., Guelph, ON, Canada, N1G 5C9.

*Corresponding author; email: dgoff@uoguelph.ca

1. Introduction

The inclusion of soluble dietary fibre in the diet results in important health benefits.[1-3] However, fibre consumption in the diet has to be at a significant concentration to be effective[4]. Soluble fibre, as a thickener or gelling agent, often does not only have nutritional functionality, but also has processing functionality. It is a significant challenge to incorporate fibre in food products in amounts nutritionally significant without resulting in over-texture, high viscosity or phase separation.

Flaxseed gum is reported to be composed of a neutral and an acidic fraction, with some proteinaceous contaminants.[5-7] The neutral fraction is mainly composed of a high molecular weight arabinoxylan, which is attributed to the pseudoplastic flow behavior of crude flaxseed gum; whereas the acidic fraction is mainly composed of a pectin-like rhamnogalacturonan of low molecular weight with low viscosity and Newtonian-like flow behavior[6,7]. This provides flaxseed gum and its fractions, especially the acidic fraction, a potential source to increase dietary fibre content in foods and drinks.

We have chosen to examine the physical functional properties of flaxseed gum also because of its significant and incremental production in Canada. As an oil crop, the yield of flaxseed or linseed (*Linum usitatissimum*) in the 2005-2006 marketing year increased to 1.16 metric tons, consisting 39% percent of the world production.[8] Flaxseeds contain approximately 45% oil, of which 60% is omega-3 fatty acids.[8,9] A portion of the seed, around 37%, is soluble fibre-rich hull, which can be produced by a patented dehulling process[10]. When soaked in water, flaxseed hulls release a mucilaginous material (gum) easily. Therefore, the current industrial extraction and fractionation process of gum starts with flaxseed hulls. However, the properties of flaxseed gum investigated in previous studies were extracted from whole seed[5-7].

The objective of this work was to extract the gum from flaxseed hulls, instead of the whole seed, and fractionate it into its two main components (acidic and neutral) using ion exchange chromatography and compare in detail the chemical (neutral sugar, uronic acid and protein contents), physical (molecular weight and its heat stability, viscosity, viscoelasticity) and functional (surface tension) characteristics of the soluble gum and its two fractions.

2. Materials and Methods.

Flaxseed hull (variety Bethune) was supplied by Natunola Health Biosciences (Winchester, Ontario, Canada).

2.1. Extraction of gum

Soaking flaxseed with a water to seed ratio of 13 at 85 to 95°C and pH from 6.5 to 7.0 for 3 h was found as the optimum condition to extract the flaxseed gum with better yield, higher apparent viscosity, and lower protein content [5]. However, Mazza and Biliaderis [11] also reported that although a better yield can be obtained at high temperatures, the solutions showed browning and contained higher protein content. Considering the browning may affect the properties of the polysaccharides, in the present work, the extraction of the gum was conducted at room temperature. A flow chart of extraction is shown in Fig. 1.

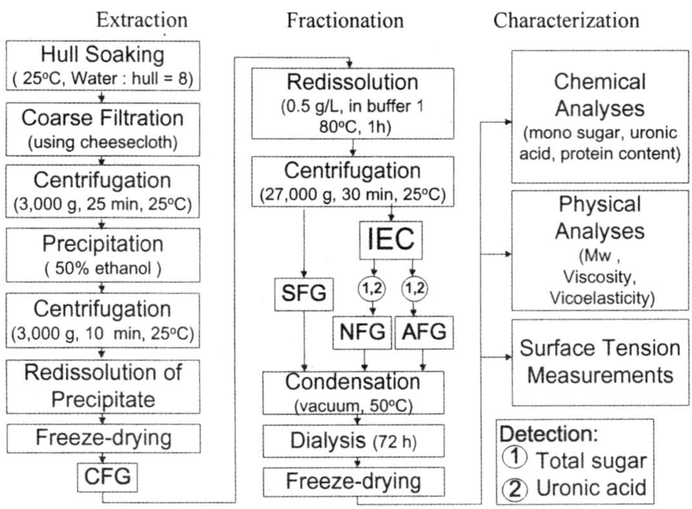

Fig.1. Flow chart of extraction, fractionation and characterization
(IEC: ion exchange chromatography; CFG and SFG: crude and soluble flaxseed gums; NFG and AFG: neutral and acidic fraction gums; buffer 1: 20 mM Tris/HCl, pH 8.)

A batch of 500 g flaxseed hull was soaked in 4 L of distilled water overnight under gentle stirring. A coarse filtration followed using cheesecloth to separate the hull. The filtered mucilage was collected and centrifuged at 3,000 g and 25°C for 25 min, using a centrifuge (Beckman Coulter, Mississauga, Ontario) to eliminate large insoluble substances. The gum was then precipitated by adding one volume of 100% ethanol to the dispersions. The precipitate was recovered by centrifugation at 3,000 g for 10 min, redissolved in distilled water at about 4-5% (w/v), and freeze-dried.

2.2. Fractionation of gum

The extracted flaxseed gum can be fractionated into two fractions: a neutral fraction and an acidic fraction using ion exchange chromatography.[6,12] The procedure of fractionation followed that of Warrand et al.[12] with minor modification: the washing solution of 0.1 M NaOH was replaced by 2 M NaCl and 1M NaOH. A flow chart of fractionation procedure is given in Fig. 1. For each batch of fractionation, 0.5 g crude flaxseed gum (CFG) was dissolved in 1 L buffer 1 (20 mM Tris/HCl, pH 8) at 80°C for 1 h, centrifuged at 27,000 g, 25°C for 30 min to remove any insolubles. The supernatant (denoted as soluble flaxseed gum, SFG) was loaded onto the pre-equilibrated column (XK 50 Column, GE Healthcare, packed with 900 mL of Q-Sephrose fast flow as matrix) at a flow rate of 10 mL/min and flushed continuously with 2 L of buffer 1. Between 25 and 150 min after the sample was loaded the eluate was collected (denoted as neutral fraction gum, NFG). The sugar content of NFG was monitored by the phenol-sulphuric acid method. The column was then washed with 1 L of buffer 2 (20 mM Tris/HCl + 1 M NaCl, pH 8). Between 60 and 110 min after buffer 2 was loaded, the eluate (denoted as acidic fraction gum, AFG) was collected. The column was successively washed with 1 L of 2 M NaCl, 0.1 M NaOH and 1 M NaOH solution, and then re-equilibrated with 2-3 L of buffer 1 for next use. The acidic and neutral fractions were then concentrated using a rotary vacuum evaporator at 50°C, dialyzed (MW cut off 3.5 kDa, Fisher Scientific) against distilled water at 25°C for 72 h, and freeze-dried.

2.3. Chemical analyses

Monosaccharide analyses of SFG, NFG and AFG were performed on a high performance anion exchange chromatography system (Dionex, DX500 Sunnyvale, CA, USA) coupled with a pulsed amperometric detector (HPAEC-PAD).[13] The hydrolysis procedure followed that of Wu et al.[14].

Uronic acid analyses were conducted using the *m*-hydroxyphenyl colorimetric method [15].

Protein analyses were conducted using the NA2100 Nitrogen and Protein Analyzer (Thermo Quest, Milan, Italy). The protein content (%) was obtained by multiplying the nitrogen content (%) by 6.25.

All chemical analyses were completed in triplicate.

2.4. Physical analyses

The molecular weight distribution and heat stability of SFG, NFG and AFG were conducted using a HPSEC system consisting of an HPLC system with degasser, autosampler, UV detector (Thermofinnigan, Mississauga, ON, Canada), and refractive index detectors (Optilab Rex, Santa Barbara, CA, USA). To eliminate baseline variations, the mobile phase (100 mM sodium nitrate + 5 mM Sodium azide) was degassed and filtered through 0.2 μm filters (Millipore, Fisher Scientific). Before analysis, the gums were dissolved in the mobile phase at 1 g/L and heated at 80 °C for 1 h under gentle stirring. Samples were filtered through disposable 0.45 μm filters (Millipore, Fisher Scientific) before injection, and aliquots (100 μL) of samples were injected onto the columns (Shodex OHpak SB-806M HQ, 7.8 mm ID×300 mm) and eluted at a flow rate of 0.6 ml/min for 60 min. Pullulan 800 (MW 788,000), 400 (MW 404,000), 100 (MW 112,000), 50 (MW 47,300), 20 (MW 22,800) were used as molecular weight markers to calibrate the column.

Both steady shear and oscillation measurements of flaxseed gum and its fractions were carried out on a strain-controlled rheometer (ARES, TA Instruments, New Castle, DE, USA) using parallel-plate geometry (50 mm diameter, 1 mm gap size). The viscosity of sample solutions of six concentrations (1, 5, 10, 15, 20, 25 g/L) was measured at five different temperatures from 4 to 50°C in shear rates ranging from 0.01 to 100 s^{-1}. Viscoelastic properties of samples were measured at 20 g/L and 25 °C.

2.5 Surface tension

The surface tension of the air-water interface was measured by the Du Nouy ring method using the Fisher Surface Tensiomat (model 21, Fisher Scientific, Nepean, ON).[16] SFG and its two fractions, NFG and AFG, were dissolved in 20 mL of distilled water at various concentrations (0.01%, 0.05%, 0.10%, 0.25% and 0.50% (w/v)), heated at 80°C for 1 h under gentle stirring and then held for more than 2 h before measurements. The surface tension (dynes/cm) of distilled water and gum solutions were measured every 5 min for at least 6 times at 25°C.

3. Results and discussion

3.1. Extraction and fractionation of gum

The yields of crude, soluble gum and its fractions are shown in Table 1. The yield of CFG extracted from the hulls was 12.67%. A higher yield of 18% was reported in a previous study.[10] The lower yield may be caused by the incomplete extraction of gum from the hulls and the precipitation with 50% (v/v) alcohol in which small molecules of polysaccharides could not be precipitated. To remove any insoluble substance before fractionation with the ion exchange column, the CFG solution (0.5 g/L) was centrifuged at 27,000 g for 30 min, during which about 23% of the crude gum was lost to produce a purified SFG. The recovery rate of the chromatographic fractionation was about 50 %:

the yield of neutral and acidic fraction was 23.18 % and 27.33 %, respectively. Almost 50 % loss of SFG was probably due to an insufficient unbinding process while using buffer 2. This result implies a sufficient cleaning procedure is necessary to remove any contaminants remaining in the column between batches of fractionation.

Table 1. Yield and chemical components of flaxseed gum and its fractions (%)

	CFG	SFG	NFG	AFG
Yield	12.67	76.71	23.18	27.33
Protein	12.39 ± 0.90	11.81 ± 0.98	-*	8.09 ± 0.42
Uronic Acid	22.52 ± 2.26	22.98 ± 0.15	1.79 ± 1.00	38.72 ± 1.10

(CFG and SFG: crude and soluble flaxseed gums; NFG and AFG: neutral and acidic fraction gums;
- *: no protein was detected; all data were calculated on a dry basis.)

3.2. Chemical composition

The protein and uronic acid contents of flaxseed gum and its fractions are shown in Table 1. The NFG was free of protein, while AFG contained less protein than the SFG. This implied that a portion of protein remained in the column during fractionation and was finally removed by the column cleaning procedure. This portion of protein might have bound with some polysaccharides and thus resulted in a low recovery rate described in section 3.1.

The uronic acid content in AFG was about 20 fold of that in NFG and twice as much as that in CFG and SFG (Table 1). This suggested a relatively pure neutral fraction was efficiently separated from the soluble fraction.

The monosaccharide composition of the flaxseed gum and its fractions is shown in Table 2. The composition of the CFG and SFG are almost identical suggesting no significant loss or change of any soluble composition during the 72 h dialysis of SFG.

Table 2. Neutral monosaccharide in flaxseed gum and its fractions (%)

	CFG	SFG	NFG	AFG
Xylose	42.34 ± 0.59	44.11 ± 0.54	70.05 ± 0.34	12.03 ± 1.04
Arabinose	13.73 ± 0.15	13.46 ± 0.24	20.00 ± 0.24	4.92 ± 0.23
Rhamnose	22.98 ± 0.35	22.38 ± 0.17	-*	51.83 ± 1.85
Galactose	17.95 ± 0.17	17.70 ± 0.10	6.52 ± 0.09	30.07 ± 0.77
Glucose	2.99 ± 0.15	2.35 ± 0.24	3.43 ± 0.09	1.14 ± 0.19

(CFG and SFG: crude and soluble flaxseed gums; NFG and AFG: neutral and acidic fraction gums;
- *: no rhamnose was detected; data are given as "mean±SD", n=3)

In addition, NFG contained mainly xylose (70.05 %), followed by arabinose (20.00 %), galactose (6.52 %) and glucose (3.43 %), but no rhamnose. In contrast, AFG contained mainly rhamnose (51.83 %) and galactose (30.07 %), with small portions of xylose (12.03 %), arabinose (4.92 %) and glucose (1.14 %). The presence of xylose and arabinose in AFG was probably due to contamination from NFG. These data are comparable with the results in previous research which showed that the neutral fraction is mainly composed of a high molecular weight arabinoxylan whereas the acidic fraction is mainly composed of a pectin-like rhamnogalacturonan of low molecular weight.[6]

3.3. Physical characterization

The molecular weight distribution and heat stability of soluble gum (SFG) and its fractions are shown in Fig. 2. NFG consisted of one fraction with high molecular weight of approximately 1,470 kDa. In contrast, AFG consisted mainly of 3 portions of molecular weight: 1,510 kDa, 341 kDa and 6.6 kDa, respectively, as shown in Fig.2.

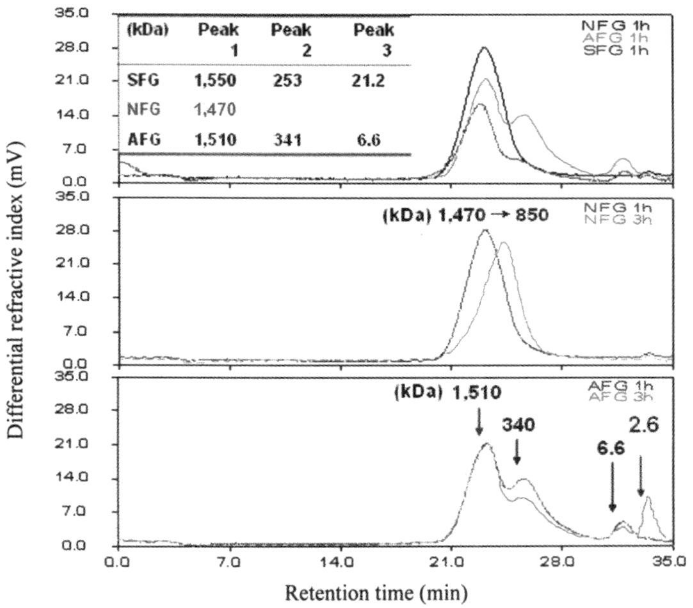

Fig. 2. Molecular weight distribution and heat stability of SFG, NFG and AFG (SFG: soluble flaxseed gums; NFG and AFG: neutral and acidic fraction gums; the column temperature was 40 °C)

When heating time at 80°C was increased from 1 h to 3 h, the molecular weight of NFG decreased from 1.47 to 0.85 kDa. As for AFG, a significant reduction in intermediate molecules (340 kDa) and the appearance of additional small molecules (2.6 kDa) also implied the degradation of this fraction.

The UV detective voltage response in Fig. 3 depicts the association of protein and polysaccharides. No protein was detected in NFG, which was consistent with the results of chemical analyses. The differences in elution time between protein and polysaccharides in both SFG and AFG suggested these two components were not covalently linked.

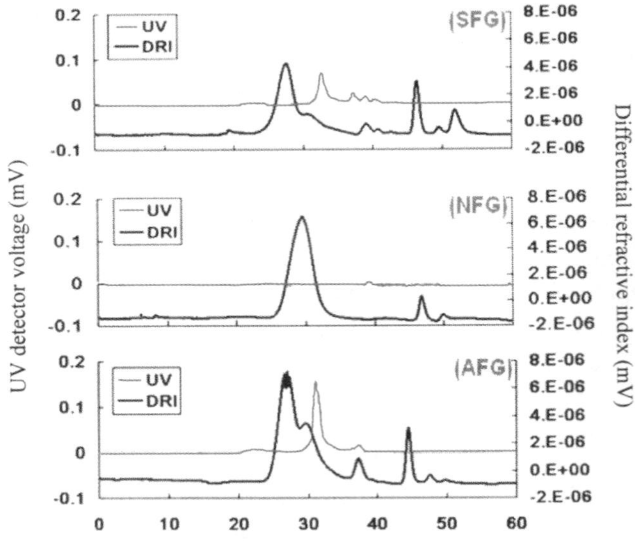

Fig.3. Protein distribution (UV line) of SFG, NFG and AFG

(SFG: soluble flaxseed gums; NFG and AFG: neutral and acidic fraction gums; the column temperature was 25 °C.)

3.4. Rheological properties

The steady shear flow curves of solutions of flaxseed gum and its fractions at 25°C are shown in Fig. 4. Overall, the solutions were very low in viscosity compared to other hydrocolloid gums at equivalent concentration. The soluble gum (SFG) exhibited pseudoplastic flow behavior over a wide range of shear rate with the increase of concentration. The neutral fraction (NFG) showed slight pseudoplastic flow behavior at concentrations above 1.0%. The steady shear flow curve of AFG was more Newtonian-like at all concentrations examined. The pseudoplastic behavior of both SFG and NFG can be attributed to the high molecular weight arabinoxylan component, that is the neutral polymer. At concentrations above 1.0%, the overall ranking of apparent viscosity was SFG>NFG>AFG, suggesting the NFG dominates the SFG viscosity at these concentrations. But at concentrations below 0.5%, the overall ranking of apparent viscosity changed to SFG>AFG>NFG, suggesting the AFG dominates the viscosity at low concentrations. This may result from differences in critical concentration between the two fractions, with the NFG contributing more to molecular entanglement than the AFG at higher concentrations.

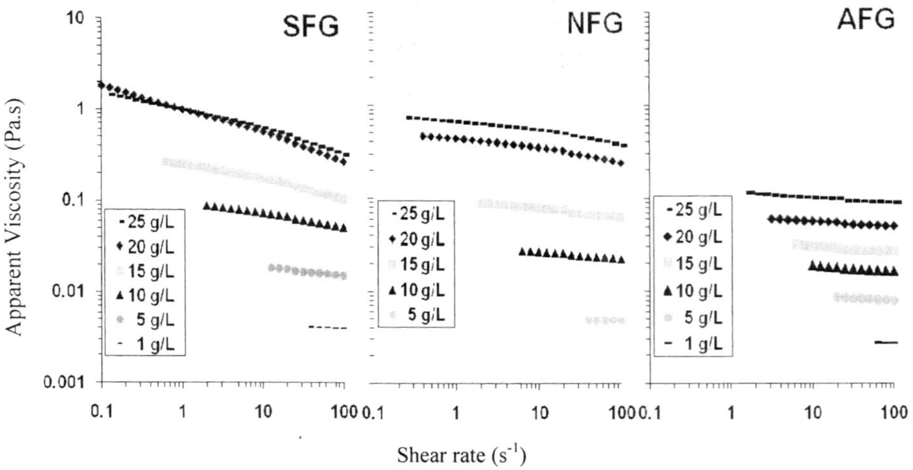

Fig.4. The steady shear flow curves of the solutions of SFG, NFG and AFG
(SFG: soluble flaxseed gums; NFG and AFG: neutral and acidic fraction gums.)

The pseudoplastic region of the flow curve was fitted to the power law model, $\eta=\kappa\gamma^{n-1}$, where K is the consistency index of the fluid and n is the flow behavior index. The concentration dependence of the flow behavior index (n) is shown in Fig.5. With the increase of concentration, the flow behavior index n of SFG and NFG decreased from 1 to about 0.7 and 0.9, respectively, whereas n of AFG only decreased slightly from 1 to 0.94 and was not sensitive to the concentration.

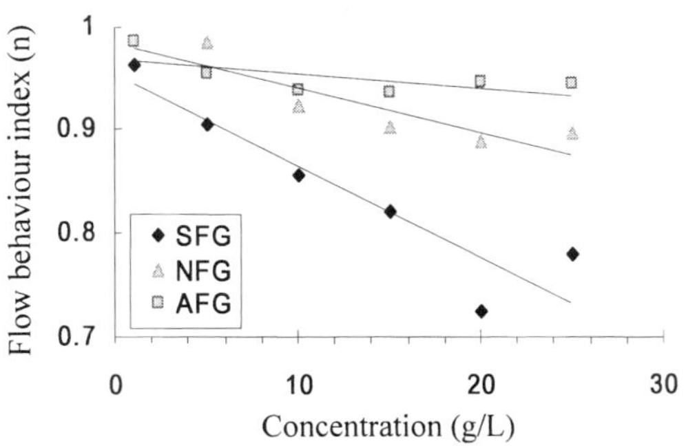

Fig.5. Concentration dependence of flow behaviour index (n) at 25°C
(SFG: soluble flaxseed gums; NFG and AFG: neutral and acidic fraction gums.)

Dispersions of soluble gum (SFG) and its fractions (NFG and AFG) all showed liquid-like behavior, as the loss modulus (G″) exceeded the storage modulus (G') over the entire frequency range investigated (Fig. 6). These results were quite different from that of Cui et al.[6], in which dispersions of crude, dialyzed flaxseed gums and neutral fractions all exhibited "weak gel" properties. This might be due to the different extraction procedures and sources of raw materials.

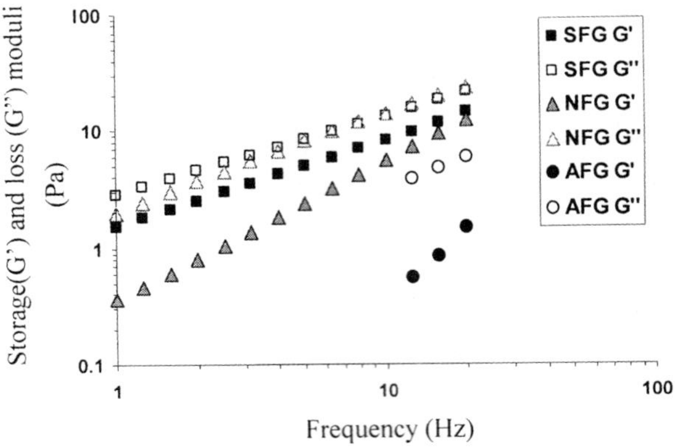

Fig. 6. Frequency dependence of storage (G') and loss (G") moduli of 20 g/L solutions of SFG,NFG and AFG

(SFG: soluble flaxseed gums; NFG and AFG: neutral and acidic fraction gums.)

3.5. Surface tension

The surface tension of water decreased rapidly with the increasing concentration of soluble gum (SFG) and its acidic fraction (AFG) and then slowly leveled off to about 53 dynes/cm at concentrations above 0.06%. The surface tension of NFG decreased relatively more slowly with the increase of concentration and then leveled off to 57.8 ± 1.0 Dynes/cm at concentrations above 0.1% (Fig. 7). The similarity of capability for surface tension reduction of SFG and AFG could be partly attributed to their similar protein content (11.81 ± 0.98 % in SFG and 8.09 ± 0.42 % in AFG). The relatively lower surface activity observed for NFG may also be partly due to the lack of protein.

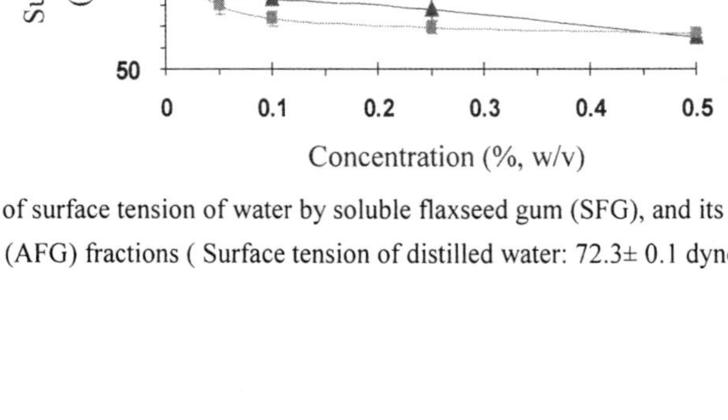

Fig. 7. Reduction of surface tension of water by soluble flaxseed gum (SFG), and its neutral (NFG) and acidic (AFG) fractions (Surface tension of distilled water: 72.3± 0.1 dynes/cm)

4. Conclusions and perspectives

Two fractions (neutral and acidic) of flaxseed gum from hulls were effectively separated using ion exchange chromatography. The neutral fraction, consisting of only one polymer with high molecular mass (1,550 kDa), was free of protein and contained <2% of uronic acid. The acidic fraction mainly consisted of polymers with different molecular weight and had associated 8% of protein, but not covalently linked. Both fractions were stable at 80°C for 1 h, however, extended heating resulted in degradation of the polymers. Surface tension decreased slightly with the increase of concentration, more so in the whole gum and acidic fraction which may due to the presence of the protein fraction. Viscosity of the dispersions was dominated by the acidic fraction at low concentrations (<0.5%) and by the neutral fraction at high concentrations. To better understand this phenomenon, further research on molecular characterization of both fractions, including conformations and elucidation of structure-function relationships, will be continued.

Flaxseed gum may become a significant source of soluble fibre, due to both its availability and its functional properties. It can be incorporated into foods at a significant concentration without excessive viscosity, making it suitable for fibre-fortified beverage application. In WPI stabilized emulsions, we have shown in other work that it can interact with the protein at the interface at acidic pH[17]. We have also shown that it is incompatible with protein at high concentration and can lead to phase separation[18]. These factors would need to be considered in product development, depending on the application.

Acknowledgements

The authors would like to thank Cathy Wang and Ben Huang for their technical assistance. Special thanks to Edita Verespej for technical assistance on the molecular weight and protein distribution measurements.

References

1. Tungland BC, Meyer D. Nondigestible oligo- and polysaccharides (dietary fiber): their physiology and role in human health and food. Comprehensive Reviews in Food Science and Food Safety 2002;1(3):1 (3) 73-92.
2. Howarth NC, Saltzman E, Roberts SB. Dietary fiber and weight regulation. Nutrition Reviews 2001;59(5):129-139.
3. Campbell JM, Fahey GC. Psyllium and methylcellulose fermentation properties in relation to insoluble and soluble fiber standards. Nutrition Research 1997;17(4):619-629.
4. Giacco R, Brighenti F, Parillo M, Ciardullo AV, Capuano M, Rivieccio AM, Rivellese AA, Riccardi G. Are the beneficial metabolic effects of fibre enriched foods preserved when consumed within a composite meal? Diabetes Nutrition & Metabolism 1998;11(2):130-135.

5. Cui W, Mazza G, Oomah BD, Biliaderis CG. Optimization of an aqueous extraction process for flaxseed gum by response surface methodology. Lebensmittel-Wissenschaft and Technologie 1994a;27(4):363-369.
6. Cui W, Mazza G, Biliaderis CG. Chemical-Structure, Molecular-Size Distributions, and Rheological Properties of Flaxseed Gum. Journal of Agricultural and Food Chemistry 1994b;42(9):1891-1895.
7. Cui W, Mazza G. Physicochemical characteristics of flaxseed gum. Food Research International 1996;29(3-4):397-402.
8. Bureau CR. The CRB commodity yearbook 2007. Hoboken, New Jersey: John Wiley & Sons, Inc.; 2007. 100 p.
9. Western Canadian flaxseed, export quality data, April 2009. Western Canadian flaxseed. Canadian Grain Commission; www.grainscanada.gc.ca.
10. Cui W, Han NF; Her Majesty the Queen in right of Canada, as represented by the Minister of Agriculture and Agri-Food, assignee. Process and apparatus for flaxseed component separation. patent US 07022363. 2006.
11. G. Mazza, Biliaderis CG. Functional properties of flax seed mucilage. Journal of Food Science 1989;54(5):1302-1305.
12. Warrand J, Michaud P, Picton L, Muller G, Courtois B, Ralainirina R, Courtois J. Large-scale purification of water-soluble polysaccharides from flaxseed mucilage, and isolation of a new anionic polymer. Chromatographia 2003;58(5-6):331-335.
13. Cui W, Wood PJ, Blackwell B, Nikiforuk J. Physicochemical properties and structural characterization by two-dimensional NMR spectroscopy of wheat beta-D-glucan - comparison with other cereal beta-D-glucans. Carbohydrate Polymers 2000;41(3):249-258.
14. Wu Y, Cui W, Eskin NAM, Goff HD. Fractionation and partial characterization of non-pectic polysaccharides from yellow mustard mucilage. Food Hydrocolloids 2009;23(6):1535-1541.
15. Blumenkrantz N, Asboe-Hansen G. New method for quantitative determination of uronic acids. Analytical Biochemistry 1973;54(2):484-489.
16. Izydorczyk M, Biliaderis CG, Bushuk W. Physical properties of water-soluble pentosans from different wheat varieties. Cereal Chemistry 1991;68:145-150.
17. Khalloufi S, Corredig M, Alexander M. Interactions between flaxseed gums and WPI-stabilized emulsion droplets assessed in situ using diffusing wave spectroscopy. Colloids and Surfaces B-Biointerfaces 2009;68(2):145-153.
18. Khalloufi S, Corredig M, Goff HD, Alexander M. Flaxseed gums and their adsorption on whey protein-stabilized oil-in-water emulsions. Food Hydrocolloids 2009;23(3):611-618.

THERMAL CHARACTERISATION OF FISH GELATIN EXTRACTED FROM SURIMI PROCESSING BY-PRODUCTS

M.H. Norziah[1*], A.B. Khairulnizam[1] and M.M Mordi[2]

[1*]Food Technology Department, School of Industrial Technology, Universiti Sains Malaysia, 11800, Penang, Malaysia. E-mail : norziah@usm.my
[2] Drug Research Centre, Universiti Sains Malaysia, 11800, Penang, Malaysia.

1 INTRODUCTION

Among commercial hydrocolloids, gelatin is regarded as special and unique, serving multiple functions with a wide range of applications in food, pharmaceutical, biomedical, and photographic industries. It is well established that the amount of gelatin used in the food industry worldwide is increasing annually.[1] Collagen is the major structural protein found in the skin and bones of animals, and gelatins are their degradation products. Most of commercial gelatins are produced from cowhide or pigskins and bones. Mammalian gelatins exhibit superior functional properties, thus are more preferred than fish gelatins.[2,3,4] However, the industrial interest in non-mammalian gelatins i.e. fish gelatin in the food industry has been influenced by the bovine spongiform encephalopathy (BSE) crisis, the growing demand for halal or Kosher gelatin by the Muslims and Jews community, respectively as well as it is also more acceptable by the Hindus. Fish gelatin production is still minor nowadays, yielding about 1% of the annual world gelatin production of 250,000 tonnes.[5] Extraction of gelatin from different fish species has been reported. Some examples are cod head,[5] tilapia,[6] shark, lungfish and carp skin,[7] snapper,[8] Alaska pollock,[9] Yellowfin tuna,[3] Sin croaker and Decapterus macrosoma.[10] Skin, bones and fins represent around 30% of fish fillet processing waste, and are produced as a consequence of the preparation of different fishery products such as fillets and surimi products. This waste is an important by-product of the fish processing industry, causing wastage and pollution.

This paper presents a study undertaken to produce gelatin from fish wastes (skins, bones, scales and fins) of different fish species, generated from a local surimi processing plant. The aim is to evaluate the potential viability of producing value-added food ingredients such as fish gelatin from these processing wastes. The extracted fish gelatin was then evaluated for its physical, rheological and thermal properties and comparisons made to several types of commercial gelatins.

2 MATERIALS AND METHODS

2.1 Materials

Fish waste consisting of bones, skins, and fins from four types of fish species, namely slender shad (*Ilisha tenualosa*), ribbon fish (*Trichiurus lepturus*), threadfin bream (*Nemipterus japonicus*) and sin croaker (*Johnius dussumieri*) was kindly supplied by a local surimi processing industry located in Perak, Malaysia. The samples were transported frozen to the laboratory and were stored below -20 °C until further use. Commercial fish gelatin was generously supplied by Croda Colloid (Chesire, UK), bovine gelatin was supplied by local company in Malaysia, while commercial porcine gelatin was provided by Gelnex Gelatin (Brazil). All other chemicals used were of analytical grade.

2.2 Gelatin Extraction Process

Fish gelatin was extracted from surimi processing wastes following a method described by Zhou and Regenstein[9] with slight modifications. Thawed fish waste was thoroughly cleaned and rinsed with excess cold water to remove dirt. The extraction was carried out by a series of soaking pretreatments with 0.2 M Ca(OH)2 maintained at a temperature below 10 °C for 1 hr using a waste to solvent ratio of 1:6 (w/v). This was followed by further soaking in 0.1 M citric acid for 3 hrs using the same ratio of waste to solvent. The final extraction was performed in distilled water at 50 °C for 3 hrs using a fish waste to water ratio of 1:2 (w/v). The mixture was then filtered with cheese cloth and further filtered using a Whatman no. 4 filter paper. The filtrate collected was oven dried overnight at 70 °C to obtain a dried fish gelatin sample which was then ground to powder of 30 mesh size. The extracted fish gelatin powder was kept in an air tight bottle for further analysis.

2.3 Determination of Gel Strength

The gel strength (bloom value) of extracted fish gelatin was determined according to the Gelatin Manufacturers Institute of America testing standard method[11]. Gelatin solution of concentration, 6.67 % (w/v) was prepared and poured into a standard glass bloom jar (150 ml capacity, Schott Duran, Germany) and chilled in a water bath at 10°C ± 0.1°C for 16 to 18 hrs. Gel strength at 8 - 9 °C was determined using TA.XT2 Texture Analyzer (Stable Micro System, Surrey, UK) with a load cell of 5 kg, cross–head speed 1 mm/s and equipped with a 0.5 inch diameter, flat bottom plunger. The gel strength was determined as maximum force (in g) when the probe proceeded to penetrate into the gel to a depth of 4 mm.

2.4 Determination of Isoelectric Point

Isoelectric point (IEP) of the extracted fish gelatins were determined following the method described by Rastogi[12]. Sodium acetate buffer solutions, 0.2 M of pH ranging from 3 to 10 were prepared. An aliquot of 10 ml of each buffer solution was placed into 15 test tubes. Gelatin powder (2 g) was completely dissolved in 48 ml distilled water by heating in a water bath at 60 °C with stirring. An aliquot of 2 ml hot gelatin solution was then added to each test tube containing 10 ml buffer solution and mixed by swirling the test tube and left to cool. Ethanol (99.5%) was added drop-wise to each test tube using a burette until a faint turbidity persists in the buffer-gelatin mixture. The volume of ethanol needed to produce permanent weak turbidity in the buffer-gelatin mixture was recorded and a plot of ml ethanol against pH was made. The pH at which most flocculation appears was taken as the IEP and this was observed as the lowest point in the curve. Triplicate runs were performed.

2.5 Electrophoresis Experiment

The molecular weight distributions of the gelatin were determined using Sodium dodecyl sulphate-polyacrylamide gel electrophoresis (SDS-PAGE) according to the method of Laemmli[13] using 4 % stacking gel and 7.5 % resolving gel. Samples were mixed with sample buffer containing 2 % SDS and 5 % β-mercaptoethanol (ratio 1:2). The mixtures were then heated at 90 °C for 5 min before loading. The samples were run at 150 V and 50 mA in the Mini-Protean Tetra Electrophoresis Cell for half an hour. Subsequently, the gel was stained with Bio-safe Coomasie Brilliant Blue and washed with distilled water. The protein fractions were identified using prestained SDS-PAGE standards broad range (catalog 161-0318).

2.6 Dynamic Viscoelastic Study

Dynamic viscoelastic studies of 6.67 % (w/v) gelatin solution was performed on a controlled stress Carri-Med CSL2 100 rheometer (TA Instruments, Surrey, England) using a cone-plate geometry (diameter, 6 cm, 2° cone angle, 64 μm gap). Temperature ramps were implemented by cooling the systems from 30 °C to 5 °C and then reheating back from 5 °C to 30 °C at a scan rate 2 °C/min, frequency 1.0 Hz and oscillating applied stress of 3 Pa. A Rheology Advantage Data Analysis software (Version 3.0.8) (TA Instruments, Surrey, England) was used for data analysis. The gelling point (T_{gel}) was taken as the point at which the phase angle (δ) falls immediately after a sharp decrease while the melting point (T_{melt}) was taken as the point at which the phase angle (δ) peaks immediately after a sharp increase.[14]

2.7 Differential Scanning Calorimetry

Thermal properties of gelatins were determined using differential scanning calorimetry (DSC-Q100, TA Instruments, Lukens Drive, New Castle). A refrigerated cooling system (RCS) was used in the instrument to achieve a temperature of −10 °C and a nitrogen DSC cell purge at 50 ml/min. Gelatin solution (6.67 % w/v) was weighed (~10 mg) in an aluminium hermetically sealed pan before being subjected to a DSC scan (cooling from 40 to -10 °C and reheating from -10 to 40 °C) at a scan rate 5 °C /min The equipment was calibrated with Indium reference material (T_m = 156.6 °C, ΔH = 28.71 J g-1). The helix-coil transition temperature (T_m) was calculated as the temperature where the endothermic peak occurs, and the enthalpy (ΔH) of the sol-gel transition was calculated as the area over the endothermic peak.

2.8 Statistical Analysis

All data presented are the means of three independent determinations with standard deviation. Data analysis was performed using SPSS V 12.0 and one-way of variance analysis was carried out using Duncan's test with confidence level as $p<0.05$.

3 RESULTS AND DISCUSSION

3.1 Gelatin Yield and Gel Strength

The yields of gelatin obtained from slender shad, sin croaker, ribbon fish and threadfin bream were 6.6 %, 3.3 %, 3.4 % and 3.4 %, respectively. Extracted fish gelatin obtained were in powder form and slightly yellowish in color. Gel strength or bloom strength is one of the most important measurements used to assess the grade and quality of a gelatin. The strength of gelatin gels depends on the intrinsic characteristics, such as amino acid content, size of protein chains, concentration, and molecular weight distribution as well as the type of extraction treatment.[15,16] The gel strength of commercial gelatin has been reported to range from 100 to 300 g. It was reported [17] that gelatin with bloom values of 250–260 g are most desired and has been used in making desserts and confectionery such as marsmallow, gummy candies and ice cream in the food industry as well as for making hard capsules for the pharmaceutical industry. Gel strength obtained for gelatin samples (6.67 % w/v) extracted from different fish species varied and ranged from 14.4 g to 49.3 g (Figure 1). Gelatin extracted from fish wastes showed lower gel strengths than commercial fish gelatin (161.0 g), commercial bovine gelatin (163.1 g) and commercial porcine gelatin (221.5 g). Gelatin from slender shad (SSG) had the highest gel strength (49.4 g) while gelatin from sin croaker (SCG) had the lowest gel strength (14.4 g). The gel strength of gelatin from ribbon fish gelatin (RFG) was 27.7 g while gelatin from threadfin bream was 16.9 g.

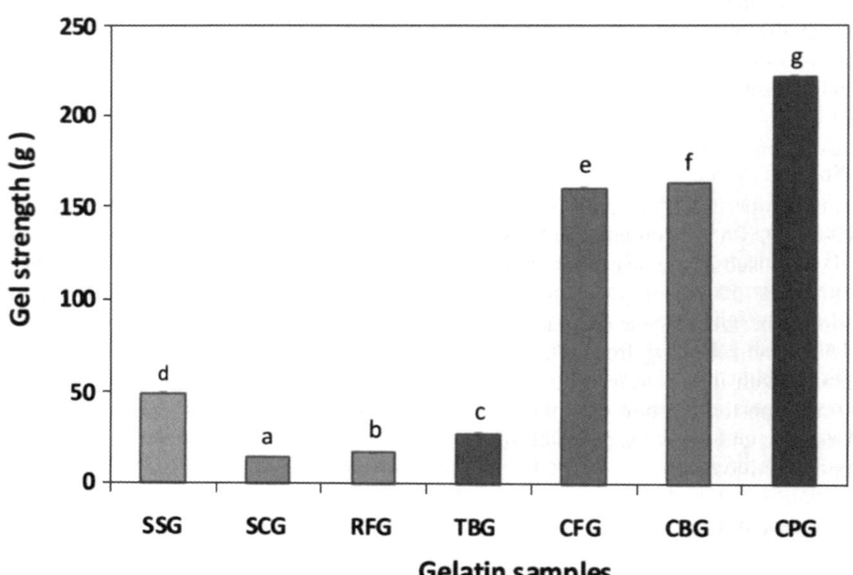

Figure 1: *Gel strength of slender shad gelatin, (SSG), sin croaker gelatin, (SCG), ribbon fish gelatin, (RFG), threadfin bream gelatin, (TBG), commercial fish gelatin, (CFG), commercial bovine gelatin, (CBG), commercial porcine gelatin (CPG). Mean values ± SD from triplicates samples.*

The lower gel strength obtained from the extracted fish gelatin may be due to the possible low content of amino acid proline and hydroxyproline resulting in less organized triple helical structures. Proline and hydroxyproline are thought to be responsible for the stability of the triple-helix of collagen structure through hydrogen bonding between free water molecules and hydroxyl group of the hydroxyproline in gelatin.[18,19] Gelatin with higher hydroxyproline content will also contribute to higher gel strength values. Furthermore, Gudmundsson and Hafsteinsson[20] also reported that gel strength was dependent on the isoelectric point which can be controlled by adjusting the pH. At or near the isoelectric point, gelatin polymers are closer to each other, stiffer and form a more compact gel. Moreover, gelatin with higher molecular weight is also known to have higher gel strength.[15] This confirms with the present result from electrophoresis studies that molecular weight obtained from the extracted fish gelatin was found to be lower compared to commercial gelatin. Low Bloom values ranging from 70 to 110 g have also been reported for cod, pollock, salmon and hake.[21,22]

3.2 Isoelectric Point of Extracted Fish Gelatin

Isoelectric point (IEP) is an important parameter of protein, which is related to the proportion of acid amino residues and base amino residues in protein.[23] Type A gelatin is produced by the acid processing of collagenous raw materials and exhibits an IEP between pH 7 and 9 whereas Type B gelatin is produced by the alkaline processing and exhibits an IEP between pH 4.6 and 5.2.[11] The IEP values obtained for slender shad gelatin (SSG), sin croaker gelatin (SCG), ribbonfish gelatin (RFG), threadfin bream gelatin (TBG), commercial fish gelatin (CFG), commercial bovine gelatin (CBG) and commercial porcine gelatin (CPG) were 8.0, 9.0, 7.8, 7.2, 8, 5.3, and 8.5, respectively (Table 1). The isoelectric points of commercial fish gelatin (CFG), commercial porcine gelatin (CPG), and gelatin extracted from fish waste (SSG, SCG, RFG, and TBG) were in the basic range which might be due to the acidic treatment before the extraction process.

3.3 Sodium dodecyl sulphate-polyacrylamide gel electrophoresis (SDS-PAGE)

Figure 2 presents the molecular weight profiles of gelatin samples determined by SDS-PAGE gel electrophoresis. All gelatin samples contained $\alpha 1$, $\alpha 2$, and β-chains, which are the main components of gelatin. It shows that extracted fish gelatin had slightly lower molecular weight chains compared to commercial fish, bovine and porcine gelatin. Extracted fish gelatin showed the presence of an intensive band with a relative molecular weight of about 115 kDa with presence of other faint bands at the lower molecular weight regions. In gelatin manufacture, the conversion of collagen to gelatin yields the molecules of varying mass, due to the cleavage of inter-chain chemical cross-links and some unfavourable breakages of inter-chain peptide linkage.[22] It was reported[16] that Nile perch skin gelatins contained low molecular weight peptides, especially, gelatins extracted with higher temperature.

3.4 Viscoelastic Properties

The dynamic viscoelastic profile of gelatin samples in the cooling scan (30-5 °C) and heating scan (5-30 °C) are shown in Figures 3 and 4. The elastic modulus, G' and viscous modulus, G'' which is a measure of the solid-like and liquid-like response of the material, respectively, and phase angle, δ (a measure of how much the stress and strain are out of phase) were obtained from these viscoelastic profiles. In general, G' values obtained were

Table 1 *Isoelectric point (IEP) of gelatin samples*

Gelatin Sample	IEP
SSG	8.0 ± 0^{cd}
SCG	9.0 ± 0^{e}
RFG	7.8 ± 0.29^{c}
TBG	7.2 ± 0.58^{b}
CFG	8.0 ± 0^{cd}
CBG	5.3 ± 0.58^{a}
CPG	8.5 ± 0^{de}

Gelatin samples: slender shad gelatin (SSG), sin croaker gelatin (SCG), ribbon fish gelatin (RFG), threadfin bream gelatin (TBG), commercial fish gelatin (CFG), commercial bovine gelatin (CBG), and commercial porcine gelatin (CPG). Values are given as mean ± standard deviation. Values with different superscripts are significantly different ($p < 0.05$).

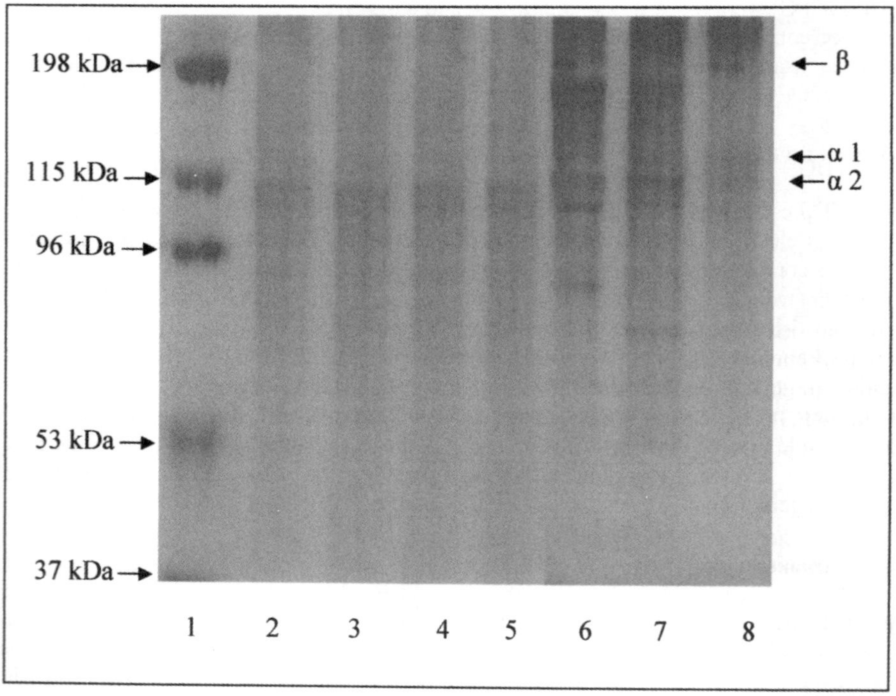

Figure 2 *Protein pattern of gelatin samples. (Lane: (1) standard marker; (2) slender shad gelatin, SSG, (3) sin croaker gelatin, SCG; (4) ribbonfish gelatin, RFG; (5) threadfin bream gelatin, TBG; (6) commercial fish gelatin, CFG; (7) commercial bovine gelatin, CBG; and (8) commercial porcine gelatin, CPG.*

higher than the viscous modulus (G'') during both cooling and heating scans, suggesting that gelatin exhibit more solid-like behaviour. The increase in elastic modulus resulted from changes in the gelatin molecule's conformation during the cooling process. Gomez-Guillen et al.[21] reported that the increase in elastic modulus of fish gelatins during gelation corresponded with an increase in the concentration of triple helical structures which is the basic unit of collagen, from which gelatin is derived. It revealed that gelatins extracted from fish waste formed gels at lower temperature than commercial fish gelatin (17.1 ± 0.21 °C), commercial bovine gelatin (19.0 ± 0.75 °C), and commercial porcine gelatin (20.8 ± 0.06 °C), suggesting that commercial gelatin had a higher concentration of helical structure than extracted fish gelatin. Gelling temperature of slender shad gelatin (SSG), sin croaker gelatin (SCG), threadfin bream gelatin (TBG), and ribbon fish gelatin (RFG) were 11.5 ± 0.1 °C, 7.1 ± 0.15 °C, 7.8 ± 0.1 °C and 9.2 ± 0.32°C, respectively as shown in Table 2. Gelling temperatures for porcine and bovine gelatins normally range from 20 to 25 °C and for fish gelatin is between from 8 to 25 °C.[24] Cheow et al.[10] has reported that the gelling temperature of gelatin from sin croaker and shortfin scad was 7.1 °C and 9.9 °C, respectively.

Table 2 *Melting and gelling temperature values of gelatin samples obtained from rheological and DSC experiments*

	Rheological experiments		DSC experiments	
Gelatin samples	Gelling point, °C	Melting point, °C	Gelling point, °C	Melting point, °C°
SSG	11.5 ± 0.1^{d}	18.7 ± 0.29^{c}	4.1 ± 0.76^{b}	18.6 ± 1.23^{c}
SCG	7.1 ± 0.15^{a}	12.9 ± 0.1^{a}	2.4 ± 0.36^{a}	15.5 ± 0.4^{b}
RFG	9.2 ± 0.32^{c}	16.2 ± 0.35^{b}	2.8 ± 0.12^{a}	16.3 ± 1.27^{b}
TBG	7.8 ± 0.1^{b}	14.7 ± 0.15^{b}	2.7 ± 0.20^{a}	14.0 ± 0.52^{a}
CFG	17.1 ± 0.21^{e}	22.6 ± 0.21^{d}	10.6 ± 0.79^{d}	22.9 ± 0.26^{c}
CBG	19.0 ± 0.75^{f}	24.2 ± 0.1^{e}	11.2 ± 1.05^{de}	25.6 ± 0.35^{d}
CPG	20.8 ± 0.06^{g}	27.9 ± 0.12^{f}	12.1 ± 0.54^{e}	26.4 ± 0.32^{d}

Values are means ± standard deviation. Values in the same column with different superscripts are significantly different ($p < 0.05$).
Gelatin samples: slender shad gelatin (SSG), sin croaker gelatin (SCG), ribbon fish gelatin (RFG), threadfin bream gelatin (TBG), commercial fish gelatin (CFG), commercial bovine gelatin (CBG), and commercial porcine gelatin (CPG).

During heating, a three-dimensional network or gel starts to lose the triple helices conformation resulting in the coil chains being able to move more freely. Gelatin gels are formed from sols when gelatin molecules are presumed partially to reform the hydrogen bonded structure of collagen, and thus the melting of gelatin gels is related to the denaturation of soluble collagens.[7] Generally, fish gelatin have a lower melting point than mammalian gelatin.[25] The melting temperatures of commercial fish, bovine and porcine gelatin samples were found to be 22.6 ± 0.21°C, 24.2 °C ± 0.1°C, and 27.9 ± 0.12 °C, respectively (Table 2). However, gelatins extracted from fish waste melted at temperatures lower than commercial gelatins which were for slender shad, SSG (18.7 ± 0.29 °C), sin croaker, SCG (12.9 ± 0.1 °C), threadfin bream, TBG (14.7 ± 0.15 °C) and ribbon fish gelatin, RFG (16.2 ± 0.35 °C). Typical melting points for porcine and bovine gelatins range from 28 to 31 °C while for fish gelatin is between from 11 to 28 °C.[24] Cheow et al.[10] reported that the melting point of gelatin from sin croaker and shortfin scad was 18.5 °C and 24.5 °C, respectively. There were several factors that affected the gelling properties and thermal stability of gelatin. As reported by Johnston-Banks[26], increase in gelling and melting temperature was directly related to higher molecular weight of gelatin.

3.5 Thermal Properties

The thermal stability of the triple helix in gelatin can be observe from differential scanning calorimetry (DSC) studies. The heat flow detected by DSC corresponds to the energy necessary to melt the junction zones and to achieve the helix-to-coil conformation.[27] DSC thermograms of cooling scans (40 to -10 °C) and heating scans(-10 to 40 °C) at 5 °C/min of gelatin samples are shown in Figure 5 (of commercial gelatin samples) and Figure 6 (of extracted fish gelatin samples). The exothermic peak observed in the cooling scans was pointed as the gelling temperature while the endothermic peak in heating scans as the melting temperature. It was observed that the melting temperature of the commercial fish gelatin (10.6 ± 0.79 °C), commercial bovine gelatin (11.2 ± 1.05 °C) and commercial porcine gelatin (12.1 ± 0.54 °C) was higher compared to the gelatin extracted from fish waste. The gelling temperature of extracted gelatin was; slender shad, SSG (4.1 ± 0.76 °C), sin croaker, SCG (2.4 ± 0.36 °C), threadfin bream, TBG (2.7 ± 0.2 °C) and ribbon fish gelatin, RFG (2.8 ± 0.12 °C) as shown in Table 2. However, these gelling points were lower than reported from the rheological studies. As compared with rheological data, the sol-gel transition was shifted to even colder temperatures due to increased cooling rate. It was observed that the melting temperature of the commercial fish gelatin (22.9 ± 0.26 °C), commercial bovine gelatin (25.6 ± 0.35 °C) and commercial porcine gelatin (26.4 ± 0.32 °C) was higher compared to the gelatin extracted from fish waste; slender shad, SSG (18.6 ± 1.23 °C), sin croaker, SCG (15.5 ± 0.4 °C), threadfin bream, TBG (14.0 ± 0.52 °C) and ribbon fish gelatin, RFG (16.3 ± 1.27 °C). Imino acids (proline and hydroxyproline) was believed to influence the rigidity of gelatin by forming hydrogen bonds that stabilize the triple-helix structure by sharp twisting of collagen structure. The higher the imino acid content the more rigid is the gelatin and the higher the melting and gelling temperatures.

4 CONCLUSION

Gelatin extracted from surimi processing by-products have been characterized. The results indicated that gelatin extracted from slender shad fish had higher gel strength, melting and gelling temperatures compared to the other extracted fish gelatins. The utilization of surimi processing by-products not only provides a potential alternative source of gelatin for the food industry especially for the halal industry in Malaysia, but it also adds value to the waste generated in abundance and presents opportunities for better utilisation.

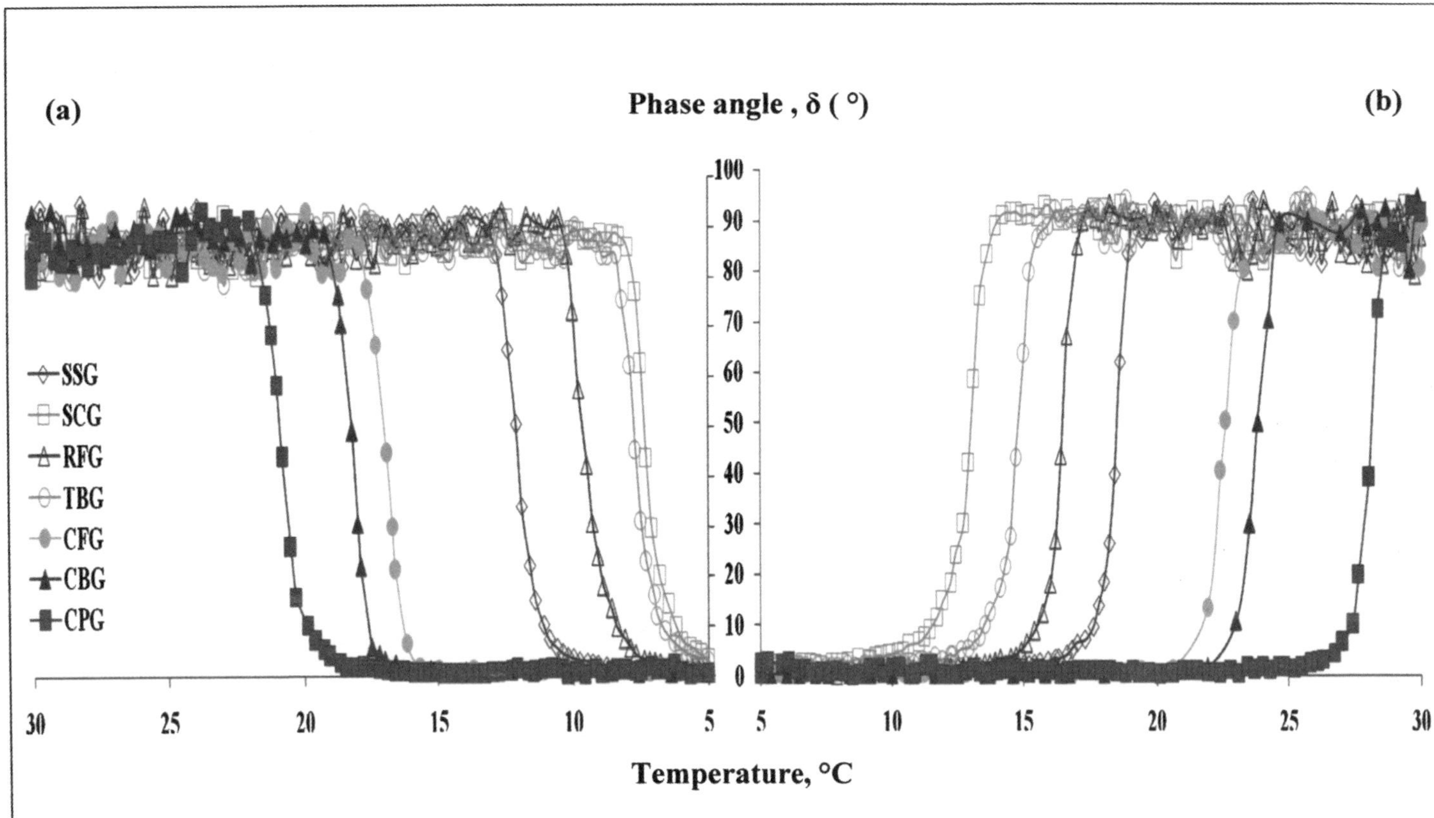

Figure 3 *Changes in the phase angle, δ upon (a) cooling (30-5 °C) and (b) heating (5-30 °C) of gelatin samples. Slender shad gelatin, SSG [◊]; sin croaker gelatin, SCG [□]; ribbon fish gelatin, RFG [Δ]; threadfin bream gelatin, TBG [○]; commercial fish gelatin, CFG [●]; commercial bovine gelatin, CBG [▲]; commercial porcine gelatin, CPG [■].*

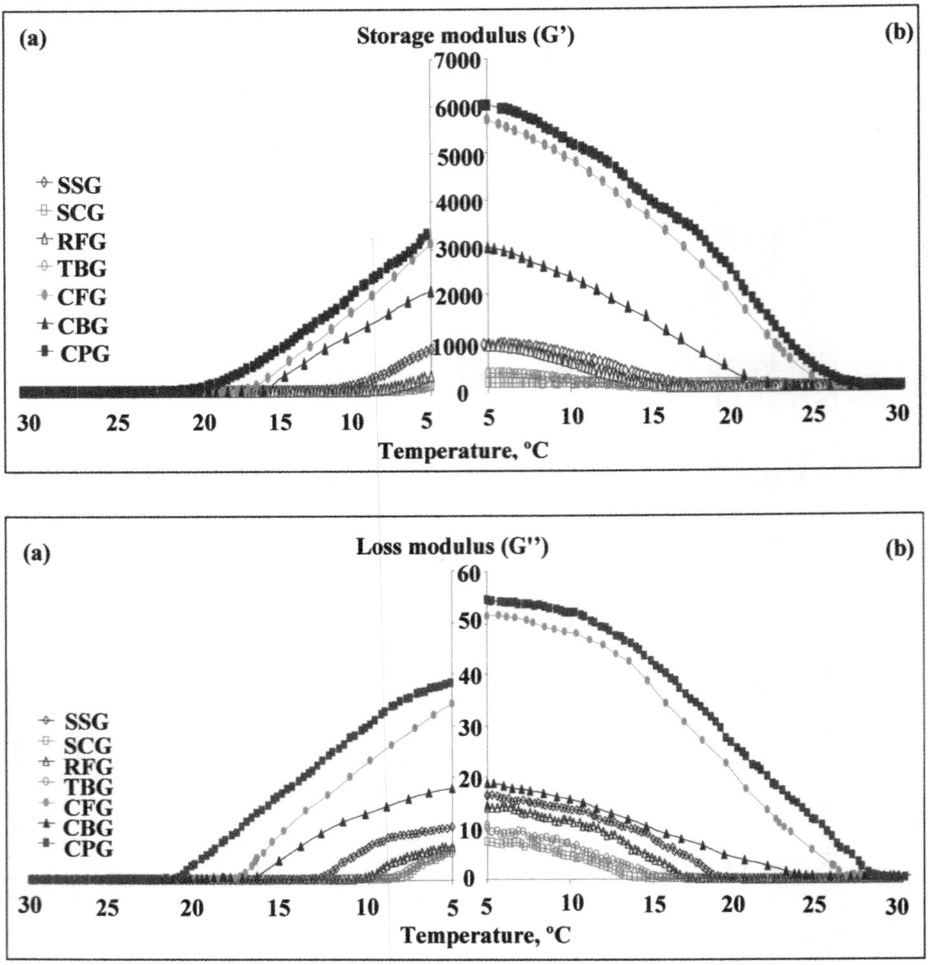

Figure 4 *Changes in the storage modulus, G' and loss modulus, G" upon (a) cooling (30 – 5 °C) and (b) heating (5 - 30 °C) of gelatin samples. Slender shad gelatin, SSG [◊], sin croaker gelatin, SCG [□], ribbon fish gelatin, RFG [Δ], threadfin bream gelatin, TBG [○], commercial fish gelatin, CFG [●], commercial bovine gelatin CBG [▲], commercial porcine gelatin CPG [■].*

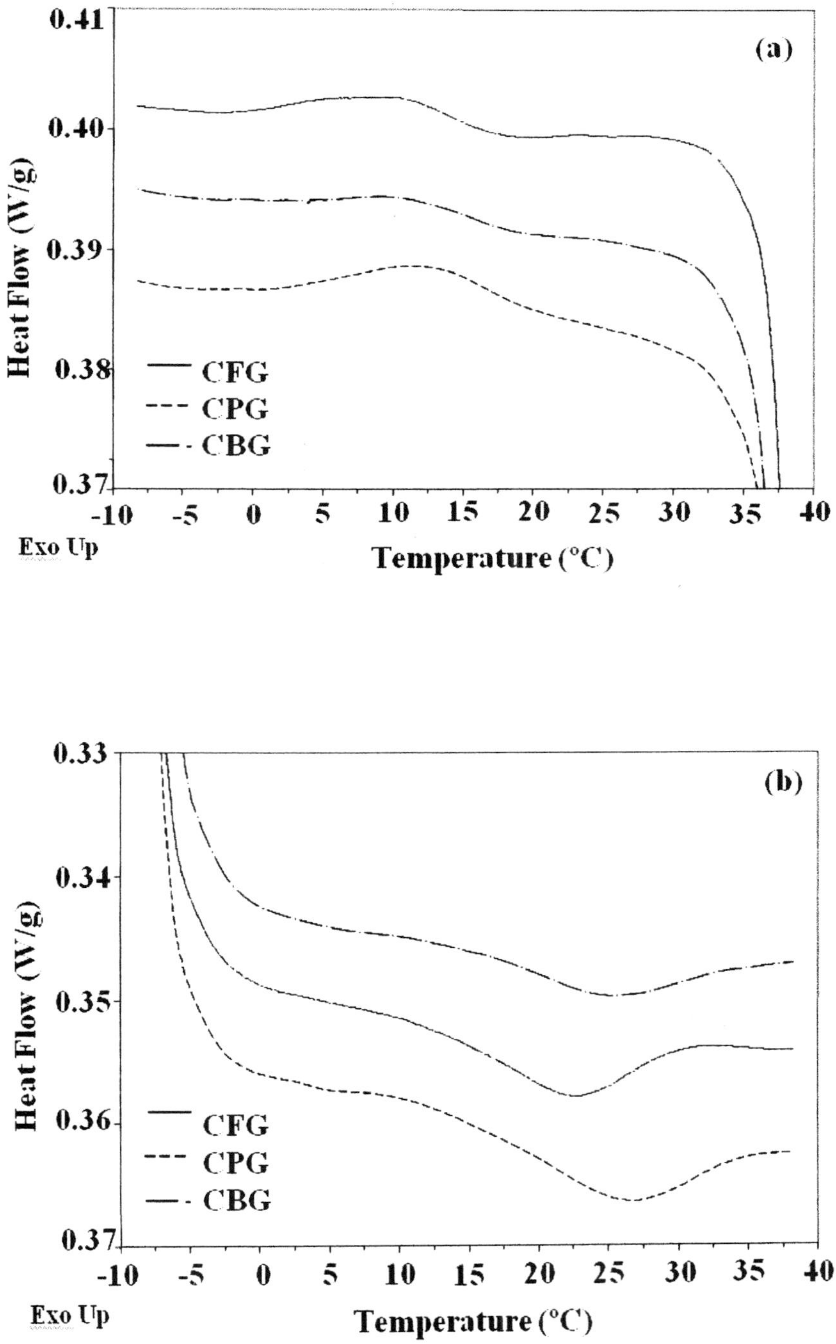

Figure 5 *DSC curves in the (a) cooling runs from 40 to -10 °C; and (b) heating runs from -10 to 40 °C for Commercial fish gelatin (CFG), commercial bovine gelatin (CBG), commercial porcine gelatin (CPG).*

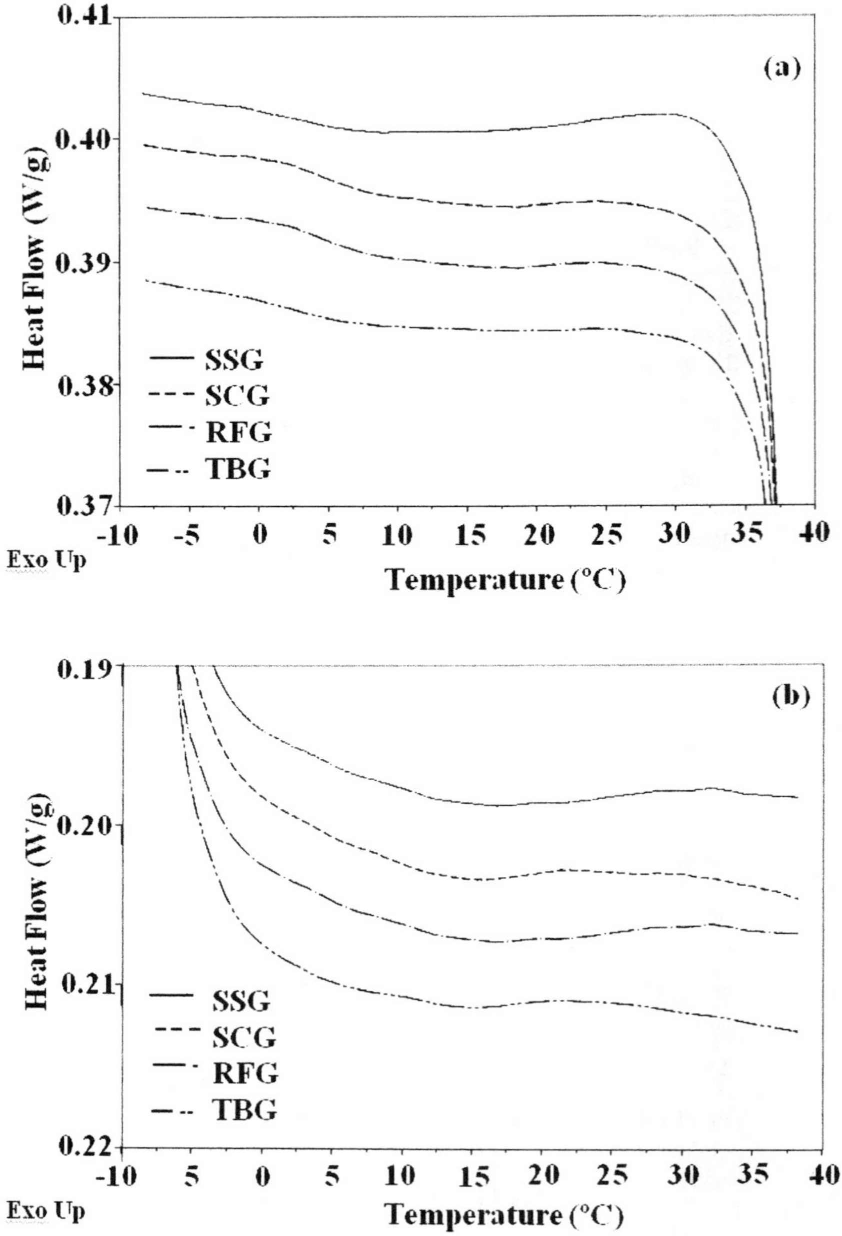

Figure 6 *DSC curves in the (a) cooling runs from 40 to -10 °C; and (b) heating runs from -10 to 40°C for Slender shad gelatin (SSG), sin croaker gelatin (SCG), ribbon fish gelatin (RFG), threadfin bream gelatin (TBG).*

Acknowledgment

The authors wish to express their thanks to Lykoon Sdn. Bhd. for the supply of fish waste. This work was funded by Short Term Research Grant from Universiti Sains Malaysia. One of the authors, A.B. Khairulnizam acknowledges the scholarship received from the National Science Fellowship from MOSTI, Malaysia.

References

1 P. Montero, and M.C. Gómez-Guillén, *J. of Food Science*, 2000, **65**(3), 434.
2 S.S. Choi and J.M. Regenstein, *J. of Food Science*, 2000, **65**, 194.
3 S.M. Cho, Y.S. Gu, and S.B. Kim, *Food Hydrocolloids*, 2005, **19**(2), 221.
4 L.M. Kasankala, Y. Xue, Y. Weilong, S.D. Hong and Q. He, *Bioresource Technology*, 2007, **98**, 3338.
5 J.A. Arnesen and A Gildberg, *Process Biochemistry*, 2006, **41**, 697.
6 B. Jamilah and K.G. Harvinder, *Food Chemistry*, 2002, **77**, 81.
7 A.G. Ward and A. Courts, in *The Science and Technology of Gelatin*. London: Academic Press, 1977.
8 A. Jongjareonrak, S. Benjakul, W. Visessanguan, T. Prodpran and M. Tanaka, *Food Hydrocolloids*, 2006, **20,** 492.
9 P. Zhou and J.M. Regenstein, *J. of Food Science*, 2005, **70**, 392.
10 C.S. Cheow, M.S. Norizah, Z.Y. Kyaw and N.K. Howell, *Food Chemistry*, 2007, **101**, 386.
11 GMIA (Gelatin Manufacturers Institute of America). (2006). Standard methods for the testing of edible gelatin. In: *Official procedure of the Gelatin Manufacturers Institute of America, Inc*. Revised September 2006 [on line].<http://www.gelatin-gmia.com> Accessed 10.06.09.
12 S.C. Rastogi, *Experimental physiology*. 2nd Edn. New Age International (P) Ltd. Publishers, New Delhi, India, (2005). [on line] <http://www.newagepublishers.com /samplechapter/000979.pdf> Accessed 10.06.09.
13 U.K. Laemmli, *Nature*, 1970, **227**, 680.
14 A.I. Sarabia, M.C. Gomez-Guillen, and P. Montero, *Food Chemistry*, 2000, **70**, 71.
15 J.A. Arnesen and A. Gildberg, *Bioresource Technology*, 2002, **82**, 191.
16 J.H. Muyonga, C.G.B. Cole and K.G. Duodo, *Food Chemistry*, 2004, **18**, 581.
17 D. Holzer, Gelatin production. US Patent 5,484,888, 1996.
18 T.V. Burjandze, *Biopolymers*, 1979. **18**, 931.
19 M.D. Fernández-Díaz, P. Montero and M.C. Gómez-Guillén, *Food Chemistry*, 2001, **74**, 161.
20 M. Gudmundsson and H. Hafsteinsson, *J. of Food Science*, 1997, **62**, 37.
21 M.C, Go´mez-Guille´n, J. Turnay, M.D. Fernandez-Diaz, N. Ulmo, M.A. Lizarbe, and P. Montero, *Food Hydrocolloids*, 2002,**16**, 25.
22 P. Zhou, S.J. Mulvaney and J.M. Regenstein, *J. of Food Science*, 2006, **71**, C313.
23 Z. Zhang, G. Li and B Shi, *J. of the Society of Leather Technologists and Chemists*, 2005, **90**, 23.
24 A.A.Karim and R.Bhat, *Food Hydrocolloids*, 2008, **23**, 563.
25 R.E. Norland, Fish gelatin, in *Advances in fisheries and biotechnology for increased profitability*, eds. M.N. Voigt, & J.K. Botta, Technomic Publishing Co, Lancaster, USA, 1990, p. 325.
26 F.A. Johnston-Banks, *Gelatin* in Food gels ed. P. Harris, London: Elsevier Applied Science Publishers, 1990, p 233.

27 D. Hickman, T.J. Sims, C.A. Miles, A.J. Bailey,M. de Mari and M. Koopmans, *J. of Biotechnology*, 2000, **79,** 245.

PRODUCTION OF SUCCINOGLYCAN BIOGUM FROM DATE SYRUP BY *Agrobacterium radiobacter* AND INVESTIGATION OF ITS RHEOLOGICAL PROPERTIES

M. Moosavi-Nasab, A. Farahnaky, H. Askari, M. Bakhtiyari

Department of Food Science and Technology, School of Agriculture, Shiraz University, Shiraz, I.R.Iran.

1. INTRODUCTION

Microbial exopolysaccharides are widely exploited in many industrial fields for their capability of forming gels, stabilizing emulsions, creating viscous solutions and, generally, of altering the rheological properties of various products. Bacteria are an important source of these biopolymers and exopolysaccharides (EPS).[1] Gums such as xanthan [2], gellan [3] and curdlan [4] have been isolated and applied to food product development. There has been a recent upsurge in interest in microbial biopolymers; several strains of lactic acid bacteria [5-8] and *Pseudomonas* spp. [8] have been screened for EPS production with novel properties. Succinoglycans are heteropolysaccharides synthesized by a variety of bacteria belonging to the family *Rhizobiaceae* (*Rhizobium, Agrobacterium*), as well as by other microorganisms such as *Alcaligenes faecalis* var. *myxogenes* and *Pseudomonas* spp.[9] They are acidic polysaccharides composed of octasaccharide repeating units in which galactose and glucose monomers occur in a molar ratio of 1 to 7. Succinate and pyruvate are commonly found as non-saccharidic substituents, whereas acetate might or might not be present, depending on the source of the polymer.[10-11] Natural and chemically modified succinoglycans show high stability even under drastic operational conditions, such as high temperature and pressure, extreme salinity and pH values, or high shear rates. Microbial production of succinoglycan on an industrial scale has been traditionally restricted to *Agrobacterium* spp. Continuous fermentation processes have also been studied extensively by several authors.[12-13]

Date fruit is becoming an important commercial crop in the producing countries with significant increase in yield by adopting advanced biotechnological approaches. However, date processing industries have not expanded at the same rate. In the work reported here, date syrup produced from low quality date fruit was inoculated with *Agrobacterium radiobacter* to produce succinoglycan. Then the rheological properties of succinoglycan solutions were investigated using a rotational viscometer.

2. MATERIALS AND METHODS

2.1 Preparation of date syrup and production of exopolysaccharide

The initial extract was prepared by soaking 200 g of stone free, low quality date fruits in 500 ml distilled water for 30 min at room temperature. The soaked dates were then blended in a blender for 1 min at low speed, and for further 3 min at high speed. The homogenized extract was filtered through a double layer of cheese cloth. The residue was then washed with hot water and solution made up to the volume required for a concentration of 20 %; subsequent dilutions to Brix 10 were made with distilled water. To remove all insoluble solids, the date extract solutions were then centrifuged (SORVALL, RC-5; USA) at 10000 g for 30 min at room temperature. *Agrobacterium radiobacter* was obtained from the Persian Type Culture Collection (PTCC) strain number 1654. The procedure for production of EPS is shown in Figure 1. Seed medium (in a final volume of 1 L) contained: sucrose, 20 g; yeast extract (Difco Laboratories, Detroit, MI, USA), 5 g; peptone (Difco), 5 g; pH adjusted to 7.0. The bacteria were inoculated into two liquid fermentation media in Erlenmeyer flasks containing sucrose or date syrup. These media (in a final volume of 1 L) contained: sucrose or date syrup (brix 10); KH_2PO_4, 1g; $MgSO_4$, 0.25g; $(NH_4)_2HPO_4$, 1g; trace elements solution, 10 ml. The trace elements solution contained in 100 ml of 0.1N HCl: $FeSO_4 \cdot 7H_2O$, 0.5 g; $MnSO_4 \cdot H_2O$, 0.2 g; $ZnCl_2$, 0.1g; $CoCl_2 \cdot 6H_20$, 0.1 g. Fermentation occurred in a Benmarin shaker (230 rpm) set at 28 °C at pH, 7. Samples were taken from fermentation media after 62, 74, 86, 98, 110, 122, 134 and 146 h. After growth and centrifugation, samples of supernatants were removed and analyzed for their EPS content. The supernatants from each treatment were pooled prior to the alcohol precipitation step. The freeze dried EPS samples were weighed, and values were converted to mg/100 ml of original medium. Exopolysaccharide was precipitated from the fermented broth using 2.5 volume of 96 % ethanol and freeze-dried, then milled to the mesh size of 53-125 μm. Yield production of exopolysaccharide was calculated as percentage of polysaccharide present in the broth.

2.2 Analysis of hydrolysis products by TLC

Purified expolysaccharide free from biomass was hydrolysed by a 10 % solution of trifluroacetic acid (TFA) at 100 °C for 4 h. Glucose, fructose, galactose, glucoronic acid and galactoronic acid were used to compare the R_f of hydrolyzed biogum samples with simple sugars (Equation 1).

$$R_f = \frac{\text{Distance traveled by the compound}}{\text{Distance traveled by the solvent front}} \times 100 \qquad \text{(Equation 1)}$$

Neutralized hydrolyzes were used for TLC analysis using TLC glass sheets (20×20 cm) precoated with 250 μm silica gel 60-F_{254} (Merck Co., Germany). The solvent system of acetone acetate, pyridine, water (55:25:20) was used to separate sugars. The chromatograms were developed by spraying with a solution containing 0.5 g thymol in 95 ml ethanol, and 5 ml 97% sulfuric acid, then heated for 15-20 min at 120°C. The background might be brightened by water vapor.

2.3 Fourier transform infrared spectroscopy (FT-IR)

FT-IR spectra of exopolysaccharides were recorded using KBr pellets on a BRUCKER spectrophotometer (EQUINOX 55, Germany) to study the structure of biogums.

2.4 Rheological measurements

Rheological measurements were carried out using a rotational viscometer (Brookfild, DV-II + Pro, USA) equipped with ÇP51 measuring geometry as a function of gum concentration (0.5, 1, 1.5 and 2 % w/w), temperature (5, 25, 40°C) and pH (2.5, 4, 7 and 10). The flow behavior index (n) and consistency index (k) values were computed by fitting the Power law model (Equation 2):

$$\tau = k\gamma^{\circ n} \qquad \text{(Equation 2)}$$

Where τ is the shear stress (mPa), γ° is the shear rate (s^{-1}), k is the consistency index (mPa.s) and n is the flow behavior index (dimensionless).

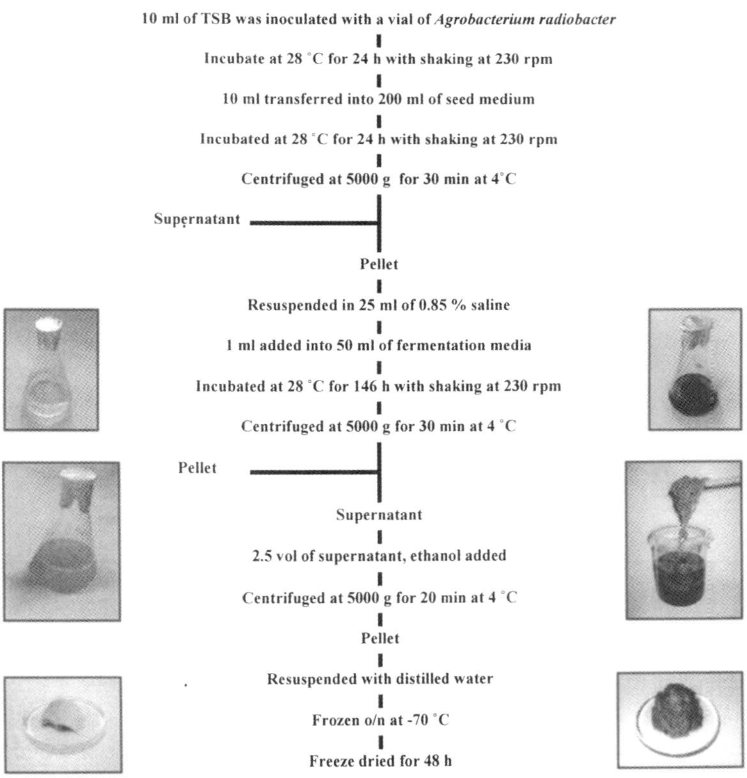

Figure 1 *Production and purification scheme of EPS by A. radiobacter.*

EPS solutions were prepared at different concentrations (% w/w) by dispersing the required amount of EPS in deionized water, under slow stirring at room temperature, and stored at 4 °C for 16 hr for complete hydration prior to assessment. The viscosity and flow properties for 1.5 % biogum solutions of EDS were measured at different pHs ranging from 2.5 to 10 (2.5,4, 7 and 10, adjusted using 0.1 M NaOH and HCl) at different shear rates and constant temperature of 25 °C.

2.5 Statistical analysis

Data were analyzed by a one-way ANOVA, followed by Duncan test to compare treatment means to the control treatment. Significant size was at $p<0.01$.

3. RESULTS AND DISCUSSION

3.1 Yield of EPS

The production of EPS in the fermentation media containing sucrose (ES) and date syrup (EDS) was monitored over a period of 146 h of incubation at 28 °C and 230 rpm and the results are given in Figure 2. The EPS production in the fermentation medium containing sucrose (ES) remained constant for the rest of the experimental period, but in the fermentation medium containing date syrup (EDS), biogum production continuously increased during fermentation. *A. radiobacter* in EDS gave a significantly higher yield of EPS than ES, and biogum production was 3 to 5 times higher than the medium containing sucrose during the fermentation period.

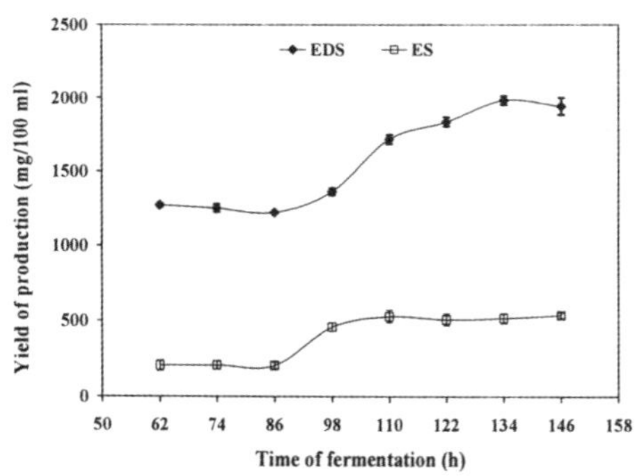

Figure 2 *Yield of Succinoglycan production in fermentation media containing either date syrup (EDS) or sucrose (ES) as a function of fermentation time.*

3.2 TLC analysis

The TLC chromatogram of the produced biogum from ES and EDS is shown in Figure 3. Glucose, fructose, galactose, glucuronic acid and galacturonic acid were used to compare the R_f of the hydrolyzed biogum samples with simple sugars. Hydrolyzed biogum showed one spot on the chromatogram with the same R_f as glucose which indicates the sample was mainly composed of glucose units. R_f for glucose, fructose, galactose, glucuronic acid, EDS, ES and galacturonic acid were 38.35, 42.46, 30.82, 12.32, 38.01, 38.21 and 6.85, respectively.

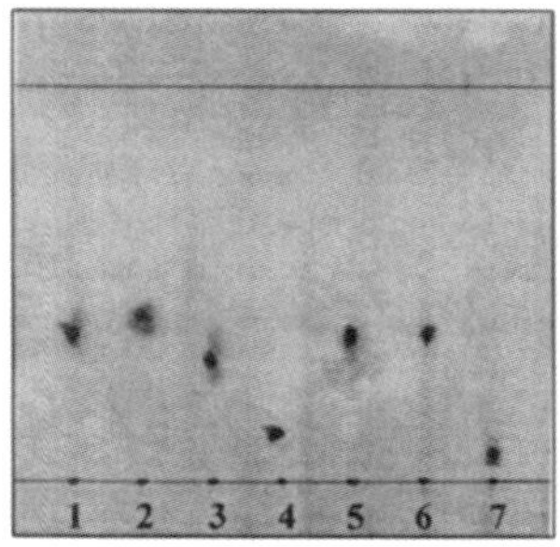

Figure 3 *The TLC chromatogram simple sugars (1) glucose, (2) fructose, (3) galactose, (4) glucuronic acid, hydrolyzed biogums from (5) EDS and (6) ES, and (7) galacturonic acid.*

3.3 FTIR spectroscopy analysis

Figure 4 shows the FTIR spectra of the EPS produced in two liquid fermentation media. The spectra are dominated by a broad band at about 3309 and 3418 cm^{-1}, which is assigned to the stretching vibration modes of -OH groups. This band tends to shift to higher wavenumbers for the spectrum of EPS isolated from EDS. The shift of the band is attributed to the weakening of hydrogen bonds.

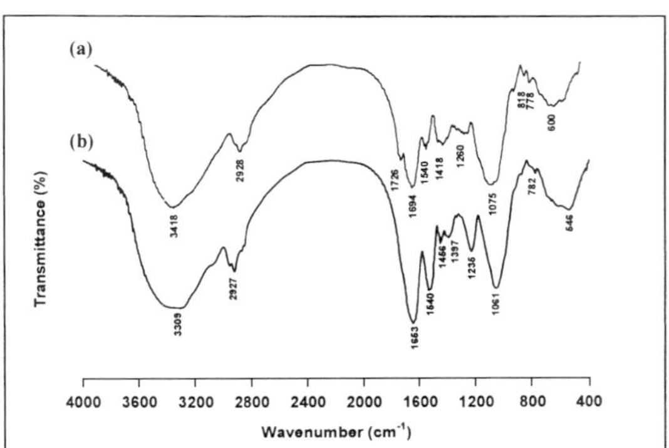

Figure 4 *FT-IR spectra of powders of (a) EDS and (b) ES.*

The peak at 1653 and 1694 cm^{-1} is due to the asymmetrical $-COO^-$ stretching vibration, whereas the band at 1418 and 1456 cm^{-1} is due to the symmetrical $-COO^-$ stretching vibration. A detailed band assignment is not presented in this work, but the main absorption peak and their tentative assignments are shown in Table 1.

Table 1 *FT-IR absorption regions and band assignments for ES and EDS*

Wavenumber (cm^{-1})		Band assignment
ES	EDS	
3309	3418	-OH stretching
2927	2928	-CH stretching of CH_2 and CH_3 groups
--	1726	C=O stretching of ester
1653	1694	COO^- asymmetrical stretching of carboxylic groups
1456	1418	COO^- symmetrical stretching of carboxylic groups
1061	1076	C-C or C-O stretching vibration

3.4 Rheological properties

The relationship between apparent viscosity (η_a) and shear rate of different concentrations of EDS biogum at 25 °C is shown in Fig. 5. It is evident that the apparent viscosity of all samples decreased with increasing shear rate. The viscosity of all solutions increased with biogum concentration.

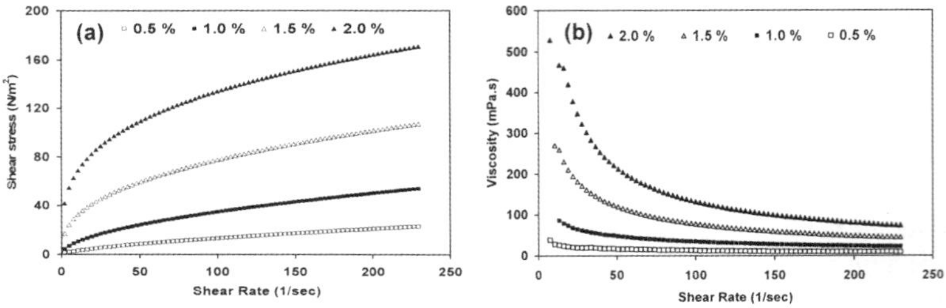

Figure 5 *Change of shear stress (a) and viscosity (b) versus shear rate as function of concentration for EDS biogum solutions at 25 °C.*

The consistency coefficient (k) and flow behavior index (n) values as a function of biogum concentration and solution temperature are shown in Table 2.

Table 2 *Compare of means n and k obtained from EDS gum solution as a function of concentration and temperature in distilled water.*

	Temperature								
	5 °C			25 °C			40 °C		
Concentration	**k**	**n**	**†R^2**	**k**	**n**	**R^2**	**k**	**n**	**R^2**
0.5 %	89.5^{a}_{B}	$0.63^{d}_{N.S.}$	0.98	65.1^{a}_{A}	$0.65^{d}_{N.S.}$	0.98	62.5^{a}_{A}	$0.61^{d}_{N.S.}$	0.98
1.0 %	455.7^{b}_{B}	0.48^{c}_{A}	0.98	303.9^{b}_{A}	0.53^{c}_{B}	0.99	267.5^{b}_{A}	0.53^{c}_{B}	0.98
1.5 %	1712.0^{c}_{C}	0.36^{b}_{A}	0.99	1311.0^{c}_{B}	0.38^{b}_{B}	0.99	1136.0^{c}_{A}	0.39^{b}_{B}	0.99
2.0 %	3783.3^{d}_{B}	0.28^{a}_{A}	1.00	3444.0^{d}_{B}	0.29^{a}_{AB}	1.00	2984.6^{d}_{A}	0.30^{a}_{B}	0.99

N.S. = Non significant

In each column and row, means supplemented by different small letters differed by Duncan test at the 1% level as function of concentration or temperature, respectively.

The values of flow behavior indices, n, were less than 1, indicating the shear thinning nature of the biogum solutions at different measuring conditions. Increasing the biogum concentration decreased the flow behavior index values and increased the consistency coefficient. The flow behavior index (n) of EDS solutions ranged from 0.656 (for 0.5 % w/w) to 0.295 (for 2 % w/w) at 25 °C, suggesting that EDS solutions with higher concentrations are more shear thinning than those with low concentrations. On the other hand, the consistency coefficient k for both of the samples increases with increasing concentration of the polysaccharide. Temperature, on the other hand, indicated an inverse effect on Power law parameters as flow behavior index increased and consistency coefficient decreased by increasing the temperature.

Table 3 shows that the rheological experimental data are well described by the Power law rheological model for the two samples ES and EDS at various concentrations at 25 °C, respectively.

Table 3 *Compare of means n and k obtained from ES and EDS gum solutions as function of concentration in distilled water at 25 °C.*

	Fermentation medium					
	EDS			ES		
Concentration	K	n	R^2	k	n	R^2
0.5 %	65.1 ± 3.75	0.65 ± 0.01	0.98	44.3 ± 2.49	0.53 ± 0.01	0.97
1 %	303.9 ± 5.27	0.53 ± 0.00	0.99	90.8 ± 4.57	0.52 ± 0.01	0.98
1.5 %	1311.0 ± 9.53	0.38 ± 0.00	0.99	200.2 ± 6.95	0.45 ± 0.01	0.99
2 %	3444.0 ± 150	0.29 ± 0.00	1.00	383.8 ± 12.53	0.39 ± 0.01	0.99

The viscosity of EPS produced from EDS was more than in the ES at different concentrations of biopolymer (Figure 6).

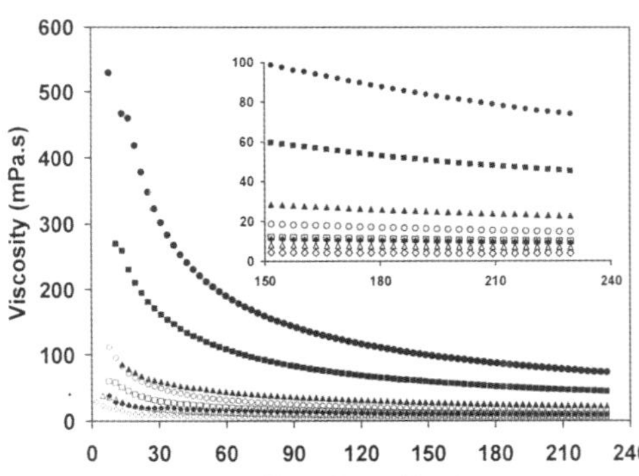

Figure 6 *Compare of viscosity ES (open symbol) and EDS (filled symbol) biogum solutions as function of concentration at 25 °C.* ●, ○ *2.0 %,* ■, □ *1.5 %,* ▲, △ *1.0 % and* ♦, ◊ *0.5 % (w/w).*

Figure 7 shows the decrease of viscosity for 1.5 % EDS gum solution as an influence of temperature.

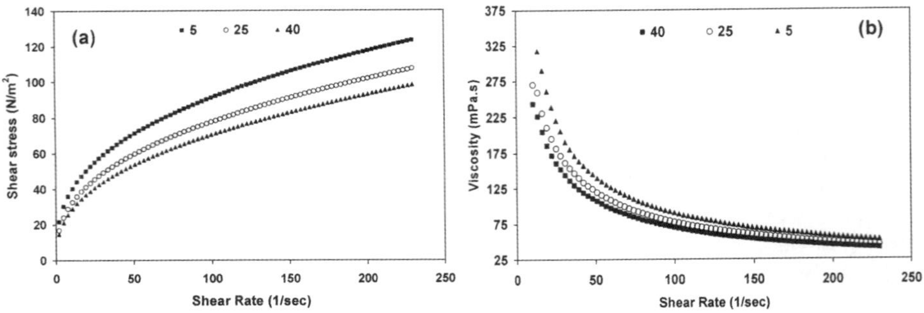

Figure 7 *Change of shear stress (a) and viscosity (b) versus shear rate as function of temperature for 1.5 % (w/w) EDS biogum solution.*

Similar results were obtained for the other concentrations. This effect, according to Garcia-Ochoa and Casas (1992), is reversible and it is due to the interactions between the polymer molecules in solution which become weaker at higher temperatures.[14]

3.5 Effect of pH on flow properties

The flow behavior index (n) and consistency coefficient (k) values, obtained by fitting the shear stress versus shear rate data to the Power law model, are given in Table 4. Similar to consistency coefficient, increasing the pH up to 7 resulted in an increase in the viscosity values and remained constant afterwards. The pH had a relatively high effect on the apparent viscosity over the pH range of 7 to 10. A number of gums such as galactomanans, CMC and gum karaya have been reported to loose their viscosity at low pH .[15] The viscosity of the succinoglycan solutions also decreased gradually at more acidic conditions that may be explained by the reduced degree of ionization of the gum carboxyl groups. The effect of pH on the viscosity of succinoglycan (EDS) at different shear rates is shown Figure 8.

Table 4 *Compare of means n and k obtained from 1.5 % (w/w) EDS gum solution as function of pH.*

pH	**2.5**	**4**	**7**	**10**
k	1071.70 ^{b}	1245.30 ^{c}	1247.00 ^{c}	912.27 ^{a}
n	0.400 ^{b}	0.386 ^{ab}	0.383 ^{a}	0.430 ^{c}
$^{\dagger}R^2$	0.99	0.99	0.99	0.99

In each row, means supplemented by different small letters differed by Duncan test at the 1% level as function of pH. *$^{\dagger}R^2$ = Regression coefficient*

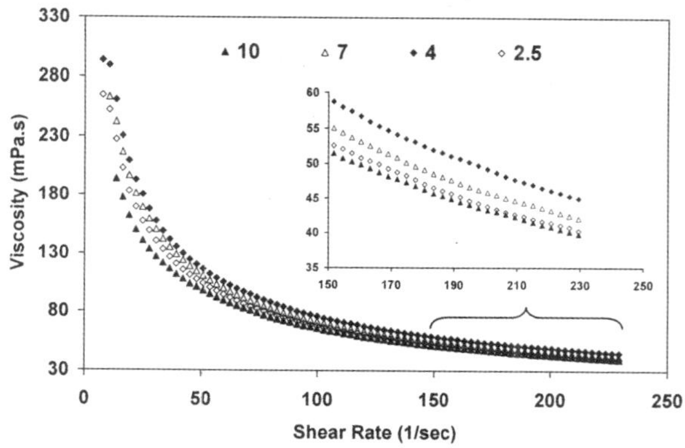

Figure 8 *Change of viscosity versus shear rate as function of pH for EDS biogum solutions at 25°C and concentration of 1.5 % (w/w).*

4. CONCLUSION

This study reported the isolation and characterization of an acidic water soluble EPS produced by a strain of *Agrobacterium radiobacter*. Water soluble EPS from other strains of this microorganism have been extensively studied[16] and up to four different EPS can be produced by a single strain.[17] There is considerable variability in the production of different EPS using various strains of *A. radiobacter*. For example, a strain designated 10C3 produces both succinoglycan and curdlan, while strain 10C3k (a mutant of 10C3) produces only curdlan in high yield.[18] A number of strains, including 1654 reported in this study (PTCC 1654), were found to show the ability to produce succinoglycan easily. Little is known about the genetic control of EPS synthesis in *A. radibacter*. In one study, the production of succinoglycan-deficient mutants was related to changes in the exogenes thought to be responsible for EPS synthesis.[19] The results of TLC revealed that the EPS contained glucose units when grown on the ES and EDS, however, there are conflicting reports on the ability of *A. radiobacter* to produce another sugar containing EPS. For example, mannose containing EPS was also produced by other strains had been reported previously to produce succinoglycan which did not contain any mannose.[20] In one of the first papers to report the isolation and identification of succinoglycan from *A.radiobacter*, strain 10C3 produced an EPS composed of 77 % glucose, 6.7 % galactose, 1.9 % mannose, and 10.6 % succinic acid when grown on a medium containing 0.1 % yeast extract.[21] It was shown in this study that the produced EPS was acidic and carrying a carboxylic group on the backbone of glucose residues. Although our research suggests that succinoglycan (the acidic polysaccharide) was produced by the strain PTCC 1654, further research is necessary to fully characterize this biogum. Succinoglycan behaved as a non-Newtonian shear thinning fluid in concentrations from 0.5% to 2% and the temperature range of 5 to 40 °C. The Power law model well described the rheological behavior of the solutions. Because of the shear thinning nature of its dispersions, succinoglycan could be suitable for application as a thickening agent. Increasing the solution concentrations increased the viscosity and pseudoplasticity while increasing temperature resulted in a decrease in viscosity and pseudoplasticity. As the temperature of the solution increased, k decreased and n increased, whereas an increase in concentration resulted in higher k increased and lower n. The apparent viscosity of succinoglycan was also affected by changes in pH of solutions and temperatures in different manner.

The results of this study could be used to achieve a good quality biogum with high yield production using a new and cheep carbon source to decrease the production costs.

References:

1 Margaritis, and G.W. pace, Oxford: Pergamon Press, 1985, Vol. 3, 1005.
2 Anonymous, San Diego, CA: Kelco Division of Merck and Co, 1988.
3 R.M. Banik, B. Kanari, and S.N. Upadhyay, World Journal of Microbiology and Biotechnology, 2000, **16**, 407-414.
4 V. Jezequel, Cereal Foods World, 1998, **43**, 361.
5 L. De Vuyst, and B. Degeest, FEMS Microbiology Reviews, 1999, **23**, 153-177.
6 G.H. Van Geel Schutten, F. Flesch, B. Ten Brink, M.R. Smith, and L. Dijkhuizen, Applied Microbiology and Biotechnology, 1998, **505**, 697.
7 K.A. Ludbrook, C. M.Russell, and R.I. Greig, Journal of Food Science, 1997, **62**, 597.
8 V.M. Marshall, A.P. Laws, Y. Gu, F. Levander, P. Radstrom, L. De Vuyst, B. Degeest, F. Vaningelgem, H: Dunn, and M. Elvin, Letters in Applied Microbiology, 2001, **32**, 433.
9 L.P.T.M. Zevenhuizen, Carbohydrate Polymers, 1997, **33**, 139-144.
10 M. Matulova, R. Toffanin, L. Navarini, R. Gilli, S. Paoletti, and A. Cesaro, Carbohydrate Research, 1994, **265**, 167.
11 T. Harada, and A. Harada, New York: Marcel Dekker, 1996, 54.
12 J.D. Linton, M. Evans, D.S. Jones, and D.N. Gouldney, General Microbiology, 1987, **133**, 2961.
13 S.G. Williams, J.A. Greenwood, and C.W. Jones, General Microbiology, 1990, **136**, 2179.
14 F.Garcia-Ochoa, and J.A. Casas, Journal of Science of Food and Agriculture, 1992. **59**, 97.
15 M. Glicksman, Food Hydrocolloids. Boca Raton, Florida: CRC Press Inc, 1982.
16 T. Haras, and A. Harada, New York: Marcel Dekker, 1996.
17 M. Hisamatsu, A. Amemura, T. Harada, T. Matsuo, H. Matsuda, and T. Harada, General Microbiology, 1982, **128**, 1873.
18 L. H. Aird, G. Brightwell, M.A. Jones, and A.W.B. Johnston, General Microbiology, 1991, **137**, 2287.
19 M. Hisamatsu, K. Sano, A. Amemura, and T. Harada, Carbohydrate Research, 1978, **61**, 89.
20 T. Harada, Archives of Biochemistry and Biophysics, 1965, **112**, 65.

EFFECTS OF L-ASCORBIC ACID AND L-CYSTEINE ON THE PHYSICOCHEMICAL CHARACTERISTICS OF WHEAT STARCH

M. Majzoobi[1], M. Radi[1], A. Farahnaky[1], J. Jamalian[1] and T. Tongdang[2]

[1] Department of Food Science and Technology, School of Agriculture, Shiraz University, Shiraz, Iran.
[2]Department of Food Science and Nutrition, Faculty of Science and Technology, Prince of Songkla University, Pattani Campus, Pattani 94000, Thailand.

1 INTRODUCTION

Starch is an important food component since it provides 70-80% of daily calories consumed by humans and also acts mainly as thickening and gelling agent in many foods.[1] Starch functionality is governed by its molecular structure. For example starch crystallinity, molecular weight of either amylose or amylopectin, degree of polymerization of starch molecules, chain length of amylopectin side chains and the ratio of amylose to amylopectin have a great impact on starch characteristics. Moreover the sources of starch and plant growing conditions also have impact on starch properties.[2] Furthermore the functionality of starch can be affected by other components in a food system. For example formation of amylose-lipid complexes and their effect on the starch gelatinization temperature and also on the type and degree of starch crystallinity have been well documented.[3] Hydrocolloids such as proteins and non-starch polysaccharides can interact with starch molecules and hence alter starch properties. [1]

Food additives may also have some effect on starch characteristics, for instance gaseous chlorine as a bleaching agent for cake flour can oxidize starch molecules and therefore change their characteristics. It has been reported that gaseous chlorine can increase the gelatinization temperature of starch. L-Ascorbic acid (AA) and L-cysteine (Cys) are two acceptable additives which are added to bread flour to improve dough and bread quality. In starch production from wheat flour, AA can be used to increase the starch and gluten yield. These chemicals are also present naturally or added in many other food products. [4]

AA and Cys are permitted up to 200 mg/kg and 90 mg/kg of wheat flour, respectively. It has been well established that AA and Cys can interact with dough proteins (gluten) by oxidizing or reducing the sulfydril or disulfide bonds, respectively.[4] Accordingly the dough may become stronger (when AA is used) or softer (when Cys is used). However the possible effects of these chemicals on starch granules, as the main component of flour, have received less attention in the literature. It is well recognized that ascorbic acid is highly reactive, easily oxidized and capable of decomposing polysaccharides. Its degradation effect on cassava starch and cereals ß-glucans have been reported.[5] It has been shown that addition of ascorbic acid to starches caused changes to their pasting properties.[6] Previous studies showed that addition of sulfites as a reducing agent can change the pasting properties of starch.[6] Accordingly the addition of Cys (a reducing agent) might have some effects on starch molecules.

The main aim of this study was to investigate the effects of AA and Cys on some physiochemical properties of wheat starch by imitating the bread dough preparation and baking conditions. The results of this study may be useful in better understanding of the improving mechanism of AA and Cys in dough and bread.

2 METHODS AND RESULTS

2.1 Materials

Pure wheat starch was purchased from Fars-Glucosin, Co. Shiraz, Iran. All chemicals were obtained from Merck, Germany.

2.2 Sample preparation

Slurries of wheat starch (30 g) were made with water (100 g) containing either L-ascorbic acid (140 mg/kg) or L-cysteine (63 mg/kg). To imitate dough preparation conditions (in terms of mixing time and temperature) samples were first mixed well for 8 min and then they were capped and left at 40 °C at a relative humidity of 80% for 45 min. Then they were dried using a cabinet drier at 50 °C for 2 hr. Then the dried starch was milled and sieved to obtain particle sizes in the range of 120-200 μm. To imitate baking conditions (in terms of mixing time and temperature) after storage of the starch slurries at 40 °C and relative humidity of 85 %, they were further stored at 100 °C for 40 min (during baking the temperature of the dough is about 100 °C). For each condition described, a control sample was also prepared. After that the samples were dried in a cabinet drier at 50 °C for 2 hr following by milling and sieving to obtain particle sizes in the range of 120-200 μm. Further experiments were performed on these samples.

2.3 Microscopic structure of starch

To study the microscopic structure of the starch samples, scanning electron microscopy (SEM) (Model 5526, Cambridge, UK) was used. To prepare the samples, a tiny amount of each sample was sputter coated with gold/palladium. Finally the samples were transferred to the microscope where they were observed at 20 kV.

2.4 Intrinsic viscosity

To determine the intrinsic viscosity of the samples, first solutions of 6 mg/ml starch in 1 M KOH were prepared and for each sample it was checked that it was fully solubilised.[7] The intrinsic viscosity was determined using a U-tube viscometer at (20 ± 0.1) °C the relative, reduced and intrinsic viscosities were determined using the Equations 1-3, respectively.[8]

$$\eta_{rel} = \frac{\eta}{\eta_0} = (\frac{t}{t_0}).(\frac{\rho}{\rho_0}) \qquad (1)$$

Where η_{rel} is relative viscosity, t and t_0 are the times required for the sample and solvent (KOH 1M) to pass through the U-tube, respectively. ρ and ρ_0 are the densities of sample solution and solvent, respectively.

Then reduced viscosity was determined according to the Equation 2:

$$\eta_{red} = \frac{\eta_{rel} - 1}{C} \qquad (2)$$

Where η_{red} is the reduced viscosity and C is the concentration of the sample.
Then intrinsic viscosity; [η], was determined using the Equation 3:

$$[\eta] = \lim_{c \to 0} \eta_{red} \qquad (3)$$

2.5 Wide angle X-ray diffraction

To determine the crystalline diffraction pattern of the starch samples, nearly 2 g of each sample equilibrated at relative humidity of 75% was pressed into a sample holder. Diffractograms were recorded using an X-ray diffractometer (Model D8 Advance, Germany). The scattered X-ray radiation was recorded by a proportional moving detector over a 4-38° (2θ) angular range with an angular velocity of 0.05°/min. The degree of starch crystallinity was determined using Equation 4. The area was calculated using the software developed and supplied by the instrument manufacturer (EVA, Version 9.0).

$$\text{Degree of crystallinity (\%)} = \left(\frac{\text{Area under the peaks}}{\text{Total curve area}}\right) \times 100 \qquad (4)$$

2.6 Thermal properties

A differential scanning calorimetry (DSC) DSC-7 (Perkin-Elmer, Beaconsfield, UK), calibrated with indium and cyclohexane, was used to analyse modified and native starch powders. An aliquot of each powder was mixed with distilled water (starch:water, 1: 3 w/w) in a stainless steel pan and left overnight for moisture equilibrium. High pressure stainless steel pans containing starch samples were scanned at a heating rate of 10 $K.min^{-1}$, from 10 to 115°C. An empty stainless steel pan was used as reference. Pyris software (Perkin-Elmer) was used to analyse the DSC traces and obtain gelatinization properties of starch samples in excess water.

2.7 Pasting properties

A Rapid Visco Analyser (RVA) (Newport Scientific Pty. Ltd., Warriewood, Australia) equipped with the software (Thermocline for windows) was used to study the pasting properties and to measure the apparent viscosity of the native and modified samples as a function of temperature.

An aliquot amount of each sample (4 g d.w.b.) was added to 25.0 g distilled water. The peddle blade was placed into the canister containing the sample and water and the canister was then inserted into the RVA. The stirring speed was 960 rpm for the first 10 seconds and then 160 rpm for the rest of the experiment. To study the pasting profile, samples were initially held at 25 °C for 2 min (to study the cold-water viscosity), then heated to 95 °C at a constant rate of 14 °C/min, held at that temperature for 3 minutes, and then cooled down to 25 °C at the same rate and held for 5 minutes.

2.8 Statistical analysis

The experiments were performed in a completely randomized design. All experiments were conducted in triplicate and the mean values and standard deviations were calculated. Analysis of variance (ANOVA) was performed and the results were separated using the Multiple Ranges Duncan's test ($\alpha < 0.05$) using statistical software of SPSS 13 (SPSS, Inc. New Jersey, USA).

2.9 Discussion

The micrographs of starch samples treated with either AA or Cys along with the control samples are shown in Figure 1. Some blister like spots can be observed on the surface of starch granules treated with AA and Cys before heating. These spots are more obvious on A-type starch granules. It has been reported that some spots can be observed on the surface of the starch granules after chemical modification. Perhaps some chemicals can cause changes to the surface of starch granules. AA and Cys have been known as an oxidative and reducing agents, respectively and hence they may have some effects on the starch granule proteins and hence change the smooth surface of the granules.

For heated samples no intact starch granules could be observed as the result of gelatinization (micrographs are not shown).

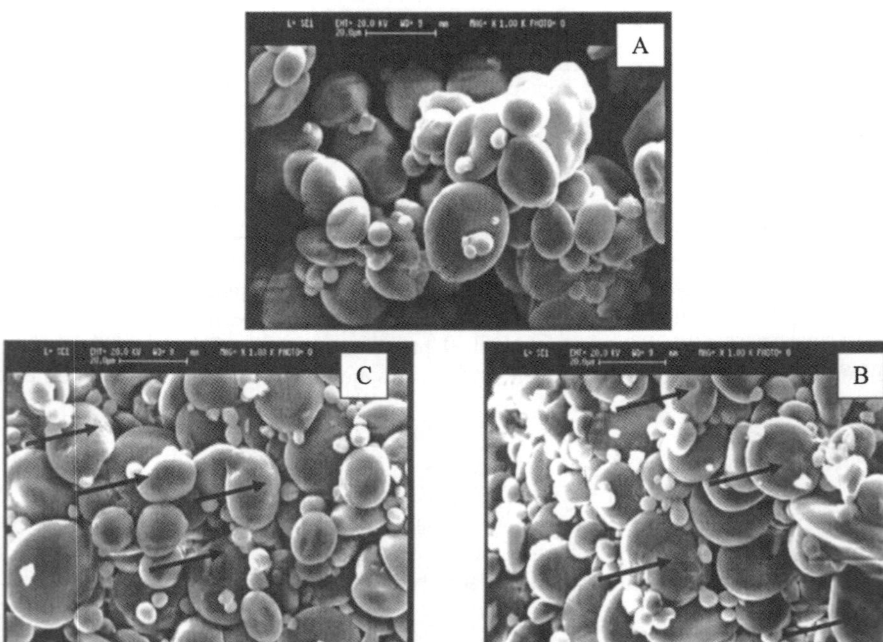

Figure 1. *Electronmicrographs of the native wheat starch (A), cysteine treated wheat starch (B) and ascorbic acid treated wheat starch (C). Each bar on the pictures represents*

X-ray diffraction patterns of starch samples along with the control and native wheat starch are shown in Figure 2. The results indicated that under the condition used for dough preparation, samples treated with AA or Cys had similar X-ray patterns to that of control and native wheat starch (typical A-pattern). Determination of the degree of starch crystallinity (shown on Figure 2) showed that all the samples had similar degree of crystallinity.

For heated samples no crystalline peak was observed (data are not shown) indicating that the amylopectin molecules had lost their crystalline structure as the result of gelatinization.

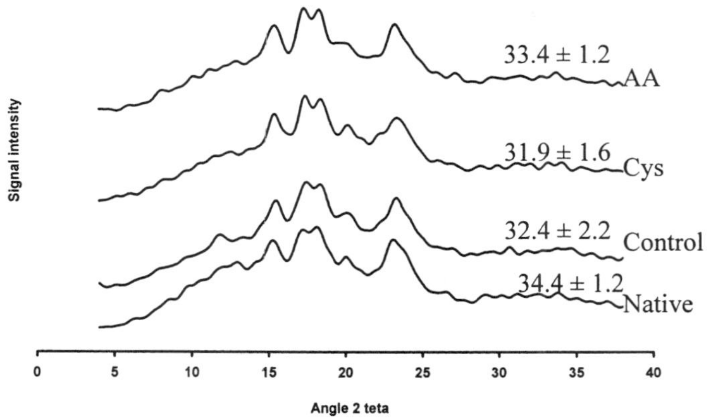

Figure 2. *The x-ray diffraction pattern of the starch samples. Values on the Figure are the degree of crystallinity ± SD.*

Determination of the intrinsic viscosity of the samples (Figure 3) showed that the intrinsic viscosity values of the samples before heating were significantly different from those of heated samples. Furthermore starch samples treated with Cys had the lower intrinsic viscosity value compared to those treated with AA and the control. Reduction in the intrinsic viscosity may indicate that AA and Cys could have resulted in starch degradation.

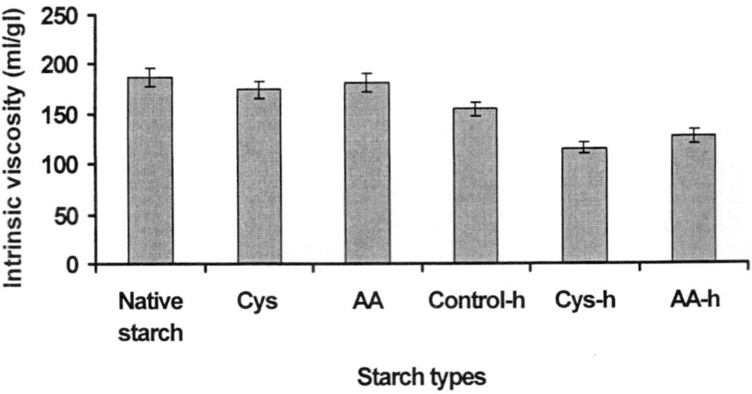

Figure 3. *The intrinsic viscosity of the starch samples as determined using a U-tube viscometer. Values are the averages of triplicates.*

Determination of the thermal properties of the samples before heating (Table 1) revealed that addition of AA or Cys had no effect on the gelatinization onset (To), peak temperature (Tp), conclusion temperature (Tc) and the gelatinisation enthalpy (ΔH) of the starch samples. Therefore it may be concluded that these chemicals did not have any effect on the crystalline amylopectin and hence on gelatinization behavior of starch. For the heated samples no gelatinization peak could be obtained from the DSC since starch was fully gelatinized during heating (data are not presented).

Table 1. *The thermal properties of starch samples as determined by DSC*.*

Sample	To[1] (°C)	Tp[2] (°C)	Tc[3] (°C)	ΔH[4] (J/g)
Control	55.62 ± 0.50	58.8 ± 0.62	69.49 ± 0.35	7.11 ± 0.65
Starch + Cys	55.24 ± 0.55	59.90 ± 0.65	68.94 ± 0.45	7.01 ± 0.51
Starch + AA	55.56 ± 0.33	58.90 ± 0.80	69.14 ± 0.70	7.11 ± 0.25

*Values are the averages of two replicates ± 1SD.
[1] Onset temperature
[2] Peak temperature
[3] Conclusion temperature
[4] Gelatinization enthalpy

Table 2. *Pasting properties of the starch samples as determined by the RVA**.

Sample	Cold visc.[1] (cP)	Peak visc.[2] (cP)	Peak time (min)	Pasting temp.[3] (°C)	Final visc.[4] (cP)
Control	13 ± 1	2103 ± 20	9.53 ± 2	25.85 ± 1	5942 ± 40
Starch + Cys	17 ± 1	2032 ± 15	9.40 ± 1	25.70 ± 1	5902 ± 32
Starch + AA	16 ± 1	2073 ± 20	9.60 ± 1	25.70 ± 1	5584 ± 35
Control (H)	97 ± 3	661 ± 10	7.73 ± 0.5	25.95 ± 1	1183 ± 20
Starch + Cys (H)	62 ± 3	507 ± 10	7.35 ± 1	25.75 ± 1	952 ± 20
Starch + AA (H)	61 ± 3	639 ± 8	7.60 ± 0.5	26.00 ± 1	1242 ± 35

* Values in the table are the averages of two replicate ± 1SD.
[1] Cold viscosity
[2] Peak viscosity
[3] Pasting temperature
[4] Final viscosity

For all the samples no cold viscosity was observed (the values obtained were very low). However all the samples had peak viscosity at 95 °C. The samples before heating had peak viscosity values greater than those of heated samples. Furthermore it seems that addition of Cys or AA did not have a great effect on peak viscosity values. Heated samples had lower peak time compared to those prepared without heating. Addition of Cys or AA had no significant effect on this value. Pasting temperature of the samples was not affected by heating or addition of either Cys or AA. Final viscosity of the samples was also affected by heating of the samples rather than each of the chemicals. For the starch + Cys the final viscosity was slightly lower than other samples. These results may indicate that addition of Cys or AA did not have any significant effect on pasting properties of the starch, however heating had some effect on pasting properties of starch which may be as a result of starch gelatinization.

3 CONCLUSION

The findings of this study suggest that addition of Cys or AA may have some effects on the functionality of wheat starch during bread dough preparation, including reduction of intrinsic viscosity and changes to the starch granule morphology.

When baking conditions were imitated, starch granules were almost fully gelatinized and hence AA and Cys could have more effects on starch molecules. For instance a significant reduction in the intrinsic viscosity was observed. Apparently Cys had greater effect on the intrinsic viscosity of starch molecules compared to AA. Reduction of the peak and final viscosities and peak time of heated samples compared to the unheated samples may also be an indication of molecular changes of starch caused by both heating and the added chemicals.

In real bread dough, the added water is mainly absorbed by the gluten, while starch granules absorb just a little water. Under this condition and based on the findings of this study, addition of AA or Cys to the dough may have a little effect on starch molecules. However during baking of the dough, the released water from the denatured gluten can be absorbed by starch granules and other hydrocolloids in the dough. At this stage AA or Cys can have greater effects on starch molecules. Further work is required to confirm such effects especially in the presence of natural minerals and other polysaccharides in the flour.

References

1 A. C. Eliasson, *Starch in Food,* Woodhead Publishing Limited, Cambridge, 2004.
2 R. T. Tester, J. Karkalas and X. Qi, *J. Cereal Sci.,* 2004, 39, 151.
3 S. N. Raphaelides and Nikolaos Georgiadis, *Carbohydr. Polyms*., 2006, 65, 81.
4 P. Sriburi, S. E. Hill and J. R. Mitchell, *Food Hydrocol*., 1999, 13, 177.
5 R. Kivela, F. Gates and T. Sontag-Strohm, *J. Cereal Sci.,* 2009, 49, 1.
6 L. Paterson, J. R. Mitchell, S. E. Hill and J. M. V. Blanshard, *Carbohyd. Res.*, *1995, 292*, 143.
7 A. Anastasiades, S. Thanou, D. Loulis, A. Stapatoris and T. D. Karapantsios, T.D. *J. Food Eng. 2002,* 52, 57.
8 S.E. Harding, *Prog. Biophys., Molec. Biol.,* 1997, 68, 207-262.

FREEZING BOUND WATER RESTRAINED BY GUM ARABIC

T. Hatakeyama[1], T. Uetake[2], Y. Inui[2] and H. Hatakeyama[2]

[1]Lignocel Research, Yatsumata, Fukui 910-3558, Japan
[2]Fukui University of Technology, Fukui 910-8505, Japan

1 INTRODUCTION

Water molecules coexisting with biopolymers, such as polysaccharides, are known to be categorized into three kinds, non-freezing, freezing bound and free water. Non-freezing water shows no first-order thermodynamic transition and the amount depends on the chemical structure of biopolymers, especially the number of hydrophilic groups. Freezing bound water shows a first order transition, however, the transition behaviour is different from that of free water[1-5]. In previous studies, it was found that freezing bound water takes an important role in the functional features of biopolymers, such as biocompatibility[3, 4, 6]. We have reported that a characteristic amount of freezing bound water is observed in polysaccharide-water systems such as xanthan gum[7], galactomannans[8], curdlan[9] and cellulose derivatives[10]. The amount of freezing bound water depends on the chemical structure of polymers and their higher order structure. Recently, the molecular conformation of polysaccharide aqueous solutions has been found to depend on the homogenization process to attain the equilibrium state[7, 11-16].
Gum Arabic (GA) is a unique polysaccharide and is widely used as thickener in food and pharmaceutical fields[17-21]. It is known that the total amount of bound water (summation of non-freezing and freezing bound water) of GA is similar to that of other functional polysaccharides, however, the amount of freezing water markedly depends on water content [22, 23], thermal history and measurement conditions. In this study, the thermal behaviour of freezing bound water restrained by GA molecules was investigated in order to elucidate the unique behaviour of GA molecular assemblies.

2 MATERIALS AND METHODS

2.1 Materials

Two kinds of gum Arabic, a commercial gum Arabic (*Acacia senegal*) (EM0) and a matured gum (Acacia (*sen*) SUPER GUM™, EM1 were provided by San-Ei Gen F.F.I., Inc, Japan.

The modification process was described in detail previously [24, 25]. i.e. the gums in the dry state are subjected to a heating process under specially controlled conditions. According to gel permeation chromatography (SEC-MALLS detection), two peaks were observed. M_w of Gum Arabic DS was 4.30 x10^{-6} g mol^{-1} and 0.48 x10^{-6} g mol^{-1} and the mass ratio of the 1st peak was 59.7 %. M_w of Gum Arabic EM1 was 5.98 x 10^{-6} g mol^{-1} and 3.90 x 10^{-6} g mol^{-1} and the mass ratio of the 1st peak was 74.5 %. The first GPC peak of GA-EM1 contains arabinoglactan protein and that of the 2nd peak contains arabinogalactan and glycoprotein.

Samples having various amounts of water were prepared. Water content of the sample was defined by W_c=(mass of water)/(mass of dry sample). g g^{-1}.

2.2 Differential scanning calorimetry

A Seiko differential scanning calorimeter DSC220C was used to measure the phase transition of the samples. Sample weight was ca. 5 mg, heating rate was 10 °C min^{-1} and an aluminium sealed type sample pan was used. Temperature was varied from -150 to 80 °C. The sample was heated from -150 to 80 °C at 10 °C min^{-1} (first heating). After maintaining the sample at 100 °C for several minutes, the sample was cooled at 10 °C min^{-1} to -150 °C (1st cooling). Again the sample was heated from -150 to 80 °C at 10 °C min^{-1} (2nd heating). The above process was repeated 5 times, and then heating and cooling curves were accumulated.

3 RESULTS AND DISCUSSION

Figure 1 shows DSC heating curves of GA(EM1)-water systems with various water contents. As shown in Fig. 1-(a), systems with W_c = 0.33 and 0.43 show no-first order phase transition, except baseline deflection due to the glass transition. In the system with W_c =0.64, a shallow exothermic peak which is attributable to cold crystallization and a melting peak are observed in addition to the glass transition. Melting peak temperature was found at a temperature lower than 0 °C, suggesting that ice formed in the system with W_c =0.64 is not normal ice. As shown in Fig. 1-(b), glass transition is scarcely observed when W_c increased. A large endothermic peak is due to melting of ice and a small shoulder peak at the low temperature side of large peak is attributed to melting of irregular ice which can be observed in the system with W_c =0.64

Phase transition behaviour of the DSC curves of GA(EM1)-water systems in a water content range from 0 to 5 can be explained as follows, i.e. (1) when the number of added water molecules is less than the number of hydrophilic groups of a GA, the main chain motion which was not observed in completely dry GA becomes detectable. Glass transition temperature decreases markedly at the initial increase of water content since water molecules break intra- and inter-molecular hydrogen bonding of GA, (2) in the above water content range, ice is not formed since each water molecule is strongly restrained by the hydrophilic group, accordingly no first order phase transition is observed by thermal analysis. By NMR relaxation study of various kinds of water restrained by polysaccharides, the average correlation time was calculated from ^{1}H longitudinal and transverse relaxation time take 10^{-8} to 10^{-7} sec depending on temperature, showing that this kind of water is categorized as a non-rigid solid[26]. When ionic groups coexist in the system, the number of strongly restrained water molecules increases. This amount of strongly bound water is categorized as non-freezing water. Figure 2 shows the relationship between the amount of non-freezing water and water content of GA (EM1), (3) when the amount of water exceeds a characteristic value (In the case of GA ca. $W_c = 0.5$), an excess amount of more mobile water molecules assembles and ice having defects is formed. This process can be detected as an exothermic event as cold crystallization. Cold crystallization accompanies simultaneous melting of irregular ice, on this account; exothermic change due to cold crystallization is masked with endothermic change of melting. This suggests that the enthalpy of cold crystallization depends strongly on experimental conditions. (4) Melting of ice is observed at a temperature lower than 0 °C. Melting temperature and enthalpy markedly depend on water content, thermal history and the chemical structure of polysaccharides, (5) from the melting enthalpy of the above ice, the amount of freezing bound water can be calculated. By NMR, relaxation times continuously decrease and no distinct change due to the formation of freezing bound water is ordinarily detected (6) when sufficient amount of free water is introduced into the system, normal ice (hexagonal packing ice) is formed and free molecular motion of GM molecules accompanied with non-freezing water is restricted and the glass transition temperature shifts to the high temperature side. (7) On this account, the lowest glass transition was observed at a W_c where freezing bound water is formed.

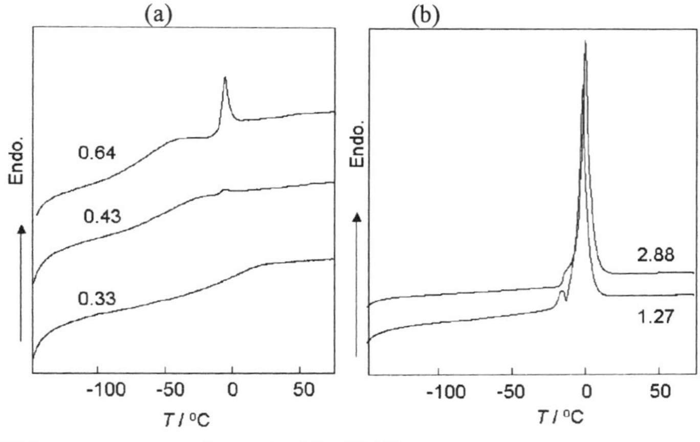

Figure 1 DSC *heating curves of gum Arabic (EM1)*
Heating rate=10 °C min^{-1} Numerals in the figures show water content

Recent studies on water-biopolymer interaction among the above mentioned three types of water, i.e. non-freezing, freezing bound and free water; the role of freezing bound water has attracted particular attention[3, 4, 6]. It has been reported that the amount of freezing bound water correlates with biocompatibility, for example protein adsorption is prevented when a sufficient amount of freezing bound water exists on the surface of membranes. In order to search for new materials having biocompatibility, it is important to find a polymer restraining a large amount of freezing bound water. Takigami and his coworkers have investigated hyaluronan-water and GA (SA)-water systems by DSC and reported that a large amount of freezing bound water was observed compared with other polysaccharides[22]. We also measured hylauronan-water systems in a similar water content range, however, we could not find the same phenomena[27]. It is thought that the thermal history of both experiments causes the above difference. Takigami measured polysaccharide-water systems in the temperature range from −50 to +50 °C, whereas in our studies the range was from −150 to100 °C. This suggests that the glassification and molecular equilibration of GA with non-freezing water seems to affect the amount of freezing bound water.

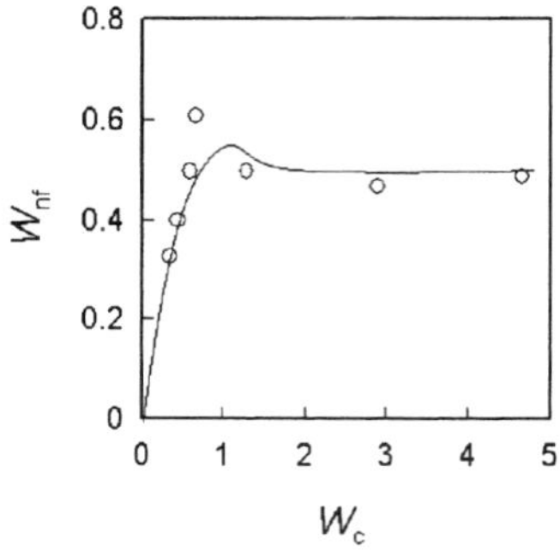

Figure 2 *Relationship between the amount of non-freezing water (W_{nf}) and water content (W_c) of GA (EM1)*

Figure 3 shows DSC heating curves of GA-water systems measured by repeated heating and cooling. W_c of GA (EM1) was selected at 0.64 where only freezing bound water is detected in the heating curves. As shown in the stacked DSC heating curves of Fig. 3, a glass transition, a shallow exotherm due to cold crystallization and a melting endothermic peak due to the melting of freezing bound water are observed. A melting peak was maintained at the same temperature, on the other hand, the peak area decreases with increasing number of heating and cooling cycles. This suggests that the amount of freezing bound water decreases by repeated thermal treatment. DSC cooling curves of GA (EM1) with W_c = 0.64 showed only a glass transition and no exothermic peak indicating crystallization was observed, although this figure is not shown. DSC cooling curves indicated that all of the water molecules are frozen in the glassy state. i.e. a homogeneous GA system and non-freezing water were formed during the freezing process. However, on heating, part of the non-freezing water assembled, and was crystallized as freezing bound water. By repeated heating and cooling the freezable portion decreases, which suggests the molecular conformation of GA molecules changed by thermal treatment.

In order to compare the effect of molecular structure on the formation of freezing bound water, GA (SD) with W_c =0.70 was measured under similar conditions as GA(EM1). Figure 4 shows DSC heating curves of GA (SD). DSC curves of GA (SD) are different from those of GA(EM1). Besides the glass transition, cold crystallization melting of both freezing bound water and free water were detected. Although the water content of both GA samples were almost the same, free water was clearly seen in GA (SD). In DSC cooling curves, two clear crystallization peaks and a glass transition were observed. In contrast to GA(EM1), DSC curves scarcely vary by repeated heating and cooling.

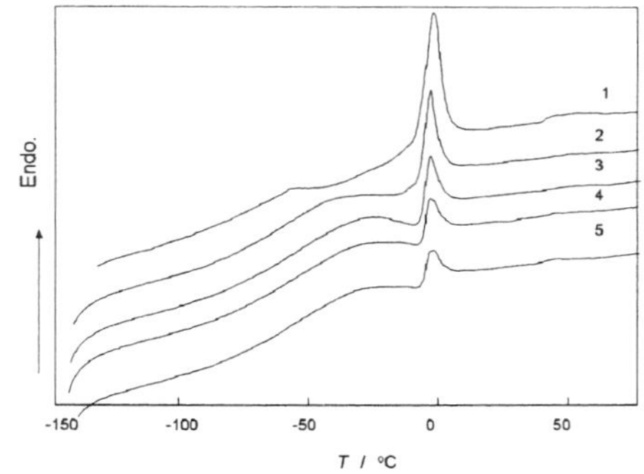

Figure 3 *DSC heating curves of gum Arabic–water system EM1 W_c =0.64*
Numerals in the figure show the number of times of heating and cooling at 10 °C min^{-1}

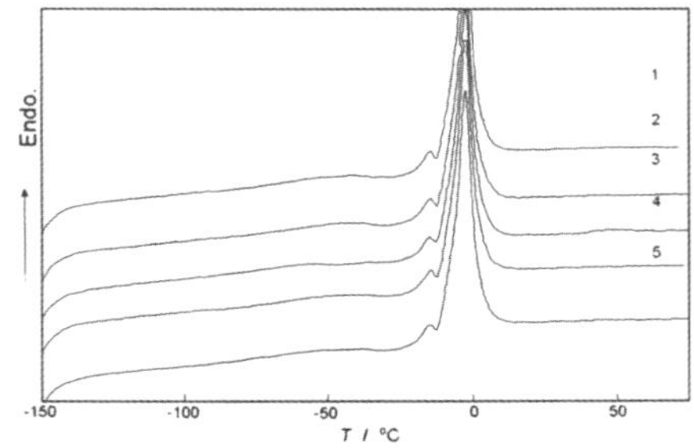

Figure 4 *DSC heating curves of gum Arabic–water system SD* W_c *=0.70*
Numerals in the figure show the number of times of heating and cooling at 10 °C min^{-1}

Figure 5 shows the phase transition temperature (T_g) of GA as a function of the number of heating and cooling cycles. The temperature of melting free water (T_m), freezing bound water (T_{ml}) of cool crystallization (T_{cc}) and T_g of GA(SD) maintain constant values regardless of the number of heating and cooling cycles. In contrast, T_g and T_{cc} of GA (EM1) shift to the high temperature side with increasing number of heating and cooling. T_{ml} of GA (EM1) is slightly higher than that of GA(SD). T_g of GA(EM1) is observed at a temperature far lower than that of GA(SD).

The enthalpy of freezing bound and free water was calculated from the endothermic melting peak using the melting enthalpy of pure water. By subtraction of freezing water (free and freezing bound water) from W_c, the amount of non-freezing water was calculated. As shown in Figure 6, in GA(EM1), the amount of freezing bound water (W_{fb}) decreases. In contrast non-freezing water content (W_{nf}) increases with increasing number of heating and cooling cycles. At the same time, it is recognized that a large difference between two types of GAs, i.e. W_{nf} values of EM1 are far larger than that of DS.

Figures 5 and 6 indicate the following facts (1) when a large amount of bound water, (non-freezing and freezing bound water) is restrained by GA molecules, molecular mobility increases and segmental motion enhances from low temperature (ca.–90 °C), (2) the higher order structure of GA molecules is not stable, (3) segmental motion is more easily enhanced for GA molecules having the higher-order structure capability to fold a large amount of freezing bound water, (4) since the total number of water molecules in the system is the same, in the initial stage, a part of the water molecules assemble and form ice which was detected as freezing bound water, (5) by repeated freezing and thawing, molecular configuration equilibrates and non-freezing water molecules are homogeneously restrained in the whole GA molecules, (6) the above process is detected as a shift of T_{cc} to the high temperature side and decrease of W_{fb}, (7) the first order structure of GA, such as molecular mass and amount of protein may affect the above process, although further investigation is necessary.

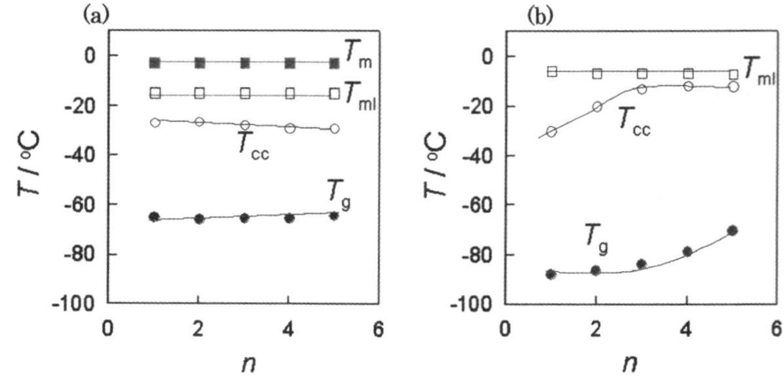

Figure 5 *Glass transition (T_g), cold crystallization (T_{cc}) and melting temperatures (T_{ml}, T_m) of gum arabic-water systems as a function of repeated heating and cooling cycles (n). (a) SD (W_c =0.70), (b) EM1 (W_c =0.64)*

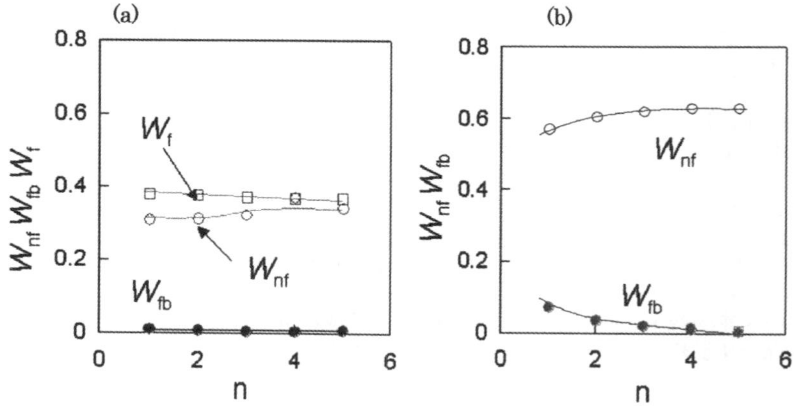

Figure 6 *The amount of non-freezing water (W_{nf}), freezing bound water (W_{fb}) and free water (W_f) of gum arabic-water systems as a function of repeated heating and cooling cycles. (a) SD (W_c =0.70), (b) EM1 (W_c =0.64)*

In our previous study, we reported that the higher order structure of aqueous solutions of polysaccharide equilibrates as a function of annealing time temperature[12-15]. It was also found that gelation of physical gels is affected by thermal treatment[7, 16, 28, 29]. W_{nf} value is thought to take the role of the index of configurational change of the polysaccharide matrix. Ordinarily W_{nf} gradually reaches equilibrium through sinusoidal vibration. GA is a complex polysaccharide and its unique functions are observed in aqueous media. Water molecules restrained by GA as bound water are thought to be important for their molecular stabilization through the variation of freezing bound water. A certain number of water molecules form irregular ice when molecular configuration allows them to assemble, at the same time, water molecules having once categorized into freezing bound water transforms to non-freezing water. By configurational change of the matrix polysaccharide, the number of non-freezing water molecules increases. It is thought that the thermal behaviour of freezing bound wateracts as an index for the above complex equilibration.

4. Conclusions

From the above results, it is considered that the higher order structure of GA molecular assemblies in the presence of water equilibrate through the structural change of freezing bound water. The amount of non-freezing water governs the molecular mobility of GA. Thermal treatment by freezing and thawing leads to a homogeneous system with stable structure

Acknowledgements
The authors are grateful to San-Ei Gen F.F.I., Inc, Japan for providing GA samples.

References

1 Hatakeyama T, Hatakeyama H (2004) Thermal properties of green polymers and biocomposites. Kluwer Academic, Dordrecht
2 Hatakeyama H, Hatakeyama T (1998) Interaction between water and hydrophilic Polymers. Thermochim Acta 308:3-22
3 Tanaka M, Motomura T, Ishii N, Shimura K, Onishi M, Mochizuk A, Hatakeyama T (2000) Cold crystallization of water in hydrated poly(2-methoxyethyl acrylate). Polym Int 49:1709-1713.
4 Tanaka M, Mochizuki A, Motomura T, Hatakeyama T (2002) Study of blood compatibility with poly(2-methoyehtyl acrylate). Relationship between water structure and platelet compatibility in poly(methoxyethylacrylate-co-2- hydroxythylmethacrylate. Biomacromol 3:36-41
5 Hatakeyama T, Kasuga H,. Tanaka M, Hatakeyama H (2007) Cold crystallization of poly(ethylene glycol)-water systems. Thermochim Acta 465:59-66
6 Mochizuki A, Hatakeyama T, TomonoY, Tanaka M (2009) Water structure and blood compatibility of poly(tetrahydrofurfuryl acrylate). J Biomat Sci 20:591–603
7 Iijima M, Shinozaki M, Hatakeyama T,. Takahashi M, Hatakeyama H (2007) AFM studies on gelation mechanism of xanthan gum hydrogels. Carbohyd Polym 68:701-707
8 Naoi S, Hatakeyama T, Hatakeyama H (2002) Phase transition of locust bean gum-, tara gum- and guar gum-water systems. J Therm Anal Cal 70:841-852
9 Hatakeyama T,. Ueda C, Hatakeyama H (2006) Structural change of water by gelation of curdlan suspension. J Therm Analy Cal 85:661-668
10 Onishi T, Hatakeyama H, Hatakeyama T (2008) DSC and AFM studies of chemically cross-linked sodium cellulose sulfate hydrogels. In Hu TQ ed. Characterization of

lingocellulosic materials. pp. 329-339. Blackwell, UK
11 Yoshida T, Takahashi M, Hatakeyama T, Hatakeyama H (1998) Annealing induced gelation of xanthan/water systems. Polym 39:119-1122
12 Fujiwara J, Takahashi M, Hatakeyama T, Hatakeyama H (2000) Gelation of hyaluronic acid through annealing. Poly. Int 49:1604-1608
13 Fujiwara J, Iwanami T, Takahashi M, Tanaka R, Hatakeyama T, Hatakeyama H (2000) Structural change of xanthan gum association in aqueous solutions. Thermochim Acta 352-353:241-246
14 Takahashi M, Hatakeyama T, Hatakeyama H (2001) Temperature and concentration dependency on equilibration of polysaccharide electrolyte hydrosols. In Kennedy JF, Phillips GO, Williams PA, Hatakeyama H eds. Recent advances in environmentally compatible polymers. pp. 145-154. Woodhead, Cambridge
15 Iseki T, Takahashi M, Hatano H, Hatakeyama T, Hatakeyama H (2001)Viscoelastic properties of xanthan gum hydogels annealed in the sol state, Food Hydrocoll 15:503-506
16 Iijima M, Hatakeyama T, Takahashi M, Hatakeyama H (2007) Effect of thermal history on kappa-carrageenan hydrogelation. Thermochim Acta 452:53-58
17 Street CA, Anderson DMW (1983) Refinement of structure previously proposed for gum arabic and other acaica gum exudates. Talanta 30:887-893
18 Whistler RL (1993) Exudate gums. In WhistlerRL, Be Miller JN eds. Industrial gums: Polysaccharides and their derivates 3rd ed. Academic Press, New York
19 Cowman MK, Funami T, Al-Assaf S, Kudasheva DS, Mohan D, Phillips GO (2006) Use of atomic force microscopy to investigate the structure of arabionogalactan proteins. FFIJ:2II:207-212
20 Al-Assaf S, Phillips GO, Williams PA (2005) Studies on acacia exudates gums. Part I:The molecular weight of *Acacia Senegal* gum exudates. Food Hydrocoll 19:647-660
21 Biswas S, Biswas, B, Phillips GO (1995) Classification of natural gums. 8. Chemometric assignment of commercial gum exudates from Africa using cluster-analysis on the protein amino-acid compositions. Food Hydrocoll 9:151–163
22 Takigami S, Takigammi M, Phillips GO (1995) Effect of preparation method on the hydration characteristics of hylan and comparison with another highly cross-linked polysaccharide, gum arabic. Carbohyd Polym 24:1-18
23 Phillips G.O, Takigami S, Takigami M (1996) Hydration characteristics of the gum exudate from *Acacia segenal*. Food Hydrocoll 10:11-19
24 Al-Assaf S, Phillips G.O, Aoki H, Sasaki Y (2007) Characterization and properties of *Acacia senegal* (L.) *Willd. var. senegal* with enhanced properties (Acacia (sen) SUPER GUM). Part 2. Mechanism of the maturation process. Food Hydrocoll 21:319-328
25 Aoki H, Katayama T, Al-Assaf S, Phillips G.O (2007) Characterization and properties of Acacia senegal (L.) *Willd. var. senegal* with enhanced properties (Acacia (sen) SUPER GUM). Part 1. Controlled maturation of *Acacia senegal var*. senegal to increase viscoelasticity, produce a hydrogel form and convert a poor into a good emulsifier. Food Hydrocoll 21:329-337
26 Hatakeyama T, Hatakeyama H (2002) Phase transition of sodium hyaluronate hylan and polyurethanes derived from hyaluronic acid in the presence of water. In Kennedy JF, Phillips GO, Williams PA, Hascall VC, eds. Hyaluronan, vol 1 Chemical, Biochemical and Biological Aspects, pp 323-328. Woodhead, Cambridge
27 Hatakeyama H, Iwata H, Hatakeyama T (1987) ^{1}H and ^{23}Na NMR studies of the interaction between water and sodium cellulose sulfate. In Kennedy JF, Phillips GO, Williams PA eds. pp. 39-46. Wood and cellulose. Ellis Horwood, Chichester
28 Hatakeyama T, Naoi S, Iijima M, Hatakeyama H (2005) Locust bean gum hydrogels formed by freezing and thawing. Macromol Symp 224:251-262
29 Hatakeyama T, Uno J,. Yamada C, Kishi A, Hatakeyama H (2005) Gel-sol transition of poly(vinyl alcohol) hydrogels formed by freezing and thawing. Thermochim Acta 431:144-148

POLYSACCHARIDE ULTRASONICATION – BEYOND DEPOLYMERIZATION

SS Wong[1], Y Sun[1], ZJ Ngiam[1], S Kasapis[2], D Huang[1]

[1]*Department of Chemistry, National University of Singapore, Block S8, Level 3, Science Drive 3, Singapore 117543*
[2]*School of Applied Sciences, RMIT University, Bundoora West Campus, Melbourne, Vic 3083, Australia*

ABSTRACT

Ultrasound depolymerization of polysaccharides has been studied extensively for several decades. Nevertheless, there are a lot of gaps in the literature in relation to the physicochemical properties of materials following ultrasonication. Thus, it is of interest to investigate the mechanism of ultrasonication and the molecular dynamics of materials generated beyond the basic concept of polymeric chain fragmentation.

An earlier study confirmed that ultrasonication of bacterial cellulose in a relatively short period of time (up to 30 min) produced an increment in crystallinity of the generated fragments[1]. The present work aims to prolong the ultrasonication period in order to promote further increase in crystallinity thus achieving levels comparable to those found for microcrystalline cellulose. Results from X-ray diffraction studies revealed that increments of crystallinity have been observed for samples sonicated at 60 and 90 min, as compared to shorter sonication-period counterparts.

Furthermore, ultrasonication has been used extensively for many years in the field of sonochemistry in an effort to aid certain chemical reactions[2]. As before, the emphasis of those undertakings in polysaccharide chemistry was to show depolymerisation and reduced molecular weight of materials. The second part of this investigation intends to explore novel applications of ultrasound in relation to the composition and functionality of pendant groups in curdlan polysaccharide. This has been achieved by varying the ultrasonication time and observing changes in the degree of substitution of modified curdlan species.

In conclusion, current findings have developed a certain understanding on the ability of ultrasonication to modify polysaccharide chain characteristics and accompanying functionality. On-going studies aim to gain further insights into the underlying mechanism of polysaccharide ultrasonication.

1. INTRODUCTION

Polysaccharide modification has been an interesting subject to pursue since decades ago in the aim of achieving novel functionalized biopolymer. Searching for desirable non-conventional modification methods is never impeding, especially in connection with the emerging concept of green chemistry. Ultrasonication has been described as a useful aid in chemical processes to achieve the aim of waste and polluting products reduction[3], thus, offers a promising alternative to modify polysaccharide.

Ultrasound is defined as sound wave having a frequency higher than the human hearing range, i.e. > 20 kHz[2]. Depolymerization of polymer via ultrasonication has been extensively studied throughout the years. For polysaccharides, studies have been done on chitosan[4], cellulose[5] and its derivatives (e.g., cellulose nitrate[6], carboxymethylcellulose[7]), starch[8], water soluble polysaccharides including guar gum[9], dextran[10], agarose and carrageenans[11]. Nevertheless, most studies were limited to depolymerization experiments.

A recent finding[1] on ultrasonication of bacterial and plant cellulose reveals interesting trends on the increment in crystallinity index of the sonicated fragments. The effect has been achieved in a relatively short period of 30 min sonication time. Thus, it is of interest to further explore the possibility of crystallinity index increment by prolonging ultrasonication time. On the other hand, Cízová and his team[12] has recently reported that by employing ultrasonication, it has significantly shortened the reaction time required for octenylsuccinoylation of carboxymethyl starch. The study provides useful fundamental information on the ability of ultrasound in assisting derivatization of polysaccharides.

Curdlan is a bacteria synthesized linear β-1,3-glucan obtained from *Alcaligenes faecalis* var. *myxogenes* strain 10C3[13]. Since its discovery by Harada in 1966[14], it has received interest in food and non food related research, due to its unique physicochemical properties. Curdlan possesses the ability to form 'high-set' thermo-irreversible gels and 'low-set' thermo-reversible gels at two distinctive temperatures, which is unique compare to other heat-set gels[15]. Meanwhile, β-1,3-glucan belongs to a group of compounds known as biological response modifiers that enhance or restore normal immune defenses[13]. Therefore, most non-food related studies were concentrated on immunomodulation and anti-tumour activity of this bacterial exopolysaccharide[16].

Since the early 1990s, studies had shown that sulfated derivatives of curdlan possess better biological activity and / or antiviral properties[17, 18]. Therefore, more subsequent work has dealt with the synthesis of curdlan sulfate and the corresponding biological activity[16]. To the best of our knowledge, most curdlan and other β-glucan sulfation processes involved the utilization of harmful chemical such as pyridine-SO_3 complex[16], pyridine-chlorosulfonic acid[19] and / or piperidine-N-sulfonic acid[18]. Moreover, most of these reactions will generate hazardous by-products, such as pyridine. It is thus, of interest to seek for alternative ways in producing curdlan sulfate from the approach of green chemistry.

In short, current work wishes to gain better insight to the ability of ultrasound in polysaccharide modification, beyond conventional depolymerization. Crystallinity modification of cellulose and functionalization of curdlan were chosen as the subject of interest.

2. MATERIALS AND METHODS

2.1. Materials

Plant cellulose (PC) (Cellulose fibrous medium, from cotton lint), powdered microcrystalline cellulose (MCC) were purchased from Sigma-Aldrich, Germany. Bacterial cellulose (BC) was a kind gift from Wong Coco(s) Pte Ltd, Singapore. Commercial curdlan samples were donated by Takeda-Kirin Foods Corporation, Tokyo Japan. Unless otherwise specified, all other chemicals were of reagent grade and were used without further purifications.

2.2. Ultrasonication of Cellulose

Plant cellulose and lyophilized bacterial cellulose were dissolved in cuprammonium hydroxide solution (CUAM) before subjecting to ultrasonication (Elmasonic S 60H, ELMA, Singen, Germany) for 5, 10, 15, 30, 60 and 90 min. Details of the procedures has been described previously[1].

2.3. X-Ray Powder Diffraction Studies (XRPD)

X-Ray powder diffraction patterns of lyophilized cellulose samples were obtained using a SIEMENS D5005 X-Ray Diffractometer (Bruker AXS, Karlsruhe, Germany) equipped with Cu-Kα (1.54 Å) radiation. Diffraction profiles of the samples were fitted with a Gaussian function to resolve into separate diffraction peaks and subsequent integration was performed on each peak to obtain the crystallinity index (CrI) of samples. The X-Ray diffraction profile of MCC was used as a benchmark for comparison.

2.4. Ultrasonication of Curdlan

Curdlan was thoroughly suspended in 50% (v/v) sulphuric acid before being subjected to ultrasonication for 360 min at 40 °C. Control samples were prepared in a similar way by suspending curdlan for 360 min at 40 °C, without subjecting to ultrasonication. The samples were subsequently neutralised, extensively dialysed and eventually lyophilized. The sonicated sample was denoted as 360_360, while the non-sonicated sample was denoted as 0_360. Samples were prepared in triplicate and the results reported are an average.

2.5. Elemental Analysis of Curdlan

Lyophilized curdlan samples were analysed for sulphur content (%S) by elemental analysis using Elementar Vario Micro (Elementar, Analysensysteme, Germany). Degree of sulfation (DS) was calculated according to the following equation[20]:

$$DS = \frac{0.162 \times (\frac{\%S}{32})}{100 - \left[(\frac{80}{32}) \times \%S\right]}$$

where %S is the sulfur content (%)

2.6. Fourier Transformed Infrared (FTIR) Spectroscopy

FTIR spectra of curdlan samples were obtained with a Spectrum One FTIR Spectrophotometer (Perkin-Elmer, Norwalk, USA) using a KBr pellet. Samples were scanned with a resolution of 4.00 cm^{-1}, within the range of 4000 cm^{-1} to 400 cm^{-1}.

3. RESULTS AND DISCUSSION

3.1. X-Ray Diffraction Profile and Crystallinity Index (CrI)

Figure 1a depicts the X-Ray diffraction patterns of bacterial cellulose sonicated at different time periods (5, 10, 15, 30, 60 and 90 min). For plant cellulose, similar diffraction profiles were observed for each time period of sonication (Figure 1b). After the deconvolution process, three major diffractions peaks at 2θ = 12.0°, 20.1° and 21.6° were obtained for both BC and PC samples. The results were in a good agreement with the previous finding, indicating all sonicated samples existed in the Cellulose II crystal conformation[1].

Even though the difference in diffraction profile for each sample were less visible, subsequent integration revealed that CrI for BC had increased over the ultrasonication time, as summarized in Table 1. This is in agreement with our hypothesis. Nevertheless, the trend of increment was not clear. Prolonged ultrasonication did not significantly increase the CrI for BC, compared to that of shorter treatment times. Therefore, further study needs to be done to confirm the underlying mechanism. Sonicated plant cellulose does not have a visible trend for the observed CrI value, which is consistent with the previous findings[1].

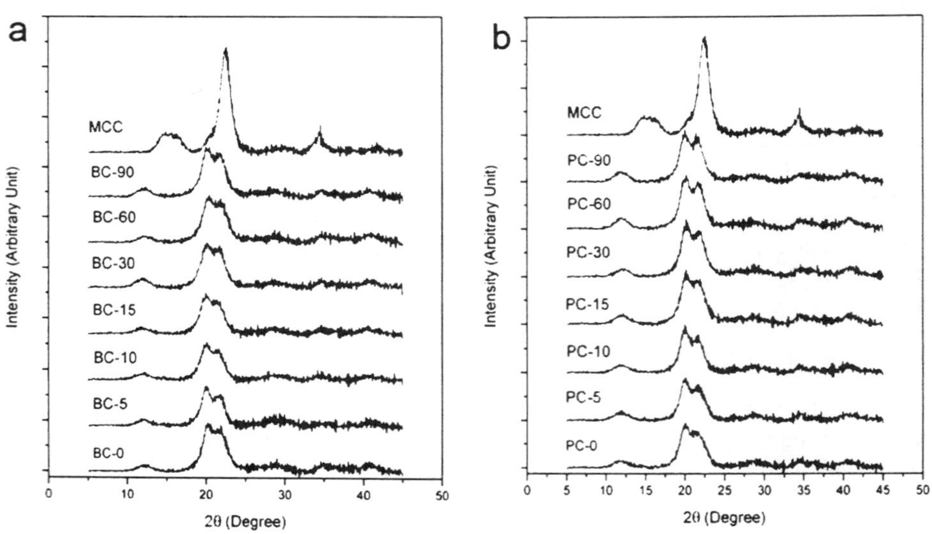

Figure 1 X-Ray diffraction patterns of (a) bacterial cellulose and (b) plant cellulose sonicated at different time period, compared to MCC.

Table 1 Crystallinity index (CrI) of bacterial cellulose (BC) and plant cellulose (PC) ultrasonicated for different times.

Sample	Ultrasonication Time (min)	CrI (%)[a]	Sample	Ultrasonication Time (min)	CrI (%)[a]
MCC	–	85.1			
BC-0	0	64.7	PC-0	0	67.8
BC-5	5	66.6	PC-5	5	69.5
BC-10	10	67.4	PC-10	10	69.5
BC-15	15	69.2	PC-15	15	67.3
BC-30	30	67.9	PC-30	30	67.9
BC-60	60	70.0	PC-60	60	67.5
BC-90	90	73.3	PC-90	90	69.1

[a] Means of readings from duplicate sample

3.2. Sulphur Content and Degree of Sulfation (DS)

Table 2 reveals the sulfur content and DS of curdlan sulfate prepared in the current study. The extent of sulfation for all samples are relatively lower compared to previous work, as deduced from the elemental analysis results. Previous studies had shown that curdlan sulfate prepared by a conventional chemical method showed a minimum %S of 5.6%[18] and a maximum of 20%[19]. This could have given a wrong impression that the conventional chemical syntheses (i.e. utilizing pyridine-SO_3 complex) are more effective. Actually, most reactions involving pyridine would require many hours of reflux and / or tedious multi-step preparations, not to mention the harmful waste (pyridine) generated after the reaction. In addition, according to a recent report published by Bae and his team[20] on the sulfation of citrus pectin using chlorosulfonic acid, the %S obtained is only 2.68. Therefore, the conventional chemical process may not be the definite way to obtain high DS values in sulfated polysaccharide.

Nevertheless, the process of ultrasonication has significantly increased the DS value of samples by at least two fold. The promising results suggested that ultrasonication is a useful method in inducing sulfation of curdlan. Moreover, it might be possible to obtain higher degree of sulfation in future by employing systematic optimization processes, such as response surface methodology, to obtain curdlan sulfate with more desirable %S.

Table 2 Sulfur content and degree of sulfation (DS) for curdlan.

Sample	Sulfur content (%)[b]	Degree of sulfation (DS)[b]
Curdlan	< 0.50	0
0_360	0.58 ± 0.02	0.030 ± 0.001
360_360	2.15 ± 0.09	0.115 ± 0.005

[b] Means of readings ± standard deviation from triplicate sample

3.3. Fourier Transform Infrared (FTIR) Spectroscopy

Figure 2 shows the FTIR spectra of curdlan samples. Both acid treated samples (0_360 and 360_360) showing an absorption peak at 1249 cm^{-1}, ascribed to the asymmetrical S=O stretching vibration[20, 21]. Meanwhile, the presence of a peak at 1060 and 1061 cm^{-1} for 360_360 and 0_360, respectively, represent the symmetrical S=O stretching vibration[21]. These confirmed that acid and ultrasonication treatment had incorporated sulfate groups into the curdlan backbone. Results from elemental analysis suggested that sulfate groups presence are relatively less, thus, it is expected that peaks attributed to sulphate groups will be less pronounced and slightly overlapped with other peaks.

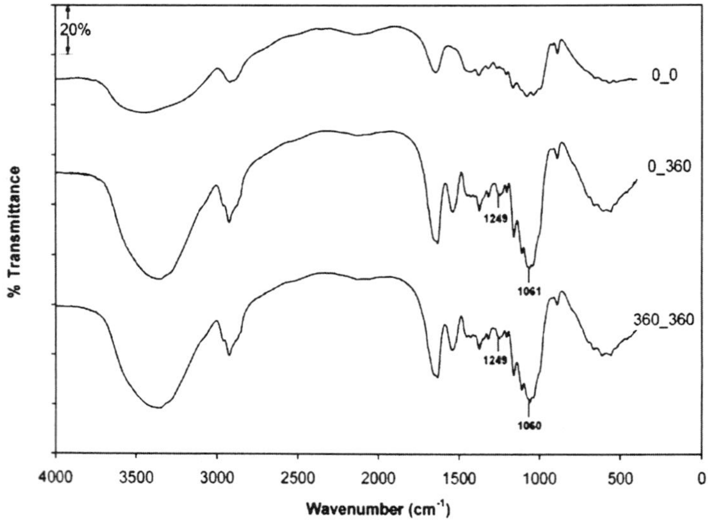

Figure 2 FTIR spectra of curdlan samples: 0_0, 0_360 (acid treatment only) and 360_360 (acid + ultrasonication).

4. CONCLUSION

Current studies expand the methodologies of polysaccharide modification. Preliminary results indicating sonicated bacterial cellulose having a higher CrI after the sonication process, yet the increment were less pronounced at longer sonication times. Sulfation of curdlan was more effective with the assistance of ultrasonication. Nonetheless, an in depth study is still needed in order to understand the detailed method to control the sonochemical processes, in order to obtain products with favourable physicochemical properties.

REFERENCES

1. Wong, S.-S.; Kasapis, S.; Tan, Y. M., *Carbohydrate Polymers* **2009,** 77, (2), 280-287.
2. Mason, T. J.; Lorimer, J. P., *Applied Sonochemistry: The uses of power ultrasound in chemistry and processing* Wiley-VCH: Weinheim, Germany **2002**.
3. Kardos, N.; Luche, J.-L., *Carbohydrate Research* **2001,** 332, (2), 115-131.
4. Kasaai, M. R.; Arul, J.; Charlet, G., *Ultrasonics Sonochemistry* **2008,** 15, (6), 1001-1008.
5. Striegel, A. M., *Biomacromolecules* **2007,** 8, (12), 3944-3949.
6. Marx-Figini, M., *Angewandte Makromolekulare Chemie* **1997,** 250, (1), 85-92.
7. Antti, G.; Pentti, P.; Hanna, K., *Ultrasonics Sonochemistry* **2008,** 15, (4), 644-648.
8. Luo, Z.; Fu, X.; He, X.; Luo, F.; Gao, Q.; Yu, S., *Starch - Staerke* **2008,** 60, (11), 646-653.
9. Tayal, A.; Khan, S. A., *Macromolecules* **2000,** 33, (26), 9488-9493.
10. Cote, G. L.; Willet, J. L., *Carbohydrate Polymers* **1999,** 39, (2), 119-126.
11. Lii, C.-y.; Chen, C.-H.; Yeh, A.-I.; Lai, V. M. F., *Food Hydrocolloids* **1999,** 13, (6), 477-481.
12. Cízová, A.; Sroková, I.; Sasinková, V.; Malovíková, A.; Ebringerová, A., *Starch - Stärke* **2008,** 60, (8), 389-397.
13. McIntosh, M.; Stone, B.; Stanisich, V., *Applied Microbiology and Biotechnology* **2005,** 68, (2), 163-173.
14. Harada, T.; Masada, M.; Fujimori, K.; Maeda, I., *Agricultural and Biological Chemistry* **1966,** 30, 196-198.
15. Nishinari, K.; Zhang, H., *Trends in Food Science & Technology* **2004,** 15, (6), 305-312.
16. Gao, Y.; Fukuda, A.; Katsuraya, K.; Kaneko, Y.; Mimura, T.; Nakashima, H.; Uryu, T., *Macromolecules* **1997,** 30, (11), 3224-3228.
17. Yutaro, K.; Osamu, Y.; Ryusuke, N.; Takashi, Y.; Masazumi, D.; Sadahiko, O.; Shigeru, S.; Yoshimasa, M.; Nobuya, N.; Yasuo, I.; Tohoru, M.; Hisashi, S.; Naohiko, Y.; Kei, M.; Toshiyuki, U.; Naoki, Y., *Biochemical Pharmacology* **1990,** 39, (4), 793-797.
18. Yoshida, T.; Hatanaka, K.; Uryu, T.; Kaneko, Y.; Suzuki, E.; Miyano, H.; Mimura, T.; Yoshida, O.; Yamamoto, N., *Macromolecules* **1990,** 23, (16), 3717-3722.
19. Yoshida, T.; Yasuda, Y.; Mimura, T.; Kaneko, Y.; Nakashima, H.; Yamamoto, N.; Uryu, T., *Carbohydrate Research* **1995,** 276, (2), 425-436.
20. Bae, I. Y.; Joe, Y. N.; Rha, H. J.; Lee, S.; Yoo, S.-H.; Lee, H. G., *Food Hydrocolloids* **2009,** 23, (7), 1980-1983.
21. Vikhoreva, G.; Bannikova, G.; Stolbushkina, P.; Panov, A.; Drozd, N.; Makarov, V.; Varlamov, V.; Gal'braikh, L., *Carbohydrate Polymers* **2005,** 62, (4), 327-332.

APPLICATION OF THE COUPLING MODEL TO THE RELAXATION DYNAMICS OF POLYSACCHARIDE/CO-SOLUTE SYSTEMS

B. Jiang[1] and S. Kasapis[2]

[1] Department of Chemistry, National University of Singapore, Block S3, Level 6, Science Drive 4, Singapore 117543
[2] School of Applied Sciences, RMIT University, Building 223, Level 1, Bundoora West Campus, Plenty Road, Melbourne, VIC 3083, Australia

1 INTRODUCTION

Gelatin has been widely used in the confectionery industry for many decades, especially at a high solid environment with sugar syrup as co-solute, for products such as wine gums and gummy bears. However, the use of gelatin is losing popularity due to reasons including religious dietary laws, vegetarianism, and health perceptions (BSE scare)[1]. Among the replacement materials, gelling polysaccharides have shown promising results in high sugar environments[1,2,3]; however, our understanding of the physics of these high solid systems is still limited due to the inherent complexity of multi-components systems.

For densely packed molecular systems such as polymers, study of the relaxation dynamics is particularly important because of the correlation between the relaxation patterns and the chemical structures of the polymers. This could give us insights to the microscopic relaxation process of the systems. The Williams-Landel-Ferry (WLF) equation in conjunction with free volume theory have been widely used to describe the softening dispersion of amorphous synthetic polymers[4], and recently this approach was proved to be suitable for amorphous biopolymer systems[5,6,7,8]. However, more and more, people argued that the free volume concept is not the most fundamental physical quantity to describe dynamics of polymer systems, and it might oversimplify the intermolecular cooperative dynamics[4]. It posed a challenge to come up with a more fundamental solution to describe the molecular dynamics in the interacting polymer systems.

The "coupling theory" has gained attention in looking for a more fundamental description of molecular dynamics of the softening dispersion. The theory argues that the relaxation of a "primitive species" is not isolated from its surroundings, and the strength of the coupling of this primitive species to its environment is determined by the chemical structures of the primitive species and the surrounding molecules[9]. Used in conjunction with the function of Kohlrausch, Williams and Watts (KWW), the coupling model could return parameters indicating the cooperativity between primitive relaxation and the surrounding materials as well as the time and temperature dependence of the relaxation properties. Although this approach has been quite well established in the field of synthetic polymers[4], to the best of our knowledge, it has only been applied to gelatin/co-solute system within the field of amorphous biopolymer systems[5]. Thus, this paper is going to explore the use of coupling theory to describe high solid polysaccharide systems in the presence of co-solute (sugars).

2 MATERIALS AND METHODS

Agarose was supplied by Sigma Chemical Co (product number A0576). Deacylated gellan gum was supplied by Kelco Biopolymers, U.K. κ-Carrageenan was obtained from Hercules, and was ion-exchanged in the potassium form using Amberlite IR-12 exchanging resin. The glucose syrup was a product of Capital Glucose Co Ltd, Thailand, with dextrose equivalent of 42, and total level of solids of 85.1%.

To prepare the high solid systems, polysaccharides were dissolved in deionised water at 90 °C with stirring for 15 min, the solutions were then cooled down and kept at 70 °C for the addition of glucose syrup. The mixtures were stirred at 70 °C until they were homogeneous and excess water was evaporated to bring the total level of solid to 80% (w/w, moisture content of glucose syrup was considered for the calculation of total solid content). $CaCl_2$ solution was added into the deacylated gellan/co-solute mixture at 70 °C and stirred to obtain 7.5 mM Ca^{2+} in the final mixture, and KCl solution was added to the κ-carrageenan/co-solute mixture at 70 °C to obtain 10 mM K^+ in the final mixture. Samples were allowed further evaporation of water to compensate for the addition of salt solution. The amount of agarose, deacylated gellan and κ-carrageenan in the mixture was 2%, 1% and 0.5% (w/w), respectively.

Mechanical behaviour of the systems was measured with the advanced rheometrics expansion system (ARES, TA Instruments, U.S.A.). Small deformation dynamic oscillation was used to record the temperature dependence of viscoelastic behaviour from 10 °C down to -50 °C at a frequency of 1 rad/s. Isothermal time dependence of mechanical relaxation was assessed by transient stress relaxation tests at a series of temperatures ranging from -50 °C to 4 °C at 3 °C intervals. Samples were allowed to relax isothermally for 8 min after application of an instantaneous strain varied from 0.02% to 1% depending on the stiffness of the sample at different temperatures.

Three samples were prepared and tested for each polysaccharide/co-solute system, three sets of data were recorded, and the average of the essentially overlapping traces was reported. The nonlinear analysis was carried out using GraphPad Prism Version 4.

3 RESULTS AND DISCUSSION

3.1 Observations of the Relaxation Process of the Polysaccharides/Co-solute Systems

Measurement of viscoelastic parameters of systems at a constant frequency as a function of temperature could give a quick glimpse of the overall viscoelastic behaviour of the system and an indication of some important pattern of behavior[13]. Figure 1 gives an example of the vitrification of a polysaccharide/co-solute system consisting of 2% agarose and 78% (dry basis) glucose syrup, as a function of temperature. The vitrification of deacylated gellan and κ-carrageenan/co-solute systems produced similar traces, thus are not presented here.

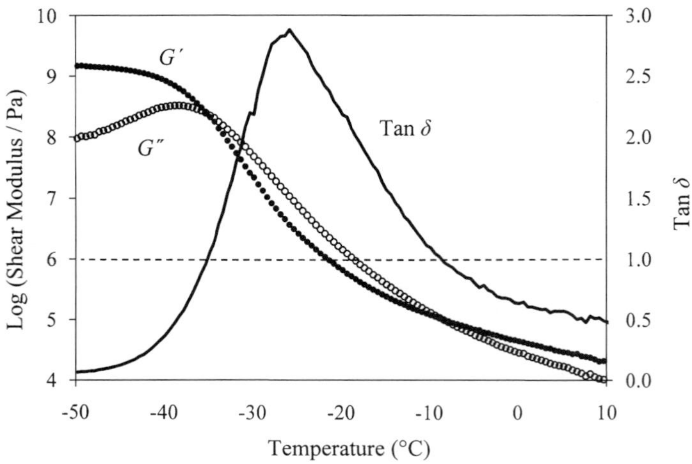

Figure 1 *Temperature dependence of shear moduli and tan δ of a system containing 2% agarose and 78% glucose syrup (scan rate, 1 °C/min; frequency, 1 rad/s; strain, 0.01 to 1%)*

Figure 1 clearly showed three sections of distinct viscoelastic behaviour. At the high temperature end (above -8 °C), the temperature pattern resembled that of rubbery plateau ($G' > G''$ with little temperature dependence of both moduli), indicating an elastic network with extensive topological entanglement or covalent cross-linking. On further cooling (below -8 °C), the elastic and viscous moduli crossed over, the viscous response became dominant ($G'' > G'$), and there was a spectacular development of both moduli from 10^5 Pa to $10^{8.5}$ Pa, the damping factor tan δ (tan $\delta = G'' / G'$) also reached a maximum far above 1, clearly indicating a glass transition zone. In this area, the long range movements of the main chain are restricted, but local relaxation processes which dissipate energy can occur. Finally, the moduli crossed over again at temperatures below -35 °C and the system entered the glassy state, with G' becoming increasingly dominant and approaching its equilibrium value. At the same time, G'' started to drop after it reached its maximum. The elastic response was more than one magnitude higher than the viscous response at -50 °C, indicating a strong solid-like behaviour of the system at glassy state.

Both calorimetry and rheology have been commonly used for the determination of glass transition temperature (T_g). However, it was found recently that a small addition of a network forming polymer to a high solid sugar preparation could accelerate the mechanical manifestation of vitrification, while the calorimetric T_g was unaltered[10]. The T_g of a system containing network forming materials was defined as "network T_g", and it was found that a mechanical study could describe better the effect of network formers on the vitrification of such a system. However, it is hard to pinpoint the T_g because vitrification takes place over a range of temperatures. Early attempts to demarcate the T_g of such a system from rheology measured the temperature dependence of shear moduli and damping factor (eg. Figure 1) and produced empirical indices. These include the temperature where the storage modulus starts to fall rapidly with increasing temperature, the temperature where the loss modulus or the damping factor reaches a maximum, or the temperature where the storage and loss moduli coincide[1]. However, in the absence of a distinct molecular process associated with these selections, the approach denotes merely an empirical indicator of convenience[14].

3.2 Modeling the Relaxation Process Using the Combined WLF/Free Volume Framework

Understanding of the relaxation process of the polysaccharide/co-solute systems should be advanced by characterizing the relaxation process as a function of time, and this was achieved by a series of isothermal stress relaxation runs at temperatures covering the range of -50 °C to 4 °C; stress relaxation modulus was recorded. The stress relaxation spectra obtained at different temperatures can be superimposed to form a master curve covering a large time scale which is not experimentally accessible (Figure 2). This was done by shifting horizontally the experimental stress relaxation curves taken at different temperatures along the logarithmic time axis to produce a smooth trace centered at an arbitrarily chosen reference temperature (T_0). The method of developing mater curves is called the time-temperature superposition principle (TTS). Thermorheological simplicity (TS) of the system is a criterion to perform TTS, which means the relaxation processes of the system should have the same temperature dependence[11]. TS was observed for the polysaccharides/co-solute systems, resulting in good superposition of stress relaxation spectra at different temperatures and smooth master curves. However, it has been reported that thermorheological complexity (TC) is more pronounced on low molecular weight materials[4]. Figure 2 shows the master curve of κ-carrageenan/co-solute system, the master curves of the high solid agarose and deacylated gellan systems are similar to Figure 2 and are not shown here. The glassy state at short times (< 10^{-1} s) and rubbery region (> 10^4 s) where the reduced stress relaxation modulus $G_p(t)$ has little time dependence can be clearly seen from Figure 2. A spectacular drop of $G_p(t)$ (from 10^9 to 10^4 Pa) during the glass to rubber transition (10^{-1} s < t / a_T < 10^4 s) is also shown.

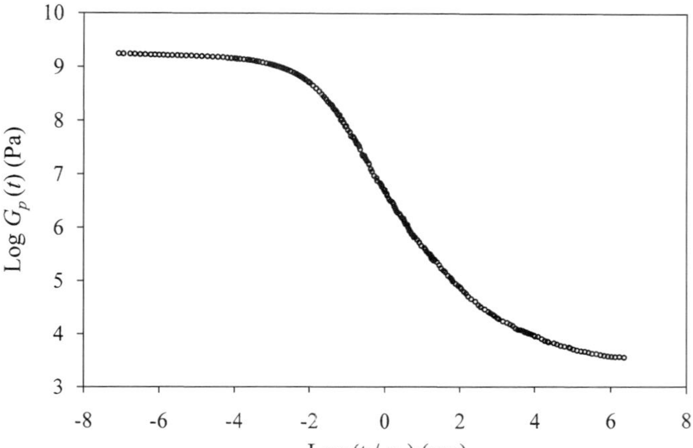

Figure 2 *Master curve of stress relaxation modulus of a system containing 0.5% κ-carrageenan, 79.5% glucose syrup, and 10 mM KCl reduced to -29 °C and plotted logarithmically against reduced time (t/a_T)*

The TTS approach generated a series of temperature shift factors, a_T, which determines how much the time scale of measurements shifts with temperature[1]. Figure 3 shows the logarithmic dependence of the shift factors as a function of temperature for the high solid agarose, deacylated gellan, and κ-carrageenan preparations.

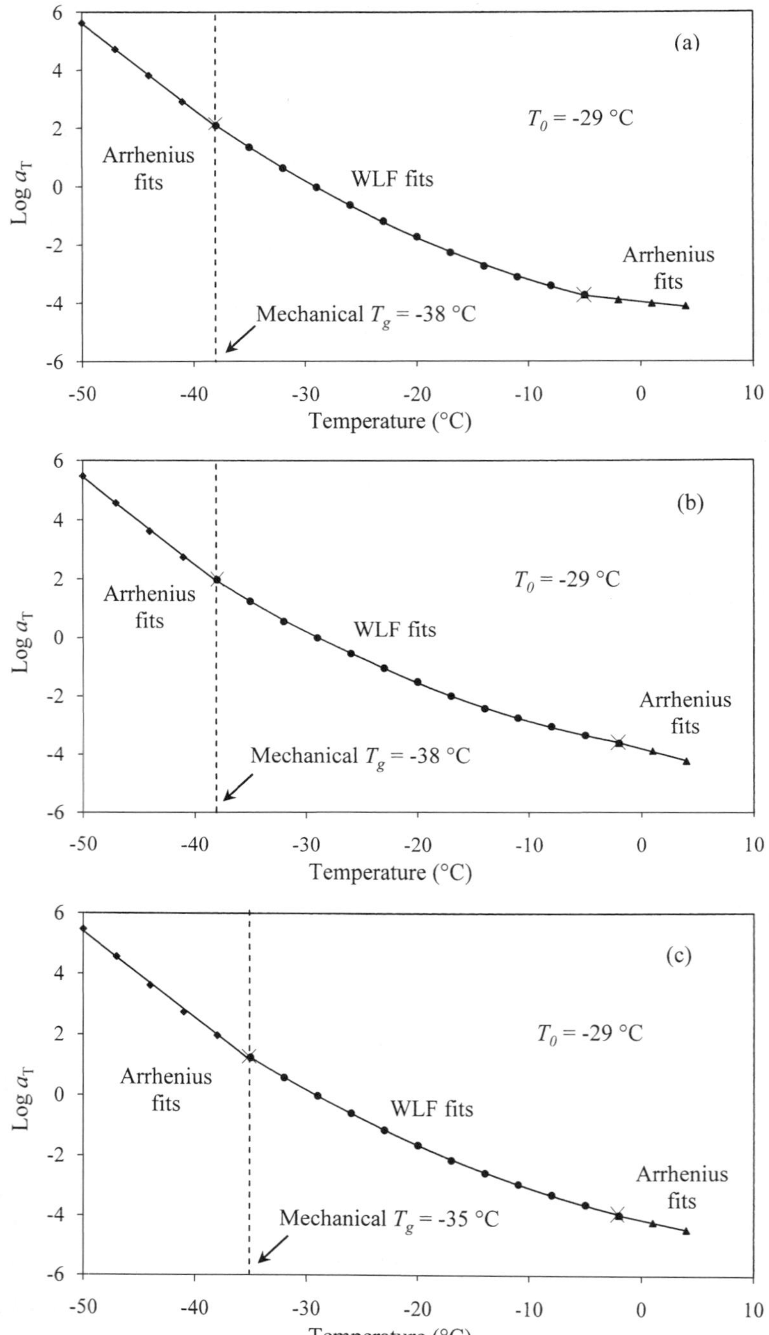

Figure 3 *Logarithmic dependence of the shift factors as a function of temperature for agarose/co-solute system (a); deacylated gellan/co-solute system (b); and κ-carrageenan/co-solute system (c), at reference temperature of -29 °C*

Clearly the shift factors show distinct temperature dependence at different temperature regions, and the points where the change of temperature dependence occurred were labelled with a cross (Figure 3). It was verified that in the lowest and the highest temperature regions, corresponding to the glassy and rubbery regions, the plot can be well described by the modified Arrhenius relationship[12]:

$$\log a_T = (E_a/2.303R)(1/T - 1/T_0) \tag{1}$$

where, T_0 is the reference temperature, R is the gas constant, and E_a is the activation energy within the relevant temperature region.

Within the intermediate temperature region where the glass transition occurs, however, the shift factors of the mechanical spectra unveil a pattern of behaviour that cannot be followed by the modified Arrhenius equation. Instead, the WLF equation could provide a good fit of the progress in mechanical properties in this temperature region:

$$\log a_T = -C_1^0(T - T_0)/(C_2^0 + T - T_0) \tag{2}$$

where, T_0 is the reference temperature, and C_1^0 and C_2^0 are WLF constants at the reference temperature. The two constants can be related to the free volume theory as follows:

$$C_1^0 = B/2.303f_0 \tag{3}$$

$$C_2^0 = f_0/\alpha_f \tag{4}$$

where, f_0 is the dimensionless ratio of free volume to total volume (v_f / (v_f + v_o)), α_f is the thermal expansion coefficient and B is usually set to 1.

The free volume theory states that the total volume of the material is the sum of the free volume v_f and the occupied volume v_o, v_o including not only the van der Waals radii but also the volume associated with vibrational motion. The free volume is therefore the extra volume required for large scale vibrational motions flexing over several atoms, that is, transverse string-like vibrations of a chain rather than longitudinal or rotational vibrations[12]. During vitrification the fractional free volume collapses from around 30% at the rubbery end to about 3% of the total volume at the glassy end of the transition[6]. Combining with the Arrhenius relationship in the glassy state, the mechanical T_g can be pinpointed at the conjunction of the glass transition region with the glassy state, as defined by the progression of the temperature dependence of shift factors[15]. Mechanistically, this coincides with the transformation from free volume derived effects in the glass transition region (transverse string-like vibrations) to an energetic barrier to rotation from one conformational state of molecular groups to another in the solid-like glassy state.

Mathematical equations were devised to determine the free volume parameters at T_g[12].

$$C_1^0 = C_1^g C_2^g/(C_2^g + T_0 - T_g) \tag{5}$$

$$C_2^0 = C_2^g + T_0 - T_g \tag{6}$$

The free volume parameters of the polysaccharides/co-solute systems obtained at T_g using these equations are listed in Table 1, together with results found in the literature from small deformation dynamic oscillation studies.

Table 1 *Current (stress relax) and literature (dynamic oscillation) free volume parameters*

Polymer	T_g (°C)		f_g (%)		α_f ($\times 10^{-4}$ °C^{-1})	
	Current	*Literature*	*Current*	*Literature*	*Current*	*Literature*
Agarose	-38	-37[6]	0.0251[6]	0.0250	3.91	6.00[6]
Deacylated gellan	-38	-26[8]	0.0308[8]	0.0311	5.64	3.67[8]
κ-carrageenan	-35	-35.5[7]	0.0269[7]	0.0290	3.88	5.90[7]

With the same composition of high-solid agarose and κ-carrageenan systems, the current study produced similar values of free volume indicators as for the literature. For the high-solid deacylated gellan preparation, however, Al-Ruqaie *et al.*[8] obtained a T_g value of 12 °C higher than the current study. This is because we used a sample of 80% total solid instead of 85.5% used by Al-Ruqaie *et.al.*. When a system of the same composition was tested, T_g of -25 °C was obtained, which is similar to that obtained by Al-Ruqaie *et al.*

3.3 Application of the Coupling Model to the Structural Relaxation Process and Utilization of the KWW Equation

For a densely packed system such as a polymer system, the softening transition is a result of a variety of motions, ranging from the Rouse modes of Guassian submolecules, to sub-Rouse, and local segmental motions[4]. The viscoelastic behaviour at the extreme short time side of the softening transition which accounts for the glass transition temperature is mainly contributed by the local segmental motion[13]. The "coupling model" states that the local segmental motions are heavily affected by the interactions between neighbouring segments of the polymer, thus the distribution of relaxation times resembled a stretched exponential function, or the function of Kohlrausch, Williams and Watts (KWW)[9]:

$$\emptyset(t) = exp[-(t/\tau)^{1-n}] \quad (7)$$

where τ is the coupled relaxation time, and *n* is the coupling constant taking values between 0 and 1.0. The equation finds its physical significance through the close relationship of both parameters with the coupling theory. The coupling constant, *n*, reflects the intensity of interactions between the underlying relaxation and the physicochemical environment of the surrounding materials[5]. A second relationship exists between the coupled relaxation time τ and the primitive relaxation time τ_0, which is governed by the interactions between neighbouring segments[13]. For stress relaxation modulus, Equation 7 recasts to the following form:

$$G(t) = (G_g - G_e)exp[-(t/\tau)^{1-n}] + G_e \quad (8)$$

where G_g and G_e are the glassy and equilibrium moduli of the local segmental motions, and t is the time after application of a fixed strain.

Master curves of stress relaxation with time were produced using the glass transition temperature as reference temperature, and the extreme short time of the glass transition region in the master curves, which is dominated by local segmental motions, was fitted using the KWW function (Equation 8). The glassy and equilibrium moduli of the local segmental motions of the agarose system were taken to be 1.61×10^9 and 1×10^8 Pa, those of the deacylated gellan system were 1.7×10^9 and 7.2×10^7 Pa, and those for the κ-carrageenan systems were 1.54×10^9 and 1×10^8 Pa. Figure 4 shows the KWW fits for the progression of the stress relaxation modulus in these systems.

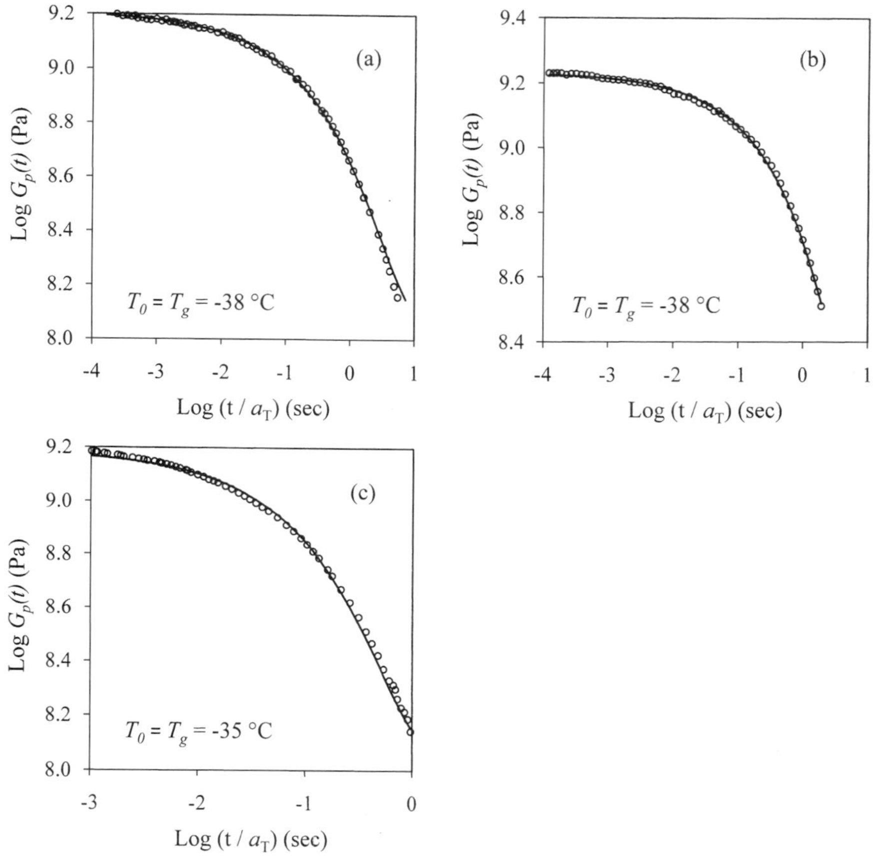

Figure 4 *Short time region of the stress relaxation master curve for the high solid agarose (a); deacylated gellan (b); and κ-carrageenan (c) preparations at the reference temperature of T_g, with the empty circle indicating experimental data, and solid line indicating the predictions of the KWW function.*

As shown in Figure 4, the stretched exponential function of KWW described well the progression of stress relaxation modulus with time at the short time region of the glass transition. Nonlinear analysis was carried out using GraphPad Prism Version 4, with the regression coefficients of all the fits being above 0.99. These fits will be used to obtain the relaxation time and coupling constant of high solid polysaccharide systems and assign physical significance to those values in relation to previously reported data for gelatin/co-solute systems.

4 CONCLUSION

Preliminary data on the application of the coupling model, in conjunction with the KWW function, to the vitrification of high solid polysaccharide/co-solute systems were obtained. In doing so, for each system, a series of isothermal stress relaxation spectra at different temperatures were recorded and a stress relaxation master curve was obtained through the method of TTS. The temperature dependence of the shift factor generated through TTS was used to pinpoint the mechanical T_g of the system using the combined WLF/free volume theory framework. Based on this prediction of the glass transition temperature, the coupling constant and relaxation times of systems will be obtained and compared with literature data for gelatin. These parameters are related to the chemical structure of the material and could provide important information on molecular dynamics of systems, which are not forthcoming with the free volume theory.

References

1 S. Kasapis, J. Mitchell, R. Abeysekera and W. MacNaughtan, *Trends Food Sci. Technol.*, 2004, **15**, 298.
2 A. Tsoga, S. Kasapis and R. K. Richardson, *Biopolymers*, 1999, **49**, 267.
3 M. Deszczynski, S Kasapis and M. W. N. Hember, *Carbohydr. Polym.*, 2003, **53**, 85.
4 K. L. Ngai and D. J. Plazek, *Rubber Chem. Technol.*, 1995, **68**, 376.
5 S. Kasapis, *Biomacromolecules*, 2006, **7**, 1671.
6 M. Deszczynski, S Kasapis, W. MacNaughtan and J. R. Mitchell, *Food Hydrocolloids*, 2003, **17**, 793.
7 S. Kasapis, *Food Hydrocolloids*, 2001, **15**, 239.
8 I. M. Al-Ruqaie, S. Kasapis, R. K. Richardson and G. Mitchell, *Polymer*, 1997, **38**, 5685.
9 J. M. Hutchinson, *Prog. Polym. Sci.*, 1995, **20**, 703.
10 S. Kasapis, *Food Hydrocolloids*, 2006, **20**, 218.
11 D. J. Plazek, *J. Rheol.*, 1996, **40**, 987.
12 R. G. C. Arridge, in *Mechanics of Polymers*, R. G. C. Arridge, Oxford University Press, Oxfors, 1975, ch. 2, p. 24.
13 D. J. Plazek and K. L Ngai, *Macromolecules*, 1991, **24**, 1222.
14 F. Badii and N. K. Howell, *Food Hydrocolloids*, 2006, **20**, 630.
15 S. Kasapis, Critical reviews in Food Science and Nutrition, 2008, **48**, 185.

NUMERICAL COMPUTATION OF RELAXATION SPECTRA FROM MECHANICAL MEASUREMENTS IN BIOPOLYMERS

Vassilis Kontogiorgos[1] and Stefan Kasapis[2]

[1]*Department of Chemical and Biological Sciences, The University of Huddersfield, Queensgate, HD1 3DH, UK*
[2] *School of Applied Sciences, RMIT University, City Campus, Melbourne VIC 3001, Australia*

1 INTRODUCTION

Stress relaxation experiments can be used as an interface to the relaxation spectrum of biomaterials. Knowledge of the relaxation spectrum is important because we obtain information about the nature and rates of molecular rearrangements and the interaction of macromolecules.[1] The Maxwell mechanical analog is used to describe a stress relaxation experiment and the stress function of the material is given by:[1]

$$\sigma(t) = \sigma_e + \int_0^{+\infty} \sigma(\tau)\exp(-\frac{t}{\tau})d\tau \qquad (1)$$

Expression (1) describes the gradual relaxation of stress, $\sigma(t)$, to the equilibrium stress (σ_e), with $\sigma(\tau)$ being the distribution function of the elements with relaxation time, τ. This is the time taken for the stress to decay to $1/e$ of its initial value. Complete material relaxation means that $\sigma_e = 0$ and the integral of the above expression can be written in terms of the generic form of the first-kind Fredholm integral:

$$g(s) = \int_a^b K(s,t)f(t)dt, \quad \alpha \leq s \leq \beta \qquad (2)$$

where, $K(s,t)$ is the kernel exp $(-t/s)$ that describes the system, $g(s)$ is the measured signal, and $f(t)$ is the unknown integral solution. The goal of the numerical calculations is to determine the spectral function $f(t)$ that represents the relaxation spectrum of the material.

Numerical evaluation of the spectral function $f(t)$ is not a straightforward task because the Fredholm integral equation is an ill-posed problem that requires a special mathematical treatment.[2] In ill-posed problems an arbitrarily small perturbation of the signal g leads to large perturbations of the solution f or in other words the solution is extremely sensitive to experimental noise.[3] To overcome such a problem, regularization methods are employed, which allow incorporation of additional information about the sought solution.[2, 4-6] In the formulation of such problems, a regularization parameter (λ) is involved that controls the trade-off between a small error in the solution (low λ values) and a smooth solution (large λ values). Successful calculation of $f(t)$ depends on the appropriate choice of parameter λ. In the present work, the L-curve method was used to evaluate the optimum regularization parameter.[7]

The aim of the investigation is to present a comprehensive methodology for numerical computation of the relaxation spectrum from stress relaxation data using state of the art algorithms so as to improve understanding of the molecular processes involved in the relaxation of three-dimensional biopolymer structures.

2 METHODS AND RESULTS

2.1 Hypothetical Exponential Decay

Numerical computation was performed in MATLAB using three different algorithms namely Hansen,[8, 9] Wendlandt,[10, 11] and Marino[12] following the names of the authors. The first two algorithms use Tikhonov's regularization with different approaches in their structure, whereas the third routine is a MATLAB implementation of the CONTIN algorithm.[5, 6]

We initially worked with hypothetical data in order to confirm the efficacy of the proposed procedure. In particular, we created an exponential decay of the function $f(t) = \exp(-t/1)$ with linear equidistantly separated points and explored the effect of regularization parameter (λ) and decay time (partially decayed or completely decayed transient) on the solution. The first step in the numerical computation to solve equation (2) is to discretize the problem i.e., to construct a system of linear equations that describes equation (2) so as:

$$A\mathbf{x} = \mathbf{b}, \; A \in \boldsymbol{R}^{nxm}, \; \mathbf{x} \in \boldsymbol{R}^{n}, \; \mathbf{b} \in \boldsymbol{R}^{m} \qquad (3)$$

where, matrix A is a discrete representation of kernel $K(s, t)$ from equation 2, **b** is a discrete vector that represents $g(s)$, **x** is the solution vector that describes $f(t)$, and **b** corresponds to the measured signal of the stress relaxation modulus (G'). To discretize kernel $K(s, t) = \exp(-t/s)$ and produce the matrix A and space *sp* where the relaxation times range, we used the MATLAB routine *discr*, which is described elsewhere in detail.[13] Optimization of the regularization parameter was performed with the routine *lcurve*, which is provided in Hansen's package.[8, 9] If the regularization parameter is smaller than the optimum the algorithms return a noisy spectrum whereas a value of λ greater than optimum yields over-smoothed solutions thus missing information from the experimental spectrum (Figure 1). As shown in Figure 1 for $\lambda = \lambda_{opt}$, computations return a peak width that does not correspond to the exact mathematical solution of $\mathbf{x}_\lambda = 1$. This is attributed to the discretization process (creation of matrix A) where there is a minimal loss of information known as the discretization error and, furthermore, round off errors during the several computation steps that contribute to further deviations from the exact mathematical solution. These errors cannot be avoided as computers work only with discrete and not a continuous array of data. Nevertheless, they are identical for all spectra, being the outcome of the same procedure, and valid comparisons among different molecular mechanisms are obtained especially for those yielding well-separated peaks. Furthermore, the important index of reference, i.e., the mean of the spectral peak is successfully computed with all scripts.

Often in stress relaxation experiments, the transient process is rarely left to decay completely owning to the prohibitive laboratory time required to observe such molecular equilibrium. In addition, biopolymer ageing during experimentation, possible slippage, and other time dependent effects contribute to decreasing signal-to-noise ratio. As a result, the transient process is commonly recorded as long as the signal amplitude exceeds the noise level.[14] This line of thought was tested presently with partially and completely decayed hypothetical transients and it was found that fully decayed transients generally give better-resolved peaks. The computational time required to calculate the solutions follows the order Hansen<Marino<Wendlandt with the Hansen algorithm being significantly faster but Wendlandt producing better resolved peaks. After ascertaining the proper application of the aforementioned routines, we proceeded to analyze experimental data of the model system of hydrated gluten.

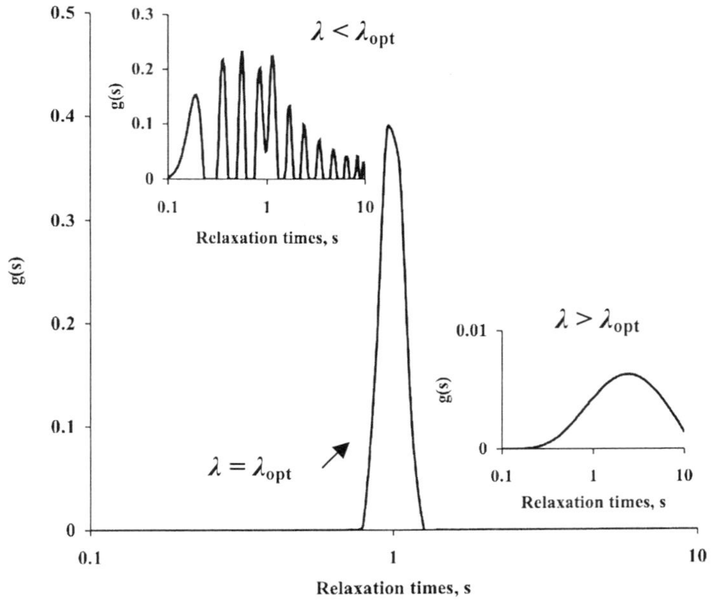

Figure 1 *Spectrum of hypothetical data decaying exponentially as f(t) = exp(- t/1) for different values of λ using the Hansen's algorithm. Top-left inset: λ is smaller than the optimum point resulting in a noisy spectrum. Lower-right inset: λ is greater than the optimum point resulting in an over-smoothed spectrum. Middle: Spectrum with the optimum regularization parameter returns the correct solution.*

2.1 Calculation of the Relaxation Spectrum of Hydrated Gluten

Analysis of stress-relaxation curves as a potential interface to the relaxation spectrum of biomaterials has received limited attention [15, 16] possibly due to the lack of a readily available methodology. In the present work, hydrated gluten (40% solids) was employed owning to well characterized viscoelastic properties that make the protein a good candidate for model studies. Furthermore, the relative compositional complexity of the system makes it akin to the intricacies found in biopolymer research, as opposed to the "simpler" synthetic polymers. Thus, the present work collected such biopolymer related data but it must be stressed that no effort was made to interpret findings at the molecular level. After ensuring that meaningful measurements are taken (LVR, sample relaxation before measurement) the stress relaxation experiment was followed for 20 sec, 30 min, 1, 5, and 10 h at 20 °C. Stress relaxation data were collected and analyzed as previously starting once more from the creation of matrix *A* and the space *sp*. Next, the optimum regularization parameter was located using the L-curve criterion. L-curves show the typical behavior of a discrete ill-posed problem with a distinct corner when plotted in a double logarithmic scale. Figure 2 shows a characteristic L-curve obtained for a gluten stress relaxation dataset. The routine *lcurve* varies the regularization parameter until a compromise is achieved between the minimization of the regularization error (residual norm) and noise error (solution norm). The vertical part of the curve corresponds to data where the solution norm is very sensitive to changes in the regularization parameter because the error in measurements dominates the regularized solution $\mathbf{x}_\lambda$. The horizontal part corresponds to solutions where the residual norm is the most sensitive to the regularization parameter because $\mathbf{x}_\lambda$ is now dominated by the regularization error. The value at the conjunction of the horizontal and vertical parts is used as the optimum regularization parameter.[7]

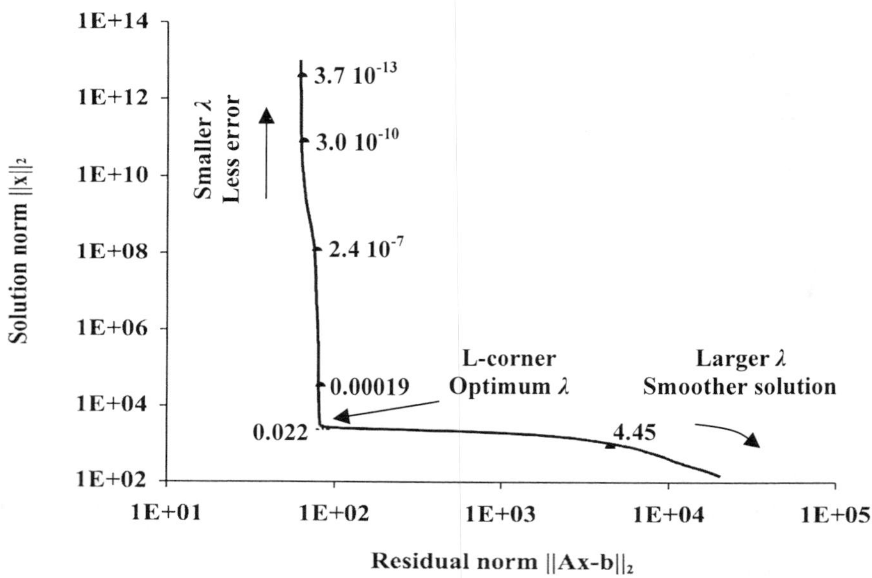

Figure 2 *Representative L-curve plot for the hydrated gluten network. The optimum regularization parameter, λ_{opt}, is located at the corner of the curve (0.022). Lower values than λ_{opt} result in noisy spectra whereas higher estimates yield over-smoothed spectra and loss of molecular information.*

Figure 3 constitutes the culmination of our efforts to treat experimental data with the different regularization algorithms of Marino, Hansen, and Wendlandt. Spectra generated by the three algorithms are in general agreement revealing a major peak at 0.02 s followed by two slower relaxing components at around 0.2 and 1.0 s. At longer relaxation times several broad peaks are obtained. Wendlandt and Hansen, using the Tikhonov regularization, give near identical traces with the former producing a sharply resolved first peak. On the other hand, Marino algorithm gives peaks with higher intensity and shifts them slightly to higher relaxation times something that is attributed to the different mathematical approach that is used in the algorithm (CONTIN implementation). The resolving power of the present protocol was tested extensively by examining data of materials left to decay for up to 10 hrs.

"Fully" or partially decayed transients return spectra that provide similar qualitative information for each algorithm maintaining the fingerprints at about 0.02, 0.2 and 1.0 sec. It appears, however, that the shortest period of observation (20 s) is not adequate to allow computations to resolve the transient at relaxation times beyond ~5 sec. For gluten, therefore, longer timescales of observation (e.g., 30 min) are required to fully resolve the dynamic range of the relaxation spectrum. Previous work has dealt with the relaxation spectrum of several gluten preparations.[16-18] However, the simple differential equation used earlier (Alfrey's rule [19]) resulted in broad spectra for various relaxing species that do not possess the resolution quality of the present methodology.

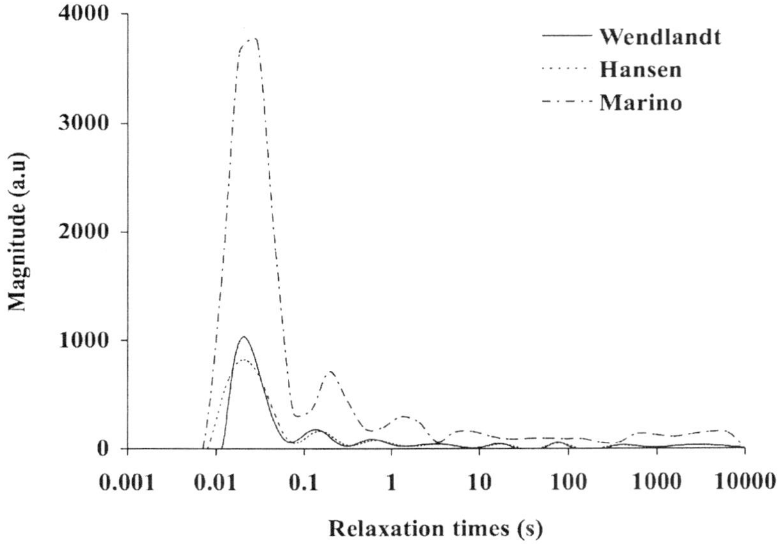

Figure 3 *Representative relaxation spectra of a hydrated gluten network at 20 °C obtained for an observation period of 5 h using the algorithms of Marino, Hansen and Wendlandt.*

2.1 Temperature Dependence of Hydrated Gluten Relaxation Spectra

The temperature dependence of hydrated gluten relaxation spectra was studied next in an effort to explore how relaxation times and molecular motion vary with changes in temperature. The stress relaxation experiment was conducted for 30 min at 0, 10, 20, 30, 40, and 50 °C and the analysis of the resulting data was performed as previously (Figure 4). The first (~ 0.0004 sec) and second (~ 0.0021 sec) peaks, which are attributed to fast relaxing species, gradually decrease in intensity as temperature increases. The fast relaxing species can be attributed to the low molecular-weight protein chains. Therefore, as temperature increases fewer low molecular-weight chains are brought out of equilibrium under the current experimental conditions indicating that they have greater mobility at higher temperatures. Species with long relaxation times (> 0.1 sec) show different behavior. Peaks that can be assigned to high molecular-weight protein fractions, appear to remain largely unaffected from changes in temperature. Thus changing the relaxation properties of gluten as a function of temperature should focus on the conformation and interactions of the high molecular-weight fractions rather than the low molecular-weight counterparts. In terms of the magnitude of the signal, the contribution of the small molecular-weight fraction dominates the relaxation behavior of the sample with the ratio of the first to the second peak being roughly 1: 4.

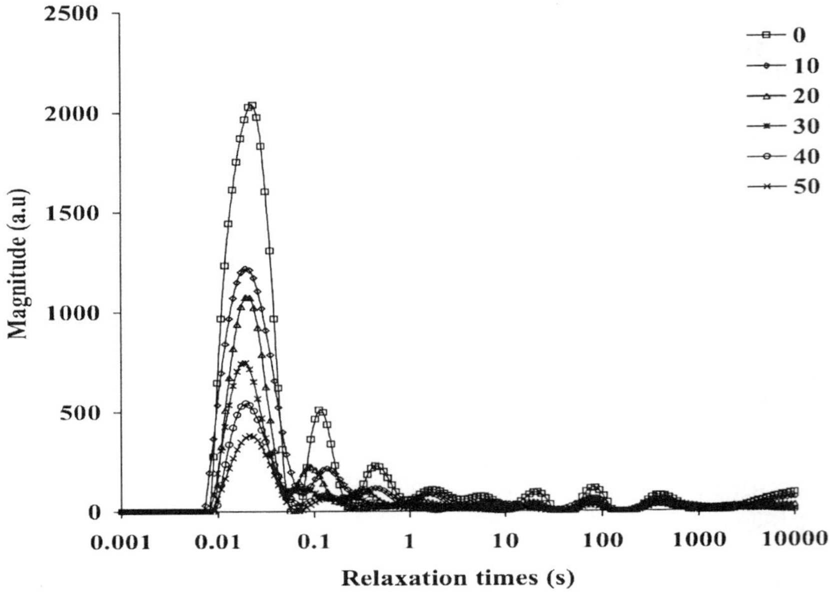

Figure 4 *Temperature dependence of hydrated gluten relaxation spectrum obtained for an observation period of 30 min using the Hansen's algorithm.*

The outcome of the present treatment yields high-resolution structure-sensitive spectra as compared to previous methods in rheological research. A novel and critical element of this framework of rheological analysis is the application of regularization algorithms in mechanical characterization of biopolymers. In the case of gluten, relaxation patterns can be utilized in the identification of molecular species responsible for specific structural properties. For example, molecular elements of long relaxation times may be associated with structural instability during processing or storage and their identification followed by removal (e.g., *via* fractionation) should improve the functional properties of ingredients and end-product formulations. Furthermore, high-resolution mechanical relaxation spectra allow introduction of the concept of *mechanical fingerprinting* that may enable pinpointing a certain structural behavior based on the relaxation spectrum of the network.

3 CONCLUSIONS

A methodology that uses MATLAB numerical algorithms for the calculation of relaxation spectra from stress relaxation measurements was developed and applied to a model biopolymer system (hydrated gluten). The L-curve criterion was used for the first time in biopolymer research to locate the optimum value of the regularization parameter permitting computation of high-resolution relaxation spectra. The three different algorithms that were employed and applied to hypothetical and "real" data successfully calculated the solution. The effect of decay (partially and "fully" decayed transient) was evaluated revealing that better resolution is obtained with "fully" decayed transients. The time required to calculate the solution follows the order Hansen<Marino<Wendlandt with the Hansen algorithm being significantly faster and Wendlandt producing better resolved peaks. Solutions calculated with the different algorithms were in general agreement and the choice would depend on the particular needs of the problem under investigation (speed or resolving power). Investigation of the temperature dependence of the spectra revealed a dramatic decrease in the contribution of the rapidly relaxing species with temperature whereas the slow relaxing chains remain unaffected.

References

1. J.D. Ferry, *Viscoelastic properties of polymers*. Wiley: New York, 1980.
2. C.W. Groetsch, *The theory of Tikhonov regularization for Fredholm equations of the first kind*. Pitman: Boston, 1984.
3. P.-C. Hansen, Inverse Problems 1992 **8,** 849
4. A.N. Tikhonov; V.Y. Arsenin, *Solutions of ill-posed problems*. Wiley: New York, 1977.
5. S.W. Provencher, Comput Phys Commun 1982 **27,** 213
6. S.W. Provencher, Comput Phys Commun 1982 **27,** 229
7. P.-C. Hansen, SIAM Review 1992 **34,** 561
8. P.-C. Hansen, Numerical Algorithms 1994 **6,** 1
9. P.-C. Hansen Regtools. http://www.mathworks.com/matlabcentral/fileexchange/loadFile.do?objectId=52&objectType=file
10. M. Wendlandt NLCSmoothReg. http://www.mathworks.com/matlabcentral/fileexchange/loadFile.do?objectId=52&objectType=file
11. M. Wendlandt; J.D. van Beek; U.W. Suter; H.B. Meier, Macromolecules 2005 **38,** 8372
12. I.-G. Marino Rilt. http://www.mathworks.com/matlabcentral/fileexchange/loadFile.do?objectId=6523&objectType=file
13. V. Kontogiorgos; B. Jiang; S. Kasapis, Foods Research International 2009 **42,** 130
14. A. Istratov; O.F. Vyvenko, Review of Scientific Instruments 1999 **70,** 1233
15. R. Mao; J. Tang; B.G. Swanson, Journal of Food Science 2000 **65,** 374
16. W. Li; B.J. Dobraszczyk; J.D. Schofield, Cereal Chemistry 2003 **80,** 333
17. Y. Song; W. Zheng; Z. Qiang, Food Hydrocolloids 2008 **22,** 414
18. V.K. Rao; S.J. Mulvaney; J.E. Dexter, J. Cereal Sci. 2000 **31,** 159
19. T. Alfrey; P. Doty, J. Appl. Phys. 1945 **16,** 700

Technofunctionality

TECHNOFUNCTIONALITY OF HYDROCOLLOIDS AND THEIR IMPACT ON FOOD STRUCTURE

TJ Foster

Division of Food Sciences, School of Biosciences, University of Nottingham, Sutton Bonington Campus, Loughborough, LE12 5RD, United Kingdom.

1 INTRODUCTION

Hydrocolloids play an important role in the structure, stability, palatability and sensory acceptance of a wide range of food materials. Such materials can be high in moisture content e.g. emulsion products, soups and sauces and drinks, or low moisture products such as baked and extruded products and confectionery items. Such functionalities have been reviewed in the past[1, 2] and there are excellent review books charting the spectrum of hydrocolloids used in the food industry[3, 4]. It is therefore a worthy argument to ask why this chapter is necessary at all. The context of the chapter however is to focus on the current needs of society of the current times (circa 2009), and addresses some of the issues pertinent to the present and future needs of society.

To set such a context then, the current pressures on the food industry are largely approached from two directions. The first is the trend which has been with us for some time, and will undoubtedly remain a key area for development and innovation in the area of hydrocolloids, and that is their effect upon general **health and wellness**. The second is an increasingly important area developing in light of societal changes, revolving around population growth, a growth in the requirements for 'western' diets, and an ever increasing need for energy, in a world of declining fossil fuel resources. This second challenge is that of achieving **sustainable food production**. Indeed, both of these areas are highlighted by Heldman[5], where food and beverage products are seen to have future scope for both disease prevention, and health enhancement, and better utilisation of resources in energy efficient processes, decreasing 'food miles' between ingredient, production and the consumer interface. Both of these areas are also central to the Strategic Research Agenda of the European Technology Platform of Food for Life[6] and the recent Royal Society of Chemistry / Institute of Chemical Engineers document 'The Vital Ingredient'[7]. The first states the need to make the healthy food choice the easy food choice for the consumer, providing a healthier diet, developing quality food products while achieving sustainable food production. The Vital Ingredient encourages the research and development community to strive for entrepreneurial changes in operation procedures, adapting technologies to reduce the amount of material, energy and water wastage. German and Watske[8] introduced new thinking about how the food production cycle might change if the view was to focus on a new added value of food 'individualisation', where the consumer will 'bioguide' all of the stages of food processing. This is a concept also picked up in 'Food for Life'[6] where consumer preference, acceptance and need has the potential to allow back engineering of food products, to ensure each individual is getting the food that they need. This is potentially futuristic and at first at odds with a sustainable approach to food production, but point-of-sale personalised production of food, could reduce the amount of energy required, introduce water only at the point of production, and have optimal use of constitutive ingredients, all over which are often in excess for the mass production, distribution and long shelf stability of the conventional food products of our times.

One of the areas of health and wellness which has received increased attention over recent years has been in the area of obesity. The well known maps of the increase in prevalence of obesity in adults of the United States are now being matched by similar looking maps on a global scale, including the worrying trend in children. Food structures, involving hydrocolloid functionality, have been suggested recently[9, 10] to combat this trend in weight gain, and some candidate structures will be reviewed later in this article.

A vision for the future is therefore to have sufficient knowledge to enable the optimisation of ingredient usage (for structural or health functionality) allied to understanding the developments of new process technologies to create designed structures which are stable and perform to requirement, enabling formulation by design. This will allow ingredient flexibility, use of greener / cleaner natural structures, utilisation of biomass (fuel and new structures), stable, naturally nutritious foods and availability when required.

2 RESULTS AND DISCUSSION

2.1 Hydrocolloid Use: What? Where? Why?

These are the questions that should be asked when deciding which hydrocolloids to use in a formulation. The decision should be determined by the hydrocolloid structure / functionality i.e. knowing the required functionality from the hydrocolloids and where it will reside in the food product microstructure. Hydrocolloid ingredients are known to be incompatible, or may interact with other ingredients. Such complexity of mixing needs to be understood in order to have controlled ingredient flexibility. It was shown, by means of an example, that locust bean gum can be replaced by guar gum in milk based systems, if the phase volume of the incompatible protein and polysaccharide phases are matched and the viscosity within the polysaccharide phase remains the same[2].

Figure 1 shows a dairy based emulsion structure, comparing that of a full fat and zero fat cream cheese. The image on the left indicates the inclusion of the fat phase inside the protein phase, which in turn is phase separated from and dispersed in the continuous polysaccharide phase. The influence on microstructure imparted by the hydrocolloids in this formulation are therefore the surface activity and gelation of the fat encapsulating polymer, and then the viscous or gelling nature of the hydrocolloid in the continuous phase. Additionally, knowledge of the microstructure of the full fat formulation enables the fat containing particles to be replaced by alternative particles in the fat free formulation. Therefore, if the phase volume of the continuous and dispersed phase remain constant, and the properties of the continuous phase remain the same, then one might expect the physical properties of the two products to be similar. This is a simple example on how fat might be physically replaced in a formulation.

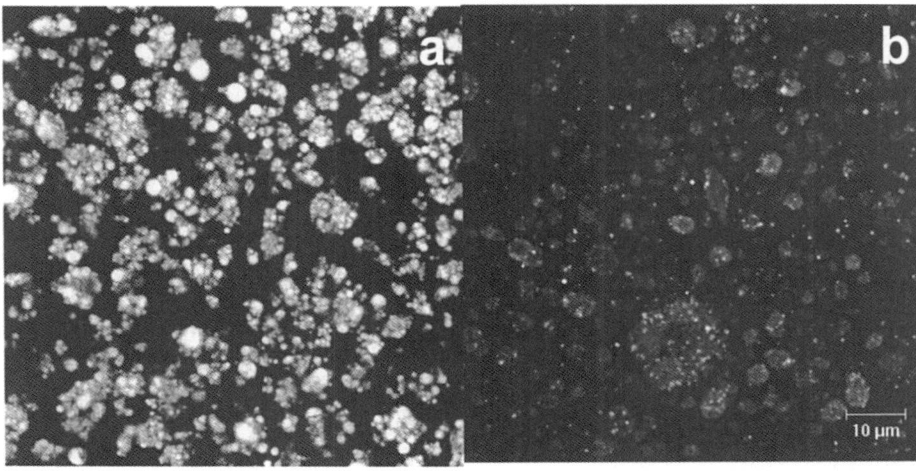

Figure 1 *Confocal micrographs of cream cheese formulations, a) full fat and b) zero fat, where polysaccharide is the dark phase, fat droplets are discernible within the protein clusters (a).*

Phase separation was also utilised in a similar way in the 1990's when there was a specific focus on fat reduction in products. Zero fat spreads for example needed to be 'spreadable' and to serve such a purpose small molecular weight crystallising hydrocolloids such as inulin and maltodextrin were employed. However, their mouthfeel wasn't always appreciated by consumers, so a need to replace the mouth-melting role of fat also needed to be incorporated. Figure 2 shows the time dependant crystallisation of the maltodextrin, and then shows the plastic phase of maltodextrin, incorporating a gelatine phase, to provide the melt-in-the-mouth sensation.

When comparing figures 1 and 2, the ingredient missing in figure 2 is fat. One might consider the dairy system of figure 1a to be an oil-in-water-in-water (O-W-W) emulsion. Therefore, with gelatin being a surface active ingredient, would it be possible to emulsify fat in the gelatin phase of figure 2? Indeed, figure 3 shows that fat is emulsified effectively by gelatin. Therefore, in both cases we have fat emulsified in a gelling protein phase, with the polysaccharide phase either a thickening or gelling polymer. A reasonable question to then ask is: Can such functionality be achieved using polysaccharides to emulsify and gel around a fat droplet? Figure 3 confirms that such functionality is possible from surface active polysaccharides.

There has been considerable interest in recent times investigating the mechanisms of polysaccharide surface activity. Dickinson has published work on hydrocolloids at interfaces[11] and as emulsifiers and emulsion stabilisers[12]. The current consensus of opinion is that the surface activity of polysaccharides results from the presence of small protein components within the sample. Indeed, for sugar beet pectin (SBP), Williams[13] and Siew[14] have shown it is possible to fractionate, via hydrophobic affinity chromatography, a protein and ferulic acid rich fraction responsible for the surface activity, and whose properties are influenced by the accessibility of the protein and ferulic acid to the interface. Surface thickness measurement have also been possible, implicating some of the pectin side chains in the stabilisation of the emulsion, and seen to be a possible Gum Arabic replacer. Kirby[15] has also confirmed the presence of protein moieties in a sample of SBP using atomic force microscopy.

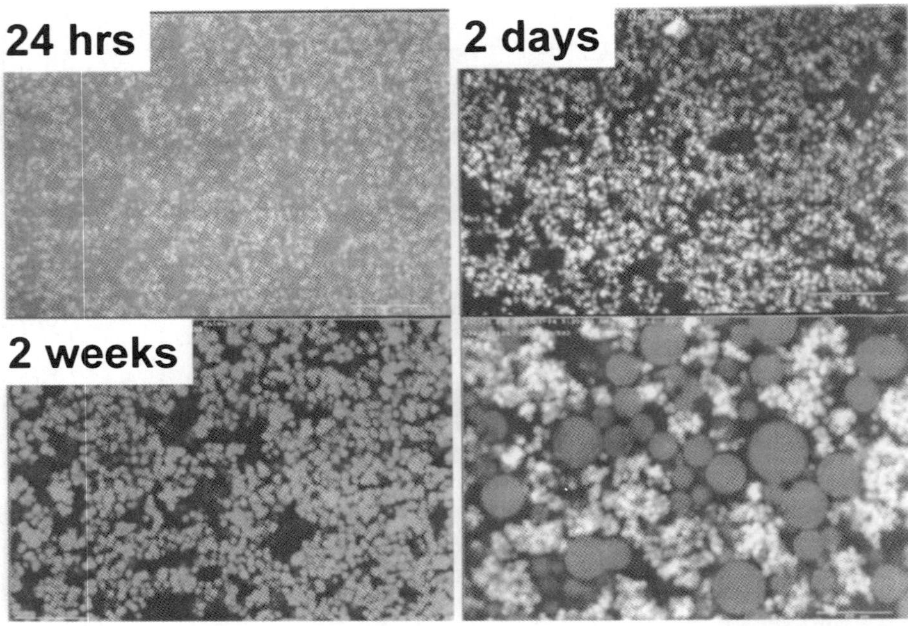

Figure 2 *Confocal micrographs of 14% DE2 maltodextrin, following crystallisation, and then after two weeks in the presence of 3% gelatine (bottom right).*

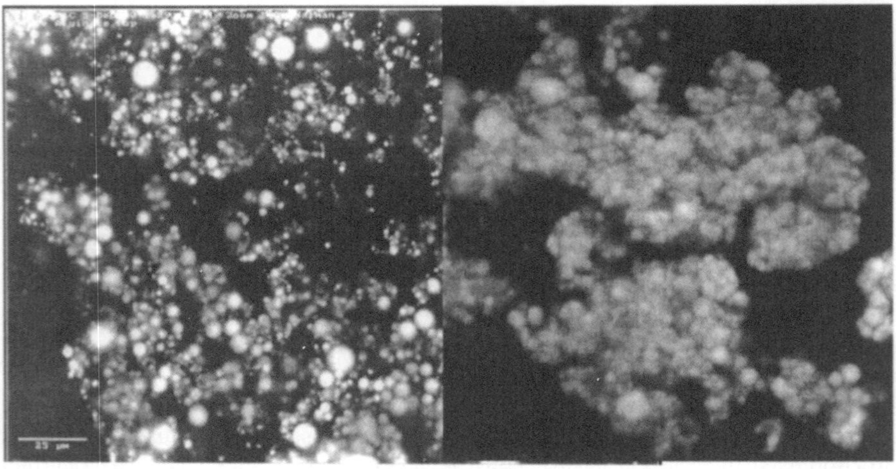

Figure 3 *Confocal micrographs of oil emulsified with sugar beet pectin, in the presence of low methoxy pectin and calcium ions (left) and a gelatin: maltodextrin mixture with the incorporation of fat. In both cases the emulsions were homogenised before the onset of gelation.*

To replace gelatin or milk protein as gelling hydrocolloids, the hypothesis tested here was to stabilise the emulsion with SBP, but then make available, to the interface, a gelling pectin and calcium ions, known to be responsible for pectin gelation. Figure 4 shows evidence that in the presence of SBP the pectin fluorescence is concentrated around the emulsion droplets, whereas in the absence of SBP the fluorescence was uniform throughout the aqueous phase. The emulsions were seen to be stable against coalescence and creaming in the presence of SBP[16, 17].

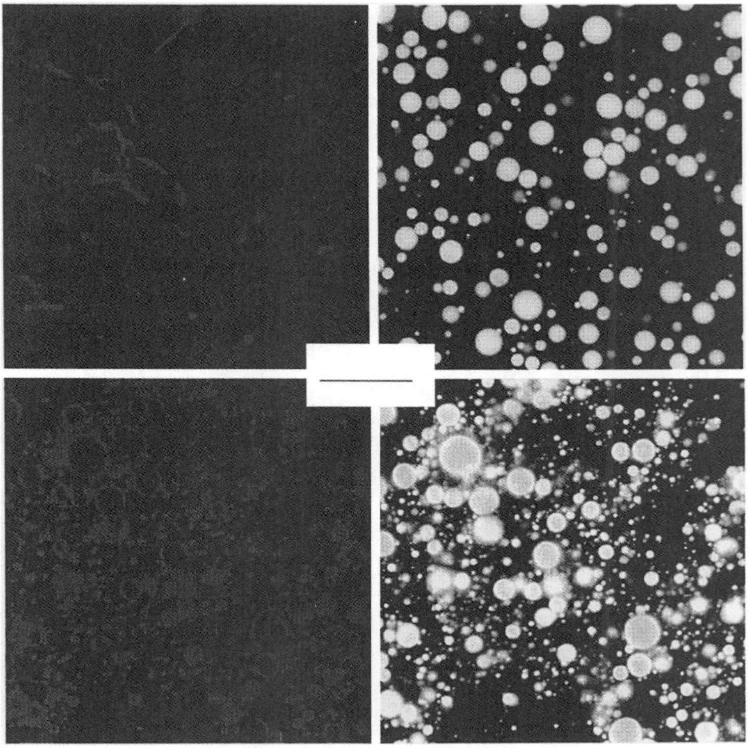

Figure 4 *Confocal micrographs showing clustering of pectin around the oil droplets (bottom images) with fat, LM pectin, SBP and calcium, and uniform structures (top images) with fat, LM pectin and calcium (Scale bar is 20μm).*

2.2 In-Body Functionality

The evidence presented thus far has been focussed on the hypothesis of structure / functionality. As mentioned in the introduction, there is an increased need to understand the promotion of health and well-being through food, and therefore new functionalities are being introduced, as more is learned about the fate of food during digestion. There is more emphasis being placed on the relationship between food structure and the way in which the food is digested [18, 19]. Such emphasis is then linked to the control of e.g. feelings of satiety[10] and delivery of nutrients[20, 21], and thankfully views are being collected by leading experts in the field[22, 23]. There is considerable knowledge in how microstructures change during processing[24] but little real evidence on how microstructures change during digestion. Figure 5 indicates differences in the rates of structure breakdown upon introduction into simulated gastric conditions for a phase separated system containing milk and galactomannan. Knowledge of the phase diagram has been used to design two starting formulations which are constant in phase composition, but moving on the tie-line are either protein or polysaccharide continuous.

As the milk system encounters low pHs the casein micelles interact and the milk phase starts to gel. Our interest was on the relative rates of gelation and proteolysis in the presence of pepsin. In the absence of pepsin at pH2, both systems produced a curdled milk structure, regardless of starting phase volume. However, in the presence of pepsin, at pH2, the protein continuous system was digested and the mixture became clear. Conversely the polysaccharide continuous system remained opaque, and the protein phase appeared to have resisted digestion. More work is required to investigate this further, but it the first sign that formulation and resultant microstructure may play a key role in the rates of digestion of simple, let alone complex food structures.

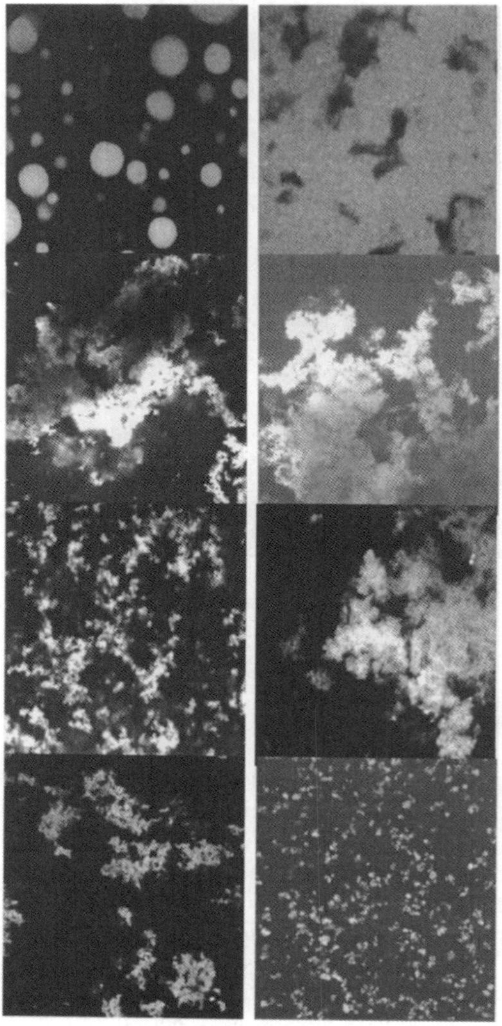

Figure 5 *Confocal micrographs of (left) polysaccharide continuous and (right) protein continuous mixtures of LBG and skimmed milk. Top image is the starting solution at neutral pH; second from top is the system at pH2 in the absence of pepsin; second from bottom at pH2 and pepsin; bottom pH2 and pepsin after 2 hours.*

Developing the concept of O-W-W into the area of digestion, with the systems described in figure 5, it is of interest to test the impact of decreasing pH in the presence of pepsin. Dickinson[18] and McClements[19] have both recently discussed the importance of interfacial design on how emulsions may be digested, influencing the way in which the interface is accessed by lipase enzymes. Figure 6 shows the polysaccharide continuous formulation with the addition of oil into the protein phase. The evidence indicates that the digestion of the protein is restricted again, this time producing the protein structures on the oil droplet interface, providing possible routes for lipolysis control.

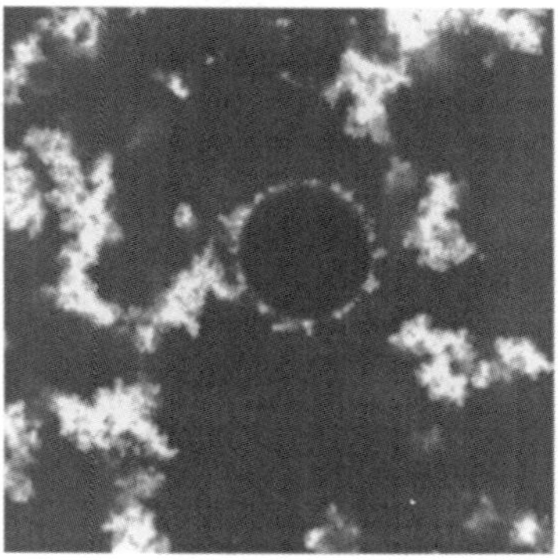

Figure 6 *Confocal micrograph showing an oil droplet in the polysaccharide continuous sample from figure 5, after one hour at pH2 in the presence of pepsin.*

2.3 Naturalness

Hydrocolloids might be seen as natural ingredients since they are derived from nature. However, they do have to be included on food product labels, and in Europe almost all have eNumbers. Most are refined and/or chemically treated and marketed as speciality ingredients, due to their specific functionality. However, such refining might provide questions for our reasoning in this paper regarding sustainability.

There have been recent moves to include fibrous material derived from citrus flesh in a range of food products[25, 26], for thickening and fat replacement. All natural ice cream has also been produced using fruit material to completely replace added hydrocolloids[27]. A recent discussion on the structure of natural materials has focussed on the structurants in such plant cell wall composites, namely pectin[28, 29, 30] and cellulosic materials in plant cell wall composites have been probed for their structural impact[31, 32].

Cellulosic structures have the potential for use in food product structures, but also in sustainable packaging. Such structuring requires the 'dissolution' of cellulose and regeneration. Figure 7 provides evidence of cellulose fibre dissolution in ***N***-Methylmorpholine-*N*-oxide and regeneration in the presence of a fluorescently tagged konjac glucomannan. Such technologies, enabled by developments in 'Green Chemistry', may allow structural flexibility to provide sustainable and 'functional' packaging alternatives to petrochemical industry derived polymers.

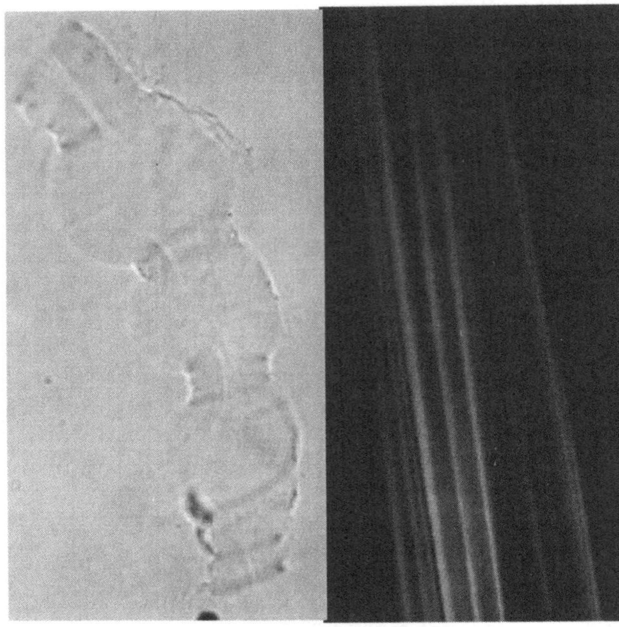

Figure 7 *Light micrograph (left) of cellulose fibrils dissolving in NMMO and a confocal micrograph (right) of a cellulose fibre regenerated in the presence of fluorescently labelled konjac glucomannan.*

3 CONCLUSIONS

In conclusion, the work presented in this paper has indicated that the field of hydrocolloid technofunctionality is developing and remains important in dealing with the current, but more importantly the future food challenges. Some of those future challenges may need a refocus on which hydrocolloids become predominant in the foods of the future, where the dual delivery against the agendas of health and wellness and sustainability will provide the focus.

Microstructure changes during simulated digestion begins to indicate that an enhanced knowledge of food digestion may provide future food structures which are optimally designed for nutrient delivery, which in turn could provide economies of production, providing only what is required.

Cellulose and natural structurants have then potential for providing new structuring opportunities, which may also provide sustainability solutions, as well as providing fibre in the diet.

What has not been covered in this paper, but which has been introduced in Mita Lad's paper in this edition is the subject of fast structuring. Such technology will have impact on the control of hydrocolloid structuring for the future opportunities of point-of-sale or distributed manufacture production.

4 ACKNOWLEDGEMENTS

The contributions of Mani Koganti, Rebecca Schofield, and Mita Lad are gratefully acknowledged.

References

1 Norton I.T. and Foster T.J., Hydrocolloids in real food systems, in *Gums and Stabilisers for the Food Industry 11* : 187-200, **2002**.
2 Foster T., Structure Design in the Food Industry, in *Product Design and Engineering Volume 2: Raw Materials, Additives and Applications*, eds. U. Brockel, W. Meier and G. Wagner, Wiley-VCH, 617-629, **2007**.
3 Handbook of hydrocolloids, eds, P.A. Williams and G.O. Phillips, Taylor Francis Ltd, ISBN 13: 9781439808207, **2009**.
4 Thickening and Gelling agents for Food, ed. A. Imeson, Chapman and Hall, **1997**.
5 Heldman D, State of the Industry, *International Food Ingredients*, 1, 32-33, **2009**
6 European Technology Platform on Food for Life, Strategic Research Agenda 2007-2020, CIAA, **2007**.
7 The Vital Ingredient: Chemical science and engineering for sustainable food, RSC / IChemE, **2009**.
8 German J.B. and Watske H.J., Personalizing food for Health and Delight, *Comprehensive reviews in food science and food safety*, 3, 145-151, **2004**.
9 Aldred D.L., van Amerongen I.A., Bodor J., Mela D.J. and Rayment P., Edible composition useful e.g. as meal replacer or other food products used in weight loss or weight control plan, contains alginate having specified L-guluronic acid content, and source of non-solubilized divalent metal ions, WO2005020717, **2005**.
10 Bialek J.M, Melnikov S.M., Winter I., Golding M.D., Pelan E.G., de Groot P.W.N. and Koppert R.J., Edible aqueous oil-in-water emulsion useful in food and beverages to treat or prevent obesity, comprises oil (having a saturated fatty acid), emulsifier (a protein) and water, WO2006053647, **2006**.
11 Dickinson E., Hydrocolloids at interfaces and the influence on the properties of dispersed systems, *Food Hydrocolloids*, 17, 25-39, **2003**.
12 Dickinson E., Hydrocolloids as emulsifiers and emulsions stabilisers, *Food Hydrocolloids*, 23, 1473-1482, **2009**.
13 Williams P.A., Sayers C., Viebke C., Senan C, Mayzoyer J. and Boulenguer P., Elucidation of the emulsion properties of sugar beet pectin, *J. Agric. Food Chem.*, 53, 3592-3597, **2005**.
14 Siew C. S, Williams P.A., Cui S.W. and Wang Q. Characterisation of the surface active components of sugar beet pectin and hydrodynamic thickness of the adsorbed pectin layer, *J. Agric Food Chem.*, 56, 8111-8120, **2008**.
15 Kirby A.R., MacDougall A.J. and Morris V.J., Atomic force microscopy of tomato and sugar beet pectin molecules, *Carbo Polym.*, 71, 640-647, **2008**.
16 Buffet G.X. and Foster T.J., Oil in water emulsions comprising a blend of biopolymers, WO 03/053151, **2003**.
17 Eckhardt G., Finlayson R.J., Foster T.J., Russell A.R. and Williams M.A.K., Multicompartment devise for preparing an emulsion, WO 03/053811, **2003**
18 Dickinson E., Interfacial structure and stability of food emulsions as affected by protein-polysaccharide interactions, *Soft Matter*, 4, 932-942, **2008**.

19 McClements D.J., Decker E.A., Park Y. and Weiss J., Designing food structure to control stability, digestion, release and absorption of lipophilic food components, *Food Biophys.*, 3, 219-228, **2008**.

20 van Amelsvoort JM, Duchateau GS, Melnikov SM and Winter I., 'Water-continuous composition for food product, e.g. beverage, dressing, soup, sauce, dip, spread or filling, comprises lipophilic physiologically active ingredient in particular fat soluble nutrient', WO2006069619, **2005**.

21 Amiot-Charlin M.J., Digestion and absorption of lipophilic food micronutrients, in *Digesting functional foods, Measuring and controlling food structure breakdown and nutrient absorption*, eds. D.J McClements and E.A. Decker, Woodhead Publishing Ltd, **2009**.

22 Digesting functional foods, Measuring and controlling food structure breakdown and nutrient absorption, eds. D.J McClements and E.A. Decker, Woodhead Publishing Ltd, **2009**.

23 Delivery and Controlled Release of Bioactives in Foods and Nutraceuticals, ed. N. Garti, Woodhead Publishing, 2008.

24 Foster TJ, Brown CRT, Norton IT, Phase inversion of water-in-water emulsions, in *Gums and Stabilisers for the Food Industry 8* : 297-306, **1996**.

25 Fischer J, Structuring of low calorie food with fruit fibres, In Gums and Stabilisers for the Food Industry 14, Eds. PA Williams and GO Phillips, RSC Publishing, 379-385, **2009**.

26 Bialek J.M., Aquino L.J.S. and Knight P.E., Edible emulsion for food product, e.g. dressing, soup, sauce, dip, spread, filling, or drink, comprises insoluble fibers, WO2005 039310, **2005**.

27 Bartkowska B., Foster T.J., Gray S.J., Regismond, S.T.A. and Underdown J., Fruit Structured Ice Cream, FR2802775A1, **2002**.

28 Jarvis M.C., Plant cell walls: supramolecular assembly, signalling and stress, *Struct Chem.*,20, 245-253, **2009**.

29 Morris V.J., Gromer A. and Kirby A.R., Architecture of intracellular networks in plant matrices, *Struct Chem.*, 20, 255-261, **2009**.

30 Voragen A.G.J., Coenen G-J., Verhoef R.P. and Schols H.A., Pectin, a versatile polysaccharide in plant cell walls, *Struct Chem.*, 20, 263-275, **2009**.

31 Foster T.J., Ablett S., McCann M.C. and Gidley M.J., Mobility-resolved C-13-NMR spectroscopy of primary cell walls, *Biopolym.* 39(1): 51-66, **1996**.

32 Kirby A.R., Ng A., Waldron K.W. and Morris V.J., AFM investigations of cellulose fibres in Bintje potato cell wall fragments, *FOBI*, 1, 163-167, **2006**.

MICROBUBBLES WITH PROTEIN COATS FOR HEALTHY FOOD: AIR FILLED EMULSIONS

Fideline L. Tchuenbou-Magaia, Ian T. Norton & Philip W. Cox
Centre for Formulation Engineering, University of Birmingham, Edgbaston, Birmingham B15 2TT, UK

1. INTRODUCTION

There is a growing trend toward consumer nutritional and health awareness. This along with obesity becoming a global pandemic has pushed the food industry to seek for novel healthy ingredients to deliver products with specific nutritional benefits. Not only must these products be shelf stable, but also they must meet consumer acceptance in terms of physical appearance, mouth-feel and flavour perception. It is believed that incorporating bubbles into food products may deliver novelty of textures and appearance, improve creaminess and make products softer to enhance the mouth-feel. Aerated structure may facilitate mastication, enzyme accessibility to substrates and enhance flavour delivering (Campbell et al. 1999; Lau & Dickinson 2005; Wildmoser, Scheiwiller, & Windhab 2004; Zuniga & Aguilera 2008). Moreover bubbles are non-fattening and we have demonstrated that air cells can be designed to resemble oil droplets in terms of physical appearance (size and shape) and rheological properties as well as their interaction with the other parts of the structure, thus offering a new and sensible approach to reducing the fat content of emulsion based products.

Bubbles can be divided in two categories: interlinked polyhedral foams and well-separated spherical bubbles where the thickness of liquid layers is similar to the gas bubble size (Exerowa & Kruglyakov 1998). Bubbles will flow whereas foam will not. Small bubbles in a range of 0.1 to 200 μm, known as microbubbles, although technologically attractive are naturally unstable because of their high Laplace pressures. However, by using hydrophobin HFBII protein, we have recently made stable air cells in a range of 1–100 μm, with approximately 50% of bubbles having a diameter less than 3 μm. Because these microbubbles suspenssions, are stable, sufficiently resilient to allow further processing and because the suspensions flow as fluid, we have termed them "Air Filled Emulsions" (AFEs).

Hydrophobins are proteins containing eight conserved cysteine residues that occur ubiquitously in mycelial fungi and will self-assemble at hydrophobic/hydrophilic interfaces (e.g. water/air, water oil and water/solid-hydrophobic-surface). Hydrophobins and specifically the HFBII form have been identified as conferring a highly viscoelastic nature to the surface of air cells, leading to excellent stability and to resistance to disproportionation and coalescence (Cox et al. 2007; Cox, Aldred, & Russell 2009).

Hydrophobins seem to be the only naturally occurring self-assembling protein so far identified, that can be used for the production of AFEs within a food context; however not only are they currently very expensive but also they are unlikely to be used for mass food production due to industrial availability. In an attempt to overcome these limitations we have successfully made AFEs with other food grade cystein rich proteins such as egg white proteins, bovine serum albumin and soy proteins. This was achieved through a sonochemistry-induced templating process around which the air cells form. The mechanism responsible for air cell formation is a combination of two acoustic phenomena: emulsification and cavitation followed by thermal cross-linking. Firstly, ultrasonic emulsification and cavitation creates the microscopic dispersion of air cells in a protein solution to form the shape of the microsphere shell; secondly, superoxides generated by sonolysis of water and in the presence of oxygen catalyses the formation of inter-protein disulphide bonds that cross-links the protein shell, thereby imparting the stability necessary for maintaining a air cell's shell architecture over time (Suslick et al. 1994).

Presented and discussed here are some results aimed at answering the following questions:

Could bulk food grade proteins be functionalised to give hydrophobins-like features in terms of bubble stability?

Could AFEs be used as an effective ingredient to reduce the fat content of high fat emulsion based products?

2. MATERIAL AND METHODS

Sonochemical preparation of "Air Filled Emulsions" (AFEs).

Hydrophobin or BSA coated AFEs were produced by irradiating protein solutions with a high-intensity ultrasonic probe (VCX 750, Sonic and Materials Inc, US) while sparging with air. The protein solution was placed in a jacketed vessel and the ultrasound horn was positioned at the air–solution interface. The solution was sonicated at 20 kHz, 50% amplitude for 3 min. The initial temperature was 49°C for BSA and 54°C for hydrophobin and the temperatures increased about 6°C during sonication as measured by a thermocouple. AFEs were concentrated (up to 75% air; by volume) by gravity overnight at room temperature in a separating funnel. Bubbles migrated to the cloudy upper layer (due to their bouyancy) and the bottom layer poor in bubbles and rich in protein was drained out away through the bottom of the funnel. Centrifugation was also used as alternative method to concentrate AFEs. In some case AFEs were centrifuged at 1000 rpm for 3min to concentrate bubbles in a top layer. Centrifugation could also be used to separate potential protein debris from the mixture.

The hydrophobin HFBII used in this study was partially purified from the *Trichoderma reesei* culture broth by foam fractionation (Tchuenbou-Magaia, Norton, & Cox 2009). The HFBII concentration in the extract was about 0.01% as estimated from the electrophoresis gel and the total proteins content of the extract was 0.02%. Lyophilised Bovine Serum Albumin (BSA), was weighed and dissolved in distilled water by stirring at room temperature for 2h, then 1M HCl solution was used to adjust the pH of the protein solution to pH 5.0.

Construction of air based triphasic emulsion (A/O/W) model salad dressing.
A/O/W emulsions were prepared using two steps:
Firstly, a standard O/W emulsion was prepared by slowly mixing sunflower oil into a Whey Protein Isolate (WPI) solution using a Silverson high shear mixer at 6000 rpm equipped with a fine screen of 19 mm diameter for 30s; food grade Xanthan (XG) solution was then added and blended for 2min. This mixture was passed through a high-pressure homogeniser at 35 Bar (FT9 homogeniser, Armfield, UK) four times. The stock solutions of XG (2%w/v) and WPI (4%w/v) were prepared by dissolving the weighed quantities of WPI and XG powders into distilled water at room temperature. The solution was stirred overnight or for 4h using a mechanical mixer or a magnetic stirrer for XG and WPI respectively. WPI stock solution was diluted with a given volume of distilled water then added sunflower oil in order to obtain an emulsion with 40% v/v sunflower oil, 1 wt% WPI and 0.2 wt% XG. Xanthan gum powder (G1253) was purchased from Sigma Chemical Company and Whey protein isolate (protein, 98 wt%) was donated by Davisco Foods International.
(2) The triphasic emulsions were made by blending a given volume of O/W emulsion and AFE together in a Silverson mixer at 3000 rpm for 7 min.
Xanthan gum was used as stabiliser/ thickening agent and was chosen along with WPI because both are used in industrial formulations of dressings and mayonnaise. The O/W was prepared so that at least 50% of air cells had a similar size with the oil droplets. Both emulsions were stable with regard to droplet size over a period of months (see table 1).
The viscosity of triphasic (A/O/W) and O/W emulsions was performed with a Bohlin rotational rheometer (Malvern Instrument Ltd, UK) under control strain mode using stainless steel cone and plates. The cone diameter was 40 mm and the cone angle 4°. Samples were sheared from 0-200 s-1. The shears in the sweep was calculated in a linear fashion. The data was plotted on linear scales for easy estimation of possible shear sensitivity of the system.
Friction measurements (tribology) were performed on a mini traction machine (MTM; PCS Instruments Ltd., UK) at 20°C (±1°C) with a steel ball (3/4 inch) on a silicone disk (46 mm diameter). The disk was freshly cut from a silicone sheet (Samco, UK). The tests were carried out with applied load 3N, a slide-to-roll ratio of 50% with entrainment speeds of 1 to 1000 mm/s and 30 data points chosen logarithmically. In order to study the time dependency properties of each sample, tribological tests were performed under the above-mentioned conditions three times, and each cycle was started without unloading the ball. The idle speed of ball was set at 1mm/s. At least three measurements per sample were carried out.

3. RESULTS AND DISCUSSION

1. Could bulk food grade proteins be functionalised to give hydrophobins-like features in terms of bubble stability?

Hydrophobin protein (HFBII) from *Trichoderma reesei* has been shown to have an outstanding ability to stabilise air bubbles by forming a strong viscoelastic film at the air/water interface thereby giving the stability necessary via an elastic restoring force for maintaining an air cell's size and shell architecture over time. However, hydrophobins are currently rare and too expensive to be used in large-scale food production. One of the purposes of our work was to find cost-effective, easily available alternative proteins that can be processed to provide all of the attractive properties of hydrophobin in terms of the stabilisation of micron sized air cells. Indeed it is possible to achieve hydrophobin-like features with various food grade proteins rich in cysteine by altering the processing conditions, subjecting those proteins to simple pre-treatments e.g. dry heating (data not shown) and by choosing the right pH for processing at. This chapter discusses the stability of suspensions of air cells made with bovine serum albumin (BSA-AFE) as compared to those made with hydrophobin HFBII (HFBII-AFE).

Sonochemical preparation of "Air Filled Emulsions" (AFEs).

For HFBII-AFE, the sonication process was chosen mainly because of the very short time scale at which cavitations occurred. Indeed, HFBII is known to aggregate very rapidly and irreversibly at the air/water interfaces thus offering little possibility for process control. However, it has been reported that mixed hydrophobin systems resulted in the reduction of this assembly rate at the interface (Askolin 2006). This reduction was also observed with HFBII contaminated with non surface active proteins (Tchuenbou-Magaia, Norton, & Cox 2009). Ultrasonic emulsification and the presence of contaminant proteins (not surface active) enabled us to overcome the hydrophobin interfacial accumulation rate problem and to construct the structures that characterise the air filled emulsions (see Figure 1(a-b)). Although the self-association in HFBII is mainly driven by the hydrophobic interaction between these small molecules (Kisko et al. 2008) we propose that the "hot spot" effect of acoustic cavitation may serve to further enhance protein aggregations, which could lead to the formation of multi-layers protein coats and reinforce the stability of hydrophobin stabilised air cells (results not shown).

In contrast BSA is a bulky globular protein; first used by (Grinstaff & Suslick 1991) to produce micrometer-sized gas-filled capsule using a sonochemical method for use as a contrast agent for *in-vivo* imaging. The formation of microbubbles with this method has been reported to be strongly inhibited by the lack of free cysteine residues in the protein (Suslick, Grinstaff, Kompany, & Wong 1994). Ostensibly BSA has only one free cysteine residue but contains a further seventeen disulphide bonds for protein stabilisation. In contrast hydrophobin HFBII has four disulphide bonds and no free cysteine; thus Suslick suggested mechanism would not account for sonochemical formation of HFBII-AFEs. However, the disulphide cross-linking mechanism behind BSA-AFEs formation, hydrophobic interaction (due to unfolding and exposure of hydrophobic amino acids) and thermal denaturation of the protein may also play an important role. Therefore the mechanisms that underpin the formation of BSA-AFEs and HFBII-AFEs are different, but result in comparable stability (see later discussion, and Figure 1 and Figure 2). Figure 1 shows light and SEM micrographs of HFBII-AFE with air cells of less than 10μm and BSA-AFE with air cells less than 5μm. It should be pointed out that HFBII-AFE also has a broader size distribution than BSA-AFE (Figure 2).

(a) (b) (c)

Figure 1. *Typical micrographs of AFE emulsion showing bubbles size: (a) brightfield image of 1day concentrated HFBII-AFE; (b) etched Cryo-SEM 5 days old HFBII-AFE; (c) phase contrast image of a 1 day concentrated BSA-AFE.*

Sonochemical preparation of BSA-AFE was found to be strongly pH and ionic strength dependent. Indeed, although Suslick and co-workers (pioneer of sonochemical cysteine-rich protein microcapsules) worked at pH 7, we have found that BSA-microbubbles were more stable and more abundant at pH 5, i.e. close to the protein's isoelectric point rather than at pH 7 with no air cells produced at all at pH 3 and 9. This might be explained by the change in secondary and tertiary structure of BSA at extreme pHs, making the molecule less favourable for adsorption. At pH values below 4 BSA expands with a decrease in its helix structure (from 55% at pH 5 around it's isoelectric point to 35% helix at pH 2.7). CascaoPereira et al. (2003) investigated the structural conformation of BSA at a single air/water interface over a range of pH, they reported that the BSA conformation in solution defined the interfacial coverage and adsorption at the air-water interface. Around BSA isoelectric point (where protein structure are most compact and stable) as well as at pH 7, but less than pH 5, this was at its greatest. Their finding were also in agreement with those of Martin-Rodriguez, Cabrerizo-Vilchez, & Hidalgo-Alvarez (1994) who reported a maximum adsorption of BSA on hydrophobic latex at pH 5. Likewise no air cells were produced with aqueous BSA in 0.2M sodium phosphate buffer at pH 7. When buffer concentration was reduced to 0.01M, air cell formation was observed but the population was far lesser than that at pH 5. It has been shown by King, Elkins, & Nelson (1999) when studying the reactivity of the cysteine of silver hake parvalbumin with DTNB (Ellman's regeant generally use to detect the chemical reactivity of sulphydryl group) that cysteine's reactivity is buffer species dependent with a substantial change in the kinetic profiles with buffers of different ionic character.

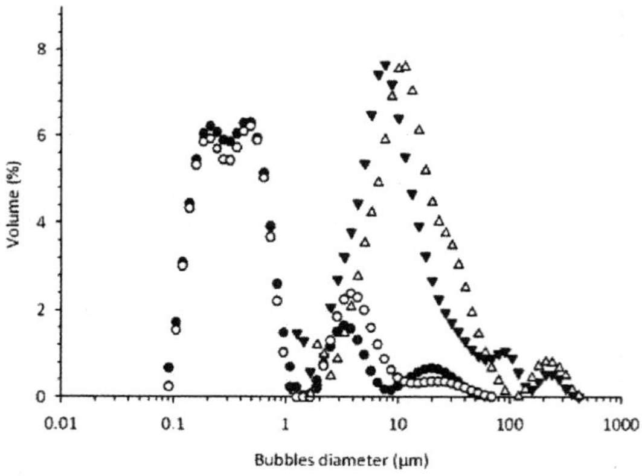

Figure 2. *Comparison of bubble volume distribution for HFBII-AFE (▼ 1day old, △ 20 days old) and BSA-AFE (● 1day old, O 20 days old).*

Comparison of the stability of HFBII-AFE and BSA-AFE

The stability study of AFEs was evaluated by monitoring the total volume, size and size distribution of air cells with time. A particle size analyser (Mastersizer Hydro 2000SM, Malvern Instrument Ltd, UK) was used for particle size analysis as described by (Tchuenbou-Magaia, Norton, & Cox 2009). It should be noted that the used of a mastersizer with AFEs or triphasic mixtures is non-trivial. Figure 2 depicts the volume percent of air cells in HFBII-AFE and BSA-AFE as function of size for 1 day and after 20 days storage at room temperature. The air cell size distribution in HFBII-AFE is broader (with a mean size ≈7 μm and a small number at ≈100 μm) than in BSA-AFE (mean size ≈ 0.5 μm and maximum ≈3 μm). This could be due in part to the lower concentration of HFBII, 0.01% W/V as compared to 5% W/V for BSA. However, it is worth mentioning that even at this high concentration BSA is still far cheaper than the hydrophobin. There was only a minor change in air cells size distribution over the period of 20 days for both emulsions. Indeed, the main peak only shifted from 7 μm to 10 μm for HFBII-AFE whereas an even smaller shift was observed for BSA. The excellent stability of BSA-AFE was also confirmed by applying a vacuum above a suspension for a period of hours to the AFEs and this treatment shows little effect on the bubbles size distribution. Likewise air cells in BSA-AFE can also survive high shear and high temperature processing steps. Figure 3 shows BSA-AFE after being subjected to a complete sterilization cycle (121°C/15 min). The BSA-AFE after this unit operation was air-dried by leaving it overnight in a fume cupboard then coated with gold and examined by SEM. Interestingly, the air cells entrapped within the protein gel (gelation due to the effect of temperature) retained their size and shape. This high stability was also observed with other bulk food grade proteins such as egg white proteins and interestingly these gave different gel properties after thermal treatment; thus offering a novel route for the production of a range of aerated protein gels with variable properties.

The protein coat of sonochemically synthesized microbubbles with BSA or heamoglobin were found to be about 25-35 nm thick, corresponding to 4 to 7 protein molecules and that some native protein remaining in the suspension also adhered to the microbubbles' shell (Suslick, Grinstaff, Kompany, & Wong 1994). Our air cells in BSA-AFEs have comparable characteristics. Also HFBII-AFEs showed interaction with other component of the system (Tchuenbou-Magaia, Norton, & Cox 2009).

These results indicate that sonochemically produced AFE with BSA and other cysteine rich food grade proteins could have the same property as AFE made with hydrophobins in terms of stability and interaction with other components in the product matrix. This may make them a suitable novel ingredient for the production of a range of aerated food such as soft drinks and low fat air based products.

Figure 3. *BSA-AFE after been subjected to heat treatment under industrial sterilisation conditions (121 °C /15min, bars): A SEM micrograph showing that bubbles within the protein gel that have kept their size and shape (Samples were air-dried, coated with gold before examining in a high vacuum scanning electronic microscope).*

Table 1. *Long time stability of triphasic emulsions: particle diameter is expressed as volume weighted diameter D [4,3] (μm)*

Emulsions	Storage time 1 day	1 month	2 months
40% O/W	2.9± 0.1	3.3 ± 0.3	3.5 ± 0.1
46% A/O/W (10% Air, 36% oil)	6.8 ± 0.1	7.8± 0.7	7.1± 0.6

2. Could AFE be used as an effective ingredient for fat reduction in traditional high fat emulsion based products like dressings and mayonnaise?

It is believed that the small, stable, air cells which were designed to resemble oil droplets in terms of their size and physical properties could be used to produce a new generation of low fat foods. This approach is strengthened by several works reporting the contribution of microbubbles to fat related attributes like creaminess (Campbell, Webb, Pondiella, & Niranjan 1999; Heuer 2009; Kilcast & Clegg 2002). In order to investigate whether AFE could be used as an ingredient for the production of low fat emulsion based products without compromising the mouth-feel and the perceived indulgence, triphasic emulsions (A/O/W) where air cells and oil droplets are dispersed in water continuous phase were constructed. Their rheological and tribological behaviours were compared to standard oil in water emulsions. This study is based on the fact that fat related attributes (smoothness, creaminess, fattiness) are known to be correlated to emulsion bulk rheology (viscosity) and the thin film rheology as studied in tribology (Dresselhuis et al. 2007). Also there is an inverse relation between lubricational properties of food and the sensorially perceived creaminess and fattiness (de Wijk & Prinz 2005; Dresselhuis et al. 2007; Malone, Appelqvist, & Norton 2003). Here, the rheological and the tribological behaviour of A/O/W and O/W emulsions model dressing are presented to illustrate a possible use of AFE for the production of low fat foods. The 46% A/O/W with 36% oil and 10% air by volume was compared to 40% O/W in order to find out if an air induced a shift in lubrication properties.

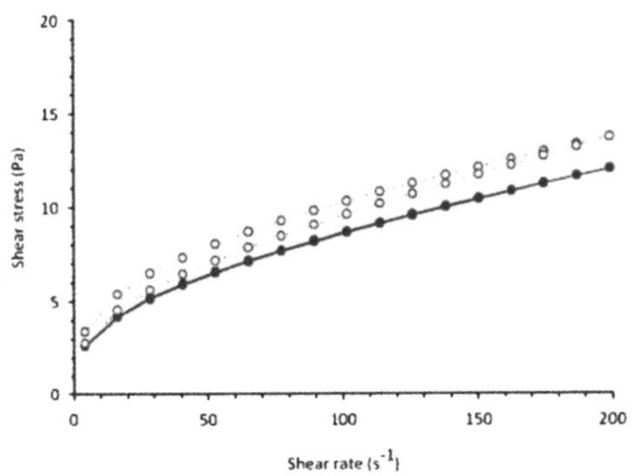

Figure 4. *Flow curves (shear stress as a function of shear rate) in linear scales ("up" and "down" curves) of 46 % A/O/W (O) and 40 % O/W (●).*

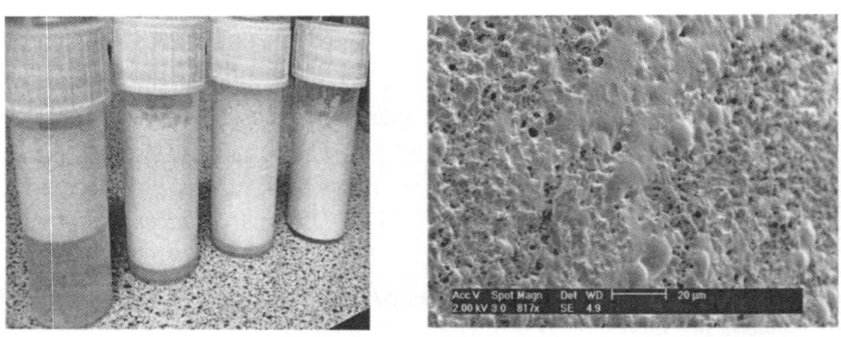

Figure 5. *(a) Effect of air phase volume on creaming stability of triphasic emulsions. Triphasic emulsions were prepared by diluting O/W, 20/80 emulsion stabilized with Tween 60 and containing 0.2% iota -Carrageenan with air HFBII-AFE (from left to right, O/W (20/80) used for the A/O/W preparation; A/O/W (10/18/72); A/O/W (20/16/64); A/O/W (60/8/32)). (b) Cryo-SEM of A/O/W (10/18/72) showing gel network (some large bubbles are visible in some of the tubes these are not to be confused with the AFE air cells).*

Comparison of the rheological and tribological behaviours of air based triphasic and standard oil/water emulsions model dressing.

Figure 4 shows the comparisons between the flow curves of a triphasic A/O/W with 46 % total phase volume (10% Air and 36% by volume) and the 40% O/W emulsion. The "up" and the "down" curves overlap for the 40% O/W emulsion whilst the triphasic A/O/W shows a hysteresis loop. Although this hysteresis could indicate an irreversible change in sample structure upon shearing, another possible and plausible explanation is the presence of weak attractive forces between particles resulting in a weak network which we attribute to a possible interactions between air cell proteins shell and the other components (XG and WPI) resulting in weak gel like behaviour (Mewis & Wagner 2009), which is usually an attractive property for such systems. In previous experiments A/O/W emulsions was made with a 20% O/W emulsion stabilised with the non-ionic surfactant, Tween 60 and contained 0.2% iota carageenan as thickening agent and these showed showed a similar behaviour (Tchuenbou-Magaia, Norton, & Cox 2009). Indeed the tri-phasic emulsion was more stable against creaming during storage than its parent O/W emulsion, and indeed creaming was observed to diminish or suppressed with increasing air phase volumes (Figure 5a). This behaviour was also thought to be due to bridging of droplets by a hydrophobin network or an interaction between the hydrophobin and the carrageen resulting in weak particulate gel (Figure5b). These results indicate that air cells (via the protein shells) in the triphasic emulsion could also help in simplifying the formulation (reduction of the amount of thickener) in a standard O/W emulsion based food to give a similar visual appearance, and as much as texture, taste and flavour as the conventional food.

The 46% A/O/W with 36% oil and 10% air by volume was compared to 40% O/W in order to find out if an air induced shift in lubrication. The lubrication properties of AFE, O/W and A/O/W emulsions are presented in Figure 6. Here commercial mayonnaise and dressings were used as reference full fat products to the low fat air based versions. It must be admitted that friction of such complex food system depends not only due to the fat and fat droplet size but also on the other components such as hydrocolloids. Moreover it has been reported that adding larger size or non-spherical particles into starch-containing custard desserts resulted in higher friction compared to smaller sizes or rounded particles (de Wijk & Prinz 2005) and commercial mayonnaise and dressings might content different type and amount of particles. Nevertheless their lubrication properties served as a good comparison and could direct the development of the air based versions with comparable lubricating behaviour and, indeed recognisable mouth-feel and consumer acceptance.

Friction measurements were performed three times with an entrainment speed varying from 1 to 1000 mm/s without unloading the ball in order to mimic the dynamic process of continuous oral breakdown of food structure and in a hope to provide an accurate picture of the lubrication properties of the product. We assumed that, if oral processing starts from the first bite till after swallowing/clearing, the friction as function of speed curve of run "n" which is equal to the following run "n+1", would indicate the in-mouth friction (in-mouth sensation) of a given product. This hypothesis was verified with other work on similar systems samples. For this reason, only data of the 1st and 2nd runs are presented in Figure 6 as there is no difference between the friction at 2nd and 3rd runs for the triphasic emulsion.

The 46% A/O/W have lower friction coefficient than the 40% O/W which means that the air based emulsion certainly contributes to the lubrication, therefore AFEs could contribute to the perceived creaminess and fattiness. The region (indicate by two vertical lines in Figure 6) at an entrainment speed between 40 and 250 mm/s has been reported by Malone, Appelqvist, & Norton (2003) as relevant to the in-mouth perception, but this might be broadened as the region might shift depending on the product type and with person specific variability in the movement of the tongue during swallowing, 10 and 305 mm/s has also been reported as a possible range (Tasko, Kent, & Westbury 2002). Moreover the role of air in lowering friction was further illustrated by the finding that AFEs has lower friction than the protein solution used for its production (Figure 6b). Although the mechanism by which AFEs lower the friction is not totally clear, we suggest that, at least one possible route would be the air cells in AFEs fill the gap between the contact zones of the tribometer resulting in lower friction. When studying the role of friction in perceived oral texture, de Wijk & Prinz (2005) found that small rounded particles contributed to the reduction of friction and to enhance the perceived creaminess of custard. Likewise Kilcast & Clegg (2002) reported that creaminess increased when small particles less than 4-7 μm were added to the system and, as showing by Figure 2 more than 50% of air cells in the BSA-AFE are within this range of size.

In addition to the above mentioned mechanism, extra coalescence might occur in triphasic air based emulsion, and consequently an extra friction reduction compared to the standard O/W emulsion. Insofar as it has been demonstrated that in-mouth aeration induces extra coalescence of oil droplets (Benjamins et al. 2009), we suggest that the partial break down of air cells upon shearing (Figure7) may also contribute to the increase in sensitivity to coalescence and subsequent perception of oil release resulting in more lubrication as observed with the triphasic A/O/W. Therefore it is likely that the A/O/W systems will have comparable if not better sensory behaviour and perceived indulgence than the full fat version at a lower fat (calorific) content.

4. CONCLUSION

We have demonstrated that cysteine rich food grade proteins can be processed to give properties comparable to hydrophobins by comparing sonochemically produced BSA-AFEs with HFBII-AFEs. AFEs in triphasic emulsion might also offer new structures and textures to the food industry due to interaction between the air cells with protein shells and other molecules resulting in weak gel system like behaviour; on the other hand the triphasic A/O/W could be manipulated to be similar to the standard oil water emulsion in terms of lubrication behaviour and sensory perception.

The next stage of this investigation is to extend the use of AFEs into a range of other food formats such as soft drinks and other emulsion based food formulations and to study the microstructure of such systems during pilot and industrial scale production.

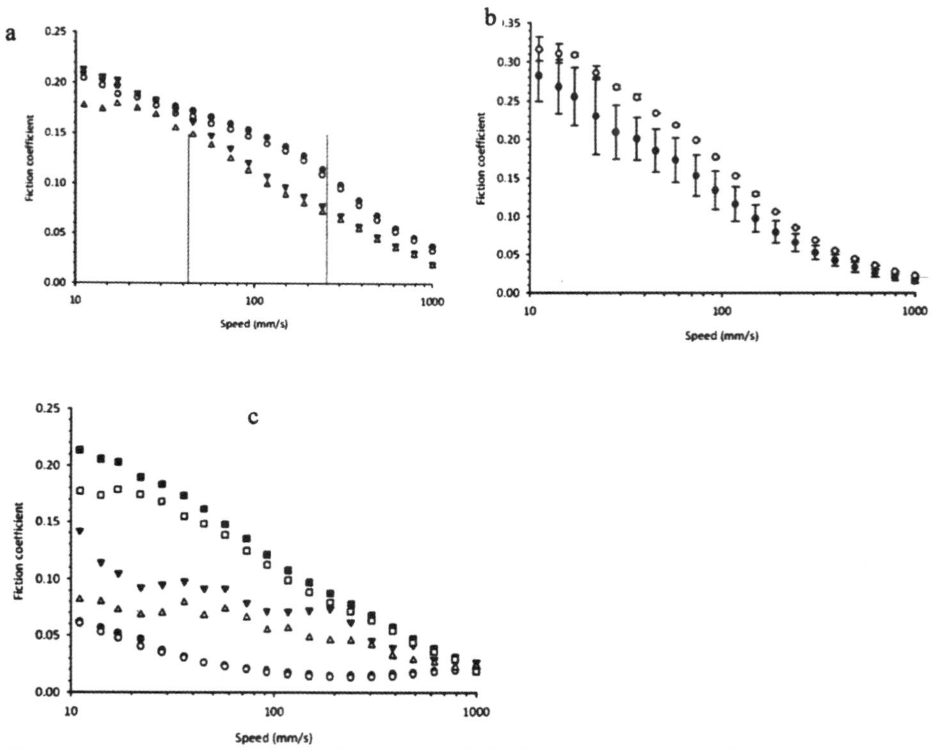

Figure 6. *Comparison of time-dependent tribological behaviour of different emulsions. a: 40% O/W (● 1st run, O 2nd run) and 46% A/O/W (10 % air and 36% oil) (▼1st run, 2nd run) emulsions at 20°C: the two vertical lines indicate the region where tribology data was reported to be correlated with the in-mouth perception ((Malone, Appelqvist, & Norton 2003); b: BSA-AFE (●) and proteins solution (O) used for its preparation; c: 46% air based triphasic emulsion (10 % air 36% oil) (■ 1st run, □ 2nd run and supermarket products (dressings, 52% fat (▼1st run, 2nd run) and mayonnaise, 79% fat (● 1st run, O 2nd run)).*

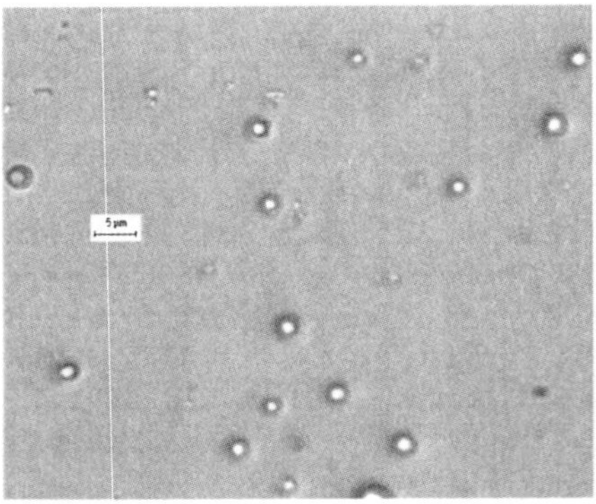

(a)

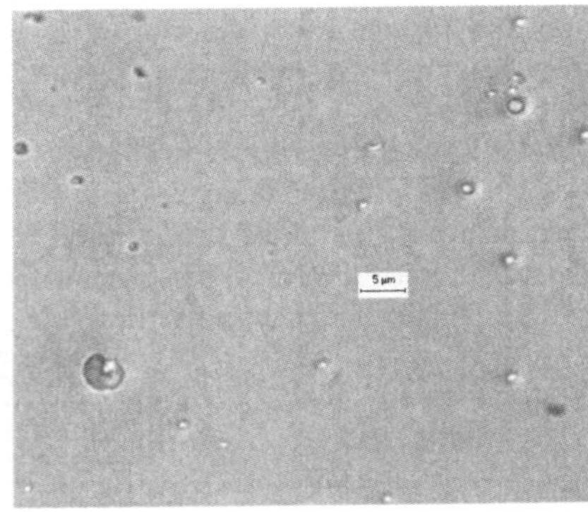

(b)

Figure 7. *Brightfield micrographs of BSA-AFE emulsion showing partial break down of air cell during tribology measurement: (a) before tribology test; (b) after tribology test.*

REFERENCES

Askolin, S. 2006, Characterisation of Trichoderma reesei hydrophobins HFBI and HFBII, Helsinki University of Technology.

Benjamins, J., Vingerhoeds, M. H., Zoet, F. D., de Hoog, E. H. A., & van Aken, G. A. 2009, "Partial coalescence as a tool to control sensory perception of emulsions", *Food Hydrocolloids*, vol. 23, no. 1, pp. 102-115

Campbell, G. M., Webb, C., Pondiella, S. S., & Niranjan, K. 1999, *Bubbles in food* American Association of Cereal Chemists.

CascaoPereira, L. G., Theodoly, O., Blanch, H. W., & Radke, C. J. 2003, "Dilatational Rheology of BSA Conformers at the Air/Water Interface", *Langmuir*, vol. 19, no. 6, pp. 2349-2356.

Cox, A. R., Cagnol, F., Russell, A. B., & Izzard, M. J. Surface Properties of Class II Hydrophobins from *Trichoderma reesei*and Influence on Bubble Stability. Langmuir 23, 7995-8002.2007.

Cox, A. R., Aldred, D. L., & Russell, A. B. 2009, "Exceptional stability of food foams using class II hydrophobin HFBII", *Food Hydrocolloids*, vol. 23, no. 2, pp. 366-376.

de Wijk, R. A. & Prinz, J. F. 2005, "The role of friction in perceived oral texture", *Food Quality and Preference*, vol. 16, pp. 121-129.

Dresselhuis, D. M., de Hoog, E. H. A., Cohen Stuart, M. A., & van Aken, G. A. 2007, "Tribology as a tool to study emulsion behaviour in the mouth," in *Food Colloids: Self-Assembly and Material Science*, E. Dickinson & M. E. Leser, eds., RSC, pp. 451-461.

Exerowa, D. & Kruglyakov, P. M. 1998, *Foam and Foam Films: Theory, Experiment, Application* Elsevier Science B.V., Amsterdam.

Grinstaff, M. W. & Suslick, K. S. 1991, "Air-Filled Proteinaceous Microbubbles: Synthesis of an Echo-Contrast Agent", *Proceedings of the National Academy of Sciences*, vol. 88, no. 17, pp. 7708-7710.

Heuer, E. A. K. 2009, *Formulation and stability of model food foam microstructures,* Engineering doctorate Thesis, School of Chemichal Engineering The University of Birmingham.

Kilcast, D. & Clegg, S. 2002, "Sensory perception of creaminess and its relationship with food structure", *Food Quality and Preference*, vol. 13, no. 7-8, pp. 609-623

King, N. M., Elkins, K. M., & Nelson, D. J. 1999, "Reactivity of the invariant cysteine of silver hake parvalbumin (Isoform B) with dithionitrobenzoate (DTNB) and the effect of differing buffer species on reactivity", *Journal of Inorganic Biochemistry*, vol. 76, no. 3-4, pp. 175-185.

Kisko, K., Szilvay, G. R., Vainio, U., Linder, M. B., & Serimaa, R. 2008, "Interactions of Hydrophobin Proteins in Solution Studied by Small-Angle X-Ray Scattering", *Biophysical Journal*, vol. 94, no. 1, pp. 198-206.

Lau, C. K. & Dickinson, E. 2005, "Instability and structural change in an aerated system containing egg albumen and invert sugar", *Food Hydrocolloids*, vol. 19, no. 1, pp. 111-121.

Malone, M. E., Appelqvist, I. A. M., & Norton, I. T. 2003, "Oral behaviour of food hydrocolloids and emulsions. Part 1. Lubrication and deposition considerations", *Food Hydrocolloids*, vol. 17, no. 6, pp. 763-773.

Martin-Rodriguez, A., Cabrerizo-Vilchez, M. A., & Hidalgo-Alvarez, R. 1994, "A comparative study on the electrokinetic behavior of bovine serum albumin molecules adsorbed onto different polymer colloids", *Colloids and Surfaces A: Physicochemical and Engineering Aspects*, vol. 92, no. 1-2, pp. 113-119.

Mewis, J. & Wagner, N. J. 2009, "Thixotropy", *Advances in Colloid and Interface Science*, vol. 147-148, pp. 214-227.

Suslick, K. S., Grinstaff, M. W., Kompany, E., & Wong, M. 1994, "Characterization of sonochemiclly prepared proteinaceous microspheres", *Ultrasonics Sonochemistry*, vol. 1, no. 1, p. S65-S68.

Tasko, S. M., Kent, R. D., & Westbury, J. R. 2002, "Variability in Tongue Movement Kinematics During Normal Liquid Swallowing", *Dysphagia*, vol. 17, no. 2, pp. 126-138.

Tchuenbou-Magaia, F. L., Norton, I. T., & Cox, P. W. 2009, "Hydrophobins stabilised air-filled emulsions for the food industry", *Food Hydrocolloids*, vol. 23, no. 7, pp. 1877-1885.

Wildmoser, H., Scheiwiller, J., & Windhab, E. J. 2004, "Impact of disperse microstructure on rheology and quality aspects of ice cream", *Lebensmittel-Wissenschaft und-Technologie*, vol. 37, no. 8, pp. 881-891.

Zuniga, R. N. & Aguilera, J. M. 2008, "Aerated food gels: fabrication and potential applications", *Trends in Food Science & Technology*, vol. 19, no. 4, pp. 176-187.

VISCOSITY DEVELOPMENT DURING COMPETITIVE HYDRATION OF STARCH AND HYDROCOLLOIDS

MD Lad, S Samanci, JR Mitchell and TJ Foster

Division of Food Sciences, School of Biosciences, University of Nottingham, Sutton Bonington Campus, Loughborough, LE12 5RD, United Kingdom.

1 INTRODUCTION

There has been interest in the past regarding the enhancement of starch properties when mixed with hydrocolloids. The description of synergism is often used in such mixtures, yet the work of Appelqvist *et al.*[1] indicated that the interaction can be viewed as a phase separation event, where the leached amylose from the starch is constrained due to incompatibility with the hydrocolloid. There has been recent interest in monitoring the starch pasting characteristics in the presence of hydrocolloids such as xanthan, guar, xyloglucan and beta glucan, amongst others [2-11]. The predominant response is an increase in peak, breakdown, final and setback viscosities with little effect on gelatinisation temperatures. The phenomenological description indicates that the hydrocolloid hydration induces a limitation of water availability, restricting the plasticisation of the starch by water and potentially reducing the total swelling of the starch granule. Other recent work has shown how such an understanding can be important in industrial application of such mixed systems [5,12-14].

The present study, however, does not focus on the swelling of starch during pasting in the presence of a hydrocolloid, but on the swelling of cold water swelling starches in competitive hydration with different hydrocolloids, upon the addition of water. The cold water starches are of waxy maize or potato origin and the hydrocolloids are either uncharged (guar gum) or anionic (alginate and xanthan). The rates of structure formation and final mix viscosity are compared using the Rapid Visco Analyzer (RVA) to determine the extent to which the hydrocolloids affect the swelling of the starch, or conversely how the swelling of the starch affects the hydration of the hydrocolloids. To further understand the developed microstructure the impact of dilution of these add-mixtures and hydration in the presence of salt is also discussed.

2 MATERIALS AND METHODS

Starches from different sources were used, a cross-linked waxy maize (CLWM) starch and a cross-linked potato starch as the cold water swelling particles. Three different types of hydrocolloids were used, xanthan, guar gum and alginate. In total four different xanthan types were used, one sourced from Sigma biological and the other three (Grindsted xanthan clear 80, Grindsted xanthan clear 200 and Grindsted xanthan clear supra) were all food grade xanthans kindly donated by Danisco. Guar gum (MEYPRODOR 30) was also supplied from Danisco and had a molecular weight of 423 kDa. The alginate used was PROTANAL HF120RB with a M:G ratio of 0.55:0.45, this was kindly supplied by FMC Biopolymer (Cork, Ireland). Sunflower oil was purchased commercially. All substances were used as supplied without further modification unless stated otherwise.

2.1 Viscosity measurements

Viscosity development and therefore sample hydration was measured using the Rapid Visco Analyzer (RVA) (Newport Scientific, Australia). Viscosity was measured as a function of constant temperature (25 °C) and shear rate (180 rpm) for a total period of 30 minutes unless stated otherwise.

2.2 Sample preparation

A range in hydrocolloid concentrations were used from 0-2% (w/w) with a constant starch concentration of 10% (w/w). The concentration of sunflower oil was kept constant at 5% unless stated otherwise. For the RVA measurements, the starch, hydrocolloid and oil were initially mixed together; to this the required amount of water was added and placed directly into in the RVA without any delay and the viscosity development measured. For measurements performed in the presence of salt three different sodium chloride concentrations were used, 0.01 M, 0.1 M and 1 M.

2.3 Confocal microscopy

Where images are presented these were taken using a Nikon Eclipse Ti inverted Confocal microscope, Supplier: Nikon UK Ltd., Kingston upon Thames. The equipment comprises lasers: Argon Ion 488 nm, Green-Helium Neon 543 nm, Blue diode 405 nm, and is fitted with a C1 detector unit (3 PMT), a C1 transmitter detector unit (transmitted light), and the data collected and analysed with EZ-C1 control software.

3 RESULTS AND DISCUSSION

3.1 Starch hydration in the presence of hydrocolloids

The use of cross linked waxy maize (CLWM) starch for rapid viscosity development at low temperatures is due to the swelling of the starch granule where the ability to maintain structural integrity provides texture. The introduction of a small amount of hydrocolloid into the mixture which is competing with the starch for the available water has shown to impact both on the initial rate of hydration and final viscosity reached after 30 minutes of mixing within the RVA.

Displayed in Figure 1 are the viscosity profiles obtained when water was introduced to samples containing 10% CLWM starch and 2% hydrocolloid either xanthan gum, guar gum or alginate with 5% oil. Also presented in the figure are the respective controls. Viscosity development for starch alone is rapid reaching equilibrium by 200 s with a final viscosity of 4422 cP. The presence of guar gum and alginate show a decrease in the initial rate compared to starch alone, taking 600 s to approach the equilibrium viscosity. Here the hydration of guar gum and alginate is reducing the rate of hydration of starch. The final viscosity in both cases is greater than that for starch alone reaching 5473 cP and 6136 cP for guar gum and alginate, respectively. In the presence of xanthan gum the initial rate of viscosity increase is further reduced, suggesting a greater competition for the available water by the xanthan. The final viscosity of the mixture recorded was reduced by over one half of starch alone indicating that the starch granules have not fully hydrated in the presence of xanthan.

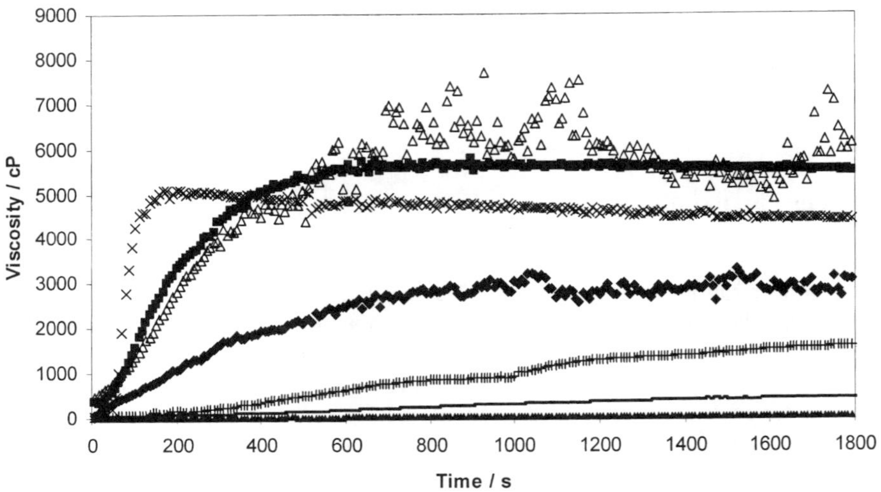

Figure 1 *Change in viscosity in a Rapid Visco Analyser for a mixture of 10% CLWM starch, 5% oil (X) + 2% xanthan (♦), 2% guar (■), 2% alginate (Δ). Control 2% hydrocolloid + 5% oil, xanthan (-), guar (▲) and alginate (+).*

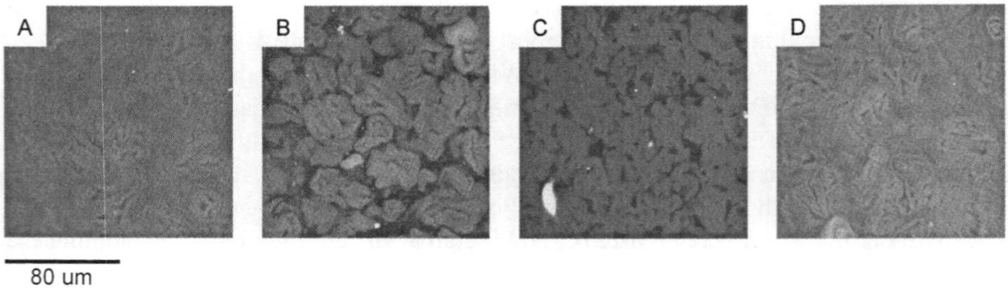

Figure 2 *Confocal micrographs of 10% CLWM starch alone (A), 10% CLWM starch + 2% xanthan (B), 10% CLWM starch + 2% guar gum (C) and 10% CLWM starch + 2% alginate.*

Further support for the partial swelling of the starch granule can be seen in the Confocal images of the samples taken after 30 minutes in the RVA (Figure 2). Starch alone shows a compact arrangement upon hydration (Figure 2 A). In the presence of alginate (Figure 2 D) this does not appear to change, there is a slight difference in the starch structure in the presence of guar gum (Figure 2 C) though this has not shown a reduction in final viscosity observed in the presence of xanthan. More significant difference in the starch granule can be observed in the presence of xanthan (Figure 2 B) where the packing volume of the starch appears to be reduced. The presence of xanthan is not preventing the starch granule from swelling, but restricts full swelling.

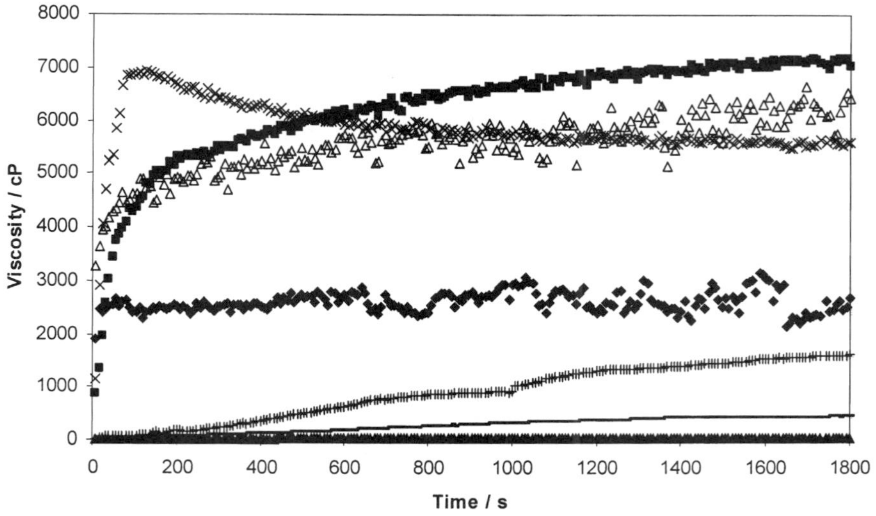

Figure 3 *Change in viscosity in a Rapid Visco Analyser for a mixture of 10% potato starch, 5% oil (X) + 2% xanthan (♦), 2% guar (■), 2% alginate (Δ). Control 2% hydrocolloid + 5% oil, xanthan (-), guar (▲) and alginate (+).*

The replacement of CLWM starch with potato starch shows the viscosity profiles display similar trends (Figure 3), again the final viscosity achieved in the presence of guar gum and alginate is above that for starch alone. In the presence of xanthan the final viscosity is significantly reduced and does not reach that observed for starch alone. The immediate difference observed between the two starches can be seen in the initial rates of viscosity increase. With the potato starch the viscosity increase and thus the hydration is much faster. Confocal images in this case (not shown) have revealed little differences in the structure of the starch granule along with the packing density, in the presence of any of the hydrocolloids which is attributed to the size difference between the two starch types. Potato starch granules swell to a larger size and therefore occupy a greater volume, and are also negatively charged, explaining the increase in rate of initial hydration.

Solutions of high xanthan concentration have been shown to form an anisotropic liquid crystalline phase[15,16] and investigations of these concentrated xanthan solutions have revealed a decreased viscosity[17]. To determine whether a concentration effect is taking place in these mixed systems, the concentrations of guar gum and xanthan were reduced. The final viscosity against hydrocolloid concentration (Figure 4) shows that the xanthan concentration has to be reduced to 0.1% to match the viscosity of starch alone. Whereas increasing guar gum concentration from 0.5 – 2% leads to a continued increase in final viscosity reducing xanthan concentration enables effective starch hydration, further supported by Confocal images (Figure 5). Starch packing density matches that of starch alone when the xanthan concentration is reduced to 0.05% (Figure 5D) which is significantly different to the starch granule in the presence of 0.5%

xanthan (Figure 5 C). This reduction in viscosity in the presence of xanthan appears to be quite unique and is not seen for the other hydrocolloids.

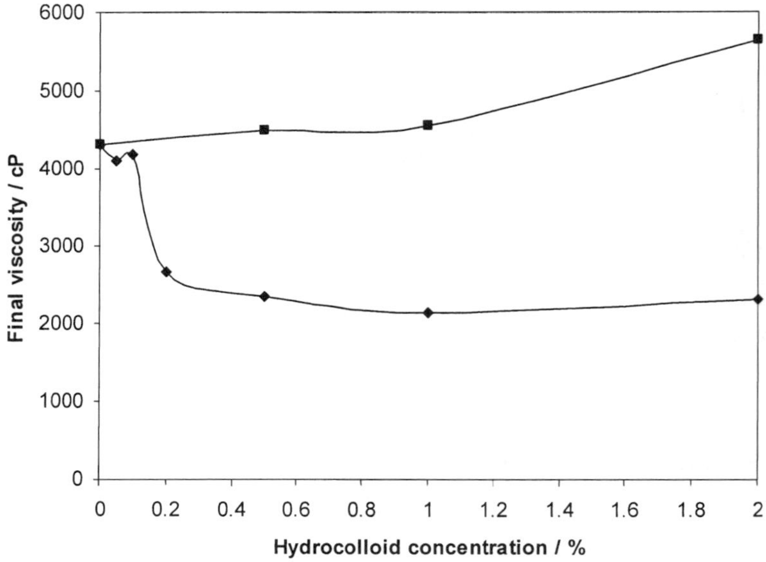

Figure 4 *Final viscosity profiles as a function of increase xanthan concentration (♦) and guar gum (■) with constant CLWM starch concentration (10%).*

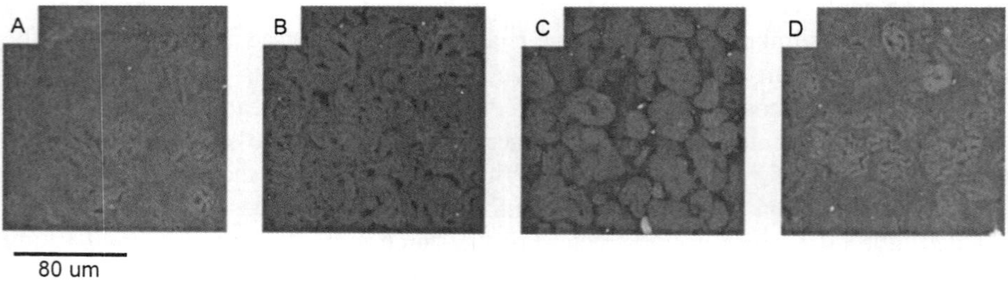

Figure 5 *Confocal micrographs of 10% CLWM starch alone (A), 10% CLWM starch + 0.5% guar gum (B), 10% CLWM starch + 0.5% xanthan (C) and 10% CLWM starch + 0.05% xanthan.*

3.2 Impact of dilution of starch-hydrocolloid admixture

Starch, in the presence of high xanthan concentrations, has shown restricted swelling with a decrease in viscosity observed akin to aqueous solutions of xanthan attributed to xanthan entering an anisotropic liquid crystalline phase. The results presented would suggest that both starch and xanthan have the ability to hydrate further. To test this, starch-hydrocolloid admixtures were diluted by 10% and 20% with viscosity development monitored for a further 15 minutes after the initial 30 minutes of hydration. The impact of dilution has shown to be dependant on starch type, concentration of hydrocolloid present within the mixture and the percentage of dilution. The findings are summarised in Table 1. The viscosity change upon dilution in the presence of guar gum and alginate always showed a decrease in viscosity (with the exception of 10% CLWM starch and 0.5% guar gum and alginate which have shown the viscosity to be maintained upon 10% dilution). Systems comprising CLWM starch and xanthan always showed an increase in viscosity. With CLWM starch being replaced by potato starch we observe a decrease in viscosity regardless of hydrocolloid type, concentration and dilution percentage. This would suggest that the starch type and the potential to swell further upon dilution go some way to explaining the increase in viscosity upon dilution. This is further supported by the Confocal images (not shown) which show no real difference in the potato starch granule before and after dilution. There is, however, a difference observed in the presence of CLWM starch and xanthan, where the starch granule appears to have swollen further upon dilution which would aid an increase in viscosity (figure not shown).

In the presence of guar gum and alginate, high equilibrium viscosity is reached before dilution (Figure 1) which either matches that of CLWM starch alone or is greater. This would suggest that the CLWM starch granule has already hydrated to its maximum and cannot hydrate further. Upon dilution a move of xanthan away from its anisotropic phase and the potential for CLWM starch to further hydrate appears to be important in determining the increase in viscosity. The viscosity does not increase in the system containing potato starch, indicating that it has fully swollen. The final viscosity of this mixture, after initial mixing, does show a decrease in viscosity when mixed with xanthan. This may indicate that it is the latent starch swelling which is the predominant factor of viscosity increase upon dilution, overriding and viscosity increase which comes from the dilution of the xanthan anisotropic phase.

Table 1 *Impact on viscosity upon dilution of starch-hydrocolloid mixtures.*

Starch type	**Hydrocolloid type and concentration**	**Viscosity change upon 10% dilution**	**Viscosity change upon 20% dilution**
CLWM	0.5% guar gum	maintained	decreases
	0.5% alginate		
	0.5% xanthan	increases	increases
CLWM	2% guar gum	decreases	decreases
	2% alginate		
	2% xanthan	increases	increases
Potato	0.5% guar gum	decrease	decreases
	0.5% alginate		
	0.5% xanthan		
Potato	2% guar gum	decreases	decreases
	2% alginate		
	2% xanthan		

3.3 Hydration of CLWM starch and xanthan as a function of increased ionic strength

Increasing ionic strength can screen the charge of the polyelectrolyte thereby decreasing the initial hydration rate of the hydrocolloid solutions. Additionally, Lee *et al.*[17,18] have shown that the concentration of xanthan required to form an anisotropic liquid crystalline phase increases with increasing ionic strength. An increase in viscosity with the CLWM starch and xanthan is indeed observed with increasing salt concentration (Figure 6) where a higher xanthan concentration is needed (1% or >), with 0.1 M salt, before the decrease in final viscosity is observed.

Evidence to support a hypothesis for a delay in xanthan hydration in the presence of increasing salt concentrations is presented in Figures 7 & 8. Here the initial viscosity increase is more rapid with increasing salt concentrations though it does not appear to reach that of starch alone. This would suggest the delay in xanthan hydration allows the starch granule to hydrate, imparting comparatively higher viscosity within the final starch-hydrocolloid mixture with added salt. This is further demonstrated when the starch-hydrocolloid mixture structure is observed via Confocal microscopy (Figure 9). As the ionic strength is increased, the individual definition of the swollen starch granule is no longer visible indicating the starch swelling potential and subsequently the packing density has increased (1B&C, 2B&C).

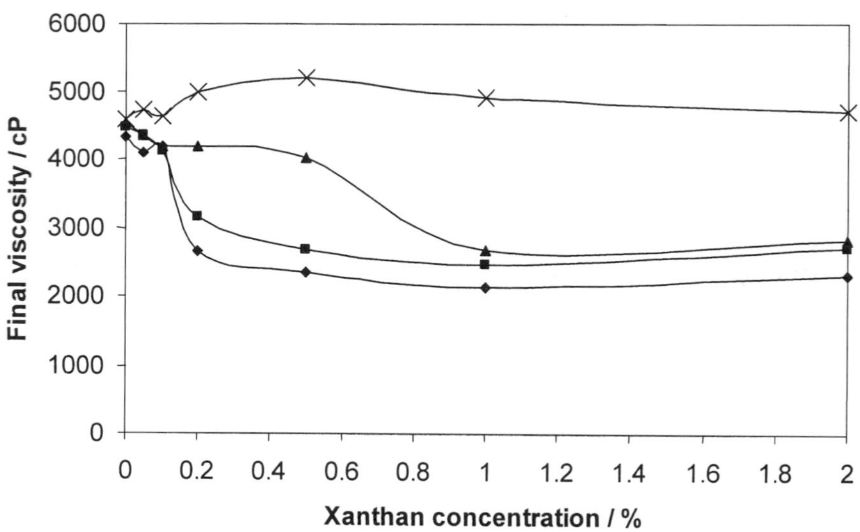

Figure 6 *Final viscosity profiles as a function of increase salt concentration 0 M (♦), 0.01M (■), 0.1 M (▲) and 1 M (×) with a constant CLWM starch concentration (10%).*

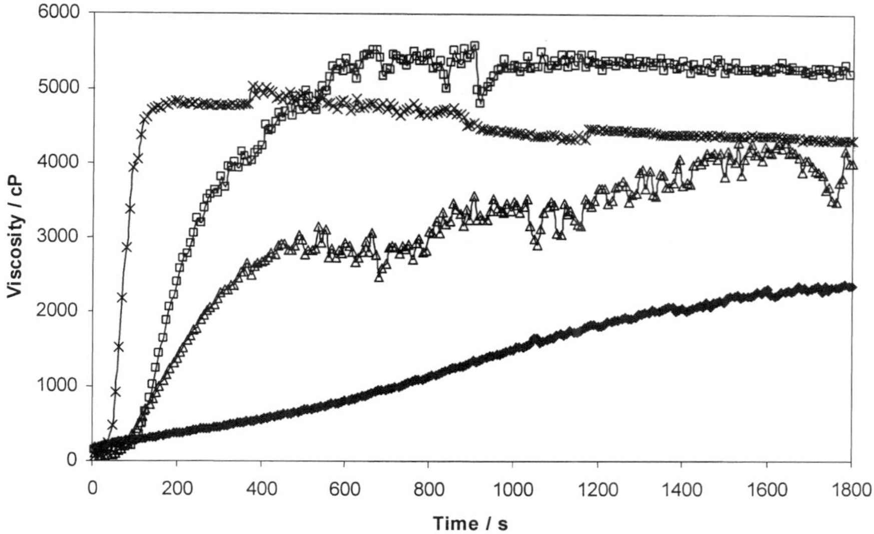

Figure 7 *Change in viscosity in a Rapid Visco Analyser for a mixture of 10% CLWM starch, 5% oil (X) + 0.5% xanthan in water (♦), with 0.1M NaCl (Δ) and 1 M NaCl (□).*

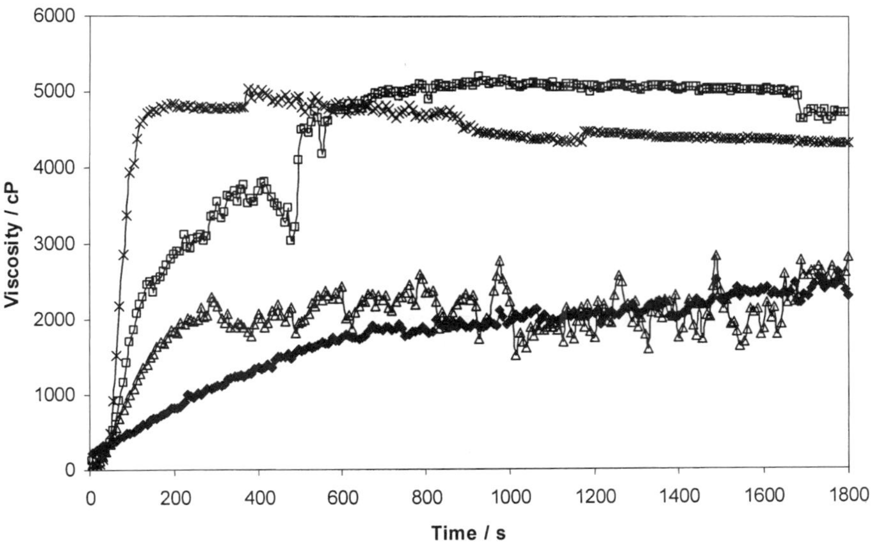

Figure 8 *Change in viscosity in a Rapid Visco Analyser for a mixture of 10% CLWM starch, 5% oil (X) + 2% xanthan in water (♦), with 0.1M NaCl (Δ) and 1 M NaCl (□).*

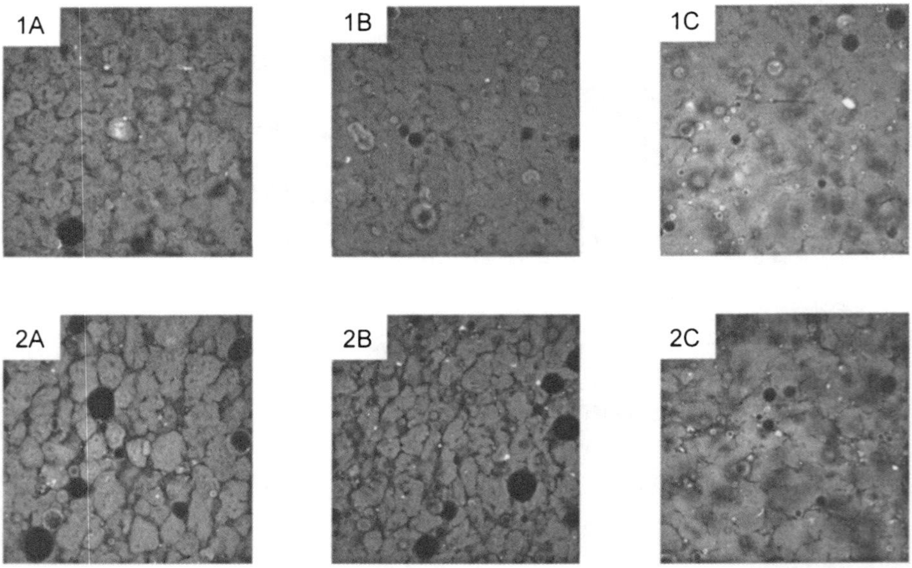

Figure 9 *Confocal micrographs of 10% CLWM starch + 0.5% xanthan 0 M salt (*1A*), 0.1 M salt (*1B*) and 1 M salt (*1C*). 10% CLWM starch + 2% xanthan 0 M salt (*2A*), 0.1 M salt (*2B*) and 1 M salt (*3C*).*

3.4 Impact of starch structure with different mixing methods

The instant and competitive hydration of hydrocolloids with cold water swelling starches in particular CLWM has shown to impact the viscosity development, final viscosity reached and ultimately the final starch granule structure. Increasing xanthan concentration in the presence of 10% CLWM starch appeared to show partial hydration of the starch granule compared to starch alone (Figures 2 B and A respectively) with a significant decrease in the initial rate of hydration. The addition of xanthan powder to fully hydrated starch paste showed an immediate decrease in viscosity. With starch powder addition to a hydrated xanthan solution the viscosity immediately increased. In both cases the equilibrium viscosity reached was close to instant hydration of CLWM starch and xanthan (Figure 10). The change in viscosity observed upon the addition of powders has shown the ability of both starch and xanthan to hydrate *i.e.* removing water from the hydrated solution/paste. The final viscosity reached for all three mixtures was similar, suggesting the system attains equilibrium to a similar end point.

Such variations in point of addition all show similar final structures (Figure 11 B, C and D), supporting the hypothesis for the hydration of both components. The hydration of xanthan certainly appears to prevent the maximum hydration/viscosity of CLWM starch to be attained. This has demonstrated that the equilibrium viscosity and end structure of starch attained is independent of mixing method.

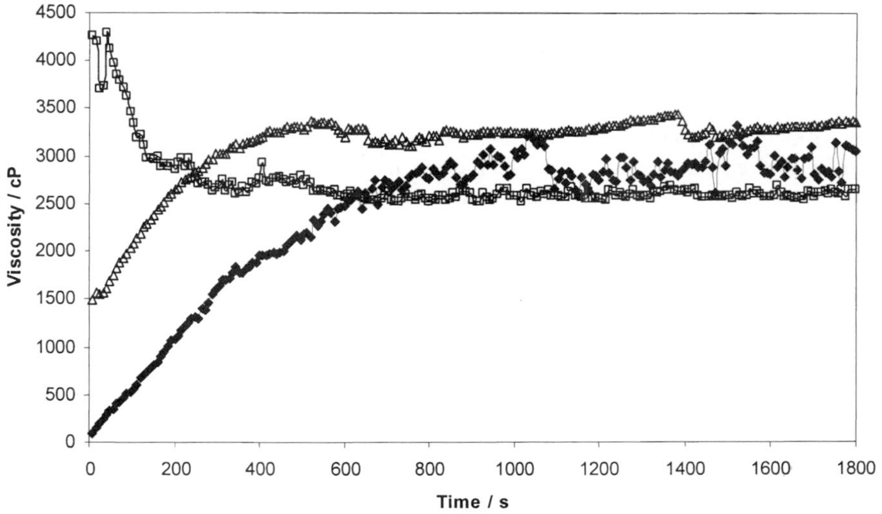

Figure 10 *Change in viscosity in a Rapid Visco Analyser for a mixture of 10% CLWM + 5% oil + 2% xanthan (instant mix) (♦), 10% CLWM starch powder to 2% xanthan paste (Δ) and 2% xanthan powder added to 10% CLWM starch paste (□).*

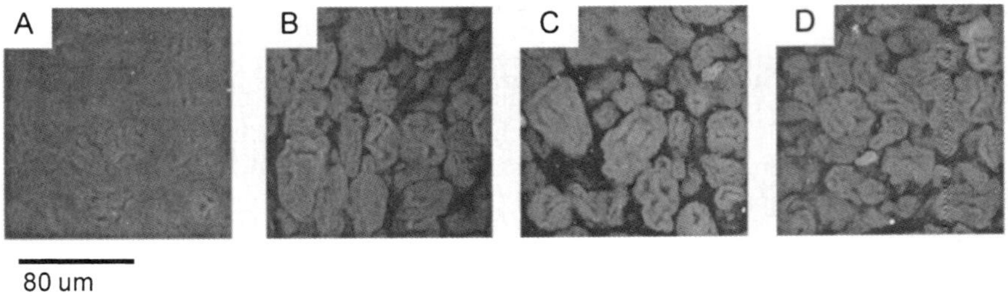

Figure 11 *Confocal micrographs of 10% CLWM starch alone (A), 10% CLWM starch paste + 2% xanthan addition as powder (B), 10% CLWM starch powder + 2% xanthan paste (C) and 10% CLWM starch + 2% xanthan addition as powder CLWM starch addition to xanthan paste(C) and 10% CLWM starch + 2% xanthan instant mix (D).*

4 CONCLUSIONS

Viscosity development of starch in the presence of hydrocolloids has shown to be dependent on starch type, hydrocolloid type and concentration. More importantly the presence of hydrocolloid has impacted the initial rate at which the viscosity increases, the final viscosity reached and the starch microstructure attained upon hydration. Differences observed have been attributed to the variation in the hydrodynamic volume of the hydrocolloid. Although further work is required, a decrease in viscosity in the case of xanthan, along with the partial swelling of starch and an increase in xanthan concentration requirement with an increased ionic strength, all appear to support the hypothesis that xanthan is in an anisotropic liquid crystalline phase in the presence of starch. Finally different mixing methods investigated demonstrate that the equilibrium viscosity and end starch structure attained is independent of mixing method. These viscosity changes observed upon dilution have the potential to aid food based formulations in the future.

5 ACKNOWLEDGEMENTS

We thank Dudley Ferdinando (Unilever Discovery, Measurement Science Unit, Colworth Science Park) for initial Confocal measurements.

References

1 Appelqvist, I. A. M.; Debet, M. R. M. *Food Reviews International* **1997**, *13*, 163-224.
2 Nagano, T.; Tamaki, E.; Funami, T. *Carbohydrate Polymers* **2008**, *72*, 95-101.
3 Weber, F. H.; Clerici, M. T. P. S.; Collares-Queiroz, F. P.; Chang, Y. K. *Starch* **2009**, *61*, 28-34.
4 Viturawong, Y.; Achayuthakan, P.; Suphantharika, M. *Food Chemistry* **2008**, *111*, 106-114.
5 Sikora, M.; Kowalski, S.; Tomasik, P. *Food Hydrocolloids* **2008**, *22*, 943-952.
6 Achayuthakan, P.; Suphantharika, M.; Rao, M. A. *Carbohydrate Polymers* **2006**, *65*, 469-478.
7 Achayuthakan, P.; Suphantharika, M. *Carbohydrate Polymers* **2008**, *71*, 9-17.
8 Banchathanakij, R.; Suphantharika, M. *Food Chemistry* **2009**, *114*, 5-14.
9 Temsiripong, T.; Pongsawatmanit, R.; Ikeda, S.; Nishinari, K. *Food Hydrocolloids* **2005**, *19*, 1054-1063.
10 Tester, R. F.; Sommerville, M. D. *Food Hydrocolloids* **2003**, *17*, 41-54.
11 Wang, F. H.; Wanf, L.-J.; Li, D.; Ozkan, N.; Li, S.-J.; Moa, Z.-H. *Carbohydrate Polymers* **2009**, *77*, 472-481.
12 Ravindran, G.; Matia-Merino, L. *Food Hydrocolloids* **2009**, *23*, 1047-1053.
13 Rojas, J. A.; Rosell, C. M.; de Barber, C. B. *Food Hydrocolloids* **1999**, *13*, 27-33.
14 Sikora, M.; Juszczak, L.; Sady, M.; Krawontka, J. *Nahrung/FOOD* **2003**, *47*, 106-113.
15 Allain, C.; Lecourtier, J.; Chauveteau, G. *Rheologica Acata* **1988**, *27*, 255-262.
16 Oertel, R.; Kulicke, W.-M. *Rheologica Acata* **1991**, *30*, 140-150.
17 Lee, H.-C.; Brant, D. A. *Macromolecules* **2002**, *35*, 2212-2222.
18 Lee, H.-C.; Brant, D. A. *Macromolecules* **2002**, *35*, 2223-2234.

EVIDENCE OF INTERACTIONS BETWEEN WHEY PROTEINS AND MUCIN: THEIR IMPLICATION ON THE ASTRINGENCY MECHANISM OF WHEY PROTEINS AT LOW pH

B Vardhanabhuti and E Allen Foegeding

Department of Food, Bioprocessing, and Nutrition Sciences, North Carolina State University, Raleigh, NC 27695-7624

1 INTRODUCTION

The sale of functional beverages worldwide is expected to reach $ 34 billion in 2010.[1] In general, high protein sports drinks make up a considerable portion of this growing market. Whey proteins are among the major protein ingredients in sports beverages due to their nutritional properties. Protein beverages made at near neutral pH (~ 6.8) or under acidic conditions (pH < 4.6 but most are at pH < 3.5); and the chosen pH has thermal processing and product characteristics implications. Beverages made at near neutral pH are generally opaque and of a consistency significantly thicker than water. In contrast, acidic beverages (pH < 3.5) are clear and have viscosities close to water. The concern over whey protein-containing beverages at low pH is their astringency,[2,3] making them less desirable to consumers.

The majority of astringency research has focused on foods containing polyphenols such as wine, tea, and fruits; however, little work has studied protein astringency. Astringency is defined as the complex sensation due to shrinking, drawing or puckering of the epithelium as a result of exposure to substances such as alums or tannins.[4] There is still debate over the mechanism(s) causing astringency. Based on electrophysiological recordings from rodent gustatory nerves, some studies argued that astringency is a taste.[5,6] It is, however, more widely accepted that the sensation of astringency is a tactile phenomenon.[7-11] Bate Smith[11] proposed that polyphenols such as tannins, form complexes with salivary proteins and/or mucopolysacharides, either precipitating them or causing sufficient conformational changes, so that they lose their lubricating power, thus making the mouth feel rough and dry. It was also proposed that the interaction between polyphenols oral tissue leads to the precipitation of tissue proteins and accompanied by the shrinkage of tissue due to loss of water and a decrease in the permeability of this tissue to water and solutes.[12] A constriction of the mucosal epithelium caused by the ability of phenols to cross-link protein chains was also suggested for generating the astringent sensation.[13] Combining both possibilities, Guinard et al.[14] proposed that the reactions with the epithelial proteins themselves may take place after the salivary proteins have been complexed and the mucus layers covering the epithelium have been stripped away. The ability of polyphenols to bind salivary proteins, particularly proline rich proteins (PrPs), has been well documented.[15-20] While polyphenols and PrPs binding is the most accepted model describing the astringency mechanisms of polyphenols, recent tribology experiments have shown that two astringent polyphenols have different effects on saliva lubricity; questioning the comprehensiveness of a solely tactile model.[21]

In contrast to polyphenols, the astringency of whey proteins at low pH is not well understood. Sano et al.[3] suggested that, when mixed with saliva (pH 7), whey proteins (pH 3.5 solution) precipitated due to the change of pH to the isoelectric point (pI) for whey proteins, ~ 5.0, causing astringency. Saliva proteins were not mentioned in the whey protein astringency mechanism. Alternatively, Beecher et al.[2] proposed that, in addition to precipitation of whey protein alone, astringency could be due to interactions between the positively charged whey proteins (pH < pI) and negatively charged saliva proteins, based on the observation that astringency of whey proteins was at maximum at pH 3.4 and then lowered at lower pH (2.6). By controlling the total acidity, Lee and Vickers[22] recently found that addition of whey proteins to acid did not increase the astringency. They suggested that astringency of whey proteins at low pH was caused by the acid and not the proteins. It was noted, however, that pH was not controlled in the study and addition of whey proteins raised the pH of the samples. This posed a challenge in interpreting the results since pH has been found to have a major effect on astringency of acids[23]. Therefore, the higher astringency of acid compared to acid+whey (having higher pH) could simply be due to the lower pH of acid sample and not because whey proteins did not contribute to astringency.

There appear to be several hypotheses regarding the astringency mechanism of whey proteins. Our proposed hypothesis is that there are direct interactions between positively charged whey proteins and negatively charged salivary proteins, contributing to the astringency of whey proteins at low pH. Though PrPs have been identified as the major salivary proteins in polyphenols astringency, salivary mucins are believed to play an important role in oral lubrication. Mucins are high molecular weight glycoproteins that are ubiquitous in most organisms.[24] Their lubrication properties are facilitated by their ability to adsorb onto surfaces of practically any chemical nature. Pig gastric mucin (PGM) consists of MUC6 and MUC5A,[25] which has large similarities to the high molecular weight human salivary mucins (MUC5B).[26] PGM has been used in formulating saliva substitutes[27] and as a model salivary mucin for investigating astringency and lubrication properties of food components.[28-33] The turbidity developed due to PGM and polyphenols complexes has been proposed as a predictive model for astringency of wine.[34] The objective of our study was to provide the evidence of interactions between β-lactoglobulin and PGM, and how this could lead to the mechanism of whey protein astringency.

2 MATERIALS AND RESULTS

2.1 Preparation of β-lactoglobulin and mucin solutions

β-lactoglobulin (β-LG) was the gift from Davisco Foods International (Le Sueur, MN) and contained 91.64% protein based on micro-Kjeldahl nitrogen analysis (N x 6.38 for β-LG). Porcine gastric mucin was purchased from Sigma (St. Louis, MO). β-LG solutions (2% w/w) were prepared by slowly dissolving the protein powder in ~ 90% of the total deionized water (>17 MΩ). The solutions were stirred at room temperature (22 ± 2°C) and left overnight in the refrigerator for complete hydration. On the next day, the solutions were warmed to room temperature, and pH (2.0 to 7.0) and the final weight were adjusted. At pH 3.8-5.8, where it was close to the pI, β-LG solutions were turbid and/or precipitated. These solutions were prepared at higher concentration and centrifuged, and the supernatant was adjusted to proper concentration and used in the study. The concentration of β-LG was determined with a bicinchoninic acid assay (Pierce Biotechnology, Rockford, IL). Mucin (0.05% w/w) was dispersed overnight in a saliva buffer (8.1 mM KH_2PO_4, 1.5 mM Na_2HPO_4, 10 mM NaCl, 1.0 mM $MgCl_2$, 2.7 mM KCl) at pH 6.7. The ionic strength and pH of this buffer was reported to mimic those of saliva.[30] Mucin solution was centrifuged to remove undissolved material at 11,872×g for 10 min at 4 °C in a Sorvall Instruments (RC-5B model) centrifuge with a F21-8×50 mL rotor (FIBERLite, Santa Clara, CA).

2.2 Turbidity as the measure of interaction between β-lactoglobulin and mucin

β-LG solutions at different pH were mixed with mucin solution at the ratio of 1:1. Turbidity (OD_{600}) and pH of the mixture were measured. As control, β-LG solutions were mixed with saliva buffer (no mucin). All β-LG solutions before mixing had the initial OD_{600} of less than 0.02. Mucin adjusted to pH 2.0 to 7.0 had an $OD_{600} \leq 0.005$ and therefore the data is not presented in Figure 1. Based on the current theory, the astringency arises from the complex formation between salivary proteins and astringent compounds, resulting in the loss of lubrication. Turbidity has been used as a measure of the interaction between astringent compounds and saliva[2, 29, 35-37] or model salivary proteins such as mucin[31] or poly-L-proline.[38]

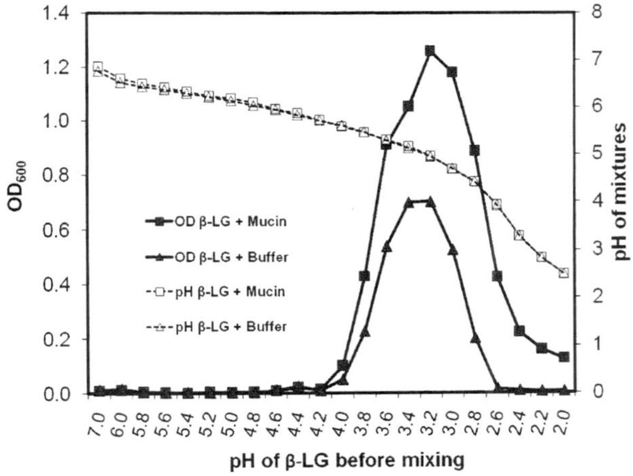

Figure 1 *Turbidity and pH of mixed β-LG and mucin or saliva buffer (without mucin.)*

Mixing β-LG solutions at pH 7.0 – 4.2 with pH 6.7 saliva buffer (no mucin) did not result in any significant change in turbidity (Figure 1). An increase in turbidity was observed when pH of β-LG solutions was between 4.0 and 2.8, which resulted in the pH of the mixtures being between 5.6-4.4. This is due to the decrease in the over-all electrostatic charge of the protein in the vicinity of the isoelectric point.[39] Increased turbidity was not observed when the pH of β-LG was further decreased (< 2.8) and the pH of the mixtures was farther away from pI. Highest turbidity of mixed β-LG and saliva buffer as well as mixed β-LG and mucin was observed when the pH of the mixtures was around 5. This result, in line with what proposed by Sano et al.[3], indicated that β-LG itself could precipitate in the oral environment due to the change of pH when mixed with saliva. The self-association and precipitation of whey proteins could induce a tactile sensation perceived by touch via mechanoreceptors.

Similarly, low turbidity was observed when mixing β-LG solutions between pH 7.0 to 4.2 with mucin. Turbidity started to increase when β-LG was $\leq$ 4.0 and reached the maximum at pH 3.4. The increase in turbidity of mixed β-LG and mucin was higher than that of mixed β-LG and saliva buffer. Since the pH of mixed β-LG/mucin and mixed β-LG/saliva buffer were the same, this higher increase in turbidity of mixed β-LG/mucin was the first evidence indicating interactions between β-LG and mucin. Furthermore, β-LG at lower pH (2.6 to 2.0) also gave an increase in turbidity when mixed with mucin compared to mixed β-LG with saliva buffer, which is another indication of β-LG and mucin interaction. It is interesting to note that the maximum turbidity was when the pH of the mixture was ~ 5 and then the turbidity decreased. If the turbidity from this mucin assay can be used as an index of astringency, the results suggest that there are certain pH values of whey proteins that will give maximum astringency. This coincides with results from Beecher et al.[2] showing model beverages containing 6% w/w whey protein isolate had higher astringency at pH 3.4 compared to pH 7 and 2.6.

We attempt to describe the decrease in astringency and turbidity as follows. Decreasing the pH of whey proteins (from 7 to 3.4) increased the net positive charge on the proteins resulting in an increase in astringency and turbidity in mixed β-LG/mucin. When the pH of whey proteins continues to decrease the proteins will be more positively charged; however, the pH of mixed β-LG and mucin (or saliva in real oral condition) will be lower than 5, resulting in less aggregation or precipitation and thus a decrease in astringency.

2.3 Verification of interactions between β-lactoglobulin and mucin using SDS polyacrylamide gel electrophoresis (SDS-PAGE)

The interactions between β-LG and mucin were verified by SDS-PAGE. β-LG solutions (1% w/w, pH 2.6) were mixed with mucin (0.05% w/w in saliva buffer) and the mixture was centrifuged. A NuPAGE Bis-Tris Electrophoresis System (Invitrogen Corp., Carlsbad, CA) was used to identify the protein composition. Samples, consisting of β-LG (pH 2.6), mucin, mixed β-LG and mucin, supernatant, and pellets, were dissolved in NuPAGE LDS Sample Buffer with NuPAGE Sample Reducing Agent (containing dithiothreitol). Proteins were separated on precast NuPAGE 12% Bis-Tris gels using NuPAGE MOPS SDS Running Buffer at a constant 200 V. Gels were stained with periodic acid-Schiff stain (PAS) using the GelCode glycoprotein staining kit (Pierce, Rockford, IL) and dried using Gel-Dry Drying Solution (Invitrogen Corp., Carlsbad, CA).

The pH of mixed β-LG and mucin was 3.6 where there should be no self-precipitation of β-LG or mucin; therefore, the turbidity arisen after mixing and the pellet formed after centrifugation was the first indication of the interactions between β-LG and mucin. SDS-PAGE patterns after PAS staining is shown in Figure 2. PAS-positive polydispersed band of high molecular weight mucin (lane 2) appeared as pink smeared band just entering the running gels. Similar results have been reported.[40-42] As expected, a pink band appeared around 18k in β-LG sample (lane 3) and no mucin band was observed. Both mucin and β-

LG bands were seen in β-LG/mucin mixtures (before centrifugation, lane 4) and the supernatant (lane 5). The presence of both β-LG and mucin bands in the pellet (lane 6) indicates that there are interactions between β-LG and mucin, leading to the complex formation and precipitation of both.

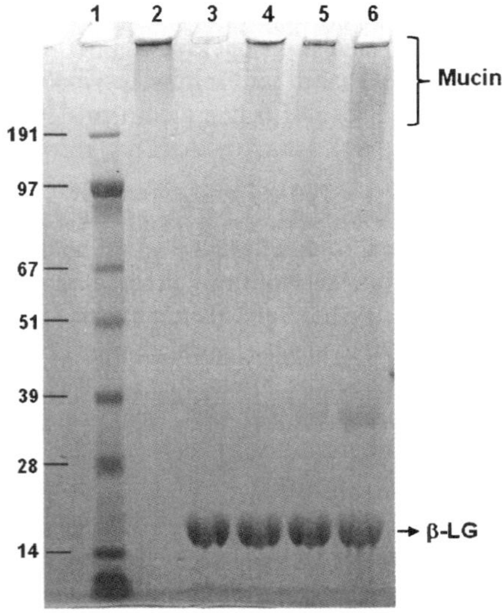

Figure 2 *SDS-PAGE of mucin (lane 2), β-LG (lane 3), mixed β-LG and mucin (lane 4), supernatant (lane 5) and pellet (lane 6) after centrifugation of mixed β-LG and mucin.*

2.4 Verification of interactions between β-lactoglobulin and mucin using Size exclusion chromatography coupled with multi angle laser light scattering (SEC-MALLS)

Interactions between β-LG and mucin were further verified by SEC-MALLS. β-LG (1% w/w, pH 2.6) was mixed with 1% w/w mucin in saliva buffer (8.1 mM KH_2PO_4, 1.5 mM Na_2HPO_4, 10 mM NaCl, 1.0 mM $MgCl_2$, and 2.7 mM KCl at pH 6.7) at 1.5 ml to 1.5 ml ratio, resulting in the mixture having the pH of 3.6. It should be noted that mucin concentration was higher in this experiment in order to allow its detection from the SEC system. The mixture was centrifuged and supernatant was diluted in saliva buffer at pH 3.6 in order to keep the pH of the supernatant constant during dilution. Pellets were resolubilized with 3 ml saliva buffer (pH 6.7 plus 200 mM NaCl). As a control, β-LG (1% w/w, pH 2.6) was also mixed with saliva buffer (pH 6.7). The β-LG/saliva buffer mixture remained clear without any precipitation and the pH of the mixture was 3.6. All samples were filtered through a 0.45-μm filter (PVDF, Millipore Corporation, Bedford, MA) before injection to the HPLC system. Sample injection volume was 75 μl, flow rate was 0.6 ml/min and separation was performed at 30 °C column temperature. Separation was carried out by size exclusion chromatography using a Shodex Protein KW-804 column (Showa Denko K.K., Tokyo, Japan) attached to a Waters HPLC system (Waters Corp., Milford, MA). Saliva buffer at pH 6.7 was used as eluent buffer for all samples except for the supernatant which used saliva buffer at pH 3.6. Peak analysis was conducted by a photo diode array (PDA) detector (Waters Corp., Milford, MA, model 2996) operating at 280 nm. A DAWN-EOS MALLS photometer (Wyatt Technology, Santa Barbara, CA) fitted with a helium–neon laser ($\lambda = 690$ nm) and a K5-flow cell, combined with a differential refractive index (RI) detector (Waters Corp; model 2414) was used for calculation of molecular mass and size of proteins. The RI detector was used as the concentration detector. The weight average molecular mass (M_w) and the mean square radius of materials were calculated with a first order Debye fit, using the refractive index increment (dn/dc) of 0.159[43] for mucin and 0.185 for β-LG.[44] Light scattering data from MALLS were processed with ASTRA software (version 5.1.5.3) (Wyatt Technology, Santa Barbara, CA).

Typical SEC chromatograms of β-LG and mucin are shown in Figure 3. High molecular weight mucin eluted in the void volume of the RI detector. The weight-average molecular mass (M_w) of mucin was 2.3 million Da and radius of gyration (R_g) was 47.6 nm, which were comparable to previously reported values.[30, 45] β-LG eluted at about 19 min and had the M_w of 29 kDa, indicating that the native proteins existed as a mixture of monomer and dimer. Mixing β-LG and mucin produced a turbid solution, while mixed β-LG and saliva buffer remained clear. After centrifugation, the supernatant chromatogram showed a decrease in β-LG (Figure 4) compared to control β-LG. The presence of both β-LG and mucin in the pellet (Figure 5) confirms the previous results that there are interactions between β-LG and mucin. It is also observed in Figure 4 that a large fraction of mucin remains soluble.

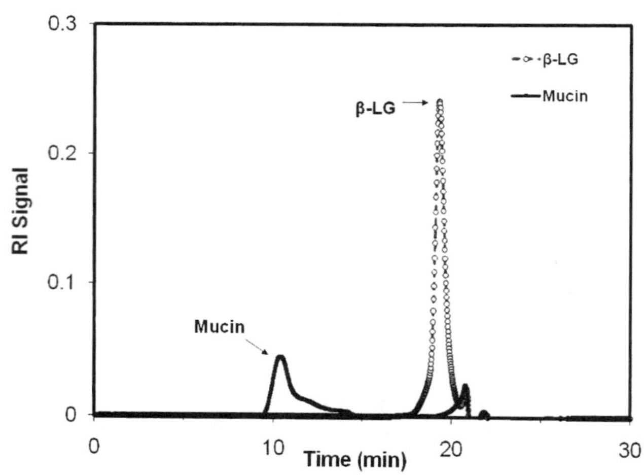

Figure 3 *Chromatograms of separate runs for β-LG and mucin.*

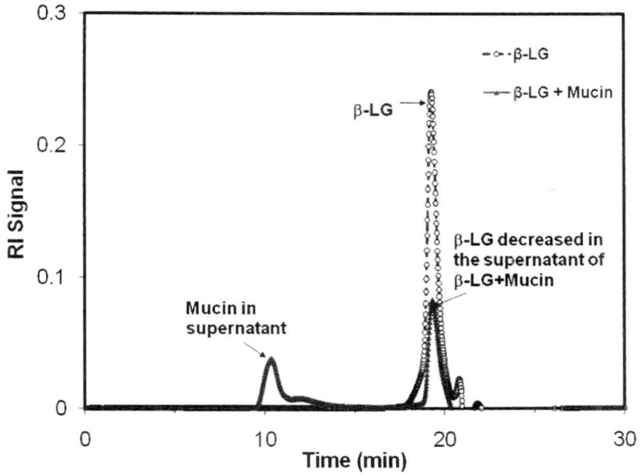

Figure 4 *Chromatogram of supernatant after centrifugation of β-LG or a mixture of β-LG and mucin. Initial concentration of β-LG was 1% (w/w).*

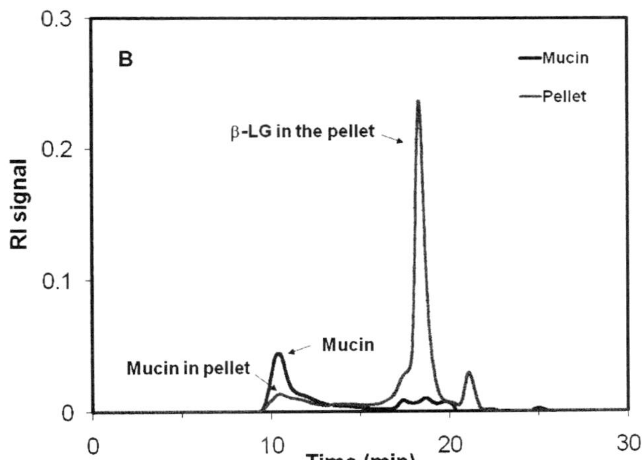

Figure 5 *Chromatograms showing composition of the pellet after centrifugation of a mixture of β-LG and mucin, and the composition of soluble mucin (no pellet formed after centrifugation. Initial concentrations of β-LG and mucin were 1% (w/w).*

Figure 6 *SDS-PAGE of saliva (lane 1), β-LG and saliva mixture (lane 2), supernatant of β-LG and saliva mixture (lane 3), and pellet of β-LG and saliva mixture (lane 4).*

2.5 Additional evidences of interactions between β-LG and salivary mucins

In addition to pig gastric mucin, the interactions between β-LG and human salivary proteins were further verified by SDS-PAGE and mass spectroscopy. Stimulated saliva was collected by having 2 volunteers chew on a piece of parafilm and expectorate for 15 min. Volunteers were instructed not to consume any food or drink (other than water) for at least 2 hr before saliva collection. Collected saliva was immediately centrifuged at 4°C. β-LG (2%, pH 3.5) was mixed with saliva at a 2:1 ratio. The mixture was centrifuged and the supernatant and the pellet were analyzed. SDS-PAGE and glycoprotein staining were conducted as previously described. Saliva (lane 1) showed the presence of glycoproteins, most likely salivary MUC 1 and MUC 5,[46] as pink smeared bands. Similar to the results using pig gastric mucin, salivary proteins and β-LG interactions were evident from the presence of salivary mucins in the pellet of β-LG and saliva mixture. In agreement with these results, LC-MS/MS analysis of also showed the presence of salivary mucins along with other salivary proteins in the pellet of β-LG and saliva mixture (results not shown).

3 CONCLUSION

An increase in turbidity caused by β-LG and mucin interactions coincided with astringency determined by sensory evaluation. Interactions between β-LG and mucin were supported by electrophoresis and chromatographic measurements. This study substantiates the possibility that the astringency of whey proteins at low pH arises from 1) self-aggregation and/or precipitation of whey proteins due to the change of pH to pI when mixed with saliva, 2) aggregation and/or precipitation of whey proteins due to the interactions between positively charged whey proteins and negatively charged salivary proteins such as mucin, and 3) loss of lubrication due to the aggregation and precipitation of salivary proteins such as mucin.

Acknowledgements

This project was supported by the National Research Initiative program of the USDA Cooperative State Research, Education, and Extension Service (CSREES), Grant number 2006-35503-17066. Support from the North Carolina Agricultural Research Service and Southeast Dairy Food Research Center are gratefully acknowledged. The use of trade names in this publication neither imply endorsement by the North Carolina Agricultural Research Service of the products named nor criticism of similar ones not mentioned. The authors are very grateful for the β-lactoglobulin donated by Davisco Foods International.

References

1 GBA, Sports and Energy Drinks: A Global Strategic Business Report, *Global Industry Analysts,* 2007.
2 J. W. Beecher, M. A. Drake, P. J. Luck and E. A. Foegeding, *Journal of Dairy Science,* 2008, **91,** 2553.
3 H. Sano, T. Egashira, Y. Kinekawa and N. Kitabatake, *Journal of Dairy Science,* 2005, **88,** 2312.
4 ASTM, 'Standard Definitions of Terms Relating to Sensory Evaluation of Materials and Products' in *Standard Definitions of Terms Relating to Sensory Evaluation of Materials and Products,* American Society for Testing and Materials, 2004 .
5 Y. Kawamura, M. Funakoshi, Y. Kasahara and T. Yamamoto, *The Japan Journal of Physiology,* 1969, **19,** 851.
6 S. S. Schiffman, M. S. Suggs, L. Sostman and S. A. Simon, *Physiology and Behavior,* 1992, **51,** 51.
7 P. A. S. Breslin, M. M. Gilmore, G. K. Beauchamp and B. G. Green, *Chemical Senses,* 1993, **18,** 405.
8 A. K. Smith, H. June and A. C. Noble, *Food Quality and Preference,* 1996, **7,** 161.
9 B. J. Lyman and B. G. Green, *Chemical Senses,* 1990, **15,** 151.
10 I. Lesschaeve and A. C. Noble, *American Journal of Clinical Nutrition,* 2005, **81,** 330S.
11 E. C. Bate-Smith, *Phytochemistry,* 1973, **12,** 907.
12 M. A. Joslyn and J. L. Goldstein, *Advances in Food Research,* 1964, **13,** 179.
13 J. P. McManus, K. G. Davis, T. H. Lilley and E. Haslam, *Journal of the Chemical Society, Chemical Communications,* 1981, 309b.
14 J. Guinard, R. M. Pangborn and M. J. Lewis, *Journal of the Science of Food and Agricultur.,* 1986, **37,** 811.
15 A. Gambuti, A. Rinaldi, R. Pessina and L. Moio, *Food Chemistry,* 2006, **97,** 614.
16 A. J. Charlton, N. J. Baxter, M. L. Khan, A. J. G. Moir, E. Haslam, A. P. Davies and M. P. Williamson, *Journal of Agricultural and Food Chemistry,* 2002, **50,** 1593.
17 C. Simon, K. Barathieu, M. Laguerre, J. Schmitter, E. Fouquet, I. Pianet and E. J. Dufourc, *Biochemistry,* 2003, **42,** 10385.
18 N. J. Baxter, T. H. Lilley, E. Haslam and M. P. Williamson, *Biochemistry,* 1997, **36,** 5566.
19 S. Kallithraka, J. Bakker and M. N. Clifford, *Journal of Sensory Studies,* 1998, **13,** 29.
20 S. Kallithraka, J. Bakker and M. N. Clifford, *Journal of the Science of Food and Agriculture,* 2001, **81,** 261.

21 J. R. Stokes, G. A. Davies, L. Macakova, G. Yakubov, J. Bongaerts and D. Rossetti, *AIP Conference Proceedings,* 2008, **1027**, 1171.
22 C. A. Lee and Z. M. Vickers, *Int. Dairy J.,* 2008, **18,** 1153-1156.
23 H. T. Lawless, J. Horne and P. Giasi, *Chemical Senses,* 1996, **21,** 397.
24 P. Roussel and P. Delmotte, *Current Organic Chemistry,* 2004, **8,** 413.
25 H. Nordman, J. R. Davies, G. Lindell, C. de Bolos, F. Real and I. Carlsted, *Biochemical Journal,* 2002, **364,** 191.
26 B. Jan-Willem Van Klinken, A. W. C. Einerhand, H. A. Büller and J. Dekker, *Analytical Biochemistry,* 1998, **265,** 103.
27 L. Wong and C. H. Sissions, *Archives of Oral Biology,* 2001, **46,** 477.
28 M. E. Malone, I. A. M. Appelqvist and I. T. Norton, *Food Hydrocolloids,* 2003, **17,** 763.
29 B. Schwarz and T. Hofmann, *European Food Research and Technology*, 2008, **227,** 1693.
30 M. H. Vingerhoeds, T. B. J. Blijdenstein, F. D. Zoet and G. A. van Aken, *Food Hydrocolloids.,* 2005, **19,** 915.
31 E. Monteleone, N. Condelli, C. Dinnella and M. Bertuccioli, *Food Quality and Preference,* 2004, **15,** 761.
32 D. Rossetti, J. H. H. Bongaerts, E. Wantling, J. R. Stokes and A.-M. Williamson, *Food Hydrocolloids,* 2009, doi:10.1016/j.foodhyd.2009.03.001.
33 G. Fia, C. Dinnella, M. Bertuccioli and E. Monteleone, *Food Chemistry,* 2009, **113,** 325.
34 N. Condelli, C. Dinnella, A. Cerone, E. Monteleone and M. Bertuccioli, *Food Quality and Preference,* 2006, **17,** 96.
35 J. Horne, J. Hayes and H. T. Lawless, *Chemical Senses,* 2002, **27,** 653.
36 M. S. Rodríguez, L. A. Albertengo, I. Vitale and E. Agulló, *Journal of Food Science,* 2003, **68,** 665.
37 K. J. Siebert and C. Euzen, *Journal of Sensory Studies,* 2008, **23,** 222.
38 C. Poncet-Legrand, A. Edelmann, J. -. Putaux, D. Cartalade, P. Sarni-Manchado and A. Vernhet, *Food Hydrocolloids,* 2006, **20,** 687.
39 C. Tanford, *Physical Chemistry of Macromolecules,* John Wiley & Sons Inc., New York, 1961.
40 J. E. Jumblatt and M. M. Jumblatt, 'Detection and Quantification of Conjunctival Mucins' in *Proceedings of the Second International Conference on the Lacrimal Gland, Tear Film, and Dry Eye Syndromes,* eds. D. D. Sullivan, D. A. Sullivan, D. A. Dartt and M. A. Meneray, Springer, New York, 1998, p. 239.
41 H. Habte, A. Mall, C. de Beer, Z. Lotz and D. Kahn, *Virology Journal,* 2006, **3,** 99.
42 M. Faure, D. Moënnoz, F. Montigon, L. B. Fay, D. Breuillé, P. A. Finot, O. Ballèvre and J. Boza, *Analytical Biochemistry,* 2002, **307,** 244.
43 A. Dėdinaitė and L. Bastardo, *Langmuir,* 2002, **18,** 9383.
44 M. B. Huglin, 'Specific Refractive Index Increments' in *Light scattering from polymer solutions,* ed., M. B. Huglin, Academic Press, New York, 1972, p. 165.
45 K. Jumel, I. Fiebrig and S. E. Harding, *International Journal of Biological Macromolecules,* 1996, **18,** 133.
46 A. Nayak and G. H. Carpenter, *Physiology & Behavior,* 2008, **95,** 290.

RHEOLOGICAL PROPERTIES OF OZONATED PECTIN DISPERSIONS

BK Tiwari[1], CP O'Donnell[1], K Muthukumarappan[2] and PJ Cullen[3]

[1]UCD School of Agriculture, Food Science and Veterinary Medicine, University College Dublin, Belfield, Dublin 4, Ireland.
[2]Agricultural and Biosystems Engineering Department, South Dakota State University, Brookings, USA.
[3]School of Food Science and Environmental Health, Dublin Institute of Technology, Dublin 1, Ireland.

1 INTRODUCTION

Hydrocolloids such as pectin are water-soluble, high molecular weight polysaccharides that are widely employed in the food industry to optimise the rheological and textural characteristics of food systems. Pectin is typically used in food formulations for optimising viscosity, creating gel-structures and enhancing structural stability[1,2,3]. Pectin is widely used in the jam industry as a gelling agent[4]. The rheological properties of polysaccharides are influenced by a number of factors such as concentration, thermal or mechanical treatments[5] and non-thermal food processing techniques such as irradiation[6], and high intensity ultrasound[7].

Ozone has been employed in water treatment applications for decades. Ozone was declared as Generally Recognized as Safe (GRAS) for use in food processing by the US Food and Drug Administration (FDA) in 1997[8]. Subsequently, ozone gained approval as a direct food additive for the treatment, storage and processing of foods both in the gaseous and aqueous phases[9]. Use of ozone has been reported for processing of various fruit juices[10,11] and washing of fruits and vegetables with ozonated water[12]. Ozone has a wide antimicrobial spectrum which combined with a high oxidation potential makes it an attractive processing option for the food industry. Relatively small quantities of ozone and short contact times are sufficient for the desired anti-microbial effect and it rapidly decomposes into oxygen leaving no toxic residues. The mechanism of ozone self-decomposition has been well documented. Degradation of water-soluble polymers such as polyethylene glycol, poly (vinyl alcohol), poly (vinyl pyrrolidone), polyacrylamide, and sodium polyacrylate have been studied by various researchers[13,14]. Previous studies have shown that ozone degrades macromolecules[15] and degrades pigments[16] due to its high oxidation potential. The objective of this study was to investigate the effect of ozone treatment on the rheological properties of pectin dispersions.

2. MATERIALS AND METHODS

2.1. Preparation of samples

Pectin (70-75% esterification) was supplied by Sigma-Aldrich Ireland Ltd, Dublin, Ireland. Pectin dispersions (2% w/w) were prepared in 500 mL distilled water and were stirred using a high shear mixer (L4RT Laboratory Mixer, Silverson Machines Ltd, Chesham Bucks, England) for 30 min at 2,500 rpm and at room temperature of 20 °C.

2.2. Ozone treatment

Experiments were carried out in a 250 mL beaker with a glass diffuser (Figure 1). Ozone was generated using an ozone generator (Model OL80, Ozoneservices, Vancouver, Canada) and recorded using an ozone gas analyzer (Model OLA-DLS, Ozoneservices, Canada). Ozone concentrations of 2.4, 4.0, 5.6 & 7.8 % (w/w) at a constant gas flow rate of 0.06 $Lmin^{-1}$ were studied. Samples were treated for 0 (control), 2, 4, 6, 8 and 10 min processing times.

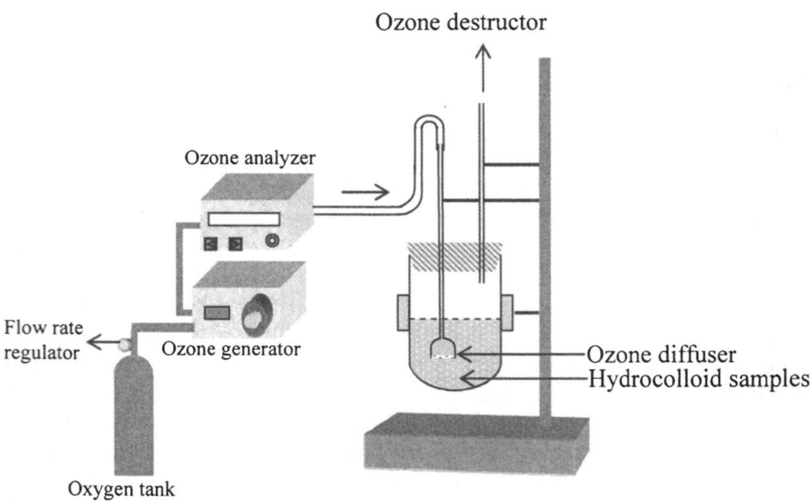

Figure 1. Experimental set up for ozone treatment of hydrocolloid samples.

2.3 Rheological analysis

Rheological analysis was carried out using a control stress rheometer (Viscoanalyzer, ATS Rheosystems, Rheologica Instruments Inc., NJ) with bob and cup geometry of 30 mm diameter with a gap of 1.25 mm. For flow curves, samples were presheared at 11 s^{-1} for 10 s followed by a relaxation time of 10 s. Shear rate was increased from 0.1 to 50 s^{-1} in 15 linear increments over a period of 30 s at 25 °C [17]. Shear stress – shear rate data was fitted to a power law model (Equation 1).

$$\eta_a = k\dot{\gamma}^{(n-1)} \qquad (1)$$

where η_a is apparent viscosity (Pa.s), k ($Pa.s^n$) is consistency index, $\dot{\gamma}$ is the shear rate (s^{-1}) and n (dimensionless) is the flow behaviour index.

For dynamic rheology, an applied stress of 0.1 Pa was found to be within the linear visco-elastic region for pectin dispersions. The dynamic rheological properties of pectin were investigated over a frequency range of 0.01 to 10 Hz in logarithmic increments of 10 per decade at 25 °C. Samples were presheared at 11 s^{-1} for 20 s. The dynamic rheological properties were characterised in terms of the elastic (G′) and viscous (*G″*) moduli, which were modelled using the power law Equation (2 & 3). Control and ozonated samples were subsequently analysed after a 24 h storage period to investigate structural recovery.

$$G' = a'\omega^{b'} \quad (2)$$

$$G'' = a''\omega^{b''} \quad (3)$$

Where a', a'' (Pa.s) and b', b'' (dimensionless) are dynamic rheological power law constants and ω (Hz) is frequency. All measurements were preformed in triplicate.

2.4 Statistical analysis

Analysis of Variance (ANOVA) and separation of means were conducted by the general linear models procedure (Proc GLM). The data were then analyzed to determine the main and interaction effects, and comparisons among treatments were analyzed using Fishers least significant differences (LSD) using SAS V.9.1 (SAS Institute, Cary, NC). Treatment means were considered significantly different at $p<0.05$. All trials were conducted in triplicate.

3 RESULTS AND DISCUSSION

3.1 Flow Curves

Flow curves for control and ozonated (2.4%w/w) pectin dispersions are shown in Figure 2. Pectin dispersion exhibited shear thinning behaviour. Apparent viscosity for all dispersions was found to decrease as a function of ozone concentration and processing time. Significant differences in consistency (k) and flow behaviour (n) indices between control and ozonated samples were observed (Figure 3). The effects of ozone concentration and processing time on the flow behaviour index (n) and consistency index (k) values of 2% pectin dispersions were significantly influenced by the processing parameters employed. A reduction of 34.4% in the pectin consistency index (k) value was observed at the lowest ozone concentration (2.4 %w/w) relative to the control. A trend towards increased Newtonian flow behaviour was observed at higher ozone concentrations and processing times. The observed reduction in viscosity due to ozonation may be caused by the depolymerisation of macromolecules. A similar decrease in viscosity of chitosan solution with ozone treatment was reported by No *et al*[18]. Seo *et al*[19] reported a 63% reduction in viscosity with ozone treatment (0.5 ppm) for 10 min compared to untreated chitosan dispersions. In this study, a reduction of 31.7% in apparent viscosity compared to control was observed for pectin dispersions at the highest ozone concentration (7.8%w/w) and longest processing time (10 min). It is evident that reduction in viscosity depends on the type of hydrocolloid and structural arrangement within the polymer.

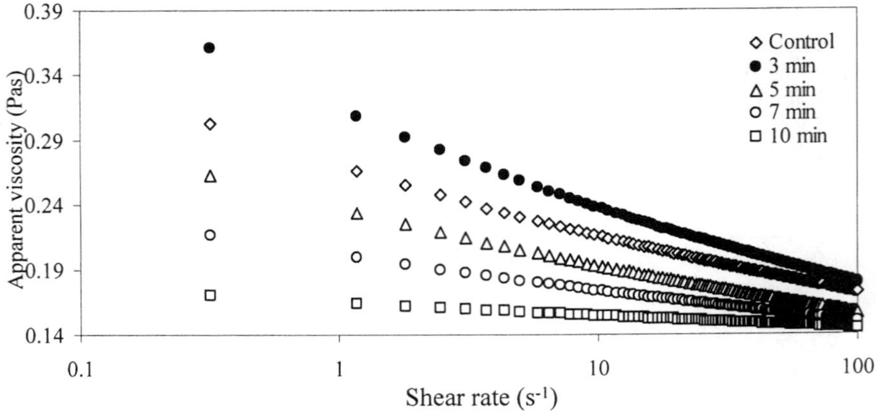

Figure 2 *Effect of ozone concentration (2.4 %w/w) on apparent viscosity of pectin dispersion with shear rate (s^{-1}) at varying treatment time*

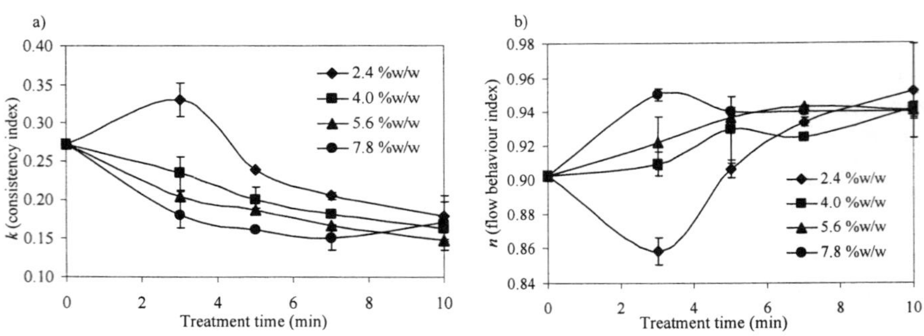

Figure 3 *Effect of ozone concentration on a) consistency index (k), b) flow behaviour index (n) of pectin dispersion at varying treatment time*

Depolymerization of a macromolecule is attributed to the oxidative destruction of β-D-glucoside bonds between units by the electrophilic attack on the C-H bond by ozone. Formation of free radicals due to ozone could also be attributed to degradation of the macromolecules. Molecular ozone reactions are selective and limited to unsaturated aromatic and aliphatic compounds. Ozone oxidizes organic matter via two pathways, by direct oxidation with ozone molecules ($O_3 + C \rightarrow C_{ox}$) and by the generation of free-radical intermediates (Figure 4), such as $OH^{\cdot}$ ($O_3 + OH^{-} \rightarrow OH + C \rightarrow C_{ox}$) a powerful, effective, non-selective oxidizing agent[20]. Direct reaction is described by the Criegee mechanism where ozone molecules undergo 1-3 dipolar cyclo addition with double bonds present, leading to the formation of ozonides (1,2,4-trioxolanes) from alkenes and ozone with aldehyde or ketone oxides as decisive intermediates, all of which have finite lifetimes. This leads to the oxidative disintegration of ozonide and formation of carbonyl compounds, while oxidative work-up leads to carboxylic acids or ketones. Ozone attacks OH radicals, preferentially to the double bonds in organic compounds (Figure 4a). Degradation of organic compounds are also reported to be due to various intermediate radical formations

(Figure 4b) leading to electrophilic and nucleophilic reactions occurring with aromatic compounds that are substituted with an electron donor (e.g. OH^-) having high electron density on the carbon compounds in ortho and para positions. Chemical reactions involve breakage of old bonds and formation of new bonds.

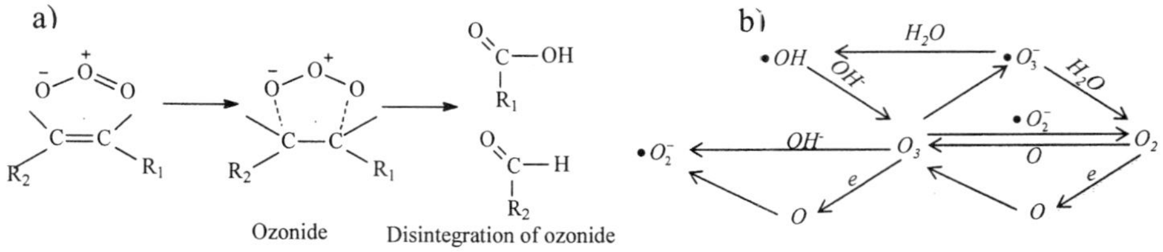

Figure 4 *a) Direct reaction of ozone with the formation of ozonide and b) formation and transformations of intermediate radicals*

3.2 Dynamic rheology

Dynamic rheological parameters a' & b' for elastic (G') and a'' & b'' for viscous (G'') moduli at 25 °C as a function of ozone concentration and processing time of 10 min are shown in Table 1 for 2% pectin dispersions. The dynamic rheological parameters for elastic moduli (a' and b') of the control pectin dispersion were lower compared to their viscous moduli parameters (a'' and b''). G' values for pectin were greatly reduced during ozone processing, indicating structural breakdown, which may be due to the oxidative cleavage of macromolecules. As for the reductions in apparent viscosity, the viscoelastic moduli for pectin were found to be highly sensitive to ozonation. This can be explained from the fact that ozone selectively depolymerises macromolecules by oxidative destruction of *β*-D-glucoside bonds between cellulose units by the electrophilic attack on the C(1)-H bond by ozone molecules[15]. Breakdown of β-1, 4 D-glucosidic linkages is also reported by Chang *et al*[21] due to formation of free radicals during ozonation. After a recovery period of 24 h no significant recovery of viscoelastic moduli was observed, indicating time independent rheological behaviour.

Table 1

Dynamic rheological parameters for control and ozonated pectin dispersion processed for 10 min at constant flow rate of 0.06 $L.min^{-1}$

Add ozone conc to table

Parameter	Control	2.4%w/w	4.0%w/w	5.6%w/w	7.8%w/w
a', Pa.s	0.04^{a}	0.019^{b}	0.020^{b}	0.023^{b}	0.021^{b}
b'	0.55^{a}	0.367^{b}	0.399^{b}	0.342^{bc}	0.428^{b}
a'', Pa.s	1.33^{a}	0.980^{b}	0.849^{c}	0.767^{d}	0.832^{c}
b''	1.00^{a}	0.986^{ab}	0.988^{ab}	0.994^{ab}	0.977^{b}

^{abc}values followed by the same letter within the same row are not significantly different ($p<0.05$).

4 CONCLUSIONS

Significant reductions in apparent viscosity and a trend towards Newtonian behaviour were observed for ozonated pectin dispersions. Dynamic rheological studies indicated that ozonation of the selected hydrocolloid dispersions reduced the gel strength as evident by changes in the viscoelastic moduli, showing no recovery over a 24 h storage period, which may be due to permanent pyrolysis of pectin dispersions. These results show the effects of ozone on rheology of pectin dispersions, which should be considered prior to incorporating pectin dispersions in food products which are subsequently ozonated.

Acknowledgements

This work was supported by a Food Institutional Research Measure (FIRM) and the Irish Department of Fisheries, Food and Agriculture.

References

1 Yaseen EI, Herald TJ, Aramouni, FM, Alavi S (2005). Food Res Intl 38:111-119.
2 Albert S, Mittal GS (2002). Food Res Intl 35:445–458.
3 Fagan CC, O'Donnell CP, Cullen PJ, Brennan, CS (2006). J Food Eng 77;261-268.
4 Quemener B, Marot C, Mouillet L, Da Riz V, Diris J (2000). Food Hyd 14:19 – 28.
5 Marcotte M, Taherian AR, Ramaswamy, HS (2001). Food Res Intl 34:695-703.
6 Kayacier A, Dogan, M (2006). J Food Eng 72:261–265.
7 Kardos N, Luche JL (2001). Carbo Res 332:115–131.
8 Graham DM (1997). Food Tech 51:121–137.
9 Khadre MA, Yousef AE, Kim J (2001). J Food Sci 6:1242–1252.
10 Tiwari BK, Muthukumarappan K, O' Donnell CP, Cullen PJ (2008). J Food Eng, 88:553-560.
11 Steenstrup, L. D, Floros, J. D. (2004). J Food Proc Preserv 28:103-116.
12 Zhang, L, Lu, Z, Yu, Z, Gao, X. (2005). Food Cont 16:279–283
13 Razumovskii SD, Zaikov GE (1984). *Studies in* Organic *Chemistry*, Elsevier, Amsterdam.
14 Cataldo F (1998). Polym Degrad Stab 35:281–291.
15 Kabalnova NN, Murinov KY, Mullagaliev R, Krasnogorskaya NN, Shereshovets VV, Monakov YB, Zaikov GE (2001). J App Polymer Sci 81:875–81.
16 Selcuk H. (2005). Dyes Pigments 64:217–22.
17 Singh G, Muthukumarappan K (2008). LWT Food Sci Tech 41:1145-1152.
18 No H$K, Kim SD, Kim DS, Kim SK, Meyers SP (1999). J Chitin Chitosan 4:177–83.
19 Seo S, King JM, Prinyawiwatkul W (2007). J Food Sci 72:522-526
20 Staehelin J, Hoigne J (1982). Environ Sci Tech 16:676.
21 Chang KLB, Tai MC, Cheng FH (2001). J Agric Food Chem 49:4845–51.

PRODUCTION AND RHEOLOGICAL PROPERTIES OF LOW-ACYL GELLAN GUM FLUID GELS INDUCED BY SODIUM AND CALCIUM IONS

J. Muñoz, M.C. García, L.M. Pérez, M.C. Alfaro

Departamento de Ingeniería Química. Facultad de Química. Universidad de Sevilla c/P. García González, 1, E41012, Sevilla, Spain

1 INTRODUCTION

Most of available literature on gellan gum reports methods of fluid gel preparation either at quiescent cooling conditions or under controlled shear in rheometers (*in situ* preparations)[1-3]. Conversely, little information is available on gellan fluid gel preparation conducted with homogenisers[4]. The latter approach is quite interesting to fill the gap between fundamental research and industrial applications of gellan fluid gels. The overall objective of this research was to explore the chances to obtain tailor-made gellan fluid gels with tuned rheological properties, concerning their applications as dispersion stabilisers. Different mechanical protocols were used to disturb the quiescent gel setting as temperature decreased once hydration and addition of gel-promoting ions were completed. A further interesting point considered was the batch reproducibility and the effect of the nature and concentration of gel-promoting ions. Gellan fluid gels can be considered as suspensions of microgel particles, whose average size is controlled by the shear stresses applied during gelation[2]. Therefore, serrated or star parallel plate geometries were used to prevent wall-depletion phenomena when conducting rheological measurements out of the linear viscoelastic range. A wide range of rheological techniques were used to gain a deeper insight into the shear flow of gellan fluid gels, involving small amplitude oscillatory shear (SAOS), creep within and out of the linear viscoelastic range (LVR), multi-step flow curves, start-up at the inception of shear and flow interrupted experiments as well as nonlinear creep/recovery/nonlinear creep series to check the structure recovery after shear.

2. EXPERIMENTAL

Kelcogel ® F ™ gellan gum kindly provided by CP-Kelco (San Diego, USA) was used as received. 99.5% purity NaCl and pro analysi $CaCl_2$ (richness > 98%) were purchased from Panreac (Barcelone, Spain) and Merck (Darmstadt, Germany), respectively. Standard commercial salt was purchased in a local shop. Fluid gels were always prepared with deionised water.

Fluid gels were prepared using a MR-D1 Ikavisc homogeniser (Ika, Staufen, Germany) with torque measuring facilities. Impellers of two different geometries were selected but keeping a similar impeller diameter / tank diameter ratio (~ 0.85). Alternatively, a T-

50/G45F Ultraturrax (Ika, Staufen, Germany) was used in some preparation procedures (ratio of diameters, 0.75). Fluid gels were produced after the hydration step of gellan (peak temperature: 80ºC) and addition of gel-promoting ions (salt, NaCl and $CaCl_2$ were used) by disturbing the gel formation. This was achieved by mechanical stirring at either 700 rpm or 1000 rpm during the whole cooling step to 20ºC. Evaporative losses were corrected for by adding appropriate amounts of deionised water. A further procedure was based on disrupting the incipient gel setting by mechanical stirring (700 rpm, for 1 minute) at 40ºC. This temperature was estimated from a temperature ramp conducted under small amplitude oscillatory shear (SAOS).
The rheological characterisation was carried out at 20ºC with several CS-rheometers: RS-100 and MARS (Thermo-Fischer Scientific/Haake; Karlsruhe, Germany) and AR2000 (TA Instruments, Crawley, England). A controlled-strain ARES LS1 rheometer (TA Instruments, Crawley, England) was also used. As a rule, serrated surface or star plate & plate geometries were used, although the double cone geometry was also used to run oscillatory tests for low- modulus systems within the linear viscoelastic range (LVR).

3. RESULTS AND DISCUSSION

3.1. Laboratory scale production of gellan fluid gels

Figure 1 shows the frequency dependence of the complex viscosity, $|\eta^*|$, derived from small amplitude oscillatory shear (SAOS) tests, for gellan fluid gels prepared with different mechanical energy-inputs. Figure 1A exhibits the results obtained with a Coulex or Saw-tooth impeller attached to a MR-D1 Ikavisc homogeniser during the hydration step, followed by the cooling step under mechanical energy. A comparison of results achieved when either the Coulex impeller or an Ultraturrax T-50/G45F homogeniser were used in the cooling step is provided. In addition, Figure 1A illustrates that the highest $|\eta^*|$ values were reached when gel setting was allowed to occur in the quiescent state and that they were not significantly different to those achieved when the Ultraturrax was only used in the last minute of the cooling step. On the contrary, the lowest $|\eta^*|$ values were obtained when keeping the Coulex stirring for 40 minutes. Intermediate results were found when the Ultraturrax was only used in the last either 5, 10 or 15 minutes of the cooling process. A similar analysis may be made on Figure 1B, where results obtained with a pitched-blade turbine are shown not to be significantly different to those of the Coulex impeller. This seems to indicate that SAOS results of fluid gels, and therefore their microstructure, are not too sensitive to a change of impeller geometry provided a similar ratio of diameters (impeller diameter/tank diameter) is used. The results obtained demonstrate that only a vertical shift factor would be required to superimpose all the mechanical spectra. In other words, the values of the elastic component as well as the viscous one can be manipulated by means of the energy input applied during processing. Figure 2 exhibits results at higher gellan concentration and provides further evidence that fluid gels with tailor-made viscoelastic properties may be produced by manipulating the energy input during processing.

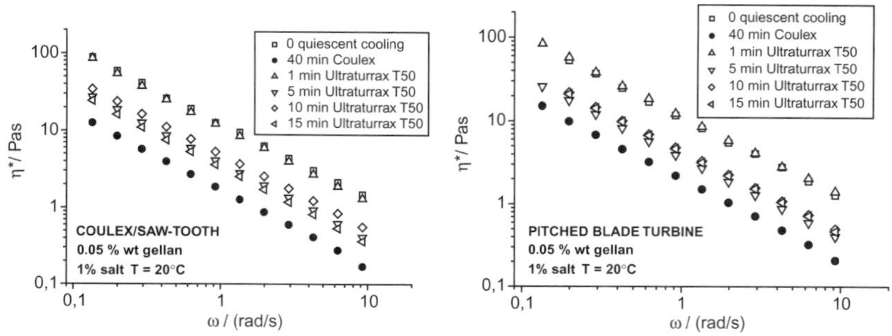

A B

Figure 1 *Influence of different mechanical* procedures on the *frequency dependence of* $|\eta^*|$ *at 20°C for 0.05 %wt gellan gum fluid gels.* ***1A****: Coulex impeller.* ***1B****: Pitched Blade Turbine. See detailed explanation in the text.*

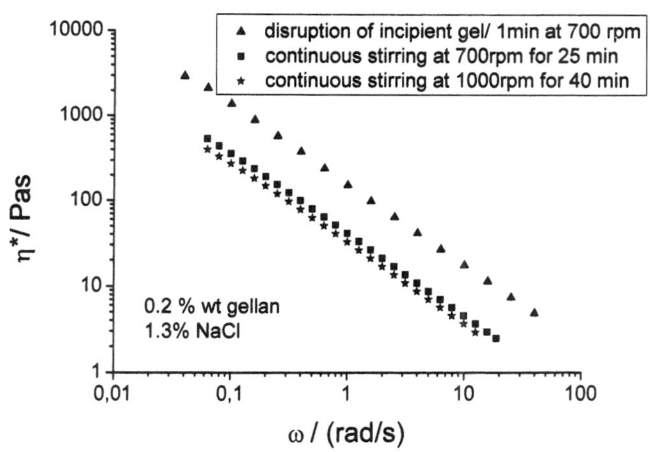

Figure 2 *Influence of the energy input level on the frequency dependence of* $|\eta^*|$ *at 20°C for 0.2 %wt gellan gum fluid gels.*

3.2 Influence of batch reproducibility on dynamic viscoelasticity

Preparation of reproducible fluid gels at laboratory scale is considered to be a rather challenging task[1]. Figures 3 and 4 show the frequency dependence of $|\eta^*|$ of samples containing the same concentration of gellan gum but different NaCl contents. The reproducibility of batches of fluid gels with lower NaCl concentration was rather poor and the wide error bands associated with standard deviation of the mean suggest that these samples were not homogeneous structured liquids (Figure 3). Low NaCl/gellan gum concentration ratios usually resulted in clear heterogeneous samples. On the contrary the fluid gel formulated with higher NaCl concentrations yielded reproducible batches with very low standard deviation bands, which demonstrated that it is vital to select a suitable content of ion-promoting gels in order to produce fluid gels and with reasonable reproducibility.

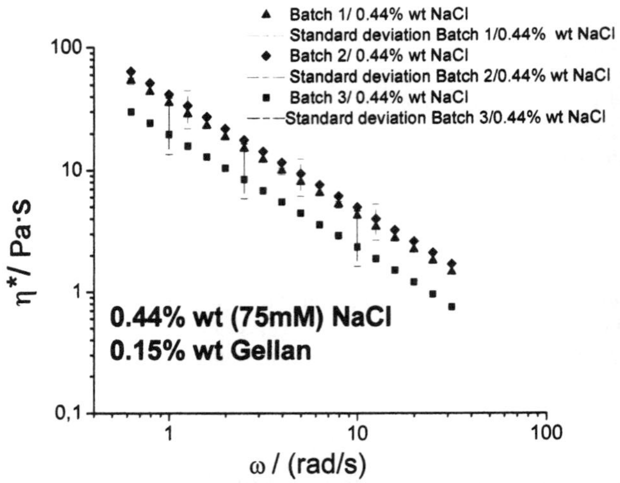

Figure 3 *Influence of batch reproducibility on the frequency dependence of* $|\eta^*|$ *at 20°C for 0.15 %wt gellan gum fluid gels containing 0.44% wt (75mM) NaCl.*

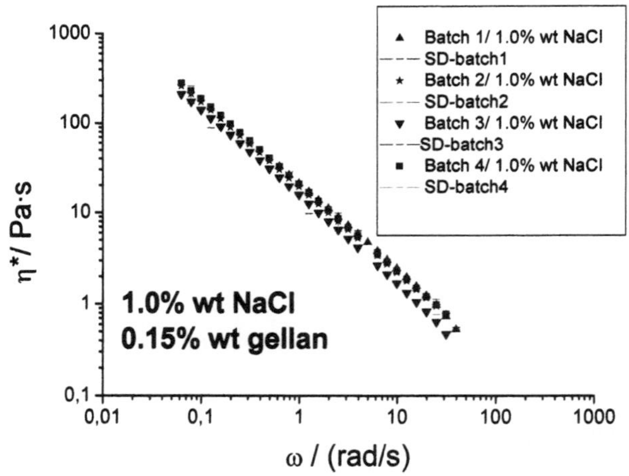

Figure 4 *Influence of batch reproducibility on the frequency dependence of* $|\eta^*|$ *at 20°C for 0.15 %wt gellan gum fluid gels containing 1.0% wt NaCl.*

3.3 Influence of $CaCl_2$ on dynamic viscoelasticity and steady shear flow curves. Comparison with NaCl-containing fluid gels.

The influence of $CaCl_2$ and NaCl concentrations was studied on fluid gels containing 0.2% wt gellan gum. Increasing $CaCl_2$ concentrations up to 0.175%wt resulted in negligible falls in G' and G'', which, in turn, did not influence the frequency dependence of the loss tangent. However, a further increase to 0.2%wt provoked a small decrease of G' but a marked rise in G'', which means that the fluid-like properties of this fluid gel were far more important, as indicated by the marked increase in the loss tangent values (data not shown). SAOS results were supported by steady shear flow curves carried out following a multi-step procedure (Figure 5). The flow properties exhibited are typical of all the fluid gels studied in this project, which can be classified as "very shear thinning materials". A dramatic fall in apparent viscosity of about 5 orders of magnitude was observed above a critical shear stress. This decreased from 0.1% to 0.175% wt $CaCl_2$ and then increased at 0.2% wt $CaCl_2$ as observed for G'' in SAOS tests. Above this sort of practical yield stress, fluid gels exhibited flow properties which may fit the Sisko equation and followed the same sequence as for the critical shear stress. Creep experiments demonstrated that a very long shearing time was required to reach real steady-state viscosity values below the critical shear stress. Steady shear flow results are supported by the slight decline in the apparent viscosity reported for gellan solutions prepared at quiescent conditions at the higher $CaCl_2$ concentrations[5] and by the similar effect found for the strength of strong gellan gum gels[6].

It is well known that the concentration range of $CaCl_2$ to form gellan fluid gels is roughly one order of magnitude lower than that of NaCl[1]. Figure 6 shows that similar complex viscosity values as well as a similar frequency dependence of $|\eta^*|$ are shown by 0.2%wt gellan gum at either 1.3%wt NaCl or 0.2%wt $CaCl_2$. However, despite G' values were rather similar, G'' and as a consequence the loss tangent of 0.2%wt $CaCl_2$ were clearly higher than for the 1.3 %wt NaCl fluid gel. The corresponding multi-step flow curves were consistent with the SAOS results, since the critical shear stresses were alike, while apparent viscosities were higher for the $CaCl_2$ gellan fluid gel. These results indicate that inorganic salt concentrations may be tuned to provide gellan fluid gels with similar rheology.

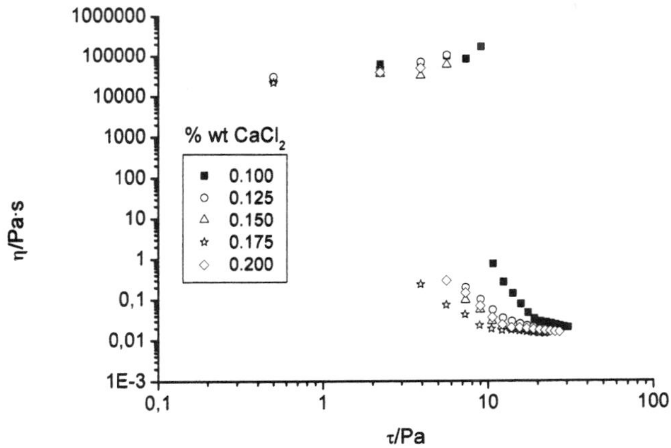

Figure 5 *Influence of $CaCl_2$ concentration on multistep flow curves at 20°C for 0.2%wt gellan gum fluid gels. Fluid gels were prepared under continuous stirring at 700rpm with a Coulex/saw tooth impeller for 40 minutes.*

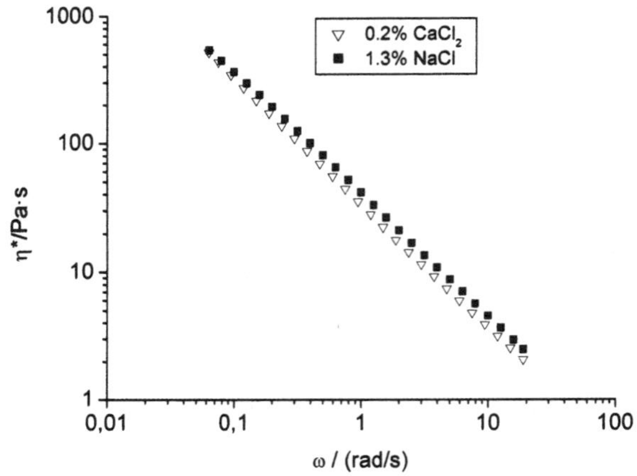

Figure 6 *Similar frequency dependence* of $|\eta^*|$ *at 20°C for 0.2%wt gellan fluid gels prepared with either 0.2%wt* $CaCl_2$ *or 1.3% NaCl. Gel preparation as in Figure 5 legend*

3.4 Time-dependent flow properties and structural recovery after cessation of shear

Most applications of gellan fluid gels rely on their role as stabilisers without requiring a high gum concentration. A good stabiliser must exhibit a high low-shear Newtonian viscosity, marked shear thinning properties and a fast viscosity (structural) recovery on cessation of shear. In addition, exhibiting viscoelastic properties is quite helpful.

Start-up at the inception of shear and flow interrupted experiments are the rheological tests of choice for a complete characterisation of structural breakdown and recovery since they preclude the influence of recent shear history. These experiments consist of several steps. First of all, the required equilibration time after the sample was loaded and reached the measuring position must be determined. Secondly a constant shear rate is instantaneously imposed for a fixed shearing time. Thirdly the stress relaxation after cessation of shear is monitored for a pre-fixed rest time and finally, the original shear rate is imposed again and the second shear stress evolution is recorded. Figure 7 provides some examples of results obtained for gellan fluid gels containing 1.3% wt NaCl. A fast and nearly instantaneous stress overshoot was achieved at a shear rate of $10s^{-1}$, supporting that nonlinear viscoelastic effects were not dominant as compared to a thixotropic behaviour. The relative fall in shear stress was more and more important as gellan gum concentration was increased from 0.025%wt to 0.25%wt gellan. It is noteworthy that the actual steady state response was only reached for the most diluted fluid gel. The comparison of the second to the first shear stress peak provided a useful tool to model the structural recovery of fluid gels after reaching a strain value of 3000. The higher the gellan gum concentration the lower relative recovery was observed, due probably to the fact that a longer relaxation time was required to rebuild the initial structure based on a more concentrated suspension of microgel particles dispersed into a polymer aqueous solution. The aforementioned information can be only provided by start of the art controlled strain rheometers, since controlled-stress rheometers fail to instantaneously impose a constant low shear rate. However, similar information can be obtained in controlled-stress rheometers by means of a series of nonlinear creep/recovery/nonlinear creep tests. Figure 8 and 9 illustrate how this approach provides information on the peak viscosity recovery in a gellan fluid gel after applying a constant shear stress.

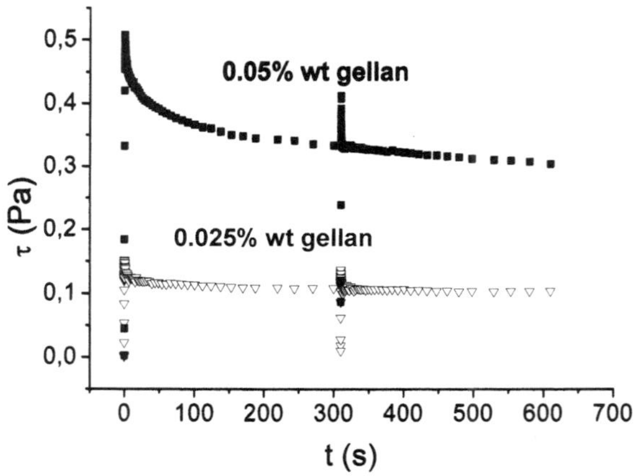

Figure 7 *Start-up at the inception of shear and flow interrupted results obtained at 20 °C in a controlled-strain rheometer at 10 s^{-1}. Relaxation time after cessation of shear: 10 s. Fluid gels prepared with 1,3% wt NaCl by disruption of the incipient gel structure.(Coulex impeller started at 40°C. 700 rpm for 1 minute).*

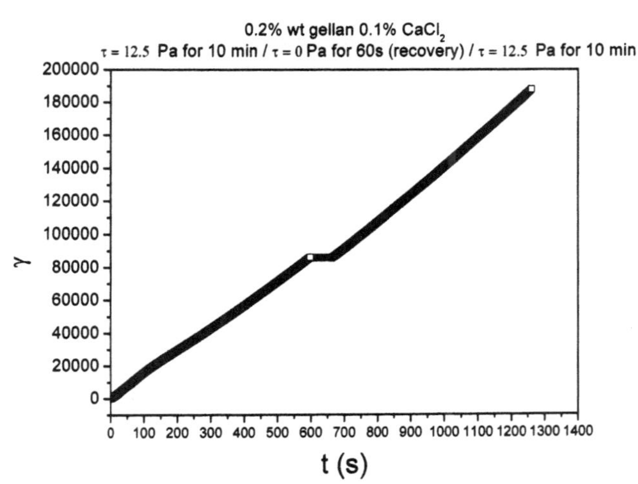

Figure 8 *Strain as a function of time* in a series of nonlinear creep (12.5Pa)/recovery/ nonlinear creep (12.5 Pa) experiments at 20°C for a 0.2% wt gellan gum fluid gel prepared with 0.1%wt $CaCl_2$ with a Coulex/saw tooth impeller. (700 rpm during 40 minutes).

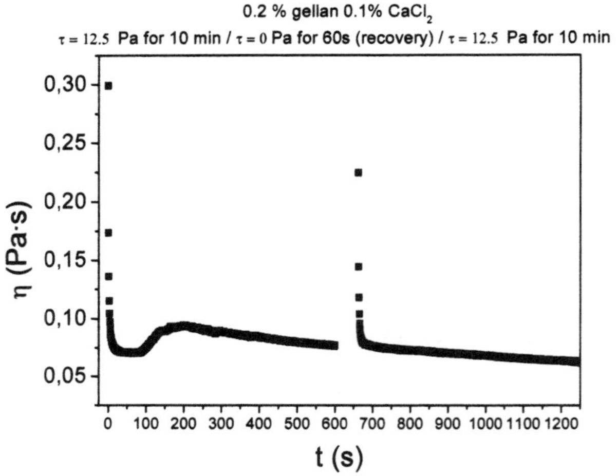

Figure 9 *Apparent viscosity values calculated from the time-dependent strain values plotted in Figure 8.*

CONCLUSIONS

Low-acyl gellan gum can form fluid gels with tailor-made values of the dynamic viscoelastic functions, which can be adjusted by selection of suitable mechanical processing as well as of the gel-promoting ion type and concentration. These factors also controlled the reproducibility of lab scale batches. Flow curves and creep experiments show that the steady shear flow behaviour of gellan gum fluid gels is consistent with a "very shear thinning" response. Start-up at the inception of shear at $10s^{-1}$ were consistent with a thixotropic response and flow interrupted experiments demonstrated that the relative recovery of the initial viscosity after a relaxation time of 10s was nearly complete below 0.05% wt gellan gum. Longer rest times were required to recover the stronger initial network formed by microgel particles at higher gellan concentrations. Similar information can be obtained from a series of nonlinear creep/recovery/nonlinear creep experiments.

ACKNOWLEDGMENTS
Financial support from the Ministerio de Educación y Ciencia of Spain (Project CTQ2007-66157/PPQ) and scholarship to M.C. García are gratefully acknowledged.

REFERENCES

1. G. Sworn, G.R. Sanderson and W. Gibson, *Food Hydrocoll.,* 1995, 9, 265.
2. M. Caggioni, P.T. Spicer, D.L. Blair, S.E. Lindberg and D.A. Weitz, *J. Rheol.,* 2007, **51**, 851.
3. A.I. Rodríguez-Hernández, S. Durand, C. Garnier, A. Tecante and J.L. Doublier, *Food Hydrocoll.*, 2003, **17**, 621.
4. L.P. Martínez-Padilla, F. López-Araiza and A. Tecante, *Food Hydrocoll.*, 2004, **18**, 471.
5. M.T. Nickerson, A.T. Paulson and R.A. Speers, *Food Hydrocoll.*, 2003, **17**, 577.
6. J. Tang, M.A. Tung and Y. Zeng, *J. Food Sci.*, 1995, 60, 748.

#THE EFFECT OF SUGARS ON LOW ACYL GELLAN GELS

V. Evageliou[1], M. Mazioti[1], G. Tseliou[1], I. Mandala[1] and M. Komaitis[1]

[1]Department of Food Science and Technology, Agricultural University of Athens, 75 Iera Odos, 11855, Athens, Greece

ABSTRACT

Low acyl gellan is a biopolymer that finds many applications in food science. It is able to gel on cooling in the presence of cations. Several sugars or their mixtures are used by the food industry in order to achieve the desired properties in sweetness and in texture. Their presence affects both the formation and the mechanical properties of various polysaccharide gels. In the present study, the effect of sucrose, glucose, fructose and inulin on the gelation of 0.5 wt% low acyl gellan in the presence of potassium chloride (100 mM) was investigated by large deformation compression experiments. The sugar concentration varied from 0-15 wt%. Samples prior to compression were refrigerated at 5°C for 24 hrs. True stress and true strain at failure along with Young's modulus were calculated from each compression curve. Stress and strain at the break point are indicative of the network's strength and elasticity, respectively, and Young's modulus is related to the network firmness. Glucose showed no contribution to the gellan network strength, as the moduli for all studied concentrations were almost identical with the sample with no sugar present. On the other hand, fructose and sucrose enhanced the strength of the network with sucrose being slightly more effective, whereas the presence of inulin led to gels with decreased strength. Regarding firmness, the general trend was that fructose and sucrose were practically not changing the firmness of the resulting gellan gels, whereas glucose and, even more, inulin showed a major drop in the modulus. Moreover, for all concentrations and all sugars studied, the values of strain remained almost unchangeable showing that the addition of cosolutes had little effect on the deformability of the resulting gels. It is very interesting that the above mentioned observations depended solely on the presence and type of each sugar and not in its concentration.

1. INTRODUCTION

Low acyl gellan is a biopolymer that finds many applications in the food area. Confectionery, jams and jellies, water-based gels and dairy products are the major ones.[1] It is able to form hard and brittle gels in the presence of cations. The three dimensional network is developed upon cooling, where double-helix structures are formed and which are further aggregated through the cations.[2]

The gel strength and the rate of gel formation are affected by the presence of cosolutes. In order to achieve the desired properties in sweetness and in texture, the food industry uses several sugars or their combinations. Sucrose is the most commonly used sugar, followed by fructose and glucose. Inulin is a fructooligosaccharide which is used by the food industry over the last years in order to formulate innovative healthy foods. The most interesting aspect of inulin is its reduced caloric value and its use as dietary fiber and an alternative sweetener.

The effect of the nature of the cosolute on the formation and the mechanical properties of various polysaccharide gels, like high methoxy pectin,[3-5] κ- carrageenan,[6,7] gelatin,[8] agarose[9] and oxidized starch[10] has been widely explored. The properties of gellan gels with various sugars (monosaccharides, disaccharides, starch hydrolysates) have also been investigated.[11-13] In addition, several researchers have studied the glass transition of gellan – high sugar systems.[14,15] They found that low concentrations of sugars (< 40 wt%) reinforced the rigidity of gellan gels, intermediate levels of co-solute (40 to 70 wt%) caused a drop in the network strength whereas high levels (>70 wt%) led to vitrification of the system.

The current work investigated the properties of the network of 0.5 wt% gellan gels in the presence of sucrose, glucose, fructose and inulin at concentrations up to 15 wt%. Potassium in the form of salt was the cation used as calcium has been used widely in previous investigations. The characterisation was based on parameters (true stress and strain, Young's modulus) obtained from compression tests. Stress and strain at the break point are indicative of the network's strength and elasticity, respectively, and Young's modulus is related to the network firmness.

2. MATERIALS AND METHODS

The deacylated gellan gum was provided by Sigma (Phytogel, P8169) and KCl was from Merck. Sucrose was normal food grade and purchased locally. Fructose and glucose were purchased from Fluca. Glucose was in its monohydrated form, which was taken into account in our calculations. Inulin was from Orafti. Distilled water was used throughout.

Samples were prepared by dissolving gellan in distilled water at 90°C using gentle agitation. Following solvation, sugars and KCl were added. The polysaccharide concentration was kept constant at 0.5 wt% whereas that of the sugars varied from 0-15 wt%. Salt concentration was constant at 100 mM.

The resultant solutions were transferred to a refrigerator at 5°C and cylindrical samples of 20 mm height and 20 mm width were taken straight for compression 24 hrs later. The samples were uniaxially compressed in an Instron Universal machine (Instron 1011, Massachusetts, USA) using a 40 mm diameter plunger at a fixed rate of 48 mm/min and up to 70 % deformation. Filter paper was placed between sample surfaces and plates in order to prevent the slippage of the gel. All measurements were made at room temperature. "True strain", "true stress" and Young's modulus were obtained from the compression curves. "True strain" shows the extent of deformation and it is given by ln (L_0/L) and "true stress" is the stress corrected for the expansion of the mean cross-sectional area of the sample during the compression. L_0 and L are the height of the sample before and during the compression. The results shown are the mean value of four compression curves.

3. RESULTS

Figure 1 illustrates the effect of different cosolutes on the stress at failure of 0.5 wt% gellan gels in the presence of 100 mM KCl, for compression tests carried out after 24 hours storage in the refrigerator. It is obvious that glucose did not affect the gellan network strength as the modulus for all studied concentrations were almost identical with the sample with no sugar present. Fructose and glucose enhanced the strength of the network with sucrose being slightly more effective. In addition, inulin resulted in a decreased gel strength. The above mentioned observations depended solely on the presence and type of each sugar and not in its concentration.

Figure 2 shows the variation of strain at failure over increasing concentration of sugars. For all concentrations and all types of sugars studied the values of strain remained almost unchanged showing that the addition of cosolutes had little effect on the brittleness of the resulting gels.

The firmness of the gels over increasing concentration of various cosolutes is shown in Figure 3. The general trend in all figures is that fructose and sucrose were practically not changing the firmness of the resulted gellan gels, whereas glucose and inulin showed a major drop in the modulus, regardless their concentration in the sample.

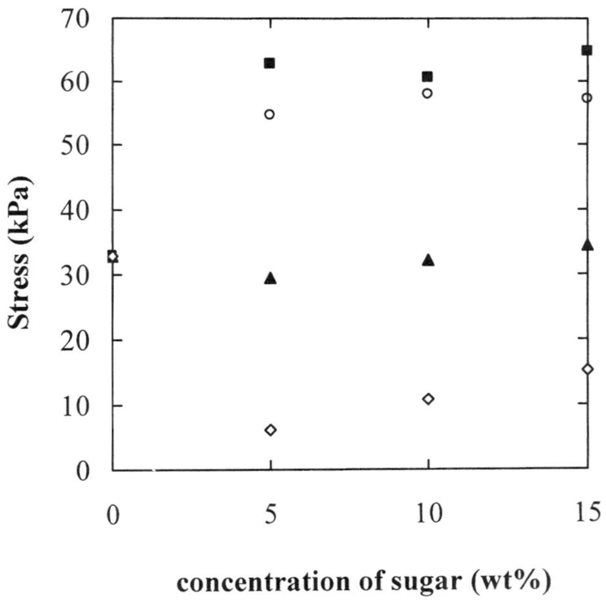

Figure 1 *Stress values at failure of 0.5 wt% gellan gum gels in the presence of 100 mM KCl with increasing concentration of sucrose (■), fructose (O), glucose (▲) and inulin (◊) after holding at 5°C for 24 hrs.*

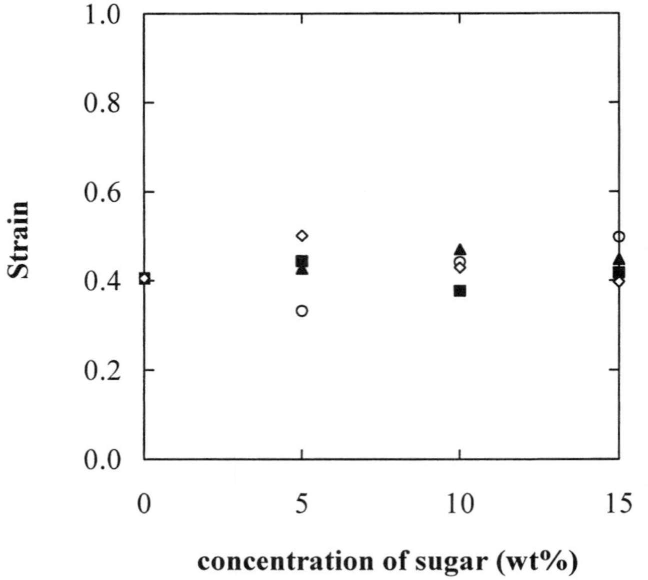

Figure 2 *Strain values at failure of 0.5 wt% gellan gum gels in the presence of 100 mM KCl with increasing concentration of sucrose (■), fructose (O), glucose (▲) and inulin (◊) after holding at 5°C for 24 hrs.*

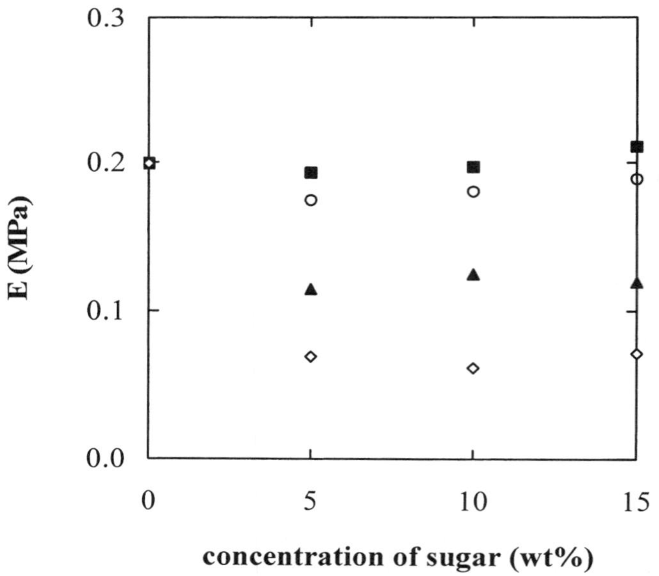

Figure 3 *Young's modulus values at failure of 0.5 wt% gellan gum gels in the presence of 100 mM KCl with increasing concentration of sucrose (■), fructose (O), glucose (▲) and inulin (◊) after holding at 5°C for 24 hrs.*

4. DISCUSSION

The behaviour of biopolymers in the presence of sugars is been widely investigated. In order for the resulted observations to be explained, Tait, et al.[16] suggested that the effectiveness of sugars is connected to the number of equatorial hydroxyl groups as their spacing is particularly compatible with the lattice structure of water. Based on that, the anticipated order for the two monosaccharides and the dissacharide studied in the present work would be fructose < glucose < sucrose. Although this order has been seen experimentally in many gelling biopolymers, like gelatine[8], agarose[9] or the initial conformational ordering of oxidised starch[10], there are also cases where there is a departure from the above mentioned order like high methoxy pectin[4] where the sequence fructose < sucrose < glucose was observed.

For both stress and Young's modulus (Figures 1 and 3), the order for the two monosaccharides and the dissacharide studied was glucose < fructose < sucrose, which departs from the order according to Tait et al.[16]. Regarding gellan, Sworn & Kasapis[11] reported different sequences for different concentrations of cosolutes but this sequence was observed for the modulus of 0.5 wt% gellan gels and for concentrations of cosolutes up to 10 wt%. In addition, the behaviour observed in Figures 1 and 3 depended solely on the type of the sugar and not on its concentration, as the values obtained were virtually the same throughout the sugar concentration range. As reported previously[11], the concentration and the nature of cosolute define the necessary gelation concentration of ion. In addition, the same researchers showed that low levels of cosolutes (20 wt%) can enhance the effect of calcium on the gelation of gellan gum which is in good agreement with previous studies[17] showing that calcium ions and sucrose complement each other. According to this suggestion, our observations can be explained by the thought that the potassium salt concentration used in our research (100 mM) can not be the necessary ion concentration for maximum gel strength for all sugars under investigation.

At the same time, inulin did not gel in any of the concentrations studied and its presence led to decreased gel strength and elasticity (Figures 1 and 3). Similar behaviour was observed when increasing concentrations of inulin were added to mixtures of low methoxy pectin in the presence of stoichiometric Ca^{2+} and was attributed[18] to excessive association of pectin into large aggregated bundles which make little contribution to network crosslinking. It is probable that phase separation occurred through volume exclusion as inulin is a large molecule.

Regarding strain the values were fluctuating around the value for the sample with no cosolute present (Figure 2). This has been seen in previous investigations by Tang et al.[12] for sucrose. Moreover, Sworn and Kasapis[11], who studied the effect of various cosolutes differing in molecular weight and size (monosaccharides, disaccharides and starch hydrolysates), among them glucose, fructose and sucrose, on the textural properties of gellan gels, found no significant effect on strain for concentration of the above mentioned cosolutes up to 20 wt%.

References

1. G. R. Sanderson in *Food gels,* ed. P. Harris, Elsevier, New York, 1990, pp 201-230.
2. H. Grasdalen and O. Smidsrod, *Carbohydr. Polym.,* 1987, **7**, 371-393.
3. J. A. Lopes da Silva and M. P. Goncalves, *Carbohydr. Polym.*, 1994, **24**, 235-245.
4. V. Evageliou, R. K. Richardson and E. R. Morris, *Carbohydr. Polym.*, 2000a, **42**, 245-259.
5. A. Tsoga, R.K. Richardson and E. R. Morris, *Food Hydrocolloids*, 2004, **18**, 907-919.

6. K. Nishinari and M. Watase, *Thermochim. Acta*, 1992, **206**, 149-162
7. V. Evageliou, S. Kasapis and M. W. N. Hember, *Polymer*, 1998, **39**, 3909-3917.
8. D. Oakenfull and A. Scott, *Food Hydrocolloids*, 1986, **1**, 163-175.
9. M. Watase, K. Kohyama and K. Nishinari, *Thermochim. Acta*, 1992, **206**, 163-173.
10. V. Evageliou, R. K. Richardson and E. R. Morris, *Carbohydr. Polym.*, 2000b, **42**, 261-272.
11. G. Sworn and S. Kasapis, *Food Hydrocolloids*, 1998b, **12**, 283-290.
12. J. Tang, R. Mao, M. A. Tung and B. G. Swanson, *Carbohydr. Polym.*, 2001, **44**, 197-209.
13. S. Bayarri, E. Costell and L. Duran, *Food Hydrocolloids*, 2002, **16**, 593-597.
14. M. Papageorgiou, S. Kasapis and R. K. Richardson, *Carbohydr. Polym.*, 1994, **25**, 101-109.
15. G. Sworn and S. Kasapis, *Carbohydr. Res.*, 1998, **309**, 353-361.
16. M. J. Tait, A. Suggett, F. Franks, S. Ablett and P. A. Quickenden, *J. Solution Chem.*, 1972, **1**, 131-151.
17. L. E. Willoughby and S. Kasapis, *Food Sci. Technol. Today,* 1994, **8**, 227-233.
18. P. Giannouli, R. K. Richardson and E. R. Morris, *Carbohydr. Polym.*, 2004, **55**, 357-365.

THE EFFECT OF K^+, Ca^{2+} AND THEIR MIXTURES ON LOW ACYL GELLAN GELS

V. Evageliou[1], M. Karantoni[1], I. Mandala[1] and M. Komaitis[1]

[1]Department of Food Science and Technology, Agricultural University of Athens, 75 Iera Odos, 11855, Athens, Greece

ABSTRACT

Low acyl gellan is a water soluble polysaccharide which forms hard and brittle gels in the presence of cations. The three dimensional network is developed upon cooling, where three fold double-helix structures are formed and which are further aggregated through the cations. Monovalent cations contribute in indirect double helices interactions whereas divalent cations in direct interactions. In the present study, the effect of a monovalent (K^+), a divalent (Ca^{2+}) cation and their mixtures on the properties of low acyl gellan gels was investigated by large deformation compression experiments. Gellan concentration was kept constant at 0.5 wt% whereas that of the salts varied from 20-80 mM for KCl and 5-80 mM for $CaCl_2$. The effect of mixtures of the two salts at a total concentration of 80 mM was also explored. Mixtures of KCl: $CaCl_2$ were prepared at molar ratios of 20:60, 40:40 and 60:20. Samples prior to compression were refrigerated at 5°C for 24 hrs. Young's modulus, true stress at failure and true strain at failure were the parameters calculated from the compression curves. Stress and strain at the break point are indicative of the network's strength and elasticity, respectively, and Young's modulus is related to the network firmness. For a given gellan concentration, the gel strength increases with cation concentration up to a certain value above which the strength is decreasing. In our study, gel strength increased with increasing concentration of potassium whereas calcium exhibited only the descending part of the curve, indicating that the maximum network strength is achieved at a concentration of 5 mM or lower. For both cations, firmness shared similar profile as strength. The values of strain did not, generally, exhibit great differences over concentration, for both cations studied. In the case of mixtures, the replacement of $CaCl_2$ with KCl had no significant effect on the strength of the network, arguing that the divalent cation defines the final strength. However, the monovalent cation determined the remaining properties as its progressive addition led to a progressive increase of both elasticity and firmness.This behaviour was attributed to the replacement of direct with indirect interactions in the double helices of gellan.

1. INTRODUCTION

Gellan gum is an anionic extracellular polysaccharide secreted by the bacteria *Sphingomonas elodea* during aerobic fermentation.[1] Its primary structure has a linear tetrasaccharide repeating unit of → 3)-β-D-Glc*p*-(1 → 4)-β-D-Glc*p*A-(1 → 4)-β-D-Glc*p*-(1 → 4)-α-L-Rha*p*-(1 →.[2,3] The native form of the polysaccharide is the high acyl form (as it is partially acylated) which by alkali treatment leads to the low acyl counterpart (deacylated).

Low acyl gellan is a water soluble polysaccharide which forms transparent gels on cooling. Formation of double helices further aggregated to form junction zones is stabilised by the presence of cations and a three dimensional network is developed. The concentration of both the polymer and the cations affect the mechanical properties of the resulting gel network.[4,5] In addition, for a given polymer concentration, the gel strength increases with cation concentration up to a certain value above which the strength is decreasing. Moreover, the valency of the cations is also important as monovalent cations are less effective in promoting gelation than the divalent ones.[6]

A possible explanation for this behaviour is that monovalent cations (e.g. K^+) promote the aggregation process by site-binding to the helices, thus suppressing the electrostatic repulsion between them,[7,8] whereas divalent cations (e.g. Ca^+) by site-binding between pairs of carboxyl groups on neighbouring helices,[7] to give structures analogous to the "egg-box" junctions proposed[9] for calcium-induced gelation of alginate or pectin. The latter are much stronger, explaining why divalent cations are more effective in promoting gelation. The decrease in gel strength at high concentrations of salts can be attributed[10] to excessive helix–helix association into very large aggregates, which leads to a reduction in the total number of junction zones in the gel network.

In the present work we have investigated the effect of a monovalent (K^+), a divalent (Ca^{2+}) cation and their mixtures on the development of low acyl gellan gels. The characterisation was based on parameters (true stress, true strain and Young's modulus) obtained from compression tests performed at different time intervals.

2. MATERIALS AND METHODS

The deacylated gellan gum was provided by Sigma (Phytogel, P8169). KCl was from Merck and $CaCl_2$ from Panreac Quimica SA. Distilled water was used throughout.

Samples were prepared by dissolving gellan in distilled water at 90°C using gentle agitation. Following solvation, appropriate amounts of KCl and $CaCl_2$ were added. The polysaccharide concentration was kept constant at 0.5 wt% whereas that of the salts varied from 20-80 mM for KCl and 5-80 mM for $CaCl_2$. The effect of mixtures of the two salts at a total concentration of 80 mM was also explored. Mixtures were prepared at KCl: $CaCl_2$ molar ratios of 20:60, 40:40 and 60:20.

The resulting solutions were transferred to a refrigerator at 5°C and cylindrical samples of 20 mm height and 20 mm diameter were taken for compression 24 hours later. The samples were uniaxially compressed in an Instron Universal machine (Instron 1011, Massachusetts, USA) using a 40 mm diameter plunger at a fixed rate of 48 mm/min and up to 70 % deformation. Filter paper was placed between sample surfaces and plates in order to prevent the slippage of the gel. All measurements were made at room temperature.

The results shown are the mean value of four compression curves. Young's modulus, true stress at failure and true strain at failure were the parameters that were calculated.

Stress and strain at the break point are indicative of the network's strength and elasticity, respectively, and Young's modulus is related to the network firmness. "True strain" shows the extent of deformation and it is given by ln (L_0/L) and "true stress" is the stress corrected for the expansion of the mean cross-sectional area of the sample during the compression. L_0 and L are the height of the sample before and during the compression.

3. RESULTS

Figure 1 shows the values of stress at failure over concentration for samples incorporating potassium, calcium ions and their mixtures compressed after 24 hrs holding at 5°C. Addition of potassium chloride progressively increased the network strength. The samples incorporating 20 mM KCl did not gel in the time scale of the experiment. Moreover, with increasing concentration of calcium, the gel strength kept decreasing. When mixtures of the two salts were used to induce gelation, it was clear that the replacement of $CaCl_2$ with KCl had no significant effect on the strength of the network, arguing that the divalent cation defines the final strength.

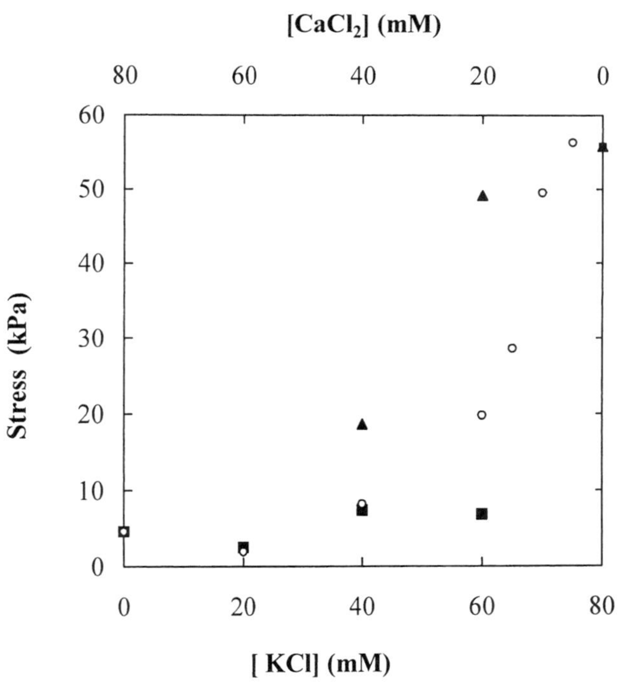

Figure 1 *Stress values at failure of 0.5 wt% gellan gum gels with (a) KCl (▲), (b) $CaCl_2$ (O) and (c) mixtures of the two salts at a total concentration of 80 mM (■) after holding at 5°C for 24 hrs.*

The elasticity of the resulting networks, expressed as the strain at the failure point, is shown in Figure 2. Gels with potassium and calcium ions, throughout their individual concentration range, did not exhibit great differences at the values of failure strain. However potassium gels showed a greater elasticity. In the mixed systems, progressive addition of potassium salt led to a progressive increase of the strain at failure which eventually reached the value for the sample containing only 80 mM KCl.

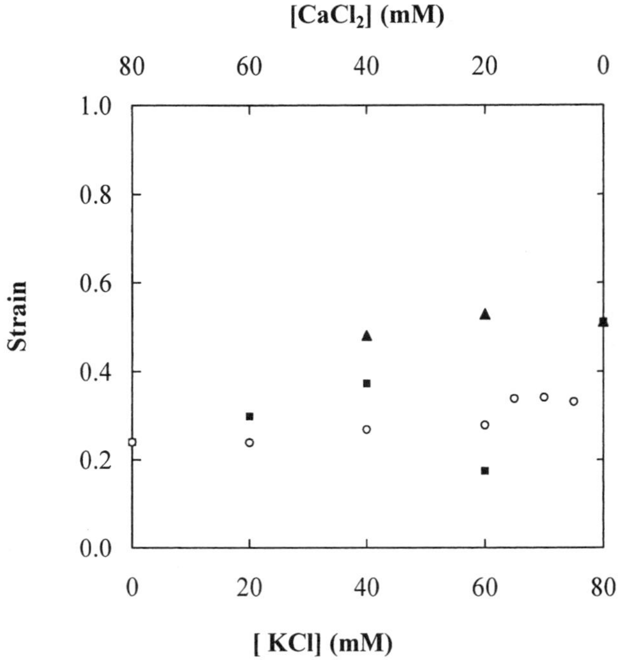

Figure 2 *Strain values at failure of 0.5 wt% gellan gum gels with (a) KCl (▲), (b) $CaCl_2$ (O) and (c) mixtures of the two salts at a total concentration of 80 mM (■) after holding at 5°C for 24 hrs.*

Figure 3 shows the values of Young's modulus for 0.5 wt% gellan gels as a function of KCl and $CaCl_2$ concentration and their mixtures compressed after holding at 5°C for 24 hrs. In the presence of potassium, the modulus, and therefore the network firmness, increased with concentration. Calcium exhibited exactly the opposite effect as the firmness was progressively decreased, in a similar profile to gel strength. However both salts shared similar maximum values of the modulus. When mixtures of the two salts were present, the modulus increased as the concentration of KCl was increasing.

4. DISCUSSION

Gellan molecules in hot water solutions exist as disordered coils. Cooling leads to ordered chains further stabilised by association into aggregates. This process is promoted by the presence of cations. Initially increasing concentrations of cations lead to increased gel strength whereas at high concentrations of cations the strength decreases. In our study, gels with potassium exhibited increased strength with increasing concentrations, whereas those with calcium decreased, indicating that the maximum network strength is achieved at a concentration of 5 mM or lower. These results are in agreement with all previous investigations on the effect of salts on gellan gels[11-13] suggesting that calcium ion is more effective than potassium as a much smaller concentration is needed in order to give a network of the same strength as potassium.

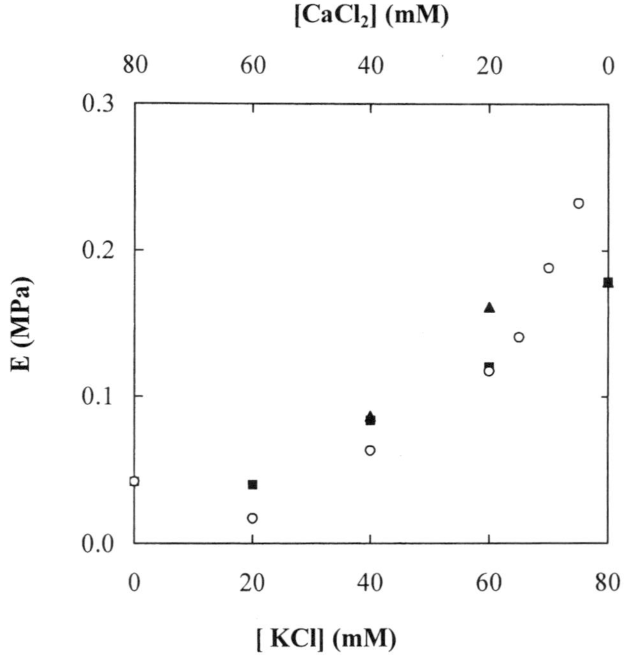

Figure 3 *Young's modulus of 0.5 wt% gellan gum gels with (a) KCl (▲), (b) $CaCl_2$ (O) and (c) mixtures of the two salts at a total concentration of 80 mM (■) after holding at 5°C for 24 hrs.*

Tang et al.[12,14] have also studied the effect of monovalent and divalent cations to compression deformation. According to them, strain initially decreased up to a certain cation level above which a standard value (~0.4 for all cations studied) was obtained. That level coincides with the critical value where the maximum strength is seen. In our study, the resulting calcium gels showed (Figure 2) only the plateau region of the curve as it was expected by the fact that all concentrations used in the present work were above the critical value. Potassium, however, shared similar behaviour throughout the concentration range. The plateau value was ~0.5 for potassium and ~0.3 for calcium. The diverse behaviour seen in our results can be attributed to different procedures used for the sample preparation and treatment prior to analysis as those can affect the kinetics of aggregation and therefore the mechanical properties[11].

When mixtures of salts were used to induce gelation, our study showed that strength was determined by the divalent cation. A possible explanation can be that having the total salt concentration of the mixtures constant at 80 mM, progressive replacement of calcium by potassium was actually similar to reducing the calcium concentration to levels lower than 80 mM. As seen from Figure 1, stress for concentrations of calcium above 40 mM exhibited very low values, close to the results for the mixtures.

These observations may be dictated by the nature of the interactions between the gellan helices and the cations. Calcium participates in direct crosslinking, which allows faster development of the gel network, in comparison with the indirect crosslinking of potassium, which is a slower procedure. Moreover, potassium also affected the deformability as the increase of its concentration in the mixtures was followed by an increase in the network elasticity. Once again, the replacement of direct with indirect interactions can be the explanation.

Firmness is another important parameter for the characterisation of a gel. For both potassium and calcium (Figure 3) firmness decreased over cation concentration in a similar profile to strength (Figure 1). Gellan gels containing mixtures of the two salts exhibited a continuous increase of the modulus on progressive replacement of calcium with potassium cation (Figure 3). The monovalent cation dominated, as the concentrations it had in the mixtures were well below 80 mM, which was the concentration where the maximum strength was seen.

References

1. G. Sworn in *Handbook of hydrocolloids, ed.* G. O. Phillips and P. A. Williams, Woodhead Publishing Ltd, Cambridge, 2000, pp 117-135.
2. P.-E. Jansson, B. Lindberg and P. A. Sandford, *Carbohydr. Res.,* 1983, **124**, 135-139.
3. M. A. O' Neill, R. R. Selvendran and V. J. Morris, *Carbohydr. Res.,* 1983, **124**, 123-133.
4. H. Grasdalen and O. Smidsrod, *Carbohydr. Polym.,* 1987, **7**, 371-393.
5. G. R. Sanderson in *Food gels,* ed. P. Harris, Elsevier, New York, 1990, pp 201-230.
6. H. Moritaka, H. Fukuba, K. Kumeno, N. Nakahama and K. Nishinari *Food Hydrocolloids*, 1991, **4**, 495-507.
7. E. R. Morris, M. G. E. Gothard, M. W. N. Hember, C. E. Manning and G. Robinson *Carbohydr. Polym.*, 1996, **30**, 163-175.
8. G. Robinson, C. E. Manning and E. R. Morris in *Food polymers, gels and colloids,* ed. E. Dickinson, Royal Society of Chemistry, Cambridge, 1991, pp 22-33.
9. G. T. Grant, E. R. Morris, D. A. Rees, P. J. C. Smith and D. Thom, *FEBS Letters,* 1973, **32**, 195-198.
10. E. R. Morris, R. K. Richardson and L. E. Whittaker, *Prog. Colloid Polym. Sci.*, 1999, **114**, 109-115.
11. M. Milas and M. Rinaudo, *Carbohydr. Polym.*, 1996, **30**, 177-184.
12. J. Tang, M. A. Tung and Y. Zeng, *Carbohydr. Polym.,* 1996, **29**, 11-16.
13. S. Kasapis, P. Giannouli, M. W. N. Hember, V. Evageliou, C. Poulard, B. Tort-Bourgeois and G. Sworn, *Carbohydr. Polym.*, 1999, **38**, 145-154.
14. J. Tang, M. A. Tung and Y. Zeng, *J. Food Sci.,* 1995, **60**, 748-752.

RHEOLOGICAL PROPERTIES OF *PLANTAGO OVATA* L. AS A FUNCTION OF CONCENTRATION, TEMPERAURE AND pH.

A. Farahnaky, H. Askari, M. Majzoobi, G.R. Mesbahi and M. Aminlari

Department of Food Science and Technology, School of Agriculture, Shiraz University, Shiraz, Iran.

1. Introduction

Psyllium is the common name used for several members of the plant genus *Plantago*. Psyllium and ispaghula husk (similar uses differ in doses) terms are interchangeably used in the text, psyllium is derived from the dried, ripe seeds of *Plantago psyllium* (synonym for *Plantago afra* L.) and from *Plantago indica* L., but ispaghula husk is derived from the ripe seeds of *Plantago ovata Forsskaol* (synonym for *Plantago isphagula Roxburgh*). [1] The seeds of psyllium are used commercially for the production of mucilage. The mucilage is obtained from the seed coat by mechanical milling/grinding of the outer layer of the seeds. It is a white fibrous hydrophilic material and forms a clear colorless mucilaginous gel by absorbing water.[2-4] The demand for hydrocolloids with specific functionality has recently increased and food scientists and technologists are always searching for new sources of polysaccharides. Accordingly, there is still place for the new sources to meet this demand.[5-6] Having pharmacological effects, and acting as a thickening agent in foods, *P. ovata* mucilage gum may have a superior technological and consumer acceptability. Although psyllium gum has been used for centuries in some countries including Iran and India no extensive scientific work has been done on its functional and rheological properties. Thus, the objectives of this study were first to extract and evaluate the thickening behavior of gum extracted from *P. ovata* and second to study the effect of gum concentration, temperature and solution pH on consistency and flow behavior indices of its solutions.

2. MATERIALS AND METHODS

The Espharze (*Plantago ovata L.*) seeds were obtained from the local market for medical herbs, Tehran, Iran. They had been collected from the Khorasan province of Iran located at between 55°, 28′-61°, 20′ longitude and 30°, 21′-38°, 17′ latitude.

2.1 Gum extraction and purification

Husks of *P. ovata* seeds were separated from the seed coats by mechanical milling of the outer layer of the seeds using a blade type grinder (Mammonlex, Taiwan) for 1 min at low speed, and for further 3 min at high speed. The seeds were then sieved using a mesh number 18 sifter and the husk was separated from seeds. Husks of seeds were dispersed in distilled water at a water/husk ratio of 20:1. The husk-water slurry was stirred continuously with a mechanical mixing paddle throughout the entire extraction period (1 h). The seed residue was discarded, and the rest of the supernatant was subjected to ethanol precipitation (97 % ethanol/mixture ratio of 3:1). The precipitate was then kept in the solvent for approximately 20 min with occasional gentle stirring. Later, the precipitate was removed using centrifugation at 10000g for 20 min at 25 °C. The gum then was dried in a conventional oven (overnight at 40 °C) and milled to a mesh size between 53 and 125 micron.

2.2 Analysis of chemical composition

Moisture, protein, fat and ash content of the mucilage were determined using the AOAC approved method. The quantity of total sugar was estimated as the remainder of the samples after substances (protein, fat, ash, and moisture) were accounted for. [7]

2.3 Fourier transform infrared spectroscopy (FT-IR)

FT-IR spectrum of purified psyllium gum was obtained at a resolution of 1 cm^{-1}. The sample was incorporated into a KBr (spectroscopic grade) and pressed into a 2 mm pellet. IR spectra were recorded in the transmittance mode from 4000 to 400 cm^{-1}, using a BRUCKER spectrophotometer (EQUINOX 55, Germany).

2.4 Rheological measurements

Rheological measurements were carried out using a rotational viscometer (Brookfield, DV-II + Pro, Brookfield Engineering, USA) equipped with CP51 geometry as a function of gum concentration (0.1, 0.2, 0.4 and 0.6 % w/w), temperature (5, 15, 25, 40, 50 and 60 °C) and pH (2.5, 4, 7and 10). For each test, 0.5 ml sample was transferred to sample compartment (cone and plate geometry) and after temperature equilibrium, was subjected to a programmed shear rate increasing linearly from 1.92-229.4 s^{-1} in 80 steps. The flow behavior index and consistency index values were computed by fitting the Power law model (Eq. 1):

$$\tau = k\gamma^{\circ n} \qquad (1)$$

Where τ is the shear stress (mPa), γ° is the shear rate (s^{-1}), k is the consistency coefficient (mPa.s) and n is the flow behavior index (dimensionless). The values of n and k were obtained by using the software (Rheocal) supplied by the manufacturer.

2.5 Statistical analysis

Rheological properties of psyllium solutions were determined by linear regression based on minimizing the sum of the squares via the software supplied by the manufacture (Rheocal, Brookfield Engineering, USA). Model equations and regression coefficients (R^2) were reported. Analysis of variance (ANOVA) test was applied to acknowledge any significant difference between the rheological parameters; n and k under the effect of temperature, pH and psyllium concentration ($P < 0.01$). The SPSS (version 13) program was used for all statistical analysis. Software of Design-Expert 6.0.2 (Stat-Ease Inc, Minneapolis, USA): D-Optimal Response Surface mode was used in order to model the experimental data and estimate any nonlinearity in the relationships between the parameters under study.

3. RESULTS AND DISCUSSION

3.1 Chemical composition

The chemical compositions of the whole seed, seed coats (husk), and the purified mucilage from *P. ovata* seeds are shown in Table 1. The moisture content of whole seed, seed coats (husk), and the purified mucilage was similar, 6.05, 6.24 and 6.13%, respectively. The protein content of the whole seed was greater than seed coats (husk), and the mucilage and the majority (i.e. about 90%) of whole seed, seed coats (husk), and the mucilage was carbohydrates. The production yield of purified mucilage from the seeds was 28.5% (w/w). This was in agreement with Fischer et al. (2004), who reported an approximate yield of 25% for *Plantago ovata* Forsk. [8]

Table1 *The chemical compositions of the whole seed, seed coats (husk), and the purified mucilage from P. ovata seeds.*

	Compositional distribution of total material (%)		
Composition	**Whole seed**	**Husk**	**Mucilage**
Moisture	6.05 ± 0.064	6.24 ± 0.016	6.13 ± 0.016
Protein	2.93 ± 0.196	0.25 ± 0.015	0.11 ± 0.004
Fat	0.30 ± 0.055	0.50 ± 0.053	0.36 ± 0.033
Ash	2.98 ± 0.234	3.59 ± 0.165	1.51 ± 0.056
Fiber	24.32 ± 0.210	2.63 ± 0.065	3.19 ± 0.130
Total carbohydrate*	87.74 ± 0.549	89.25 ± 0.143	91.81 ± 0.062

* *Calculated from total (Moisture, Protein, Fat and Ash)*

3.2 FT-IR spectroscopy

Figure 1 shows the FT-IR spectrum of dry powder of purified psyllium gum. The spectrum are dominated by a broad band at about 3447 and 3207 cm^{-1}, which is assigned to the stretching vibration modes of -OH groups and –OH of alcohol, respectively. The peak at 1654 cm^{-1} is due to the asymmetrical $-COO^-$ stretching vibration, whereas the band at 1402 cm^{-1} is due to the symmetrical $-COO^-$ stretching vibration.

3.3 Rheological measurement

3.3.1 Effect of concentration and temperature on flow properties

The consistency coefficient (k) and flow behavior index (n) values obtained by fitting the shear stress versus shear rate data to the Power law model as a function of gum concentration and temperature are shown in Table 2. The regression coefficient (R^2) was 0.98 or higher for all tested samples indicating the appropriateness of the Power law model to describe the flow properties of the mucilage solutions. Examination of flow properties as a function of shear rate also designated the non-Newtonian behavior for different concentrations and temperatures. The values of flow behaviór indices, n, were less than 1, indicating the shear thinning nature of the gum at different measuring conditions. Increasing the gum concentration decreased the flow behavior index values and increased the consistency coefficient. The flow behavior index of gum solutions ranges from 0.74 (for 0.1 % w/w) to 0.50 (for 0.6 % w/w) at 25 °C and pH 7, suggesting that mucilage solutions with higher concentrations are more shear thinning than those with low concentrations. On the other hand, the consistency coefficient (k) of samples increases with increasing concentration of the polysaccharide. Temperature, on the other hand, indicated an inverse effect on Power law parameters as flow behavior index increased and consistency coefficient decreased by increasing the temperature. These phenomena are in agreement with the work done by Marcotte et al. (2001 a,b) on xanthan gum, carrageenan, pectin and starch.[9-10] The resultant data were analysed using the Design-Expert 6.0.2 and the best models were obtained. All variables and their interactions with significant effect in each model were kept and any variable or interaction without significant effect was removed as performed using the software. Graphs of consistency index, flow behavior index as obtained from the Design Expert are presented in Figure 3.

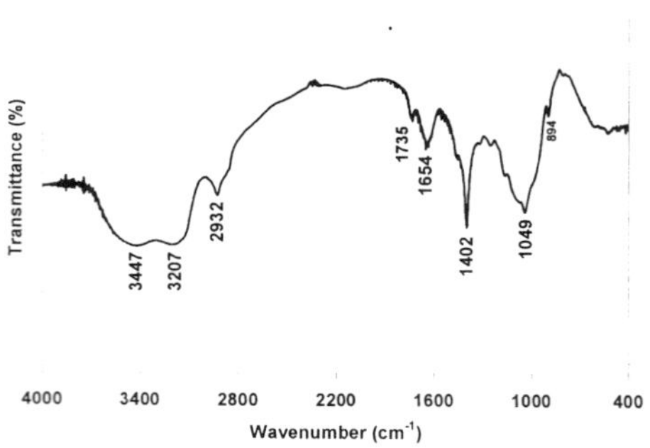

Figure 1 *FT-IR spectrum of psyllium gum*

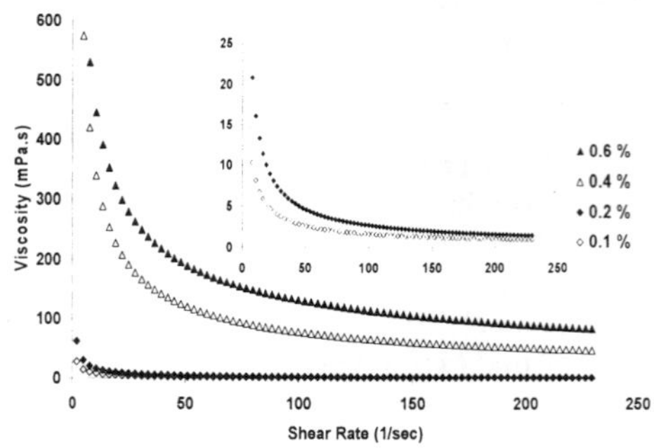

Figure 2 *The change in apparent viscosity of psyllium solutions at different concentration, pH 7 and 25 °C*

A decrease in consistency coefficient was observed with increasing temperature, indicating a decrease in apparent viscosity at higher temperatures (Table 2, Fig. 3). In addition consistency coefficient increased with the concentration of psyllium at all temperatures. For the prepared samples under acidic conditions, the increase in temperature resulted in considerable changes in viscosity values (Fig 3, Table 2). This behavior can be explained by intermolecular forces restricting the molecular motion of the fluid. These forces, which determine the intermolecular spacing, are significantly affected by the change in temperature. As temperature increases, the viscosity of the samples decreases except at 40 °C under alkaline conditions, where it increased. Analysis of variance (ANOVA) ($p<0.01$) showed that flow behavior index (n) and consistency

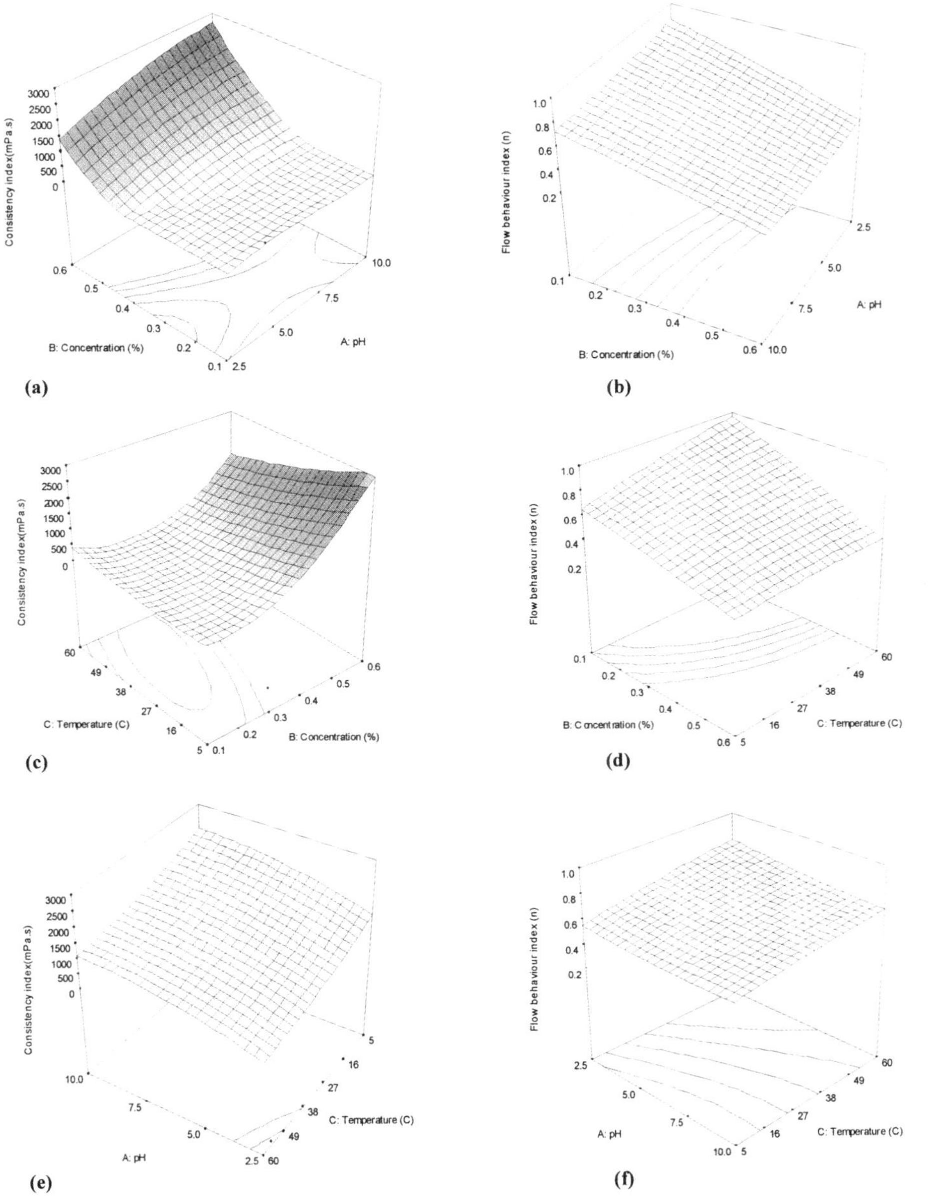

Figure 3 *Response Surface Methodology (given by Design Expert software): effect of pH, concentration and temperature on (a,c,d) consistency index and (b,d,f) flow behavior index of psyllium gum solutions.*

Table 2 *The Power law parameters for psyllium gum solutions at different concentration, pH and temperature.*

Concentration	pH	Temperature (°C) 5 k	5 n	15 k	15 n	25 k	25 n	40 k	40 n	50 k	50 n	60 k	60 n
0.1 %	**2.5**	12.20_{C**}^{a*}	$0.73_{A}^{n.s.}$	4.95_{B}^{a}	0.83_{B}^{d}	4.19_{B}^{a}	0.87_{B}^{d}	1.60_{A}^{a}	0.98_{C}^{b}	1.54_{A}^{a}	0.99_{C}^{c}	1.30_{A}^{a}	$1.0_{C}^{n.s}$
	4	15.40_{D}^{a}	$0.69_{A}^{n.s}$	8.00_{C}^{b}	0.77_{B}^{c}	5.34_{B}^{a}	0.81_{B}^{c}	2.41_{A}^{a}	0.92_{C}^{b}	1.65_{A}^{a}	0.97_{C}^{c}	1.43_{A}^{ab}	$1.0_{C}^{n.s}$
	7	16.25_{E}^{a}	$0.68_{A}^{n.s.}$	10.73_{D}^{c}	0.72_{AB}^{b}	7.12_{C}^{b}	0.74_{B}^{b}	5.145_{BC}^{b}	0.79_{C}^{a}	2.32_{AB}^{b}	0.93_{D}^{b}	1.60_{A}^{b}	$0.99_{E}^{n.s.}$
	10	24.30_{E}^{b}	$0.64_{A}^{n.s.}$	13.40_{D}^{d}	0.67_{AB}^{a}	8.90_{C}^{c}	0.70_{B}^{a}	5.53_{B}^{b}	0.77_{C}^{a}	4.02_{B}^{c}	0.83_{D}^{a}	1.82_{A}^{c}	$1.0_{E}^{n.s}$
0.2 %	**2.5**	19.05_{C}^{a}	0.67_{A}^{b}	5.780_{B}^{a}	0.83_{B}^{d}	5.06_{B}^{a}	0.81_{B}^{c}	5.45_{B}^{a}	0.83_{B}^{b}	2.78_{A}^{a}	0.98_{C}^{b}	1.12_{A}^{a}	0.98_{C}^{c}
	4	40.90_{E}^{b}	0.53_{A}^{a}	11.52_{C}^{b}	0.73_{B}^{c}	8.71_{BC}^{b}	0.73_{B}^{b}	7.37_{ABC}^{a}	0.82_{C}^{b}	4.05_{AB}^{a}	0.97_{D}^{b}	2.62_{A}^{b}	0.96_{D}^{bc}
	7	42.25_{D}^{b}	0.52_{A}^{a}	21.90_{C}^{c}	0.62_{B}^{b}	12.35_{B}^{c}	0.68_{B}^{a}	12.17_{B}^{b}	0.76_{C}^{ab}	6.15_{A}^{b}	0.88_{D}^{a}	4.77_{A}^{c}	0.90_{D}^{ab}
	10	65.50_{C}^{c}	0.55_{A}^{a}	31.85_{B}^{d}	0.54_{A}^{a}	12.55_{A}^{c}	0.68_{B}^{a}	12.050_{A}^{b}	0.70_{B}^{a}	7.68_{A}^{b}	0.83_{C}^{a}	5.76_{A}^{c}	0.88_{C}^{a}
0.4 %	**2.5**	477.80_{D}^{a}	$0.49_{A}^{n.s}$	166.90_{C}^{a}	0.48_{A}^{a}	99.45_{B}^{a}	0.50_{A}^{a}	21.15_{A}^{a}	0.71_{B}^{a}	14.85_{A}^{a}	0.71_{B}^{b}	8.59_{A}^{a}	0.76_{C}^{b}
	4	594.55_{D}^{b}	$0.49_{A}^{n.s}$	340.50_{C}^{b}	0.65_{A}^{b}	228.50_{B}^{b}	0.64_{A}^{b}	329.15_{C}^{b}	0.51_{C}^{b}	184.75_{A}^{b}	0.51_{C}^{a}	171.60_{A}^{b}	0.58_{B}^{a}
	7	627.75_{E}^{b}	$0.49_{A}^{n.s.}$	366.60_{C}^{b}	0.64_{C}^{b}	322.0_{B}^{c}	0.62_{C}^{b}	441.60_{D}^{c}	0.51_{A}^{b}	272.7_{A}^{c}	0.55_{B}^{a}	239.15_{A}^{c}	0.56_{B}^{a}
	10	638.80_{D}^{b}	$0.49_{A}^{n.s.}$	373.10_{B}^{b}	0.64_{C}^{b}	340.7_{B}^{d}	0.63_{C}^{b}	492.85_{C}^{d}	0.50_{A}^{b}	336.35_{B}^{d}	0.51_{A}^{a}	260.15_{A}^{c}	0.55_{B}^{a}
0.6 %	**2.5**	2528.50_{E}^{a}	0.35_{B}^{b}	1499.0_{D}^{a}	$0.35_{B}^{n.s}$	1044.60_{C}^{a}	0.33_{A}^{a}	492.95_{B}^{a}	0.33_{A}^{a}	165.40_{A}^{a}	0.44_{C}^{b}	47.50_{A}^{a}	0.59_{D}^{b}
	4	3794.0_{C}^{b}	0.27_{A}^{a}	2596.50_{B}^{b}	$0.34_{B}^{n.s}$	1148.50_{A}^{b}	0.52_{D}^{b}	2752.5_{B}^{b}	0.40_{C}^{b}	2392.0_{B}^{b}	0.38_{C}^{a}	1088.70_{A}^{b}	0.38_{C}^{a}
	7	4074.50_{D}^{b}	0.27_{A}^{a}	2988.0_{C}^{bc}	$0.33_{B}^{n.s}$	1411.50_{A}^{c}	0.50_{D}^{b}	3176.5_{C}^{c}	0.41_{C}^{b}	2426.0_{B}^{b}	0.38_{C}^{a}	1511.50_{A}^{c}	0.37_{C}^{a}
	10	4366.50_{D}^{b}	0.26_{A}^{a}	3249.0_{C}^{c}	$0.32_{B}^{n.s.}$	1815.0_{A}^{d}	0.49_{E}^{b}	3205.0_{C}^{c}	0.41_{D}^{b}	2512.50_{B}^{b}	0.38_{C}^{a}	1707.50_{A}^{c}	0.36_{C}^{a}

N.S.= Non Significant
In each rows and columns, means supplemented by different letters differed by Duncan's test at the 1% level.
** : Differed at different pH.*
***: Differed at different temperatures*
R^2: Average 0.98

coefficient (k) were significantly influenced by the change in temperature. These results also are in agreement with other studies by Vardhanabhuti and Ikeda (2006), Farhoosh and Riazi (2007) and Marcotte et al. (2001 a,b) who also reported that temperature had a large effect on the consistency coefficient of different gum and the consistency coefficient increased with the concentration of starch, pectin, carrageenan, and xanthan gum.[9-12] Gomez-Diaz and Navaza (2003) related the increase of k values to the increase of water binding capacity.[13] Wancho et al. (1996) reported that the coefficient k is a strong function of the concentration of the solution and the temperature, whereas index n does not have a strong dependency on the concentration and temperature of the polymeric solutions.[14] The relationship between apparent viscosity (η_a) and shear rate of different gum concentrations at pH 7 and 25 °C is shown in Fig 2. It is evident that the apparent viscosity of all samples decreased with increasing shear rate as well as apparent direct dependency of the gum concentration. After a sharp reduction the viscosity change became less pronounced at high shear rates. The values of viscosity at the weak shear rate could permit to appreciate the consistency of the product in the mouth [15], while the values of viscosity at the high shear rate allow understanding the viscosity of the product during certain processing operations such as pumping and spray drying.[16] The concentration of polysaccharide in solution is known to affect directly the viscosity and the degree of pseudoplasticity. [17] Increasing the psyllium concentration increased the viscosity as shown in Fig 2. This is due to the higher solid contents which generally causes an increase in the viscosity resulting from mainly molecular movements and interfacial film formation.[18]

3.3.2 Effect of pH on flow properties

The flow behavior index (n) and consistency coefficient (k) values, obtained by fitting the shear stress versus shear rate data to the Power model, are given in Table 2. Increasing the pH of the solution, from 2.5 to 10, augmented the pseudoplasticity and consistency coefficient. These occurrences have been explained by the electrostatic repulsion of functional groups that tends to keep the molecules in an extended form, thus producing a high viscous solution and thus the k values increases.[19-20] The consistency coefficient will be at a maximum when its molecular chains are in a state close to the rod conformation in the solution.[21] A number of gums such as CMC and karaya lose their viscosity at low pH.[22]

4. Conclusion

Plantago ovata L. seeds are covered by a polysaccharide rich layer known as psyllium, which is able to thicken aqueous solutions well even at low concentrations. To study structural aspects, the psyllium gum was characterized using FT-IR and the results showed that psyllium gum contains carboxylic groups on the backbone of polymer. The flow characteristics of psyllium (*plantago ovata* L.) solutions at different concentrations (0.1-0.6 %), temperatures (5-60 °C) and pH (2.5-10) analyzed in a wide range of shear rates indicated that psyllium solutions exhibited a typical non-Newtonian shear thinning behavior known for most polysaccharides and the Power law equation was able to model the rheological data. A dependence of viscosity on pH (from acidic to alkaline regions) was also observed. Two types of interactions play an important role in aqueous psyllium solution, namely the charge repulsion between carboxylic groups and the hydrogen bonding formed between carboxylic acid and ionized carboxyl groups. Decrease of pH causes, the number of dissociated carboxylic groups in psyllium chains to decrease, and the psyllium molecules lose their hydrophilicity. When some dissociated carboxylic groups in psyllium chains are gradually protonated, the hydrophobic segments appear in psyllium chains. The reduction in viscosity at low pH, could also be attributed to biopolymer degradation due to the cleavage of glycosidic bonds within the polysaccharide structure. Overall the results indicated that psyllium gum has a good potential to be used in foods and pharmaceuticals as a new source of thickening agent.

References:

1. M. Blumenthal, A. Goldberg, and J. Brinkmann, American Botanical Council, Integrative Medicine, Austin, TX. 2000.
2. J.F. Kennedy, J.S. Sandhu, and D.A.T. Southgate, Res., 1979, **75**, 265.
3. J.S. Sandhu, G.J. Hudson, and J.F. Kennedy, Carbohydrate Research, 1981, **93**, 247.
4. R.A. Laidlaw, and E.G.V. Purcival, Journal Chemical Society, 1950, 528.
5. P.A. Williams, G.O. Phillips, in Handbook of Hydrocolloids eds., G.O. Phillips, P.A. Williams, CRC Press, New York, 2000, 1–19.
6. M.P. Yadav, D.B. Johnston, A.T. Hotchkiss, and K.B. Hicks, Food Hydrocolloids, 2007, **21**, 1022.
7. A.O.A.C. Association of Official Analytical Chemists, Washington, DC, 1984.
8. M.H. Fischer, N.X. Yu, G.R. Gray, J. Ralph, L. Anderson, and J.A. Marlett, Carbohydrate Research, 2004, **339**, 2009.
9. M. Marcotte, A.R. Taherian, H.S. Ramaswamy, Food Research International, 2001a, **34**, 695.
10. M. Marcotte, A.R. Taherian, H.S. Ramaswamy, Journal of Food Engineering, 2001b, **48**, 157.
11. B. Vardhanabhuti, and S. Ikeda, Food Hydrocolloids, 2006, **20**, 885.
12. R. Farhoosh, and A. Riazi, Food Hydrocolloids, 2007, **21**, 660.
13. D. Gomez-Diaz, J.M. Navaza, Journal of Food Engineering, 2003, **56**, 387.
14. R.K. Wanchoo, S.K. Sharma, and R. Bansal, Journal of Polymer Materials, 1996, **13**, 49.
15. E.R. Morris, and L.J. Taylor, Progress in Food and Nutrition Science, 1982, **6**, 285.
16. S.W. Race, Improved Food Technology, 1991, **45**, 86.
17. I.W. Sutherland, Biotechnology Advances, 1994, **12**, 393.
18. M. Maskan, and F. Gogus, Journal of Food Engineering, 2000, **43**, 173.
19. B. Launay, I. Doubiler, and G. Cavalier, in Functional Properties of Food Macromolecules, Mitchell, J.R. ed., Elsevier Applied Science Publishers, New York, 1986, 1–78.
20. J.C. Onweluzo, Z.A. Obanu, and K.C. Onuoha, Nigerian Food Journal, 1994, **12**, 1.
21. T. Feng, Z.B. Gu, and Z.Y. Jin, Food Science and Technology International, 2007, **13**, 55.
22. M. Glicksman, Food Hydrocolloids, Vols. 2 and 3. CRC Press Inc, FL, 1982.

EFFECT OF NaCl, $CaCl_2$ AND Na_2HPO_4 ON RHEOLOGICAL PROPERTIES OF SEED HUSK OF *Plantago ovata* L.

A. Farahnaky [1], H. Askari [1], M. Majzoobi [1], G.R. Mesbahi [1] and M. Aminlari [1,2]

[1] Department of Food Science and Technology, School of Agriculture, Shiraz University, Shiraz, Iran.
[2] Department of Biochemistry, School of Veterinary Medicine, Shiraz University, Shiraz, Iran.

1. INTRODUCTION

Plantago ovata Forsk belongs to the family *Plantaginaceae*. Out of about 200 species of the genus only two, namely *P. ovata* and *P. psyllium*, have been extensively used for the production of seed husk, a thin white membrane covering the concave side of the seed. Chemical and physical studies of the active fraction of psyllium mucilage show that it has arabinose 22.6%, xylose 74.6%, molar basis; only traces of other sugars. With about 35% of non-reducing terminal residues, the polysaccharide is highly branched. The data are compatible with a structure consisting of a densely substituted main chain of β-(1→4)-linked d-xylopyranosyl residues, some carrying single xylopyranosyl side chains at position 2, others bearing, at position 3, trisaccharide branches having the sequence L- Araf-α-(1→3)-d-Xylp-β-(1→3)-l-Araf.[1] The seed collected after threshing the harvested plants is transported to factories/mills for processing. The seeds are passed through sieving pans to separate out dust particles, and shriveled and deformed seeds. The ground seed material is again sieved and winnowed to separate out the husk from coarse remains of the kernel. The best grade Isabgol is 70 mesh. Husk free full or broken seeds are used as cattle feed.[2] Psyllium has also been incorporated into breakfast cereals, ice cream, instant beverages, bakery and other dietary products.[3-4] Furthermore, the rheological behavior of psyllium is of special importance when they are used to modify the textural attributes of foods. It has been recognized that the rheological properties of hydrocolloids in solution depend on many factors: concentration of the active compound, temperature, degree of dispersion, dissolution, electrical charge, previous thermal and mechanical treatments, presence or absence of other lyophilic colloids, age of the lyophilic solution and the presence of electrolytes and non-electrolytes.[5-7] Because of its pharmacological effects, foods fortified with *P. ovata* mucilage gum may have a superior consumer acceptance. Although psyllium gum has been used for centuries in Iran no extensive scientific work has been reported on its functional and rheological properties. Thus, the objective of this study was to evaluate the flow behavior of psyllium gum solutions in the presence of different salts at different pH's.

2. MATERIALS AND METHODS

Gum extraction and purification, Fourier transform infrared spectroscopy (FT-IR) and scanning electron microscopy (SEM) were performed as explained in the manuscript entitled "Rheological properties of *Planago ovata* L. as a function of concentration, temperature and pH" published in this issue.

2.1 Rheological measurements

Rheological measurements of 0.4 % (w/w) aqueous solutions of the mucilage isolated from Plantago ovata seeds were carried out using a rotational viscometer (Brookfild, DV-II + Pro, Brookfield Engineering, USA) equipped with CP51 measuring geometry as a function of salt concentration and pH at 25 °C. Effect of different salts such as NaCl (0.5, 1, 2 and 3 % w/w), $CaCl_2$ (0.5, 1, 2 and 3 % w/w) and Na_2HPO_4 (0.2, 0.4 and 0.6 % w/w) at different pHs (2.5, 4, 7 and 10) were investigated. For each test, 0.5 ml sample was transferred to sample compartment, then subjected to a programmed shear rate increasing linearly from 1.92-229.4 s^{-1} in 80 steps. The flow behavior index (n) and consistency index (k) values were computed by fitting the Power law model (Eq.1):

$$\tau = k\gamma^{\circ n} \qquad (1)$$

Where τ is the shear stress (mPa), γ° is the shear rate (s^{-1}), k is the consistency coefficient (mPa.s) and n is the flow behavior index (dimensionless).

2.2 Statistical analysis

The SPSS (version 13) program was used for all statistical analysis. Software of Design-Expert 6.0.2 (Stat-Ease Inc, Minneapolis, USA): D-Optimal Response Surface mode was used in order to model the experimental data and estimate any nonlinearity in the relationships between the parameters under study.

3. RESULTS AND DISCUSSION

3.1 Analysis of chemical composition

The proximate analysis of the whole seed, seed coats (husk), and the purified mucilage from *P. ovata* seeds is shown in Table 1. The moisture content of whole seed, seed coats (husk), and the purified mucilage was similar, 6.05, 6.24 and 6.13%, respectively. The protein content of the whole seed was greater than seed coats (husk), and the mucilage and the majority (i.e. about 90%) of whole seed, seed coats (husk), and the mucilage was carbohydrates.[8] The production yield of purified mucilage from the seeds was 28.5% (w/w).

Table 1 *The chemical compositions of the whole seed, seed coats (husk), and the purified mucilage from P. ovata seed.*

	Compositional distribution of total material (%)		
Composition	**Whole seed**	**Husk**	**Mucilage**
Moisture	6.05 ± 0.064	6.24 ± 0.016	6.13 ± 0.016
Protein	2.93 ± 0.196	0.25 ± 0.015	0.11 ± 0.004
Fat	0.30 ± 0.055	0.50 ± 0.053	0.36 ± 0.033
Ash	2.98 ± 0.234	3.59 ± 0.165	1.51 ± 0.056
Fiber	24.32 ± 0.210	2.63 ± 0.065	3.19 ± 0.130
Total carbohydrate*	87.74 ± 0.549	89.25 ± 0.143	91.81 ± 0.062

* *Calculated from total*

3.2 Rheological measurement

3.2.1 Effects of pH on flow properties

The consistency coefficient (k) and flow behavior index (n) values obtained by fitting the shear stress versus shear rate data to the Power law model as influence by pH is shown in Table 2. Examination of flow properties as a function of shear rate also designated the non-Newtonian behavior for different pH's. The values of flow behavior indices, n, were less than 1, indicating the shear thinning nature of the gum at different measuring conditions. Increasing the pH increased the consistency coefficient. The consistency coefficient, k for all of the samples increases with increasing pH of the mucilage solutions.

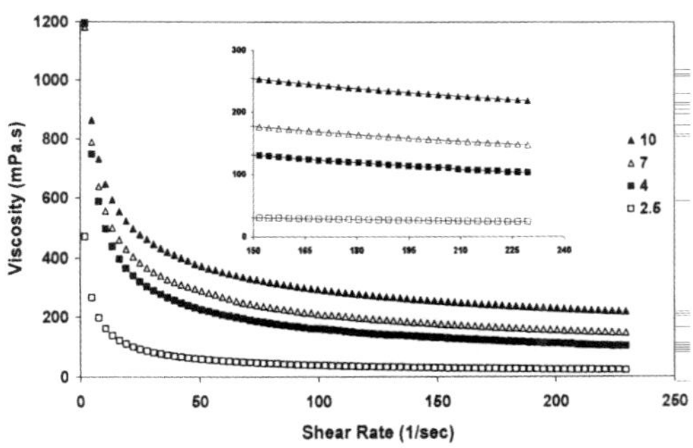

Figure 1 *Effect of different pH on apparent viscosity of psyllium solution at concentration of 0.4 % and 25 °C.*

Analysis of variance (ANOVA) (α = 0.01) showed that flow behavior index (n) and consistency coefficient (k) were significantly influenced by the change in pH. The relationship between apparent viscosity (η_a) and shear rate for different pH's at 0.4 % (w/w) gum concentration at 25 °C is shown in Fig. 1. It is evident that the apparent viscosity of all samples decreased with increasing shear rate. Direct dependency of viscosity on the solution pH was also apparent, as shear rate increased, after a sharp reduction the viscosity change was less at high shear rates. Shear thinning is the result of an orientation effect. As shear rate increases, the long chains of polymer molecules and randomly positioned chains become increasingly aligned in the direction of flow resulting in less interaction between adjacent polymer chains. The values of viscosity at low shear rates is representative of the consistency of the product in the mouth [9], while the values of viscosity at the high shear rates allow understanding the viscosity of the product during certain processing operations such as pumping, spraying and drying. Accordingly, as the viscosity of the solution decreases with shear rate, pumping efficiency increases as pump flow rate increases.[10] Increasing the pH of solutions, from 2.5 to 10, augmented the shear thinning and consistency coefficient. These occurrences have been explained as induction of the electrostatic repulsions by functional groups that tend to keep the molecules in an extended form, thus producing a highly viscous solution and thus the k values increases.[11-12] The consistency coefficient will be at a maximum when its molecular chains are in a state close to the rod conformation in the solution.[13] As shown in Fig. 1 increasing the pH up to 10 resulted in an increase in the viscosity. The viscosity of the psyllium seed gum also decreased gradually at more acidic conditions and may be explained by the lower ionization of the mucilage carboxyl groups.

3.2.2 Effects of salts on flow properties

The effects of salts on the viscosity of various gums have been established. Xanthan gum, galactomannans, and hydroxypropyl methyl cellulose maintain their viscosity with the addition of salts.[14] In this study, the effect of NaCl, Na_2HPO_4 and $CaCl_2$ on the flow behavior of psyllium (0.4 %) was investigated.

NaCl is the most popular salt used in foods as a flavoring and preserving agent. Its primary effects on polyelectrolytes are considered to shield electrostatic repulsions among charged groups on the polymer chains. This can lead to a reduction in viscosity due to less expansion of the molecules or a decrease in viscosity due to intermolecular associations. As shown in Fig. (2 a,b), sodium chloride greatly affected the flow behavior of psyllium. Addition of NaCl from 0.5 to 3 % (w/w) decreased the consistency coefficient (k) and increased the flow index (n) of hydrocolloid solutions (Table 3). Also, adding NaCl at different concentrations decreased the viscosity of the psyllium at different pH. The results revealed the negative effect of NaCl on psyllium gum, therefore psyllium seems to be suitable to thicken foods containing low concentrations of NaCl.

The effect of different concentrations of $CaCl_2$ on the flow properties of psyllium gum solution is given in Table 2 and Fig. 2. The flow behavior indices changed progressively when the salt concentration increased from 0.5 to 3 % (w/w) for $CaCl_2$. The k value increased and flow behavior index (n) of hydrocolloid solutions decreased with addition of $CaCl_2$ (Fig. 2 c, d). A rapid decrease in apparent viscosity was observed with addition of 0.5 % salt. Increase of viscosity as affected by $CaCl_2$ addition, may result from interactions of cations with gum macromolecules producing junctions zones. The junction zones would not easily be deformed under small strains due to their strength and rigidity as mentioned earlier, and only the network chains were stretched at this initial stage. When the deformation was large, the junction zones were also deformed, contributing to stress increase. Then, the stress resulted not only from the viscosity, but also from the deformation of junction zones.

Sodium phosphates (SPs) are the most popular form of salts used in the meat industry.[15-17] SPs exhibit a more pronounced poly-ionic character compared to other salts such of naturally occurring orthophosphates in meats, which have virtually no effect on water binding.

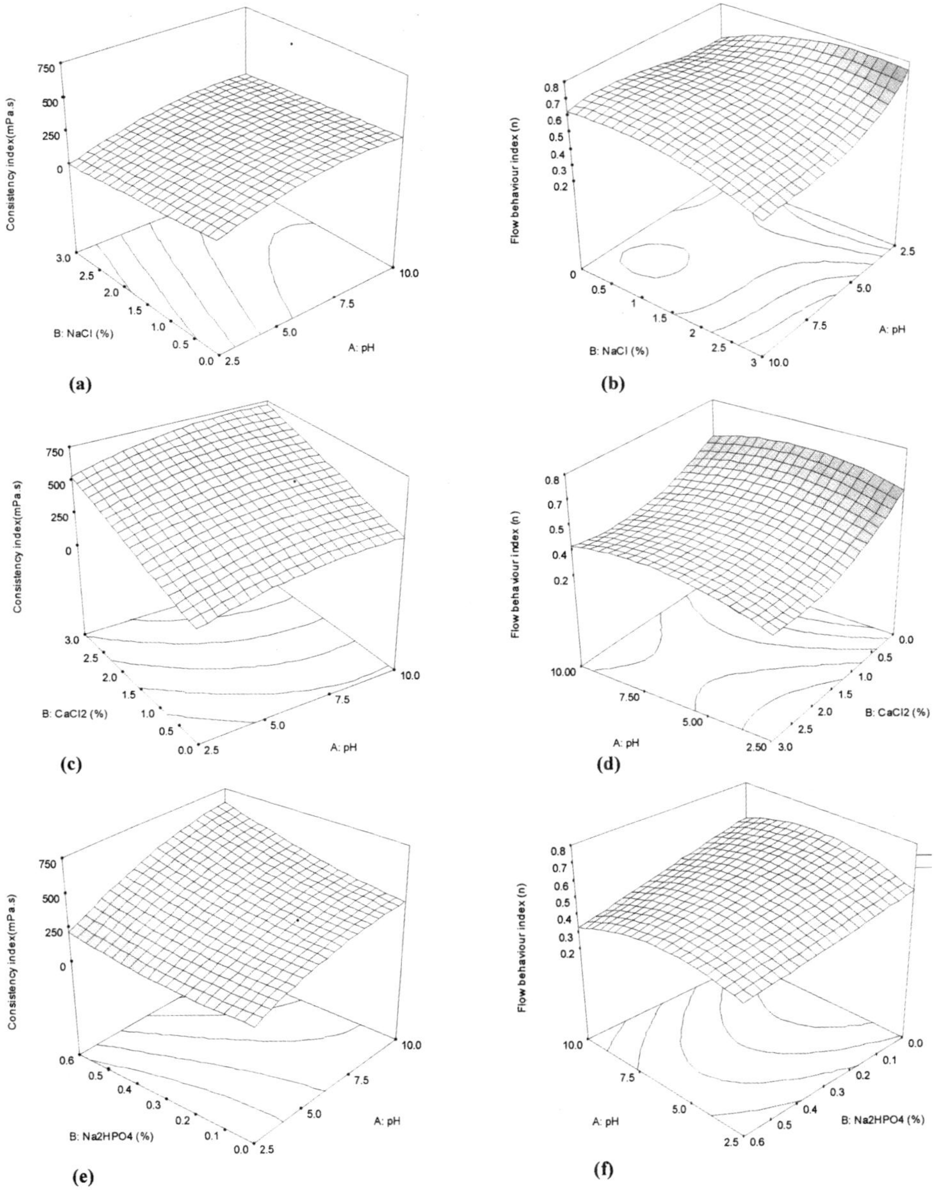

Figure 2 *Response Surface Methodology (given by Design Expert software): effect of NaCl (a,b), $CaCl_2$ (c,d) and Na_2HPO_4 (e,f) on consistency index and flow behavior index of 0.4 % (w/w) psyllium gum solutions at different pH and 25 °C.*

Table 2 *The Power law parameters for psyllium gum solutions at different concentration of salts and pH.*

Salt	pH	2.5			4			7			10		
	Concentration	**k**	**n**	R^2	**k**	**n**	R^2	**k**	**n**	R^2	**k**	**n**	R^2
NaCl	**Control**	32.70_{A}^{c}	0.505_{A}^{a}	0.98	228.50_{B}^{c}	0.645_{B}^{a}	0.98	322.00_{C}^{e}	0.625_{B}^{b}	0.98	340.70_{D}^{b}	0.630_{B}^{b}	0.98
	0.5 %	32.70_{A}^{b}	0.680_{B}^{b}	0.96	259.40_{B}^{d}	0.635_{A}^{a}	0.98	252.95_{B}^{d}	0.645_{A}^{b}	0.96	261.50_{B}^{a}	0.635_{A}^{b}	0.97
	1.0 %	20.85_{A}^{a}	0.745_{B}^{c}	0.97	217.00_{B}^{c}	0.650_{A}^{a}	0.99	238.75_{C}^{c}	0.625_{A}^{b}	0.97	251.60_{C}^{a}	0.630_{A}^{b}	0.96
	2.0 %	19.20_{A}^{a}	0.755_{C}^{cd}	0.98	181.15_{B}^{b}	0.620_{B}^{a}	0.96	188.45_{C}^{b}	0.620_{B}^{b}	0.98	270.50_{D}^{a}	0.590_{A}^{b}	0.98
	3.0 %	17.90_{A}^{a}	0.765_{C}^{d}	0.98	67.25_{B}^{a}	0.730_{B}^{b}	0.98	139.30_{C}^{a}	0.510_{AB}^{a}	0.96	265.90_{D}^{a}	0.420_{A}^{a}	0.96
$CaCl_2$	**Control**	99.45_{A}^{a}	0.505_{A}^{e}	0.98	228.50_{B}^{a}	0.645_{B}^{d}	0.98	322.00_{C}^{b}	0.625_{B}^{d}	0.98	340.70_{D}^{a}	0.630_{B}^{c}	0.98
	0.5 %	110.55_{A}^{a}	0.435_{A}^{d}	0.97	234.15_{B}^{a}	0.580_{B}^{c}	0.96	256.70_{C}^{a}	0.565_{B}^{c}	0.96	329.00_{D}^{a}	0.435_{A}^{b}	0.98
	1.0 %	361.10_{A}^{b}	0.330_{A}^{c}	0.97	359.15_{A}^{b}	0.460_{C}^{b}	0.98	444.35_{B}^{c}	0.420_{BC}^{b}	0.96	497.25_{C}^{b}	0.400_{B}^{a}	0.96
	2.0 %	383.30_{A}^{b}	0.270_{A}^{b}	0.97	569.25_{C}^{c}	0.390_{B}^{a}	0.96	541.65_{BC}^{d}	0.380_{B}^{ab}	0.98	522.05_{B}^{b}	0.395_{B}^{a}	0.96
	3.0 %	435.00_{A}^{c}	0.230_{A}^{a}	0.96	687.00_{B}^{d}	0.380_{B}^{a}	0.96	652.60_{B}^{e}	0.365_{B}^{a}	0.96	764.20_{C}^{c}	0.385_{B}^{a}	0.98
Na_2HPO_4	**Control**	99.45_{A}^{b}	0.505_{A}^{c}	0.98	228.50_{B}^{a}	0.645_{B}^{b}	0.98	322.00_{C}^{a}	0.625_{B}^{b}	0.98	340.70_{D}^{a}	0.630_{B}^{c}	0.98
	0.2 %	82.40_{A}^{a}	0.505_{B}^{c}	0.98	259.65_{B}^{b}	0.605_{C}^{b}	0.96	302.05_{C}^{a}	0.590_{C}^{b}	0.97	411.70_{D}^{b}	0.455_{A}^{b}	0.97
	0.4 %	130.45_{A}^{c}	0.390_{A}^{b}	0.99	314.45_{B}^{c}	0.530_{C}^{a}	0.96	388.15_{C}^{b}	0.445_{B}^{a}	0.97	557.45_{D}^{c}	0.390_{A}^{a}	0.96
	0.6 %	202.35_{A}^{d}	0.345_{A}^{a}	0.96	369.25_{B}^{d}	0.510_{C}^{a}	0.98	532.65_{C}^{c}	0.425_{B}^{a}	0.96	633.50_{D}^{d}	0.365_{A}^{a}	0.96

N.S. = Non significant
In each column, means supplemented by different small letters differed by Duncan test at the 1% level as a function of salt concentration.
In each row, means supplemented by different capital letters differed by Duncan test at the 1% level as a function of pH.
[†]*R^2 = Regression coefficient*

Increases in poly-ionic properties enable phosphates to attach to positive sites on protein molecules leading to improved protein solubility and an enhanced water binding.

3.3 Effect of concentration and pH on microstructure of psyllium gum

Although rheology provided an important analytical tool for understanding the gelling properties of psyllium polysaccharides, it did not provide the complete story. The microstructure of psyllium gel was also analyzed using SEM for a better understanding of the properties of the psyllium gum gel. Gels are a challenge to prepare for electron microscopy by conventional methods because they have high water content. Artifacts can be easily introduced by conventional sample preparation. Macro-porous hydrogels can be examined using scanning electron microscopy (SEM) after freeze drying, when the material has an adequate modulus to avoid the structural collapse during dehydration. In the current study, freeze-dried samples were used to remove the artifacts. The prepared solutions were rapidly frozen at -70 °C and then dried for 48 hr in a freeze drier. Fig. 3 shows SEM of a range of concentrations (2, 2.5 and 3 % w/w) of psyllium gels. The SEM of 2 % shows some large flat sheets with a limited collapse of gel structure, confirming that only weak gels formed at this concentration. A lath sheet-like structure was formed within the 2.5 % sample and sheets were more numerous, larger with inclusions of macro pores.

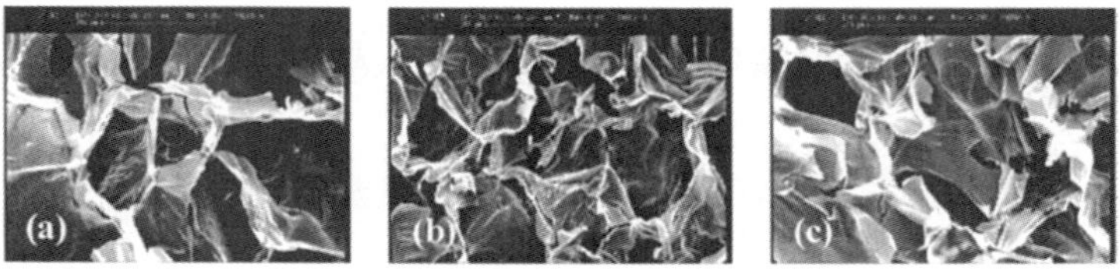

Figure 3 *Scanning electron micrographs of psyllium gels. Concentration of 2 % (w/w) (a), 2.5 % (b), and 3% (c) at distilled water (magnification 200).*

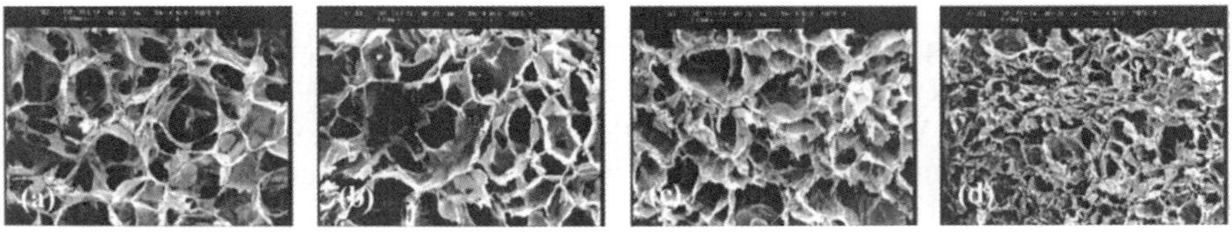

Figure 4 *Scanning electron micrographs of psyllium gels. Concentration of 2.5% at (a) pH 2.5, (b) pH 4, (c) pH 7, (d) pH 10 (magnification 40).*

There was almost no interconnection between laths and it is possible that the large sheets were torn. However, a significantly stronger gel was formed which did not collapse upon freeze drying when the concentration of psyllium was further increased to 3 % (w/w). The SEM images showed a sheet-like or membrane structure with significant interaction forming a 3-dimensional network. From Fig 3-c, it can be seen that the gel network was getting denser with increase of concentration of psyllium gum in solution. Furthermore, the microstructure of psyllium gel investigated by SEM at different pH (2.5, 4, 7 and 10) is presented in Fig. 4. The microstructure of psyllium gels at high pH showed that the polymer structure was porous, the pores were small in size and large in number. This phenomenon can be explained by the protonization of carboxyl groups under alkaline conditions.

4. CONCLUSION

The impact of different salts (NaCl, $CaCl_2$ and Na_2HPO_4) on rheological changes of psyllium solutions were different. NaCl had a negative effect on viscosity while the other two salts had a positive effect. The effect of ion concentration on viscosity is important not only for determination of polyelectrolyte behavior of macromolecules but also to estimate their functional and rheological properties in the presence of ions.
The IR spectra of the psyllium evidenced the presence of carboxyl groups which may serve as binding sites for divalent cations. The carboxyl groups may also work as functional moieties to generate new and/or modified polysaccharide variants using different approaches including polymer engineering or novel formulation designing by linking this polysaccharide with starch and/or other synthesized polymers. Overall, salts either naturally present or added to food formulations can have profound effects on functional properties of psyllium solutions, even at low concentrations, and each of which may behave differently.

References:

1. M. H. Fischer, N. Yu, G. R. Gray, J. Ralph, L. Anderson and J. A. Marlett. Carbohydrate Research. 2004. **339**, 2009.
2. A.K. Koul, and S. Sareen, in Supplement to Cultivation and Utilization of Medicinal Plants. eds., S.S. Handa, and M.K. Kaul, Jammu, India: Regional Research Laboratory (CSIR), 1996, 477–495.
3. J.K.C. Chan, and V. Wypyszyk, Cereal Foods World. 1988, **33**, 919.
4. M.K. Dhar, S. Kaul, S. Sareen, and A.K. Koul, Plant Genetic Resource, 2005, **3**, 252.
5. M.A. Rao, Food Technology, 1992, **46**, 116.
6. M.A. Rao, and J.F. Kenny, Canadian Institute Food Science Technology Journal, 1975, **8**, 142.
7. M. Rao, M.A. Rao and S.S.H. Rizvi, Engineering Properties of Foods, Marcel Dekker, New York, 1986, 1-47.
8. A.O.A.C. Official Methods of Analysis, Association of Official Analytical Chemists, Washington, DC, 1984.
9. E.R. Morris, and L.J. Taylor, Progress in Food and Nutrition Science, 1982, **6**, 285.
10. S.W. Race, Food Technology, 1991, **45**, 86.
11. B. Launay, I. Doubiler and G. Cavalier, in Mitchell, J.R. ed. Functional Properties of Food Macromolecules, Elsevier Applied Science Publishers, New York, 1986, 1–78.
12. J.C. Onweluzo, Z.A. Obanu, and K.C. Onuoha, Nigerian Food Journal, 1994, **12**, 1.

13. T. Feng, Z.B. Gu, and Z.Y. Jin, Food Science and Technology International, 2007, **13**, 55.
14. P.A. Williams and G.O. Phillips, in, G.O. Phillips and P.A. Williams, eds., Handbook of Hydrocolloids, CRC Press, New York, 2000, 1–19.
15. J.D. Dziezak, Food Technology, 1990, 44, 80.
16. L.E. Lampila, Journal of Aquatic Food Product Technology, 1992, **1**, 29.
17. V. Tenhet, G. Finne, R. Nickelson II, and D. Toloday, Journal of Food Science, 1981, **46,** 344.

RHEOLOGY OF BALANGU SHIRAZI (*Lallemantia royleana*) SEED GUM: A HIGH VISCOSITY THICKENING AGENT

A. Farahnaky, H. Askari, M. Bakhtiyari

Department of Food Science and Technology, School of Agriculture, Shiraz University, Shiraz, Iran.

1. INTRODUCTION

Gums from seeds are important food additives in the food industry.[1] They are widely used as thickening, gelling, stabilizing and suspending agents in foods.[2] Some plants, growing in different regions of Iran, have valuable polysaccharides, and their seed extracts can be used as novel food hydrocolloids. For example, the seed of plants known as Madwort seed (*Alyssum homolocarpum* Fitch), Basil seed (*Ocimum basilicum*), Balangu seed (*Lallemantia royleana (walla)* benth), cress seed (*Lepidium sativum*), are commonly used as pharmaceutical sources and have valuable medical effects and their seeds have large amount of mucilage.[3-5] *Lallemantia royleana* grows extensively in different regions of European and Middle East countries especially in Turkey, Iran, and India. This seed is a good source of fiber, oil and protein and has some medicinal, nutritional and human health properties.[4] A factor in its physiological fiber behavior is its high viscosity and gel-like character in water. This property, in turn, is related to the functions associated with its high-molecular-weight polysaccharides.[6] Balangu seeds absorb water quickly when soaked in water and produce a sticky, turbid and tasteless liquid.[7] Balangu seed gum has been used for centuries in Iran for different purposes but no scientific work has been published on its functional and rheological properties. Thus, the objectives of this study were to evaluate the rheological behavior of gum extracted from *Lallemantia royleana* and to study the effects of concentration, temperature and pH.

2. MATERIALS AND METHODS

The Balangu Shirazi (*Lallemantia royleana*) seed was obtained from a local medical plant market, Tehran, Iran. The seeds were manually cleaned to remove all foreign matter such as dust, dirt, stones, chaff and immature seeds. All chemicals used in this study were of analytical grade and purchased from Merck (Darmstadt, Germany) company.

2.1 Gum extraction and purification

The whole seeds of Balangu Shirazi were mixed with 20 times their weight of water and allowed to stand and hydrate for 4 h at 25 °C. The suspension was then mixed for 1 min at low speed, and for further 3 min at high speed using a blender. The mixture of seed and mucilage was centrifuged at 10000 g for 20 min at 4 °C. The residual of seeds were discarded, and the rest of the supernatant was subjected to ethanol precipitation (97 % ethanol/mixture ratio of 3:1). The gum was then dried in a conventional oven at 40 °C and milled to mesh size 53-125 micron.

2.2 Analysis of chemical composition

Protein, fat, ash and moisture content of the whole seeds were determined according to the approved methods.[8]

2.3 Fourier transform infrared spectroscopy (FT-IR)

The FT-IR spectrum of purified Balangu seed gum was obtained at a resolution of 1 cm^{-1}. the sample was incorporated into KBr and pressed into a 2 mm pellet. Spectra were recorded in the transmittance mode from 4000 to 400 cm^{-1}, using a BRUCKER spectrophotometer (EQUINOX 55, Germany).

2.4 Rheological measurements

Rheological measurements were carried out using a rotational viscometer (Brookfield Engineerings, DV-II + Pro, USA) equipped with CP51 measuring geometry as a function of gum concentration, temperature and pH. For each test, 0.5 ml sample was transferred to sample compartment (cone and plate geometry) and subjected to a programmed shear rate increasing linearly from 1.92-229.4 s^{-1}. The flow behavior index (n) and consistency index (k) values were computed by fitting the Power law model (Equation 1) using the software (Rheocal, Brookfield Engineerings, USA) supplied by the manufacturer:

$$\tau = k\gamma^{\circ n} \quad (1)$$

Where τ is the shear stress (mPa), γ° is the shear rate (s^{-1}), k is the consistency coefficient (mPa.s) and n is the flow behavior index.

The viscosity and flow properties were measured for all Balangu seed gum solutions at different pH's ranging from 2.5 to 10.

2.4.1 Effect of concentration on flow properties

There is usually a direct nonlinear relationship between concentration of a solute and viscosity at a constant temperature.[9] In most foods, it is often possible to identify the components that play an important role in the rheological properties. The effect of concentration, C, (soluble or insoluble solid) on the consistency index of the Power law model, is described by either exponential (Equation 2) or power functions (Equation 3) [10] as,

$$k = k_{c_1}.e^{b_1 C} \tag{2}$$

$$k = k_{c_2}.C^{b_2} \tag{3}$$

Where, K_{C1}, K_{C2}, b_1 and b_2 are experimental constants.[10]

2.4.2 Evaluation of temperature dependency of gum solutions

Gum solutions were prepared at different concentrations by dispersing the required amount of gum in distilled water, under slow stirring at room temperature, stored for 16 h at 4 °C for complete hydration prior to assessment, maintained for 3 min at temperature of 85 °C and then rapidly cooled. Prepared· samples were loaded into the viscometer at measurement temperatures of 5, 25, 40 °C. The temperature dependency of the consistency coefficient (indicator of the viscous nature of the sample) was assessed by fitting the Arrhenius model (Equation 4) as suggested by Sengul et al. (2005):

$$k = k_t.e^{(\frac{E_a}{R.T})} \tag{4}$$

Where k_t is the proportionality constant (or consistency coefficient at a reference temperature, Pa.s), E_a the activation energy (J/mol), R the universal gas law constant (J/mol.K), and T the absolute temperature (°K).[11]

2.4.3 Compare of viscosity with other gums

Tragacanth gum, carboxy methyl cellulose (CMC), Na-alginate and Arabic gum were used for comparison of their viscosity at concentration 0.4 % (w/w) with Balangu seed gum.

2.5 Statistical analysis

Analysis of variance (ANOVA) test was applied to acknowledge any significant difference among rheological parameters; n and k under the effect of temperature, pH and gum concentration ($P < 0.01$). The SPSS (version 13) program was used for all statistical analysis and Design-Expert 6.0.2 (Stat-Ease Inc, Minneapolis, USA): D-Optimal mode was used to analyze the resultant data.

3. RESULTS AND DISCUSION

3.1 Analysis of chemical composition

The chemical composition of the whole seed from *Lallemantia royleana* seed is shown in Table 1. The extraction yield of mucilage was 19.65% (w/w) of the seed.

Table1 *Chemical composition of whole Balangu shirazi seed (Lallemantia royleana)*

Composition	Distribution (%)
Moisture	6.05 ± 0.06
Protein	2.93 ± 0.20
Fat	0.30 ± 0.06
Ash	2.98 ± 0.23
Fiber	24.24 ± 0.11
Total carbohydrate*	87.74 ± 0.55

* *Calculated from total (Moisture, Protein, Fat and Ash)*

3.2 FT-IR spectroscopy analysis

Figure 1 shows the FT-IR spectrum of purified Balangu seed gum. The spectrum is dominated by a broad band at about 3395 cm^{-1}, which is assigned to the stretching vibration modes of the -OH groups. The peak at 1613 cm^{-1} is due to the asymmetrical $-COO^-$ stretching vibration, whereas the band at 1419 cm^{-1} is due to the symmetrical $-COO^-$ stretching vibration.

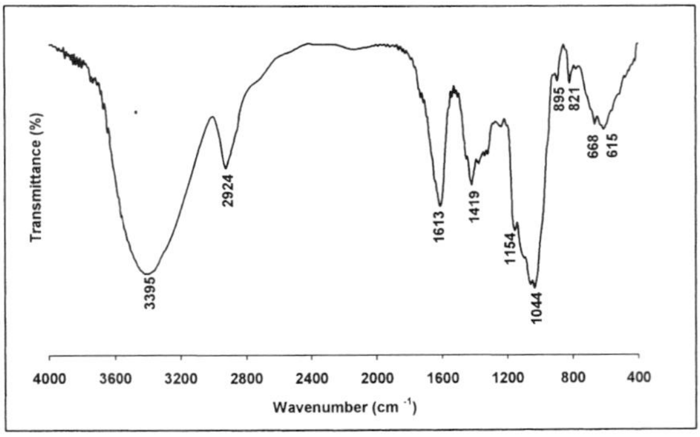

Figure 1 *FT-IR spectrum of Balangu seed gum powder*

3.3 Rheological measurements

3.3.1 Effect of concentration and temperature on flow properties

The consistency coefficient (k) and flow behavior index (n) values obtained by fitting the shear stress-shear rate data to the Power law model as influenced by gum concentration and temperature are shown in Table 2. Examination of flow properties as a function of shear rate also designated the non-Newtonian behavior for different concentrations and temperatures. The values of flow behavior indices, n, were less than 1, indicating the shear thinning nature of the gum at different measuring conditions. Increasing the gum concentration decreased the flow behavior index values and increased the consistency coefficient. The flow behavior index, n, of gum solutions ranges from 0.62 (for 0.1 % w/w) to 0.31 (for 0.6 % w/w) at 25 °C and pH 7, indicating that mucilage solutions with higher concentrations are more shear

Table 2 *Compare of means n and k obtained from Balangu gum solution as a function of concentration, temperature and pH .*

Concentration (w/w)	pH	Temperature 5 °C k	5 °C n	5 °C R^2	25 °C k	25 °C n	25 °C R^2	40 °C k	40 °C n	40 °C R^2
0.1 %	2.5	37.20_{B}^{a}	0.54_{A}^{a}	0.97	18.15_{A}^{a}	0.55_{AB}^{a}	0.97	12.70_{A}^{a}	0.58_{B}^{a}	0.97
	4	81.60_{C}^{b}	0.55_{A}^{ab}	0.99	42.95_{B}^{b}	0.55_{A}^{a}	0.97	13.20_{A}^{a}	0.73_{B}^{c}	0.97
	7	138.05_{B}^{d}	0.57_{A}^{b}	0.98	62.85_{A}^{c}	0.62_{B}^{b}	0.98	53.10_{A}^{b}	0.63_{B}^{b}	0.98
	10	114.05_{C}^{c}	0.62_{A}^{c}	0.98	71.00_{B}^{d}	0.65_{B}^{c}	0.98	58.10_{A}^{c}	0.62_{A}^{b}	0.99
0.2 %	2.5	38.95_{C}^{a}	0.54_{A}^{c}	0.97	27.75_{B}^{a}	0.56_{A}^{b}	0.97	18.15_{A}^{a}	0.62_{B}^{c}	0.97
	4	99.35_{B}^{b}	0.61_{A}^{b}	0.97	35.55_{A}^{a}	0.73_{C}^{c}	0.98	41.50_{A}^{a}	0.67_{B}^{d}	0.98
	7	399.50_{B}^{c}	0.52_{A}^{b}	0.99	289.30_{A}^{b}	0.54_{B}^{ab}	0.99	225.10_{A}^{b}	0.54_{B}^{b}	0.98
	10	653.15_{C}^{d}	0.43_{A}^{a}	0.98	306.60_{B}^{b}	0.50_{B}^{a}	0.98	239.45_{A}^{b}	0.49_{B}^{a}	0.97
0.4 %	2.5	74.45_{C}^{a}	$0.60_{N.S.}^{c}$	0.97	38.05_{B}^{a}	$0.60_{N.S.}^{c}$	0.97	24.05_{A}^{a}	$0.61_{N.S.}^{c}$	0.97
	4	205.05_{C}^{b}	0.68_{B}^{d}	0.97	142.00_{A}^{b}	0.67_{B}^{d}	0.99	168.65_{B}^{b}	0.61_{A}^{c}	0.98
	7	1477.50_{C}^{c}	$0.40_{N.S.}^{b}$	0.98	1269.00_{B}^{c}	$0.39_{N.S.}^{b}$	0.98	1091.50_{A}^{c}	$0.40_{N.S.}^{b}$	0.97
	10	2731.50_{B}^{d}	0.32_{A}^{a}	0.98	1477.00_{A}^{d}	0.36_{C}^{a}	0.98	1412.50_{A}^{d}	0.34_{B}^{a}	0.98
0.6 %	2.5	86.60_{C}^{a}	0.69_{B}^{d}	0.99	58.35_{B}^{a}	0.64_{A}^{d}	0.97	25.50_{A}^{a}	0.72_{C}^{d}	0.98
	4	894.60_{A}^{b}	0.52_{C}^{c}	1.00	879.10_{A}^{b}	0.47_{B}^{c}	0.98	1021.80_{B}^{b}	0.42_{A}^{c}	0.97
	7	5435.00_{C}^{c}	0.25_{A}^{b}	0.99	3397.00_{B}^{c}	0.31_{B}^{b}	0.98	2914.00_{A}^{c}	0.31_{B}^{b}	0.98
	10	7029.00_{B}^{d}	0.21_{A}^{a}	0.99	3770.00_{A}^{d}	0.29_{B}^{a}	0.98	3515.50_{A}^{d}	0.28_{B}^{a}	0.98

In each column and row, means supplemented by different small letters differed by Duncan test at the 1% level as a function of pH and temperature, respectively.

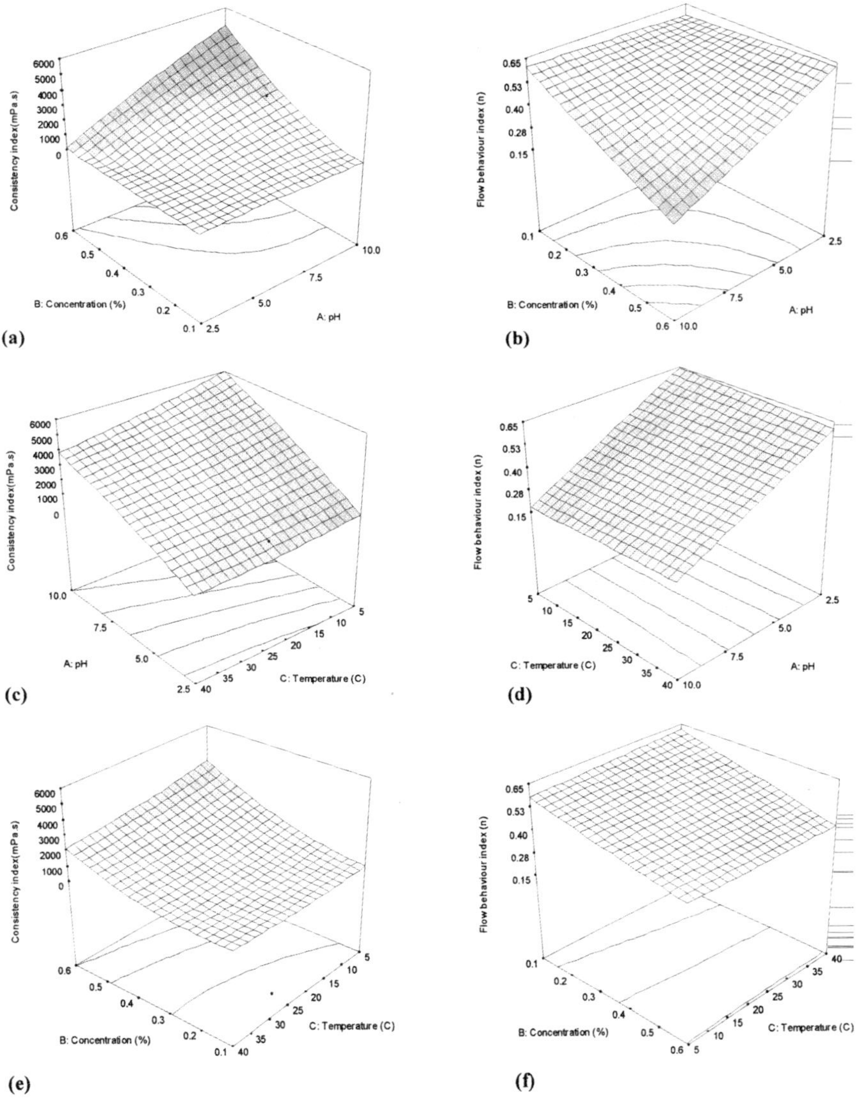

Figure 2 *Response Surface Methodology (given by Design Expert software): effect of pH, concentration and temperature on consistency index (a,c,d) and flow behavior index (b,d,f) of Balangu seed gum solutions.*

thinning than those with low concentrations. On the other hand, the consistency coefficient k for all of the samples increased with increasing concentration of the polysaccharide. Temperature, on the other hand, indicated an inverse effect on Power law parameters as flow behavior index increased and consistency coefficient decreased by increasing the temperature. These phenomena are in agreement with the work done by Marcotte et al. (2001 a,b) on xanthan gum, carrageenan, pectin and starch.[12-13] The resultant data were analyzed using the Design-Expert 6.0.2 and the best models were obtained. Models of consistency index, flow behavior index as obtained from the Design Expert are presented in Figure 2 and Table 3. A decrease in consistency coefficient was observed with increasing temperature, indicating a decrease in apparent viscosity at higher temperatures. In addition, consistency coefficient increased with gum concentration at all temperatures. For the prepared samples at all conditions, the increase in temperature resulted in considerable changes in viscosity values. This behavior can be explained by intermolecular forces restricting the molecular motion of the fluid. These forces, which determine the intermolecular spacing, are significantly affected by the change in temperature. Analysis of variance (ANOVA) ($\alpha = 0.01$) showed that flow behavior index (n) and consistency coefficient (k) were significantly influenced by temperature.

Table 3 *Equations obtained in the D-Optimal mode of Response Surface Methodology (given by Design Expert software) in terms of actual factors: effect of pH, concentration and temperature on each measured rheological parameters in Balangu Shirazi seed gum solutions.*

Parameter	Equations in terms of actual factors: Effect of pH, concentration and temperature	$(R^2)^{\dagger}$
$(K)^{\dagger}$ (mPa.s)	$= -55.85 + 147.68 \times pH - 8227.41 + 18.65 \times Temperature(°C) - 15.95 \times pH^2 + 10042.32 \times Concentration(\%)^2 + 0.56 \times Temperature(°C)^2 + 1320.48 \times pH \times Concentration(\%) - 5.53 \times pH \times Temperature(°C) - 82.08 \times Concentration(\%) \times Temperature(°C)$	**0.920**
$(n)^{*}$	$= +0.53 + 0.011 \times pH + 0.46 \times Concentration(\%) + 2.427E-004 \times Temperature(°C) - 0.12 \times pH \times Concentration(\%) + 2.02E-004 \times Temperature(°C) \times pH - 2.66E-003 \times Concentration(\%) \times Temperature(°C)$	**0.754**

† K: Consistency coefficient

* n: Flow behavior index

The temperature dependence of the viscosity at 0.4 % concentration and pH 7, shown in Table 4, was assessed by applying the Arrhenius-type model.[14-15] Activation energy (E_a) for the flow process is related to chain flexibility and high R^2 values indicate that the apparent viscosity of solutions in relation to temperature obey the Arrhenius type equation.[16] It can be seen that E_a values for Balangu seed gum decreased from 20.35 J/mole for 0.1 % concentration to 6.18 J/mole at 0.4 % concentration, meaning that the solution with lower concentration had the greater viscosity dependency to the temperature. A higher E_a value signifies a more rapid change in viscosity.[17] The outcomes imply that temperature control will be critical for Balangu seed gum solutions at low concentrations.

Table4 *Parameters of Arrhenius model for the effect of temperature on consistency coefficient at pH 7.*

Concentration (w/w)	K_t (mPa.s)	E_a (J/mol)	$^{\dagger}R^2$
0.1	0.0197	20.35	0.94
0.2	2.4317	11.81	0.99
0.4	102.68	6.184	0.98
0.6	17.909	13.16	0.97

3.3.2 Effect of pH on flow properties

Increasing the pH of the solution, from 2.5 to 10, augmented the pseudoplasticity and consistency coefficient (Fig. 3). These occurrences have been explained by the increased electrostatic repulsion by functional groups that tend to keep the molecules in an extended form, thus producing a high viscosity solutions.[18-19] The viscosity of Balangu seed gum also decreased gradually at more acidic conditions due to a lower degree of ionization of the mucilage carboxyl groups.

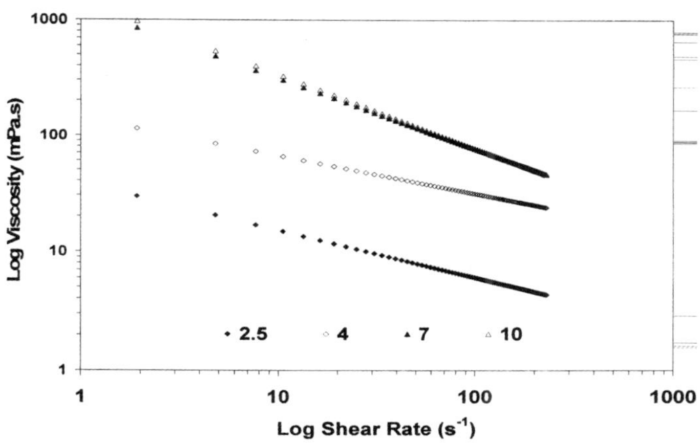

Figure 3 *Change of viscosity versus shear rate as function of pH for 0.4 % gum concentration solution at 25 °C.*

3.3.3 Effect of concentration on consistency coefficient

The effect of gum concentration on the consistency coefficient was described by exponential and power functions. Linearized forms of Equations 2 and 3 were plotted and the corresponding model parameters for pH 7 are provided in Table 5. The analysis of correlation coefficients of models showed that the exponential and power models successfully describe the relationship between concentration and consistency coefficient. For both models, the analysis of the experimental constants reveals that the effect of gum concentration on the consistency coefficient was more pronounced at lower temperatures. This can be explained by the increase in viscosity thus the consistency coefficient due to the combined effect of increasing Balangu seed gum concentration and decreasing temperature.

Table 5 *Effect of gum concentration on consistency coefficients at pH 7 and different temperatures.*

Exponential Model (Eq.3)			Power Model (Eq.4)		
A_1	d_1	$^{\dagger}R^2$	A_2	d_2	$^{\dagger}R^2$
3.5077	-1.6544	0.99	1.3296	-0.3629	0.98

3.3.4 Comparison of viscosity of Balangu gum with other gums

The rheological property of Balangu seed gum solution was compared with Tragacanth gum, CMC, Na-alginate and Arabic gum at 0.4 % (w/w) concentration. The highest consistency coefficients were for both Balangu seed gum and Tragacanth gum. The relationship between apparent viscosity (η_a) and shear rate of all tested gums at 25 °C are shown in Fig 4. It is evident that the apparent viscosity of all samples decreased with increasing shear. Balangu seed gum created solutions with a viscosity similar to CMC and Tragacanth gum and higher than Na-alginate and Arabic gum. According to the results and because of the shear thinning nature of its dispersions, Balangu seed gum could be suitable for application as a thickening agent.

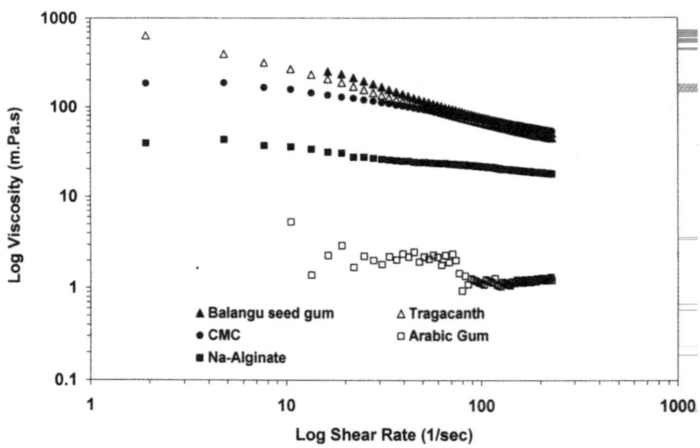

Figure 4 *Comparison of viscosity of Balangu gum solution with other gums at 0.4 % Concentration and 25 °C.*

4. CONCLUSION

The flow characteristics of Balangu *Lallemantia royleana*) seed gum were analyzed in the shear range between 1.92 and 229.4 s^{-1} in 80 steps. Steady shear viscosity measurements showed a typical non-Newtonian shear thinning behavior, which could be correlated to the shear rate by the Power law model. Temperature dependence was found for mucilage solutions, which its extent was concentration dependent. The pH of gum solutions had an important effect on viscosity and from acidic to alkaline regions, viscosity increased sharply. The viscosity level of Balangu gum was comparable with tragacanth gum and higher than other gums such as CMC and Na-alginate at concentration of 0.4 % (w/w) and 25 °C. Based on the viscosity data Balangu gum would be classified as a strong viscosifying and thickening agent and would be suitable for applications where thickening is the main functional factor.

References:

1. Glicksman, M., 1969. A Series of Monographs: Gum Technology in the Food Industry. London: Academic Press, Inc.
2. De Man, J.M., 1999. A Series of Monographs: Gum Technology in the Food Industry. London: Academic press, Inc.
3. A. Mohammad Amini, M.H. Haddad Khodaparast, and R. Farhoosh, MSc thesis, Ferdowsi University of Mashhad, Iran, 2007.
4. F. Naghibi, M. Mosaddegh, S. Mohammadi Motamed, and A. Ghorbanii, Iranian Journal of Pharmaceutical Research, 2005, **2**, 63–79.
5. S. M. A. Razavi, R. Farhoosh, and A. Bostan, Research project no.1475, Iran: Ferdowsi University of Mashhad, 2007.
6. S. Al-Assaf, P.A. Williams, and G.O. Phillips, 2002. The molecular characterization of hyaluronan and hylan. In Hyaluranan 2000. Vol. 1, Chemical, Biochemical and Biological Aspects, Pp. 55-74.
7. S. M. A. Razavi, T. Mohammadi Moghaddam, and A., Mohammad Amini, International Journal of Food Engineering, in press, 2008.
8. A.O.A.C., 1984. *Official Methods of Analysis*. (14). Washington, DC: Association of Official Analytical Chemists.
9. Bourne, M. C., 1982. Food Texture and Viscosity: Concept and Measurement. Academic Press, Inc., USA.
10. Rao, M.A., Anantheswaran, R.C., 1982. Food Technology, **36**, 116–126.
11. Sengul, M., Ertugay, M.F., Sengul, M., Food Control, 2005, **16**, 73–76.
12. M. Marcotte, A.R. Taherian, H.S. Ramaswamy, Food Research International, 2001a, **34**, 695.
13. M. Marcotte, A.R. Taherian, H.S. Ramaswamy, Journal of Food Engineering, 2001b, **48**, 157.
14. Barbosa Canovas, G.V., Peleg, M., 1983. Journal of Texture Studies 14, 213–234.
15. N.I. Singh, W.E. Eipeson, Journal of Texture Studies, 2000, **31**, 287.
16. Nielsen, L.E., 1977. Polymer Rheology. Marcel Dekker Inc., New York.
17. L. Medina-Torres, E. Brito-De La Fuente, B. Torrestiana-Sanchez, and R. Katthain, Food Hydrocolloids, **14**, 417.

18. Launay, B., Doubiler, I., Cavalier, G., 1986. Flow properties of aqueous solutions and dispersions of polysaccharides. In: Mitchell, J.R. (Ed.), Functional Properties of Food Macromolecules. Elsevier Applied Science Publishers, New York, pp. 1–78.
19. J.C. Onweluzo, Z.A. Obanu, and K.C. Onuoha, Nigerian Food Journal, 1994, **12**, 1.

FORMATION OF "WEAK GEL" NETWORKS BY KONJAC GLUCOMANNAN IN MIXTURES WITH SUCROSE, GLUCOSE OR FRUCTOSE

Amy Sweetnam, Antonio Sullo and Edwin R. Morris

Department of Food and Nutritional Sciences, University College Cork, Cork, Ireland

1 INTRODUCTION

Konjac glucomannan (KGM) is a water-soluble polysaccharide obtained from the tubers of *Amorphophallus konjac* and related species.[1] It gives solutions of high viscosity at low concentrations, and is therefore used as a thickener, particularly in food products. KGM has a (1 $\rightarrow$ 4)-diequatorially-linked backbone, containing both β-D-glucose and β-D-mannose residues in the approximate ratio 1 : 1.6, with no long blocks of either.[2,3] Solubility is conferred by some slight branching, and the presence of *O*-acetyl groups on ~5-10 % of the sugar residues. Removal of the solubilising acetyl substituents (by alkaline hydrolysis) allows the polymer chains to associate and form a gel network.[1] We now report that network formation can also occur when unmodified, native KGM is mixed with high concentrations of sugar (sucrose, glucose or fructose).

2 EXPERIMENTAL PROCEDURE

KGM (Deoxy Ltd., Cork, Ireland) was used at a fixed concentration of 1.0 wt %. The concentration of sucrose, glucose or fructose was varied from 0 to 60 wt % (in 10 % increments). Mixtures were characterised by oscillatory rheological measurements at 25°C (CarriMed CSL-100 rheometer; 4 cm cone-and-plate geometry; 2° cone angle; 1 % strain).

3 RESULTS AND DISCUSSION

3.1 Effect of sugar concentration

As would be expected, the mechanical spectrum recorded for 1.0 wt % KGM in water, with no added sugar (Figure 1a), has the form typical[4] of a solution of entangled polymer coils. At low frequencies, where there is sufficient time for substantial disentanglement to occur within the period of oscillation, liquid-like response, characterised by the loss modulus, G", predominates over elastic response (storage modulus, G'), both moduli increase steeply as the frequency of oscillation (ω) is increased, with G' showing a greater frequency-dependence than G", and there is little change in complex dynamic viscosity [$\eta^* = (G'^2 + G''^2)^{1/2}/\omega$]. At higher frequencies, however, where there is progressively less time for coils to disentangle before the direction of applied strain is reversed, G' rises above G" and η* falls, paralleling the "shear thinning" observed[4-6] in rotational measurements of viscosity (η) with increasing shear rate ($\dot{\gamma}$).

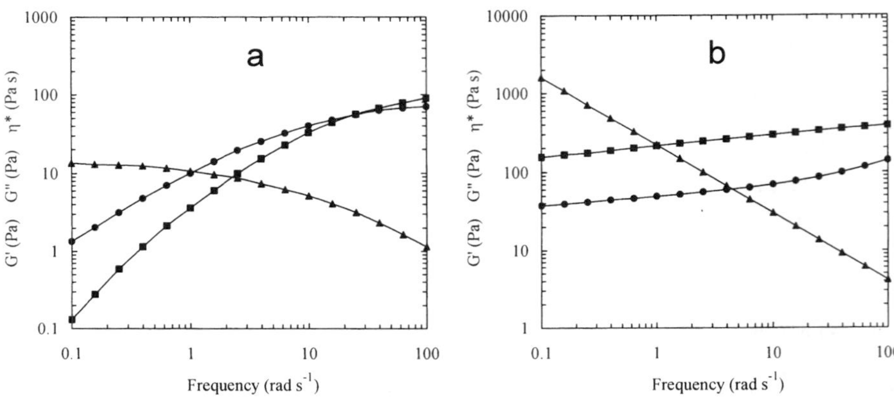

Figure 1 *Mechanical spectra (1 % strain) showing the frequency-dependence of G' (■), G" (●) and η* (▲) for 1.0 wt % KGM at 25°C (a) alone and (b) with 60 wt % sucrose.*

The same general form of frequency-dependence is evident in double-logarithmic plots of η* versus ω for solutions of 1.0 wt % KGM incorporating sucrose at concentrations up to 50 wt % (Figure 2), although the onset of the steep decrease in η* occurs at progressively lower frequency as the concentration of sucrose is increased, and the maximum value (η^*_o) of η* in the flat, "Newtonian plateau" region at low frequency increases by about a factor of 10 between 0 and 50 wt % sucrose.

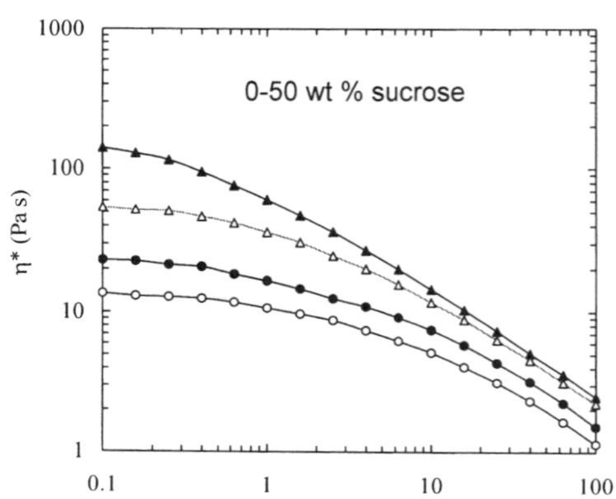

Figure 2 *Frequency-dependence of η* (1 % strain) for 1.0 wt % KGM at 25°C in the presence of sucrose at concentrations (wt %) of 0 (O), 10 (●), 30 (Δ) and 50 (▲).*

On further increase in sucrose concentration to 60 wt %, however, there is a radical change in rheological response. The mechanical spectrum (Figure 1b) now has the form typical[4] of a gel network, with G' >> G", little frequency-dependence of either modulus, and a linear dependence of log η^* on log ω, with no indication of a Newtonian plateau at low frequency. The mixture, however, was still sufficiently fluid to be loaded onto the rheometer. Preparations that give predominantly solid-like response to low-amplitude oscillation, but that can also be poured or spread, are known as "weak gels".[4]

Figure 3a shows the effect of sucrose concentration on values of G', G" and tan δ (G"/G') at a fixed frequency of 1 rad s^{-1} for 1.0 wt % KGM at 25°C. The transition from an entangled network to a "weak gel" can be seen as a sharp increase in solid-like character (G') as the concentration of sucrose is raised from 50 to 60 wt %, with an accompanying sharp decrease in tan δ.

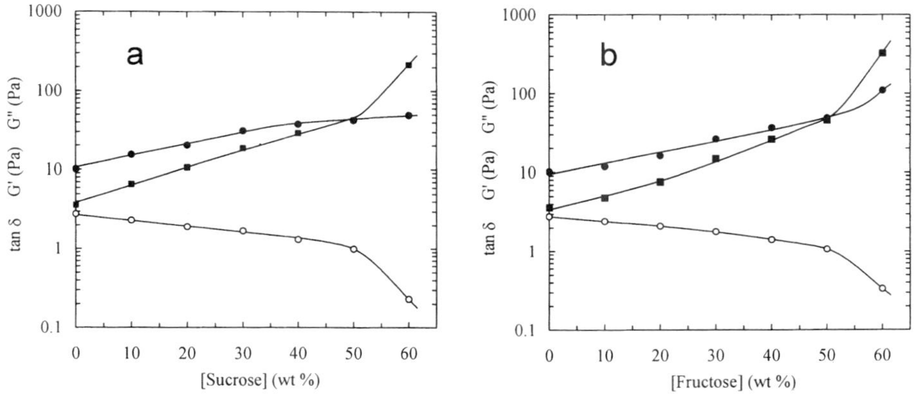

Figure 3 *Variation of G' (■), G" (●) and tan δ (O) with concentration of (a) sucrose and (b) fructose in mixtures with 1.0 wt % KGM at 25°C. Measurements were made at 1 rad s^{-1} and 1 % strain.*

A broadly similar pattern of response is evident (Figure 3b) with fructose as cosolute, but the variation of G', G" and tan δ with increasing concentration of glucose (Figure 4a) is substantially different. G' passes through a maximum at ~50 wt % glucose and then decreases as the glucose concentration is increased to 60 wt %, with an accompanying minimum in tan δ, and a reduction in G". Our proposed interpretation is that association of KGM chains into a continuous "weak gel" network becomes optimum when the concentration of glucose reaches ~50 wt %, and that further increase in glucose concentration causes excessive association (incipient precipitation), with consequent collapse of network structure. As shown in Figure 4b, the values of G' observed with sucrose or fructose as cosolute at the highest concentration studied (60 wt %) are still building up towards the maximum obtained with 50 wt % glucose.

As a simple indication of the strength of the network formed at optimum crosslinking, Figure 5 shows a direct comparison of the mechanical spectrum obtained for 1.0 wt % KGM with 50 wt % glucose (Figure 5a) and a spectrum (Figure 5b) reported previously[7] for high-methoxy pectin under conditions typical of its use in jams and jellies (0.5 wt % pectin; 65 wt % sucrose; pH 3.1). The two spectra are similar in form, and the moduli of the KGM network are only slightly lower (by about a factor of 2) than those of the pectin gel, indicating a texture similar to that of a comparatively weak, easily spreadable, jam.

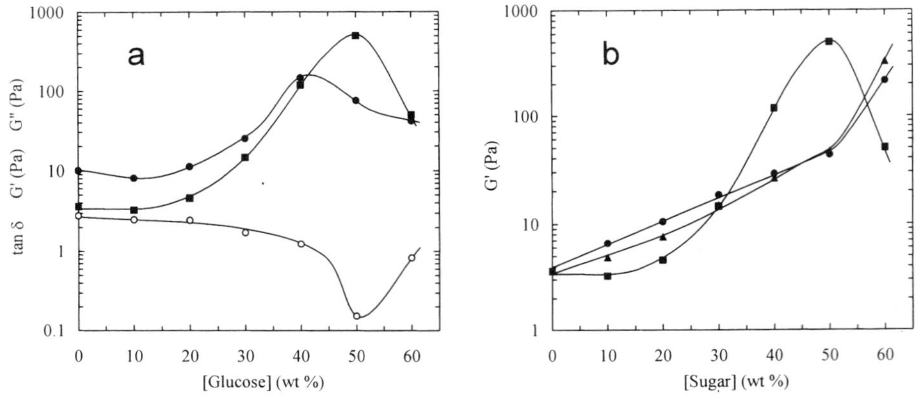

Figure 4 *(a) Variation of G' (■), G" (●) and tan δ (O) with concentration of glucose in mixtures with 1.0 wt % KGM. (b) Effect of glucose (■), sucrose (●) and fructose (▲) concentration on elastic character (G') of 1.0 wt % KGM. Measurements were made at 1 rad s^{-1}, 1 % strain and 25°C.*

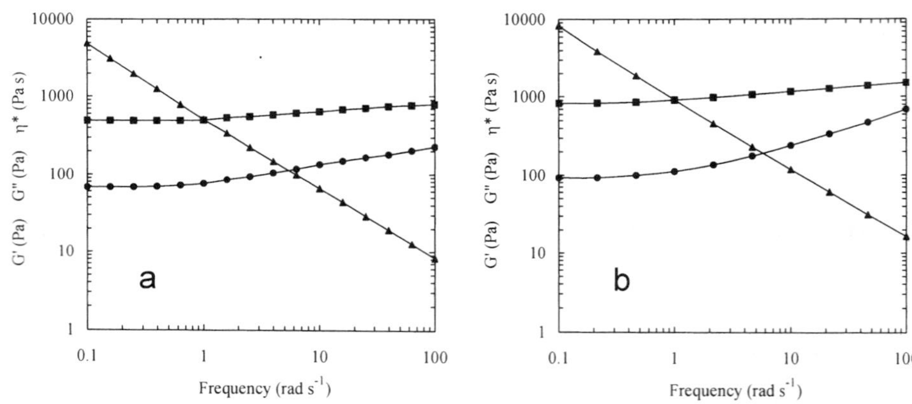

Figure 5 *Mechanical spectra (1 % strain) showing the frequency-dependence of G' (■), G" (●) and η* (▲) for (a) 1.0 wt % KGM with 50 wt % glucose, and (b) 0.5 wt % high-methoxy pectin with 65 wt % sucrose at pH 3.1.*

3.2 Quantitative analysis of "shear thinning"

Solutions of disordered polysaccharides, at concentrations where the individual coils are forced to interpenetrate one another and form an entangled network, show a general form of shear thinning,[5] which can be expressed quantitatively[6] by:

$$\eta = \eta_o/[1 + (\dot{\gamma}/\dot{\gamma}_{½})^{0.76}] \quad (1)$$

where η is viscosity, $\dot{\gamma}$ is shear rate, η_o is the maximum viscosity in the Newtonian plateau region at low shear rates, $\dot{\gamma}_{½}$ is the shear rate at which η is reduced to $\eta_o/2$, and 0.76 corresponds to the maximum (negative) slope of log η versus log ω at high shear rates.

Solutions of entangled coils[4] show close "Cox-Merz[8] superposition" of η and η^* at equivalent values of shear rate ($\dot{\gamma}$ /s^{-1}) and frequency (ω/rad^{-1}). Thus, for oscillatory measurements, Equation 1 can be expressed as:

$$\eta^* = \eta^*_o/[1 + (\omega/\omega_{1/2})^{0.76}] \qquad (2)$$

which can be re-arranged[6] to a linear equation:

$$\eta^* = \eta^*_o - (1/\omega_{1/2}^{0.76})\, \eta^*\omega^{0.76} \qquad (3)$$

Plotting η^* versus $\eta^*\omega^{0.76}$ should therefore give a straight line of gradient $-1/\omega_{1/2}^{0.76}$ and intercept η^*_o, provided that disentanglement is all that is required to allow flow to occur.

When plotted in this way, the frequency-dependence of η^* for 1.0 wt % KGM in the presence of sucrose at concentrations in the range 0-40 wt % shows good linearity (Figure 6a), as expected for entangled networks. There is, however, a systematic increase in intercept (η^*_o) as the concentration of sucrose is increased. The same effect is evident in the double-logarithmic plots of η^* versus ω shown in Figure 2. A likely interpretation is that increasing concentration of sucrose from 0 to 40 wt % raises the effective molecular weight of the KGM by promoting progressive association of the polymer chains into small "microgel" clusters, which interact with one another solely by physical entanglement.

The corresponding plots of η^* versus $\eta^*\omega^{0.76}$ with 0-40 wt % fructose as cosolute show a similar standard of linearity (Figure 6b), but the intercepts (η^*_o) are consistently lower than observed (Figure 6a) at the same concentrations of sucrose, indicating that fructose is less effective than sucrose in promoting self-association of KGM.

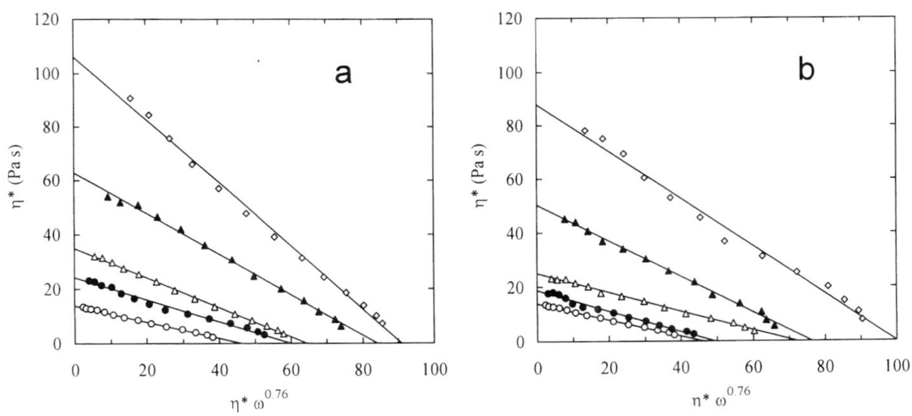

Figure 6 *Variation η^* (25°C; 1 % strain) with frequency (ω) for 1.0 wt % KGM in mixtures with (a) sucrose and (b) fructose at concentrations (wt %) of 0* (O), *10* (●), *20* (Δ), *30* (▲) *and 40* (✧)*, plotted according to Equation 3.*

As shown in Figure 7, the plots remain linear when the concentration of sucrose (Figure 7a) or fructose (Figure 7b) is increased to 50 wt %, with sucrose still giving a higher intercept than fructose. However, on further increase in concentration of cosolute to 60 wt % the plots are no longer linear, and their direction is reversed, demonstrating that, as would be expected from the gel-like mechanical spectra observed at this concentration of sucrose or fructose, the samples no longer show the frequency-dependence typical[6] of entangled coils.

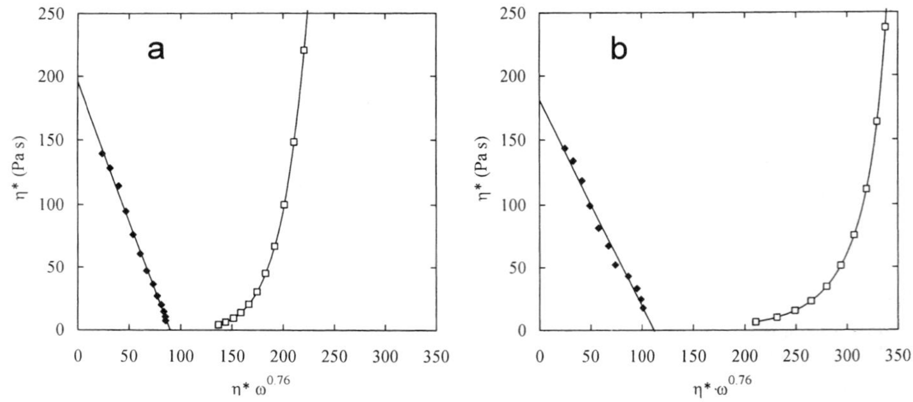

Figure 7 *Variation η* (25°C; 1 % strain) with frequency (ω) for 1.0 wt % KGM in mixtures with (a) sucrose and (b) fructose at concentrations of 50 (♦) and 60 (□) wt %, plotted according to Equation 3.*

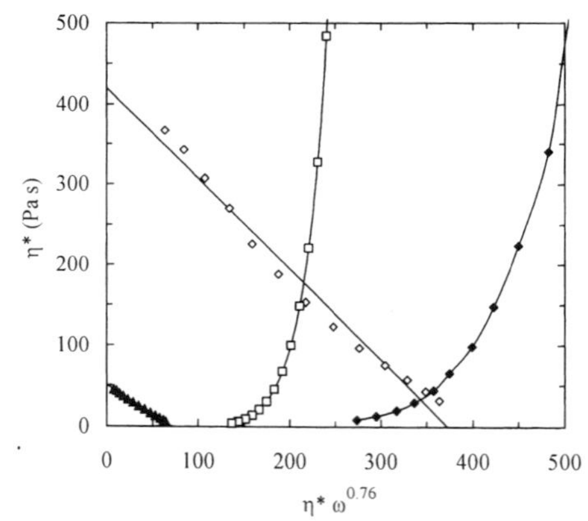

Figure 8 *Variation η* (25°C; 1 % strain) with frequency (ω) for 1.0 wt % KGM in mixtures with glucose at concentrations (wt %) of 30 (▲), 40 (✧) 50 (♦) and 60 (□), plotted according to Equation 3.*

With glucose as cosolute (Figure 8) loss of solution-like response (linear plots of η^* versus $\eta^*\omega^{0.76}$) comes earlier, at 50 wt %, which is again consistent with the gel-like mechanical spectrum (Figure 5a) observed at this concentration of glucose. The order of effectiveness in promoting self-association of KGM (Figures 6-8) is therefore glucose > sucrose > fructose.

3.3 Interpretation

One obvious way in which sugars (and other cosolutes) can promote self-association of polysaccharides, and other macromolecules, is by replacing part of the water needed to maintain the polymer in the fully-hydrated (solution) state. However, different cosolutes differ widely in their ability to stabilise macromolecular assemblies, and indeed some (notably urea) act as "denaturants", causing loss of stability. These differences are often attributed to changes in the "solvent quality" of water by interaction with the cosolute molecules. For sugars, particular attention has been paid to equatorial hydroxyl groups, whose spacing is compatible with the structure of liquid water.[9] Solute-induced changes in water structure, however, are difficult to quantify, or even to define.[10]

An alternative approach,[11] which is tractable to quantitative analysis, is to consider the concentration of cosolute molecules around the macromolecules, where they can compete with self-association of the polymer chains. The origin of enhancement ("adsorption") or depletion ("desorption") of cosolute at the macromolecular surface, relative to the concentration in the bulk of the solution, and the effect on self-association of the polymer, were explored in an elegant theoretical and experimental study,[12] using a lattice model to address explicitly all six possible pairwise interactions in polymer–water–cosolute mixtures (water–water; polymer–polymer; polymer–water; cosolute–cosolute; cosolute–water; and cosolute–polymer). Experimentally, the temperature (T_o) at the onset of self-association (formation of double helices on cooling) was measured for agarose (1.0 wt %) alone, and in the presence of various concentrations of a wide range of cosolutes (including glucose and urea). The same cosolutes (as 2 M solutions) were passed down a column of crosslinked agarose (Superose from Pharmacia). T_o was found to decrease linearly with increasing retention time (elution volume) on the column, demonstrating that preferential adsorption of cosolute molecules at the agarose–water interface causes inhibition of polymer–polymer association (double-helix formation). One of the main conclusions from application of the lattice model was that cosolute–water interactions have an important role in determining enhancement or depletion of cosolute at the surface of the polymer. Adsorption of cosolute molecules decreases their contact with water, and is therefore promoted if the cosolute–water interactions are thermodynamically-unfavourable; conversely, thermodynamically-favourable interactions with water promote desorption of cosolute. Self-association of the polymer chains, however, is dictated by the absolute concentration of surrounding cosolute, irrespective of whether that is higher (adsorption) or lower (desorption) than in the bulk of the solution.

In addition to cosolute–water interactions, the concentration of cosolute at the polymer surface is, of course, also dependent on direct interactions between cosolute and polymer. For uncharged polysaccharides (such as KGM) in aqueous mixtures with neutral sugars these interactions are likely to be dominated by dipolar attraction between –OH groups. As noted previously,[7] the number of (highly polar) primary alcohol groups per residue (Figure 9) in the sugars studied follows the order: fructose > sucrose > glucose, which is the reverse of the observed order of effectiveness in inducing self-association of KGM. It therefore seems possible, at least, that dipolar attraction between polymer and cosolute has a dominant role in increasing the concentration of sugar around the KGM chains, with consequent progressive restriction of macromolecular association as the polarity of the sugar increases.

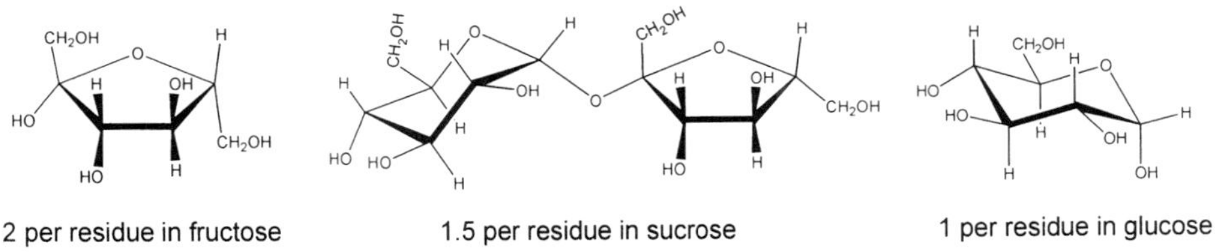

Figure 9 *Acidic primary alcohol (CH_2OH) groups in fructose, sucrose and glucose.*

4 SUMMARY AND CONCLUSIONS

The main conclusions from this investigation can be summarised as follows.

- Sugars promote self-association of KGM by depriving it of solvent.
- As the sugar concentration is increased, the initial effect of intermolecular association is to give small "microgel" cluster that interact with one another by physical entanglement.
- On further increase in sugar concentration, 1.0 wt % KGM forms a continuous "weak gel" network (at ~60 wt % fructose or sucrose and ~50 wt % glucose).
- Increasing glucose concentration from 50 to 60 wt % promotes excessive aggregation, causing network collapse and consequent reduction in moduli (Figure 4a).
- Self-association of KGM is restricted by sugar molecules around the polymer chains.
- The concentration of surrounding sugar increases with polarity, in the order fructose > sucrose > glucose.
- Self-association of KGM, with consequent enhancement in rheology, therefore follows the opposite order: glucose > sucrose > fructose.
- The "jam-like" texture (Figure 5) of KGM at optimum concentrations of sugar might have practical applications (e.g. in "fruit preps" for incorporation in ice cream or yogurt).

References

1 W.R. Thomas, in *Thickening and Gelling Agents for Food*, ed. A. Imeson, Blackie Academic and Professional, London, 1997, p. 169.
2 V. Davé and S.P. McCarthy, *J. Environmental Polymer Degradation*, 1997, **5**, 237.
3 K. Nishinari, P.A. Williams and G.O. Phillips, *Food Hydrocolloids*, 1992, **6**, 199.
4 S.B. Ross-Murphy, in *Biophysical Methods in Food Research*, ed. H.W.-S. Chan, *Critical Reports on Applied Chemistry*, SCI, London, 1984, p. 195.
5 E.R. Morris, A.N. Cutler, S.B. Ross-Murphy, D.A. Rees and J. Price, *Carbohydr. Polym.,* 1981, **1**, 5.
6 E.R. Morris, *Carbohydr. Polym.,* 1990, **13**, 85.
7 V. Evageliou, R. K. Richardson and E.R. Morris, *Carbohydr. Polym.,* 2000, **42**, 245.
8 W.P. Cox and E.H. Merz, *J. Polym. Sci.*, 1958, **28**, 619.
9 M.J. Tait, A. Suggett, F. Franks, S. Ablett and P.A. Quickenden, *J. Solution Chem.*, 1972, **1**, 131.
10 A. Holtzer and M.F. Emerson, *J. Phys. Chem.*, 1969, **73**, 26.
11 J.A. Schellman, *Biopolymers*, 1987, **26**, 549.
12 S. Nilsson, L. Piculell and M. Malmsten, *J. Phys. Chem.*, 1990, **94**, 5149.

Mixed Hydrocolloid Systems

PROTEIN – POLYSACCHARIDE COMPLEXES: FROM BASICS TO FOOD APPLICATIONS

Christophe Schmitt and Eric Kolodziejczyk

Department of Food Science and Technology, Nestlé Research Center, P.O. Box 44, CH-1000 Lausanne 26, Switzerland. Email: christophe.schmitt@rdls.nestle.com

1 INTRODUCTION

Three different types of phase behaviours have been described in mixtures of proteins and polysaccharides: co-solubility, thermodynamic incompatibility and complex coacervation.[1,2] The type of phase separation depends on the intrinsic physicochemical properties of the macromolecules (charge, charge density, molecular weight, chain flexibility), but also on the extrinsic physicochemical conditions in which the mixture is prepared (pH, ionic strength, concentration in biopolymers, weight mixing ratio, temperature).[3]

Complex coacervation is mostly observed in mixtures of proteins with anionic polysaccharides below the isoelectric pH (IEP) of the proteins. In this case, the biomacromolecules start forming soluble electrostatic complexes with increasing size and decreasing charge. When these complexes reach a critical size/charge ratio, they insolubilize and form discrete liquid droplets called coacervates. In a final stage, the coacervates coalesce, leading to macroscopic liquid-liquid phase separation.[4]

In this chapter, we will describe the bulk behaviour of the specific system β-lactoglobulin/acacia gum in acidic conditions leading to complex coacervation. The functional properties of these electrostatic complexes and coacervates at the air/water and oil/water interface will be then shortly described. Finally, some food applications will be reviewed, including the preparation of ice cream or sherbet with improved heat-shock stability and creaminess, the manufacturing of stable acid-based mousses without the use of gelatine or the coemulsion concept that was applied for the production of a chocolate-based confectionery exhibiting shape retention properties at 40°C.

2 BULK BEHAVIOUR OF BETA-LACTOGLOBULIN/ACACIA GUM MIXTURES

The major whey protein, β-lactoglobulin (β-lg), was characterized by a negative electrophoretic mobility above its IEP of 5.2 for the dimer and a positive value below this critical pH (Fig. 1a). This behaviour was mainly due to the balance of the charges carried by the amino acid side-chains from protein at the various pHs.[5] The electrophoretic mobility of acacia gum was mainly negative on the whole pH range. It decreased slightly at acid pH, *i.e.* close to the pK_a value of the carboxylic groups present on the gum backbone (Fig. 1a).[6] From the electrophoretic mobility behaviour of the single biopolymer dispersions, it can be predicted that electrostatic complexes should form upon mixture in acidic conditions. Such complex formation can be observed in figure 1b, which is reporting the changes in the hydrodynamic diameter of the biopolymers after mixing at a weight ratio of 2:1 and pH 4.2, corresponding to optimum conditions of charge neutralisation for this system.[7] The initial hydrodynamic diameter of the β-lg was around 6 nm and the bimodal distribution of the acacia gum accounted for the AG and AGP major fractions (average diameter of 24 nm).[6] Upon mixing, soluble complexes with a diameter of about 120 nm and a narrow polydispersity were formed. Over time, these complexes started aggregating to reduce their overall charge, leading to larger intermolecular complexes of about 600 nm (Fig. 1b). Ultimately, these complexes became insoluble, leading to discrete coacervate droplets as shown in figure 2b. It was recently shown that the whole complex coacervation coarsening mechanism could be described by a nucleation and growth model.[8]

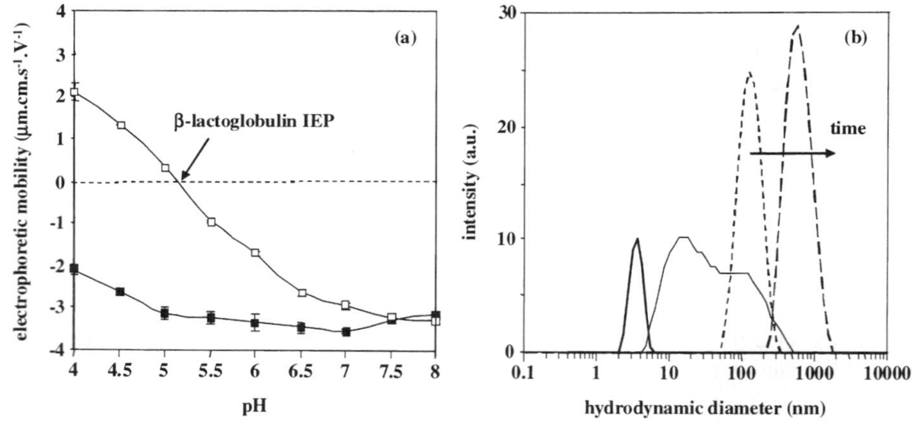

Figure 1 *(a) Variation of the electrophoretic mobility as a function of the pH for 0.1 wt% dispersion of β-lactoglobulin and acacia gum measured at 25°C. The dashed line represents electroneutrality. Vertical bars represent standard deviation. (b) Intensity-weighted distribution of the hydrodynamic diameter of 0.1 wt% dispersion of β-lactoglobulin (—), acacia gum (—) and mixtures (---) 10 s after mixing, (– –) 4 h after mixing at a weight mixing ratio of 2:1 measured at pH 4.2 and 25°C using dynamic light scattering.*

When β-lg/acacia gum mixtures were prepared at various initial biopolymer concentrations and after the equilibrium phase separation conditions were reached, the phase diagram shown on figure 2a was obtained. Interestingly, the very sharp corner of the 2-phase region was a sign that the critical mixing concentrations leading to phase separation were very low, *i.e.* in the order of ~ 0.1 wt%. This showed that the affinity between the two biopolymers in these pH conditions was very high as demonstrated by an affinity constant K_{eq} = 1.3 x 10^7 $mol^{-1}.L.$[4] Mixing conditions inside the biphasic region delimited by the binodal line led almost instantaneously to the formation of coacervates, as for example for mixture M, corresponding to 2.5 wt% β-lg mixed with 1.25 wt% acacia gum (Fig. 2b). Coacervates appeared as mono- or multi-vacuolated droplets enriched with the two biopolymers and some hydration water (around 70-80% at thermodynamic equilibrium).[7,9] Biopolymer mixtures leading to complex coacervation should be therefore considered as non-equilibrium systems exhibiting various degrees of structure and, possibly, of functional properties.

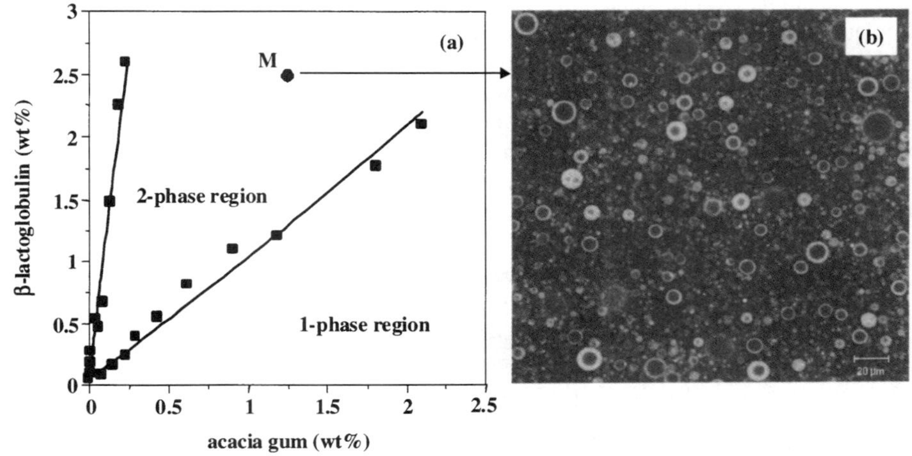

Figure 2 *(a) Phase diagram obtained at pH 4.2 and 25°C for the β-g/acacia gum/water system. The full line represents the experimental binodal determined by chemical analyses of the equilibrium phases after centrifugation. Squares represent the corresponding experimental binodal points. The M point corresponds to a mixing of 2.5 wt% β-lg with 1.25 wt% acacia gum dispersion. (b) Confocal scanning laser micrograph of the mixture corresponding to point M of the phase diagram. The grey areas correspond to regions enriched with biopolymers whereas the dark areas correspond to water-rich regions. β-lg was covalently labelled with fluoresceine isothiocyanate (FITC). Scale bar is 20 μm.*[10]

The electrostatic complexes formed initially, as well as the coacervates appearing thereafter were shown to exhibit surface activity both at the oil/water and at the air/water interface.[11] This behaviour was mainly due to the amphiphilic nature of the proteins and the hydrophilic nature of the polysaccharide, with structural modulations induced by the electrostatic interactions.

3 INTERFACIAL BEHAVIOUR OF BETA-LACTOGLOBULIN/ACACIA GUM COMPLEXES

β-lactoglobulin/acacia gum complexes or coacervates were able to diffuse to interfaces and to re-organize in order to form a mixed interfacial film. Such surface activity was observed in figure 3, where either a layer of fused coacervates or even discrete coacervates were visible at the interface of an air bubble generated in a drop containing β-lg/acacia gum mixture at a ratio of 2:1. In this experiment, the pH of the bulk phase was progressively reduced from 7.0 to 4.2 by *in situ* acidification using glucono-δ-lactone (GDL). This enabled observation of the different steps of the complex coacervation and the migration of the complexes and coacervates to the interface. Once adsorbed, the electrostatic complexes formed a stable viscoelastic layer at the air/water interface, exhibiting a dilatational modulus E' of 60 $mN.m^{-1}$ (at a frequency of 0.1 Hz) *vs.* 40 $mN.m^{-1}$ for pure β-lactoglobulin film.[12-14] Similarly, the same type of structure was obtained at the interface of oil droplets,[15,16] as this was the case in the sunflower emulsion presented in figure 3b.

The stability against disproportionation of air bubbles and foams stabilized by 0.1 wt% β-lactoglobulin/acacia gum complexes was shown to be much higher than for a bubble stabilized only by β-lg. For example, figure 4 reported that the shrinkage rate of an air bubble could be retarded by more than 30%. This was mainly due to the reduced gas permeability of the β-lg/acacia gum film which decreased from 0.521 $cm.s^{-1}$ for β-lg to 0.021 $cm.s^{-1}$ for the complexes.[14]

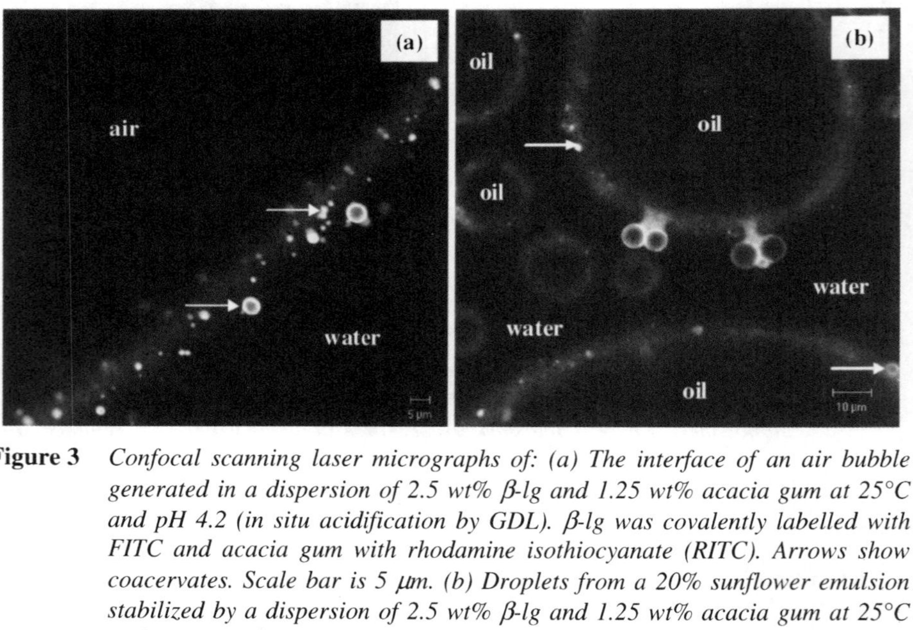

Figure 3 *Confocal scanning laser micrographs of: (a) The interface of an air bubble generated in a dispersion of 2.5 wt% β-lg and 1.25 wt% acacia gum at 25°C and pH 4.2 (in situ acidification by GDL). β-lg was covalently labelled with FITC and acacia gum with rhodamine isothiocyanate (RITC). Arrows show coacervates. Scale bar is 5 µm. (b) Droplets from a 20% sunflower emulsion stabilized by a dispersion of 2.5 wt% β-lg and 1.25 wt% acacia gum at 25°C and pH 4.2. β-lg was covalently labelled with FITC and acacia gum with rhodamine isothiocyanate (RITC). Arrows show coacervates. Scale bar is 10 µm.*

The use of the diminishing bubble method highlighted that the structure of the thin-film obtained at the air/water interface after 12,500 s was significantly different from the protein alone when β-lg/acacia gum complexes were used (Fig. 4a & b). The β-lg formed a homogeneous black film of about 10 nm thickness (Fig. 4a) whereas this thickness increased to > 20 nm in the presence of complexes that were retaining a large amount of water (white reflecting areas on fig. 4b).[14]

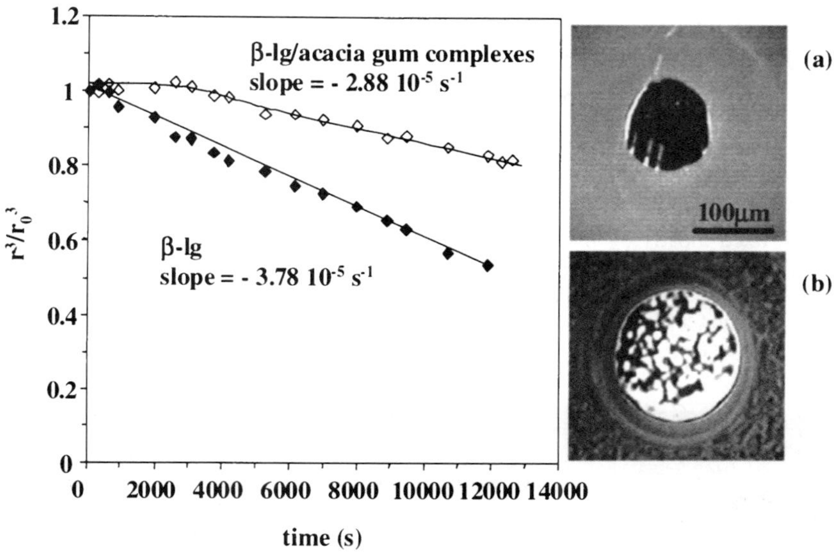

Figure 4 *Variation of the normalized radius of an air bubble stabilized by 0.1 wt% β-lg or β-lg/acacia gum complexes at pH 4.2 and 25°C. Corresponding light micrographs of the thin film formed at the top of the bubble after 12,500 s of shrinkage (a) with β-lg, (b) with β-lg/acacia gum complexes. Scale bar is 100 μm.*[14]

Based on these results, the formation of complexes and coacervates between β-lg and acacia gum was used in order to stabilize air/water and oil/water interfaces in aerated or emulsified food systems like ice cream, whipped cream or chocolate-based confectionery. A proper control of the kinetics of phase separation of the complexes to coacervates was applied in order to take advantage from the functionality of the different species formed during the complex coacervation process. To achieve these trials, β-lg was replaced by a whey protein isolate that is known to behave very similarly to β-lg in presence of acacia gum.[17]

4 SOME EXAMPLES OF FOOD APPLICATIONS USING BETA-LACTOGLOBULIN/ACACIA GUM COMPLEXES

4.1 Strawberry ice cream

Whey proteins/acacia gum complexes were used to stabilize the air bubble interface of a model strawberry ice cream containing 9% fat. Careful control of the formation of the complexes and coacervates was achieved by preparing two ice cream mixes, one enriched with whey protein isolate, the other one enriched with acacia gum, at pH 4.2 using a conventional ice cream process (mixing, homogenisation, pasteurisation and ageing). After ageing at 4°C, the two mixes were combined using a specific aeration device in order to form electrostatic complexes concomitantly with the air/water interface before entering the freezer. Doing so, complexes were mainly formed and were able to optimally stabilize the interface by *in situ* formation of coacervates.[18] The final contents of whey protein isolate and acacia gum in the ice cream were 2.5 and 1.25 wt%, respectively. The final product overrun was 100% and the freezing step was achieved at -6°C followed by hardening at -40°C.

It could be observed that the air bubble interface was properly stabilized by coacervates in the finished product as shown in figure 5a. Microscopic observation of the control ice cream mix (without β-lg/acacia gum complexes or with only β-lg) showed the air bubbles were stabilized mostly by fat clusters[19] and that the proteins were dispersed evenly in the product, without specific concentration at the interface, as was the case with complexes. We can therefore conclude that the mechanism of air bubble stabilisation was completely different in the products.

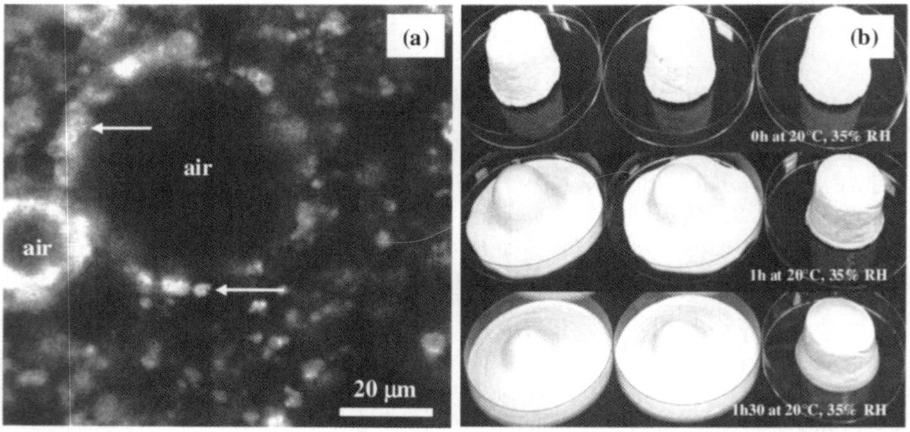

Figure 5 *(a) Confocal scanning laser micrograph (CSLM) of a 100% overrun strawberry ice-cream stabilized by whey protein isolate/acacia gum complexes and coacervates at pH 4.2 after 7 days of storage at -40°C. Protein was labelled non-covalently with the rhodamine 6G reagent. Arrows indicate coacervates adsorbed at the interface of the air bubbles. The scale bar is 20 µm. (b) Pictures showing the melting behaviour of 100 mL ice cream samples at 20°C for 1h30. Samples from left to right: control ice cream, ice cream with 2.5 wt% whey protein isolate, ice cream with 2.5 wt% whey protein isolate/1.25 wt% acacia gum.*

Interestingly, this microscopic structure of the product had a macroscopic effect on the melting properties of the ice cream (Fig. 5b). If the product based on a standard recipe or stabilized only with β-lactoglobulin were melting at room temperature in the timeframe of 1h30, no sign of melting was visible on the product that was stabilized by whey proteins/acacia gum complexes. This product was therefore keeping its aerated structure without loosing its shape, even after complete melting of the water, therefore enabling the design of a new texture. Indeed, sensory analysis performed on this product revealed that it was harder, smoother and creamier than the control. In addition, after a series of 7 days of heat shocks (-20/-8°C every 12 hours), the air bubble size distribution of the ice cream stabilized with complexes did not change, whereas it strongly shifted to larger air bubbles for the control sample (due to air bubble coalescence and disproportionation).

4.2 Lemon sherbet

The interfacial stabilization by whey proteins/acacia gum was further tested in a lemon sherbet containing 2 wt% fat. This product is therefore known to be really sensitive to heat shock.[20] The initial 2 wt% fat-recipe was used as a control (NS) and a modified recipe containing 2.5 wt% whey protein isolate/1.25 wt% acacia gum to produce complexes and coacervates (CS) was tested. The final product overrun was about 100% and the pH 3.8.

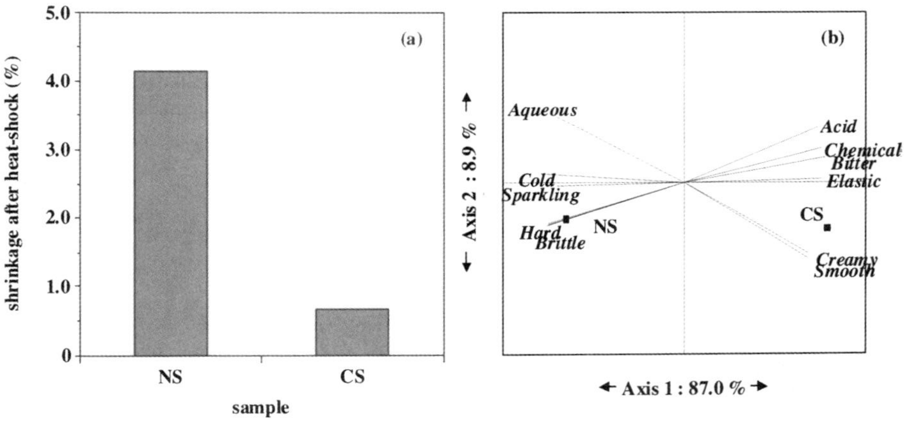

Figure 6 *(a) Relative shrinkage of the lemon sherbet after 7 days of heat shock (-20/-8°C every 12h). (b) Principal component analysis map of the sensory attributes generated by a trained panel upon testing the two lemon sherbets after 7 days of storage at -40°C. NS: control recipe; CS: complex-stabilized recipe.*

After a series of 7 days of heat shock, the shrinkage of the control product was about 4% of the initial sample volume, which can be considered to be a sign of instability of the air bubbles. For the sherbet produced in the absence of fat but with whey protein isolate/acacia gum complexes, the shrinkage was only 0.5%, *i.e.* 8 times lower than for the fat containing product (Fig. 6a). Microscopic investigation clearly highlighted the presence of complexes and coacervates at the interface of the air bubbles, whereas only fat was observed in the control sample.

In addition, the two sherbets were evaluated by a trained sensory panel in order to find the most discriminating attributes. Figure 6b revealed that 96% of the sensory differences between the two samples could be described by the reported 11 attributes. The non-fat sherbet stabilized by whey protein isolate/acacia gum was perceived smoother and creamier than the control, even if the latter contained 2 wt% fat. The air bubble stabilisation by whey protein isolate/acacia gum complexes therefore enabled designing a creamier product without using fat.

4.3 Sour cream mousse

An acidic chilled whipped dessert was also tested for air bubble interfacial stabilisation using whey protein isolate/acacia gum complexes. In this type of dessert, air bubbles are generally stabilized by the partial coalescence of the milk-fat spreading at the air/water interface during the whipping step in the presence of low molecular weight surfactants.[21] The shelf-life stability of the product is thereafter mainly achieved by increasing of the viscosity of the continuous phase by addition of a cold-gelling hydrocolloid such as gelatine in order to reduce drainage and bubble coalescence. The presence of the hydrocolloid generally leads to a thick texture. In the present trial, a sour cream recipe based on 43 wt% liquid cream (35% fat) and 5.7 wt% skim milk powder was used. The control recipe contained 0.7 wt% gelatine and pH was adjusted to 4.2 by addition of citric acid. For the product stabilized with whey protein isolate/acacia gum complexes, the skimmed milk powder was partly replaced by 2.5 wt% whey protein isolate and 1.25 wt% acacia gum and no gelatine was used. The liquid mixes were UHT treated and whipped to a final overrun of 100% using a rotor/stator mixer at 4°C and the stability of the air bubble was followed by microscopy analysis during a period of 28 days of storage at 4°C.

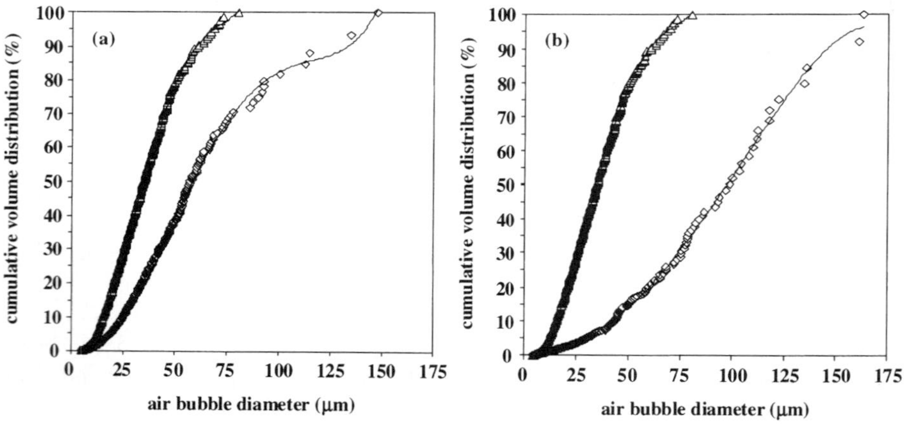

Figure 7 *(a) Cumulative air bubble size distribution obtained by image analysis of micrographs of sour cream mousse after 1 day of storage at 4°C. (b) Cumulative air bubble size distribution obtained by image analysis of micrographs of sour cream sample after 28 days of storage at 4°C. Triangles: whey protein isolate/acacia gum-based sample. Diamonds: gelatine-based sample.*

Observation of the sour cream stabilized by gelatine using CSLM after labelling of the fat phase using Nile red revealed that air bubbles were mainly stabilized by coalesced fat. On the contrary, when whey protein isolate/acacia gum complexes were used, the fat droplets were rather homogeneously distributed in the bulk, the air/water interface being stabilized by the complexes. The analysis of the cumulative air bubble size distribution showed that the product without gelatine contained smaller bubbles ($D_{v0.5}$ = 30 μm) than the control sample ($D_{v0.5}$ = 60 μm) after 1 day of storage (Fig. 7a). More interestingly, the air bubble size distribution was not affected upon storage for 28 days at 4°C in presence of whey protein isolate/acacia gum complexes, whereas it shifted to a $D_{v0.5} > 100$ μm in the control sample (Fig. 7b). Therefore, it was possible to design a sour cream recipe with air bubbles stable against disproportionation and coalescence keeping a low bulk viscosity of 100 mPa.s^{-1} compared to 340 mPa.s^{-1} when gelatine was used. Finally, sensory analysis of the samples after 7 days of storage at 4°C revealed that the sour cream stabilized with whey protein isolate/acacia gum complexes was perceived with more buttery/milky notes and with a more aerated texture with a spoon compared to the gelatine control.

4.4 Temperature resistant chocolate-based confectionery

In a last set of trials, a concept based on the electrostatic flocculation of oil droplets by whey protein isolate/acacia gum complexes was tested for the preparation of a temperature resistant chocolate-based confectionery that should keep its shape around 40°C, *i.e.* higher than the melting temperature of cocoa butter (around 35-37°C).

When a concentrated o/w emulsion ($\phi = 0.66$) stabilized with β-lactoglobulin was mixed with a similar emulsion stabilized with acacia in conditions of complex formation (ratio 2:1, pH 4.2), instantaneous electrostatic flocculation occurred due to charge neutralisation, leading to a mayonnaise-like o/w emulsion called coemulsion (Fig. 8a).[18] This coemulsion was characterized by an interfacial structure of the droplets composed of complexes and coacervates (Fig. 8b), leading to strong cohesiveness between the droplets.

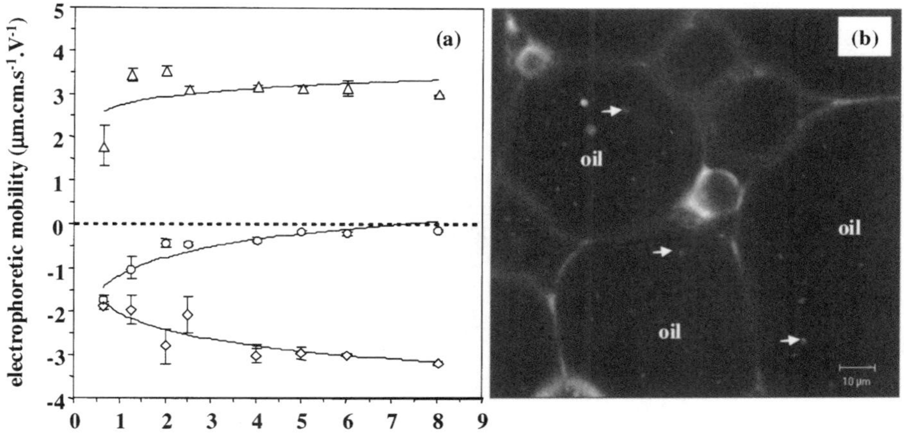

Figure 8 *(a) Variation of the electrophoretic mobility of concentrated (ϕ = 0.66) sunflower emulsions stabilized by various biopolymer concentration at pH 4.2 and 25°C. Emulsions were diluted 500 times before measurement. Triangles: β-lactoglobulin-stabilized o/w emulsion. Diamonds: acacia gum-stabilized o/w emulsion. Circles: β-lg/accaia gum coemulsion at a weight mixing ratio of 2:1. Vertical bars represent standard deviation. (b) CSLM of a sunflower oil coemulsion (ϕ = 0.66) stabilized by 2.5 wt% β-lactoglobulin and 1.25 wt% acacia gum at pH 4.2. β-lg was labelled with FITC and acacia gum with RITC. Grey areas are rich in electrostatic complexes and coacervates. Arrows show some labelled protein inclusions dispersed in the oil phase. Scale bar is 10 μm.*

The coemulsion concept was used in order to increase the heat resistance of a model chocolate confectionery, replacing a fraction of the cocoa butter with coemulsified cocoa butter in order to build up a network of droplets within the fat phase at high temperature. The control recipe was based on 50 wt% sucrose, 20 wt% skim milk powder, 26 wt% cocoa butter and 0.4 wt% lecithin. When the coemulsion was used, it contained 66 wt% cocoa butter stabilized by 2.5 wt% whey protein isolate and 1.25 wt% acacia gum at pH 4.2. The coemulsion was partially replacing weight by weight 10, 15, 20 and 30% of the cocoa butter. The melting resistance of the confectionery was evaluated after storage overnight at 40°C by performing an uniaxial compression test (to 80% deformation of the sample) in order to determine the maximum stress before rupture of the sample (or firmness). Table 1 showed that a higher firmness than the control sample was obtained after replacement of 10% of the cocoa butter by the coemulsion. Replacement of 15% led to a much higher modulus of ~ 32 kPa, whereas further increase of cocoa butter substitution reduced again the firmness.

Table 1 *Stress at rupture (or firmness) obtained upon uniaxial compression of chocolate-based confectionery samples at 40°C.*

Sample name	*Stress at rupture (kPa)*
Control	10
Coemulsion 10%	25
Coemulsion 15%	32
Coemulsion 20%	15
Coemulsion 30%	5

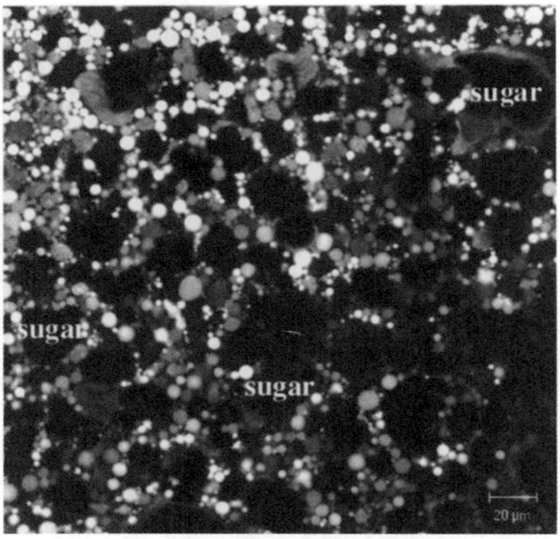

Figure 9 *CSLM of a chocolate-based confectionery samples containing 15% cocoa butter coemulsion at 40°C. Fat was labelled with Nile red and appears as white/grey areas. Sugar particles appear as dark areas. Scale bar is 20 µm.*

Microscopic observation of the samples using CSLM after staining of the fat phase by Nile red revealed that for 15% replacement of cocoa butter by the coemulsion, a homogeneous fat droplet network was formed within the melted cocoa butter and the sugar particles (Fig. 9). This led to the measured increased firmness of the product at 40°C and to its macroscopic shape retention properties. Use of 10% coemulsion did not allow homogenous network formation. On the contrary, higher yield of cocoa butter replacement induced recrystallisation of the sugar particles due to the increased water content brought by the coemulsion. These large sugar crystals were creating sample defects leading to an early rupture and to a grainy texture as reported by the sensory panel.

5 CONCLUSIONS

In this overview, we showed that the formation of β-lactoglobulin/acacia gum complexes in optimum charge neutralisation conditions was a dynamic phenomenon. Aggregation of the electrostatic complexes formed initially led to the formation of discrete coacervate droplets and ultimately to macroscopic liquid-liquid phase separation. These complexes did exhibit important surface stabilizing properties as they were able to form a viscoelastic film at the air/water or oil/water interface upon *in situ* re-organization. This film was shown to significantly improve the stability or air bubbles against coalescence or promote strong electrostatic flocculation of emulsion droplets. Such properties were used to design heat shock stable ice cream or sherbet with increased creaminess. As well, whey protein isolate/acacia gum complexes could stabilise chilled sour cream mousse without the use of gelatine. Finally, electrostatically bridged oil droplets were used to produce a chocolate-based confectionery exhibiting shape retention at temperatures above 40°C.

Acknowledgements

Authors would like to acknowledge C. Bovay, V. Clément, R. Cartier, S. Rami-Shojaei, L. Sandoz, C. Vaccher, P. Rousset and A. Fracheboud from Nestlé Research Center Lausanne for their scientific and technical support.

References

1 J.-L. Doublier, C. Garnier, D. Renard, D. and C. Sanchez. Protein-polysaccharide interactions. *Curr. Opin. Coll. Interf. Sci.*, 2000, **5**, 202.
2 S.L. Turgeon, M. Beaulieu, C. Schmitt and C. Sanchez. Protein–polysaccharide interactions: phase-ordering kinetics, thermodynamic and structural aspects. *Curr. Opin. Coll. Interf. Sci.*, 2003, **8**, 401.
3 C. Schmitt. C. Sanchez, S. Desobry-Banon and J. Hardy. Structure and technofunctional properties of protein-polysaccharide complexes: A review. *Crit. Rev. Food Sci. Nutr.*, 1998, **38**, 689.
4 C. Schmitt, L. Aberkane, and C. Sanchez. 'Protein-polysaccharide complexes and coacervates', in *Handbook of hydrocolloids – Second edition*, eds., G.O. Phillips and P.A. Williams, Woodhead Publishing Ltd, Cambridge, 2009, pp. 420-476.
5 F. Fogolari, L. Ragona, S. Licciardi, S. Romagnoli, R. Michelutti, R. Ugolini and H. Molinari. Electrostatic properties of bovine β-lactoglobulin. *Proteins: Struct., Func. Gen.*, 2000, **39**, 317.

6 D. Renard, L. Lavenant-Gourgeon, M.-C. Ralet and C. Sanchez . Acacia senegal gum: continuum of molecular species differing by their protein to sugar ratio, molecular weight, and charges. *Biomacromolecules*, 2006, **7**, 2637.
7 C. Schmitt, C. Sanchez, A. Thomas and J. Hardy. Complex coacervation between β-lactoglobulin and acacia gum in aqueous medium. *Food Hydrocoll.*, 1999, **13**, 483.
8 C. Sanchez, G. Mekhloufi and D. Renard. Complex coacervation between β-lactoglobulin and Acacia gum: A nucleation and growth mechanism. *J. Coll. Interf. Sci.*, 2006, **299**, 867.
9 C.L. Cooper, P.L. Dubin, A.B. Kayitmazer and S. Turksen, S. Polyelectrolyte-protein complexes. *Curr. Opin. Coll. Interf.* Sci., 2005, **10**, 52.
10 C. Schmitt, C. Sanchez, A. Lamprecht, D. Renard, C.-M. Lehr, C.G. de Kruif and J. Hardy. Study of β-lactoglobulin/acacia gum complex coacervation by diffusing-wave spectroscopy and confocal scanning laser microscopy. *Coll. Surf. B*, 2001, **20**, 267.
11 E. Dickinson. Interfacial structure and stability of food emulsions as affected by protein-polysaccharide interactions. *Soft Matter*, 2008, **4**, 932.
12 C. Schmitt, T. Palma da Silva, C. Bovay, S. Shojaei-Rami, P. Frossard, E. Kolodziejczyk and M.E. Leser. Effect of time on the interfacial and foaming properties of β-lactoglobulin/acacia gum electrostatic complexes and coacervates at pH 4.2. *Langmuir*, 2005, **21**, 7786.
13 C.C.C. Liz, V. Petkova, J.J. Benattar, M. Michel, M.E. Leser and R. Miller. X-ray reflectivity studies of liquid films stabilized by mixed β-lactoglobulin-acacia gum systems. *Coll. Surf. A*, 2006, **282-283**, 109.
14 C. Schmitt, E. Kolodziejczyk and M.E. Leser. 'Interfacial and foam stabilisation properties of β-lactoglobulin-acacia gum electrostatic complexes', in *Food Colloids: Interactions, Microstructure and Processing*, ed., E. Dickinson, Royal Society of Chemistry, Cambridge, 2005, pp. 284-300.
15 R.A. Ganzevles, K. Zinoviadou, T. van Vliet, M. Cohen-Stuart and H.J. de Jongh. Modulating surface rheology by electrostatic protein/polysaccharide complexes. *Langmuir*, 2006, **22**, 10089.
16 F. van de Velde, F. Weinbreck, M.W. Edelman, E. van der Linden and R.H. Tromp. Visualisation of biopolymer mixtures using confocal scanning laser microscopy (CSLM) and covalent labelling techniques. *Coll. Surf. B.*, 2003, **31**, 159.
17 F. Weinbreck, R. de Vries, P. Schrooyen and C.G. de Kruif. Complex coacervation of whey proteins and gum Arabic. *Biomacromolecules*, 2003, **4**, 293.
18 E. Kolodziejczyk and C. Schmitt. Interface stabilisation of a product with 2 or more phases with a protein-polysaccharide complex. *World patent application WO2004/028281 A1*, Nestec S.A., 2004.
19 H.D. Goff. Colloidal aspects of ice cream - A review. *Int. Dairy J.*, 1997, **7**, 363.
20 U.K. Dubey and C.H. White. Ice cream shrinkage: a problem for the ice cream industry. *J. Dairy Sci.*, 1997, **80**, 3439.
21 G.A. van Acken. Aeration of emulsions by whipping. *Coll. Surf. A*, 2001, **190**, 333.

MILK PROTEINS AND PECTINS – COMPATIBILITY AND CONJUGATION

Ulrike Einhorn-Stoll[a], Christopher Guyot[b], Carsten Rittweger[a]

[a] Food Science and Food Chemistry, Food Quality and Material Science, Technische Universitaet Berlin, Koenigin-Luise-Strasse 22, D-14195 Berlin, Germany
[b] Chair for Food Process Engineering and Dairy Technology, Technische Universität München, Weihenstephaner Berg 1, D-85354 Freising, Germany

1 INTRODUCTION

Thermodynamic compatibility and conjugate formation of proteins and polysaccharides have been described in many publications during the last 30 years.[1,2,3,4,5,6] Milk proteins are widely used not only in classical milk products but also in baked goods, meat processing and other fields, and pectins are well known as stabilisers in different food systems. Therefore, combinations of milk proteins and pectins were also chosen as substrates for investigations of compatibility and conjugation.[7,8,9,10] Former experiments using these materials showed a considerable and similar influence of protein type, degree of methoxylation (DM) of the pectin component and pH on the thermodynamic compatibility[11] as well as on the emulsifying properties of conjugates formed from these components.[12] Mixtures of whey protein isolate (WPI) and high methoxylated pectin (HMP) were completely compatible. The conjugates made from those components were excellent emulsifiers and a higher pH and degree of methoxylation (DM) improved the emulsifying properties. Sodium caseinate (SC) was only to a certain extent compatible with pectin, a higher pH and a lower DM favoured the compatibility. Incubation products of sodium caseinate-pectin mixtures had medium emulsifying properties, which were also higher in case of the application of low DM pectins. It seemed that the higher compatibility of caseinate and low DM pectin allowed a better conjugation process with formation of an improved emulsifier. Might it be possible to enhance the compatibility of caseinate and pectin also by reducing the molecular weight of the latter as described for pectin-gelatine systems[13] and would this improve the conjugation process too? Is there even any dependence of conjugation on thermodynamic compatibility because of a closer contact of compatible components during conjugate preparation? This is examined by using mechanically degraded HMP for compatibility tests as well as conjugation with SC. For comparison, the conjugation was made also with WPI as protein component.

2 MATERIALS AND METHODS

2.1 Materials

The milk proteins used were commercial whey protein isolate Lactalbumin 80 (MILEI, Germany) and sodium caseinate (DENA, Germany). The proteins were dissolved in distilled water, dialysed for five days against distilled water at 5 °C and lyophilised. The ash content was not reduced in case of WPI (2.1 % before and after dialysis) and only slightly decreased for the caseinate (3.2 to 2.9 %, respectively).
For further experiments, the casein was enzymatically hydrolyzed using the pH-stat method of Adler-Nissen with thermitase at pH 8.5 and 50 °C, as described elsewere.[14] The degrees of hydrolysis (DH) are given in Table 1.

Table 1 *DH of the caseins*

Casein	**CAS 0**	**CAS 1**	**CAS 2**	**CAS 3**	**CAS 4**
DH [%]	0	2.7 ± 0.23	4.2 ± 0.40	5.5 ± 0.13	9.1 ± 1.67

The mechanically disaggregated pectins HMP 1 to HMP 3 were made from high methoxylated citrus pectin HPM 0 from Copenhagen Pectin A/S (Lille Skensved, Denmark) using a vibration mill as described earlier.[15] An unmodified high methoxylated citrus pectin HMP HF was kindly provided by Herbstreith and Fox (Neuenbuerg, Germany). The molecular parameters of the pectins used are given in Table 2. The DE was determined photometrically and the intrinsic viscosity [η] with an Ubbelohde viscosimeter as described elsewere.[11] From the latter, the molecular weight MW and the hydrodynamic (rotation) radius HR were calculated.

Table 2 *Molecular parameters of the pectins*

Pectin	**HMP 0**	**HMP 1**	**HMP 2**	**HMP 3**	**HMP HF**
DE [%]	78.9	74.4	74.5	73.5	72.1
[η] [cm^3/g]	387 ± 2	220 ± 8	95 ± 1	40 ± 3	624 ± 18
MW [Da]	87,000	40,000	12,000	4,000	168,000
HR [nm]	17.52	11.53	5.93	3.03	25.55

2.2 Methods

Compatibility tests and construction of binodal curves were described earlier[11]. Proteins and pectins were dissolved in phosphate buffer, mixed, stirred for 1 h and centrifuged for 1 h at 5000 g. The binodal curves were constructed using the volume-ratio method. The area below the binodal curve is the region of compatibility, the area above that of incompatibility. Because of the complete compatibility of pectin with WPI, the tests of the influence of the pectin MW were made only with caseinate and pectins. The preparation of protein-pectin conjugates and examination of the reaction products by SDS electrophoresis were described elsewere[12]. The incubations were made in a climate chamber at 60 °C and 65 % relative humidity for up to 10 days. The emulsifying properties of the conjugates were determined as emulsifying activity EA (turbidity at 550 nm) in a modified Pearce and Kinsella photometric method[16]. In case of the hydrolysed proteins, the compatibility tests were made using only the soluble part of the samples (which was not completely comparable to the unseparated material in SDS electrophoresis) and for the conjugation process the complete hydrolysate was used.

3 RESULTS AND DISCUSSION

3.1 Influence of pectin molecular weight

With decreasing molecular weight of the pectin component, the thermodynamic compatibility with sodium caseinate increased and already at pectin MW 12,000 all mixtures were completely compatible (Figure 1). This is in agreement with results of Antonov et al.[13] and can be explained mainly by the reduced excluded volume effect of smaller biomacromolecules.[17] High molecular pectins are assumed to be relatively stiff linear chains and have a relatively high excluded volume (which is not accessible for the SC). The depolymerised pectins have a lower MW, reduced hydrodynamic radius and higher molecular flexibility. This means more available solution volume for the protein molecules, even at higher pectin concentrations, and a better thermodynamic compatibility.

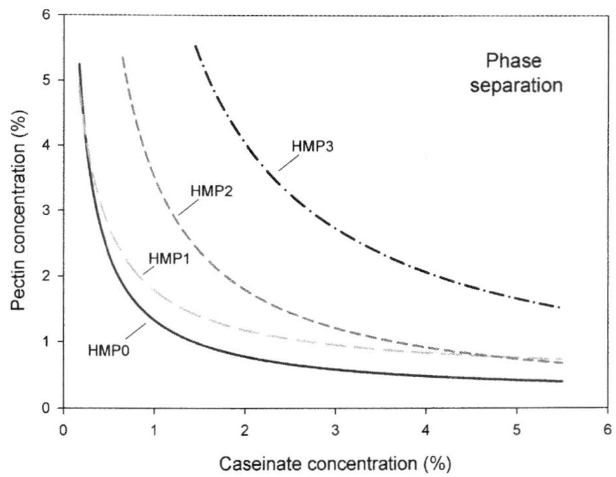

Figure 1 *Binodal curves and region of thermodynamic compatibility of SC with pectins of different molecular weight*

The emulsifying properties of the pure caseinate and the caseinate-pectin mixtures were decreasing dramatically after incubation. The incubation of whey protein alone as well as in combination with pectins, however, increased the emulsifying activity significantly, nearly independent on the degree of pectin degradation and the conjugation time (Figure 2).

The question was whether there were really conjugates formed of the two components during incubation. The SDS electrophoresis results (Figure 3) showed conjugates as high molecular substances (dark stripes on the top of the stacking gel as well as the resolving gel) in case of whey protein-pectin reaction products. These conjugates have excellent emulsifying properties. The incubation of pure WPI (not shown) caused a certain formation of higher molecular components, too, which can explain the increase of the emulsifying activity of the pure protein. In this case, however, the low molecular components (MW < 29,000) were found nearly unchanged, even after 10 d incubation.

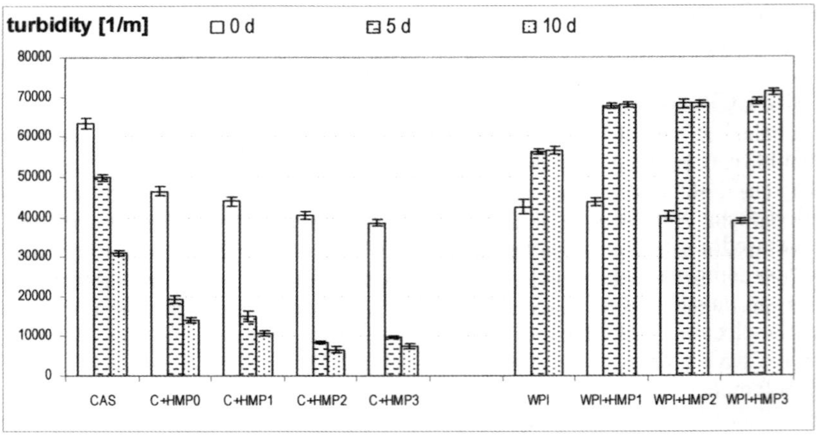

Figure 2 *Influence of pectin molecular weight on emulsions made with incubated SC-pectin and WPI-pectin mixtures. 0 d, 5 d, 10 d incubation time.*

Only very small amounts of such high molecular material, however, could be found on the top of the stacking gels after incubation of casein and pectin. In contrast, degradation of the major fractions of casein (MW about 23,000) occurred (Figure 3). These degraded proteins were not able to give enough short term stability to the emulsions, and the smaller the MW of the pectins, the lower was their stabilising contribution.

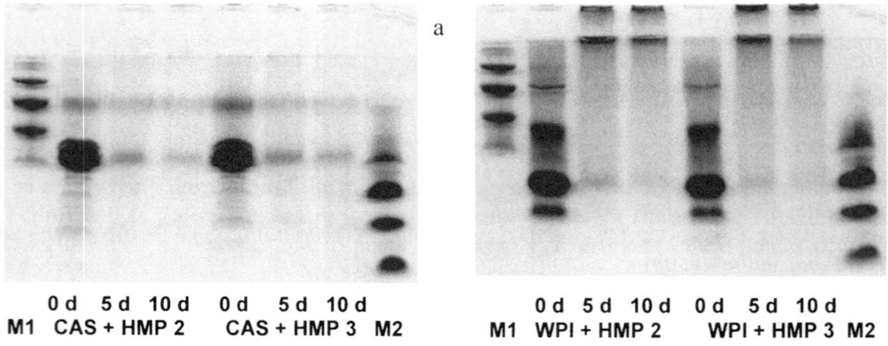

Figure 3 *Conjugate formation detection in SDS electrophoresis: casein-pectin mixtures (a) and WPI-pectin conjugates (b); M1/2 are marker protein mixtures with MW 29,000 to 97,000 (M1) and 6,500 to 29,000 (M2).*

The limited thermodynamic compatibility of caseinate and pectin can be not the reason for the different behaviour of the two proteins and their conjugates. Though HMP 2 and HMP 3 were completely compatible with caseinate at the tested concentrations, no conjugates were formed. A possible reason might be the different structure of the two proteins. Whereas whey proteins behave like dissolved polymers of low MW, caseins often form colloidal particles[2] and the processing of the sodium caseinate might additionally influence the behaviour.[17]

3.2 Influence of protein molecular weight

The assumed influence of the compact caseinate structure on the interactions with pectins was further examined. Therefore, the next step was to test whether degradation of the sodium caseinate could break up the colloidal particles and increase the ability of the protein for conjugate formation. The sodium caseinate was hydrolysed with the endoprotease thermitase. Afterwards, the thermodynamic compatibility was increased for all hydrolysed proteins (Figure 4) but the influence was not as systematic as in the case of pectin degradation, and none of the systems were completely compatible. On the contrary, the number of incompatible mixtures seemed to be increasing with higher protein DH.

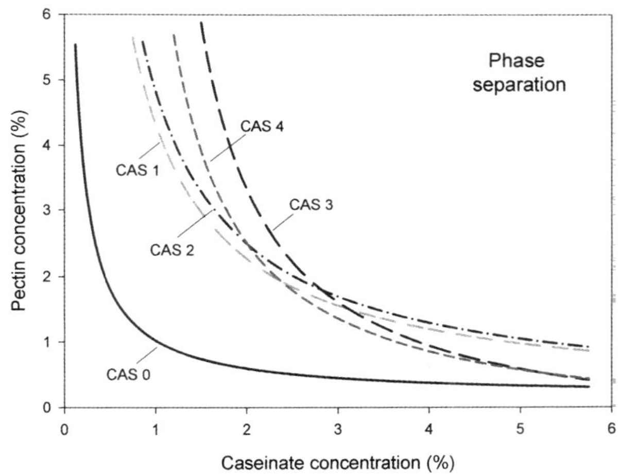

Figure 4 *Binodal curves and region of thermodynamic compatibility of hydrolysed SC with pectin*

The emulsifying properties of the mixtures of degraded SC and unmodified high molecular pectin HMP HF after incubation were considerably different from those containing high molecular caseinate and degraded pectins. Generally, hydrolyses of caseinates decreased the emulsifying activity, possibly because of an increasing bridging flocculation. At higher protein DH, the emulsifying activity was slightly improved by incubation with pectins, though it was on a relatively low level for DH below 5 %. The casein hydrolysis seemed to modify the protein structure in a way that it was able to undergo interactions with the pectin to a certain extent (Figure 5).

A comparison with the emulsifying behaviour of the incubated pure caseinates, however, gave another result. The incubated pure hydrolysates had nearly the same emulsifying ability like the incubated mixtures with pectin (CAS 1 and 2). Pure CAS 3 could not stabilise emulsions (the molecules were too small) but only in this case the incubation process seemed to produce a kind of conjugate.

The results of SDS electrophoresis allow a possible explanation for the effects (Figure 6). Not only the mixtures of casein hydrolysates and pectin (left side) but also the pure casein hydrolysates (middle and right side) formed higher molecular components after incubation. These reaction products might be able to stabilise the emulsion droplets to a certain extent and increase the emulsifying activity.

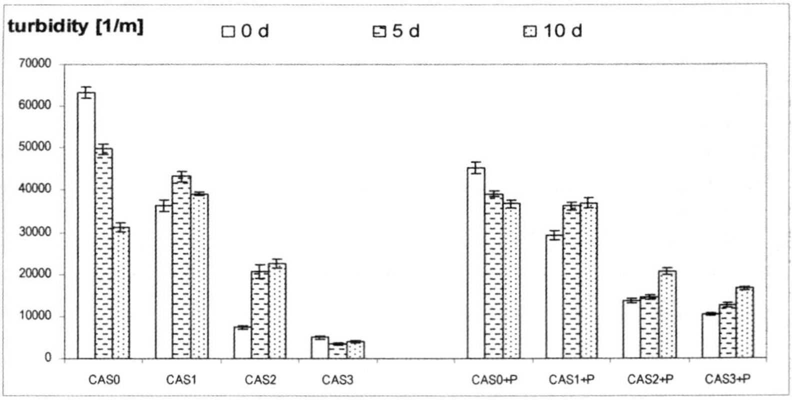

Figure 5 *Influence of protein molecular weight on emulsions made with incubated SC-pectin mixtures and with pure incubated SC. 0 d, 5 d, 10 d incubation time.*

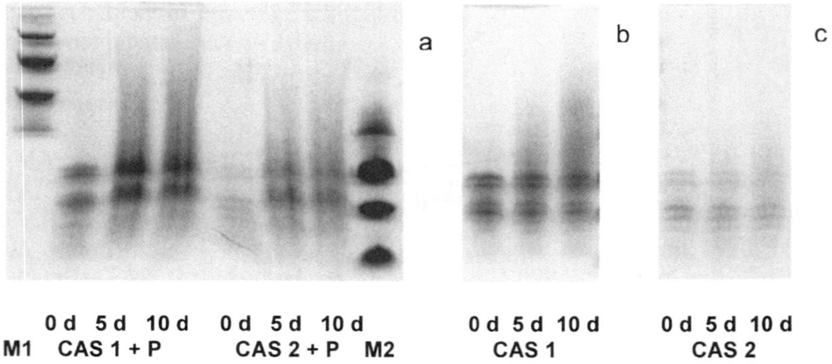

Figure 6 *SDS electrophoresis of hydrolysed CAS 1 and 2 after incubation with pectin (a) and CAS 1 (b) and CAS 2 (c) incubated without pectin.*

4 CONCLUSIONS

Mechanically degraded high methoxylated pectins were able to form conjugates with whey protein isolate that had excellent emulsifying properties. The degree of degradation had no significant influence.

In case of sodium caseinate and high methoxylated pectin, reduced molecular weights of the pectin as well as the protein component caused a higher thermodynamic compatibility. In the case of pectin degradation, this is mainly a result of the reduced excluded volume. The partial casein hydrolysis may have a comparable effect on compatibility, but other reasons can be assumed, too. The initial excluded volume of SC is much lower than that of pectin[17] and a reduction has a much smaller influence. Therefore, in this case the Flory-Huggins theory might be a better explanation for the increased compatibility. A higher hydrophilicity of the soluble part of the hydrolysate, used for compatibility tests, can play a role because of the lower hydrophobicity difference compared with pectin ($\Delta\chi$-effect).

The poor formation of conjugates and the resulting low emulsifying activity after incubation of sodium caseinate and depolymerised pectin can be explained with the compact protein structure that inhibits the necessary interactions and can be changed only for instance by protein hydrolysis. Instead, the single components SC and pectin potentially undergo separate degradation reactions which result in an increasing brown colour formation. The low emulsifying activity of casein hydrolysate-pectin mixtures is a result of the poor EA of the small casein molecules that can be improved to a certain extent by polymerisation during incubation. A limited conjugate formation seems to be possible at high protein DH but mostly the presence of pectins has no significant influence on this process.

Generally, the thermodynamic compatibility of sodium caseinate and pectin is not sufficient as an indicator for a formation of casein-pectin conjugates with good emulsifying properties. A close contact of the components in compatible systems is a necessary but not the only criterion for the success of the Maillard-type reaction. The positive effect of pectin demethoxylation on the compatibility with caseinate has different reasons than that found after pectin depolymerisation: Demethoxylation caused a higher compatibility because of an adjusting molecular structure of both components that also allowed a better formation of casein-pectin conjugates with improved emulsifying properties. In contrast, the pectin depolymerisation mainly changed the excluded volume but the molecular structure is nearly unchanged and the emulsifying properties after incubation decreased. It seems that the mechanism of increased compatibility is more important for the formation of protein-pectin conjugates than its extent.

References

1 V.Y. Grinberg and V.B. Tolstogusov, *Food Hydrocolloids*, 1997,**11**, 145.
2 A. Syrbe, W.J. Bauer and H. Klostermeyer, *Int. Dairy J.*, 1998, **8**, 179.
3 V.B. Tolstogusov, *Food Hydrocolloids*, 2003, **17**, 1.
4 A. Kato, Y.Sasaki, R. Furuta, K. Kobayashi, *Agric. Biol. Chem.*, 1990, **54**, 107.
5 E. Dickinson, V.B. Galazka, *Food Hydrocolloids*, 1991, **5**, 281.
6 A. Maroziene, C.G. de Kruif, *Food Hydrocolloids*, 2000, **14**, 391.
7 N. Neirynck, P. van der Meeren, M. Lukaszewicz-Lausecker, J. Cocquyt, D. Verbeken, K. Dewettinck, *Colloids and Surfaces A: Physicochem. Eng. Aspects*, 2007, **298**, 99.
8 R. Lutz, A. Aserina, Y. Portnoy, M. Gottlieb, N. Garti, *Colloids and Surfaces B: Biointerfaces,* 2009. **69**, 43.
9 O.G. Jones, E.A. Decker, D.J. McClements, *Food Hydrocolloids*, 2009, **23**, 1312.
10 B.L.H.M. Sperber, H.A. Schols, M.A. Cohen Stuart, W. Norde, A. G.J. Voragen, *Food Hydrocolloids*, 2009, **23**, 765.
11 U. Einhorn-Stoll, T. Salazar, B. Jaafar, H. Kunzek, *Nahrung/Food*, 2001, **45**, 332.
12 U. Einhorn-Stoll, M. Ulbrich, S. Sever, H. Kunzek, *Food Hydrocolloids*, 2005, **19**, 329.
13 Y. Antonov, N. Lashko, Y. Glotova, A. Malavikova, O. Markovich, *Food Hydrocollods,* 1996, **10**, 1.
14 U. Einhorn-Stoll, U. Kretzschmar, E. Lippert, *Acta Biotechnol.* 1994, **14**, 379.
15 U. Einhorn-Stoll, H. Kunzek, G. Dongowski, *Food Hydrocolloids*, 2007, **21**, 1101.
16 U. Einhorn-Stoll, M. Weiss, H. Kunzek, *Nahrung/Food*, 2002, **46**, 294.
17 V.B. Tolstogusov, *Nahrung/Food*, 2000, **44**, 299.

EFFECT OF WHEY PROTEIN TYPE AND XANTHAN GUM ON THE RHEOLOGICAL PROPERTIES AND EMULSION STABILITY OF THE FINAL MIXTURES

G. Panaras, G. Moatsou, I. Mandala
Department of Food Science & Technology, Agricultural University of Athens, 75 Iera Odos, 118 55 Votanikos, Greece, e-mail: imandala@aua.gr

ABSTRACT

Commercial whey protein concentrates (WPCs), coded as WP1-WP5 with different protein contents (i.e. 65 or 80%) and origin were studied. Their functional properties such as their solubility and oil holding capacity were determined. Furthermore, their protein profiles, rheological properties and storage stability in emulsions containing also olive oil and xanthan gum were investigated. Emulsion stability was measured by a technique based on multiple light scattering, using a near-infrared light source, over a week cold storage period. An exponential first order regression model was applied for comparing the emulsification ability of the different WPCs used. The phase separation kinetics differed in relation to a) the total protein concentration of the commercial samples and to b) the source of WPC. The presence of xanthan in emulsions was important, since a sample without xanthan separated completely after 8 hours. Neither the solubility nor the oil holding capacity was related to the emulsion stability. WP2 had a high oil-holding capacity and presented a better emulsion stability. A high lactose content impaired the protein emulsifying ability. Differences in protein denaturation and aggregation kinetics were revealed accordingly to the change in storage modulus (G') at increasing temperature. Gelation occurred over a temperature range from 71-76°C. The gelation temperature depended also on the qualitative composition of samples and specifically on a) the β-lactoglobulin ratio and b) the CMP (caseinomacropeptide) %. The greatest increase rate of G' was observed in WP5 samples and the lowest in WP3. The greatest, final G' values at 85°C were noticed in WP5 and the lowest in WP3. Similar final G' values to WP3 were observed in WP4 and WP2, which had the same initial protein content.

1 INTRODUCTION

Whey protein concentrates (WPCs) are widely used in the food industry, because they are considered highly functional and nutritional ingredients. Their functionality is related to their protein content; mainly β-lactoglobulin (β-lg) and α-lactalbumin (α-la).[1] However, their sensitivity to environmental conditions, such as pH, ionic strength, and/or thermal processing can limit their applications.[2] In many foods proteins and polysaccharides (PS) coexist. The parallel use of different PS could enhance protein functionalities through different protein-polysaccharide interactions.[3, 4]

Among different polysaccharides, xanthan gum (XG) is widely used in many products as a thickener, stabilizer, emulsifier and foaming agent. XG is an anionic polyelectrolyte mainly considered to be non-gelling, which is used for the control of viscosity, presenting weak-gel shear-thinning properties. Its relatively low viscosity at high shear rate makes it easy to mix, pour, and swallow.[5, 6]

The interactions between WPCs-XG depend among other factors on the iso-electric point (Ip) of whey proteins. Above Ip both protein and XG are negatively charged causing an electrostatic protein-polysaccharide repulsion, while below Ip oppositely charged polymers cause an electrostatic protein-polysaccharide attraction.

Thus, the investigation of WPCs-XG mixtures is of high interest especially regarding the stability and texture modification that they may provide in the final products. The objectives of this study were to investigate: a) the functional properties of WPCs having different protein content and origin regarding milk kind and b) their interactions with xanthan, specifically their emulsification capacity in its presence.

2 MATERIALS & METHODS

2.1 Whey Protein Concentrates (WPCs)

Commercial WPCs produced by ultrafiltration and spay drying were used. Samples were coded as follows and their chemical composition is presented in Table 1.

- WP1 (Lacprodan®-80 with 78±2% protein content, Arla Foods Ingredients amba-Denmark)
- WP2 (Nutrilac®DR-7015 with 65% minimum protein content, Arla Foods Ingredients amba-Denmark)
- WP3 (WHEYPRO65 produced from sheep and goat's whey with 65% protein content on dry matter, Hellenic Protein S.A.)
- WP4 (PROMIL WPC 65 produced from sheep and goat's whey with 65±1% protein content, Velco-Industrial Ingredients, Greece)
- WP5 (PROMIL WPC 80LF with 80±1% protein content, Velco-Industrial Ingredients, Greece)

Table 1 *Chemical composition of the WPCs used in the study, according to the manufacturers*

Sample	Protein %	Fat %	Ash %	Lactose %
WP1	78±2	max. 8	max. 3.5	7±2
WP2	min. 65	11-17	max. 5	max. 10
WP3	65 on d.m.	3.8	3.3	23
WP4	65±1	max. 4	max. 3.5	min. 23
WP5	80±1	max. 8	max. 3.5	max. 4

2.2 Evaluation of the protein profiles of WPCs

Aqueous solutions of WPCs were analyzed in duplicate by RP-HPLC on a Vydac C4 214 TP 5415 column (Separation group, Hesperia, CA 92345, USA). The elution times of native α-lactalbumin (α-la), β-lactoglobulin (β-lg) and caseinomacropeptide (CMP) were verified by analysis of purified standard proteins; their quantification was based on the area of the relevant peaks.[7]

2.3 Oil-holding capacity & solubility

A mixture of 5 ml olive oil and 0.25g WPCs was centrifuged at 4000 rpm for 5 min after vortexing for 30 s and left undisturbed for 5 min. The oil layer was carefully separated from the top of the tube using a syringe. The difference between the weight of oil added and the weight of oil separated at the top of the tube was calculated as a percentage to give oil-holding capacity.

For the determination of WPC solubility, about 0.5 g of WPC was dissolved in a tube containing 5 ml of phosphate buffer pH 7.5. After vortexing for 10 s, the tube containing the sample solution was left undisturbed for 5 min. Then, the tube was centrifuged at 4000 rpm for 10 min and the liquid phase with the components solubilized in it was removed with a syringe.[8] The sample tube was dried in an air drying oven for 2 h at 105°C and weighed. Solubility was determined by Equations 1, 2:

$$\text{Insolubility (\%)} = (\text{weight of dried insoluble sample} \times 100)/\text{sample weight} \tag{1}$$

$$\text{Solubility (\%)} = 100 - \text{Insolubility (\%)} \tag{2}$$

2.4 Emulsification capacity in the presence of xanthan

Stock solutions of WPC were prepared, dispersing WPC powder into phosphate buffer solution. The prepared samples (2% w/w) were stirred for 2 h, centrifuged at 1000 rpm for 30 min to eliminate air bubbles and insoluble substances. Supernatants were collected and stored at 4°C overnight to ensure full hydration of the biopolymers.[9, 10] Xanthan stock solutions (0.1% w/w) were also prepared following the same procedure as above. The pH of the final solutions was 7.5.

Emulsions containing WPC (2% w/w), xanthan (0-0.1% w/w), olive oil (20% v/v) were prepared by mixing all ingredients in a homogenizer at 11000/min for 2 min. Sodium azide (0.02% w/v) was added as an antimicrobial agent into the final emulsions. Emulsion stability was measured by a technique based on multiple light scattering, using a near-infrared light source, λ_{air}=850nm, over a week under cold storage. The optical characterization of the samples was evaluated using a Turbiscan MA 2000 (Formulation, Toulouse, France). The samples of 6 ml each were placed in tubes. Two detector devices were used in order to monitor light transmitted through the sample (180° from the incident light, transmission sensor), and backscattered by the sample (45° from the incident radiation, backscattering detector) along the height of the tube. The transmitted and backscattered light was presented, as a function of the sample height. Change in the thickness of the bottom phase (e.g. evolution change of the separated bottom phase) was observed. A non-linear fitting regression model was used for recording the bottom phase volume and backscattering change. The models used were of first order (Equation 3, 4).

$$y = A_1 + y_1 \cdot (1 - e^{-k_1 t}) \tag{3}$$

$$y = A_0 + y_0 \cdot e^{-kt} \tag{4}$$

where A_0, A_1, y_0, y_1, k, k_1 constants of the kinetics.

2.5 Rheological properties in the presence of xanthan

Temperature sweep tests (from 25 to 85°C) of WPC solutions (20% w/w) at small oscillatory shear tests were performed. A plate-plate rheometer (SR-5, Universal Stress Rheometer/Rheometrics Scientific, Inc., NJ) was used. Samples were placed between the plates, allowed to rest to equilibrate for 5 min before beginning experiments. A water trap was used in order to avoid water evaporation. Temperature sweep tests were performed from 25 to 85°C at a heating rate of 5°C /min until 60°C and 1°C /min until 85°C to simulate the effect of heating causing the denaturation of proteins. A strain in the range of linear viscoelastic region was selected. Values of storage modulus (G′) were recorded. All the tests were replicated three times.

3 RESULTS AND DISCUSSION

3.1 Protein profile

The quantification results of RP-HPLC analysis based on the chromatographic area of the relevant peaks are shown in Table 2. It is evident that the protein composition of WPCs from sheep and goat's whey (WP3 and WP4), differs with regard to high CMP and low α-la contents. On the other hand, the quantification results regarding the WPCs from cow's milk were similar regardless of their total protein content, i.e. 65 or 80%. Thus, the native protein profile of WPCs was different according to the source of whey.

Table 2 *Quantification of native whey proteins of WPC samples by means of RP-HPLC*

Sample	Protein %	CMP[1] %	α-la[2] %	β-lg[3] %	β-lg/α-la	β-lg/CMP	α-la/CMP
WP1	80	8.5	26.2	65.3	2.5	7.7	3.1
WP2	65	8.5	27.1	64.4	2.4	7.6	3.2
WP3	65	14.2	19.7	66.1	3.4	4.7	1.4
WP4	65	12.7	21.4	66.0	3.1	5.2	1.7
WP5	80	8.6	27.2	64.2	2.4	7.5	3.2

[1] caseinomacropeptide, [2] α-lactalbumin, [3] β-lactoglobulin

3.2 Oil-holding capacity & solubility

According to the data presented in Table 3 the greatest oil-holding capacity was observed in WP2 solution, which had the lowest solubility value. The solubility depends on the level of protein denaturation and aggregation of the samples.[11] Thus, it depends strongly on the source and treatment of the whey used for their preparation, i.e. cheese making conditions and thermal treatment of the whey.[12]

Table 3 *Oil-holding capacity and solubility of whey protein concentrates*

Samples	Oil-holding capacity %	Solubility %
WP1	14.5 (0.10)	97.8 (0.16)
WP2	18.3 (0.20)	82.8 (0.15)
WP3	14.0 (0.11)	96.8 (0.14)
WP4	14.3 (0.12)	97.7 (0.13)
WP5	14.9 (0.13)	99.2 (0.15)

3.3 Emulsification capacity in the presence of xanthan

The phase separation kinetics and specifically the volume of the down phase (or thickness change) presented in Figure 1 and Figure 2 differed in relation to a) the total protein concentration of the initial commercial samples and to b) the source of WPC used for the same initial concentration.

WPCs from the same producer, but of different initial protein content (i.e. 65 or 80%), exhibited differences probably due to the fact that the fraction and quantity of protein macromolecules they contain influenced their emulsification capacity in a different way. Thus, the higher the protein content of WPC (i.e. 65 or 80%), the slower the separation of the down phase (e.g. comparisons between WP1-WP2 or WP4-WP5) (Fig. 1). Furthermore, proteins from different production companies but of similar protein content presented different emulsification ability in the final emulsion. EWP2 presented the greatest stability and the EWP4 the lowest.

Factors that affect emulsification properties of WPCs are protein solubility, β-lg concentration, salt content, free sulfydryl content, fat content, pH and presence of other solutes.[11,13,14,15] Taking into consideration that the composition of the samples varied considerably (Table 1), their emulsifying behaviour can be attributed to several factors.

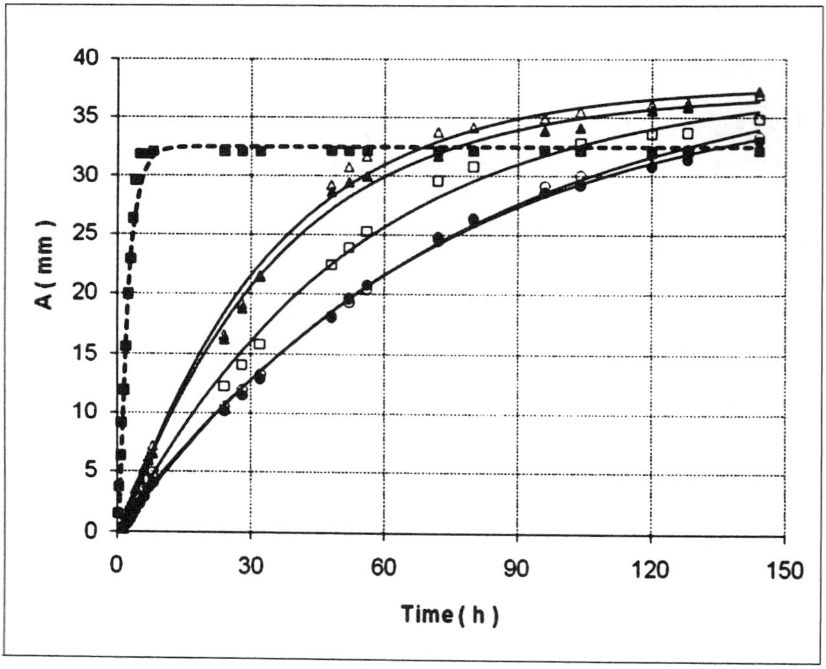

Figure 1 *Volume change kinetics of pseudoemulsions (EWPi) containing olive oil, xanthan gum (0.1% w/w) and different WPCs. (● WP1, ○ WP2, □ WP3, Δ WP4, ▲ WP5, ■ WP1a)*

Regarding samples WP1, WP2 and WP3 that had the same protein content and ash content, it seems that the high lactose content of WP3 and WP4 (23%) impairs their emulsifying ability compared to the low content of WP2 (10%). Furthermore, the α-la, which shows better emulsifying properties than β-lg[15], is present in higher quantity in WP2. WP2 had a high oil-holding capacity and seems to be more effective in emulsion stabilization. All other samples had similar oil-holding capacity values and thus differences in their emulsification ability were not evident.

The equilibrium volume of control sample (EWP1a) prepared without the stabilizer used (xanthan) and containing WP1 was reached in different time intervals (from about 2 to 5 days) for all samples, depending on the WPC sample used. The fastest phase separation was observed in EWP4, the slowest in EWP1. Equilibrium in the presence of xanthan was not reached in any of the samples investigated for the storage period of 6 days. Furthermore, the volume of the separated phase was 65.4% for EWP1a, since it was 61% for the respective sample with xanthan (EWP1).

According to the separation kinetics as expressed by turbidity change (backscattering values), EWP2 was the most ''stable'' sample (slower volume change) (Figure 2). Kinetic constants of destabilization during storage are presented in Table 4. k_1 values indicate that a lower rate of bottom phase thickness increase occurs in samples EWP2 and EWP1, whereas approximately a double thickness increase rate was noticed in EWP4 and EWP5 samples. The presence of a stabilizer seems to be quite important, since k_1 is only 3% of that without xanthan (comparison between EWP1 and EWP1a). Backscattering kinetics were similar to those of thickness change during storage. However, the k value of EWP1 was 5% of that of the control samples without xanthan, e.g. a slightly greater value was observed than that mentioned above for thickness change. Furthermore, differences of k values among samples investigated were more limited than those of the k_1 values among the respective samples. This could mean that backscattering values changed a little slower than thickness values. The slower change in backscattering could be linked to the remaining macromolecules in the continuous phase, e.g. the movement of xanthan or unabsorbed whey proteins indicates that they move slower than oil droplets. As the pH of the prepared samples was above Ip, protein-solvent interactions increase. At pH 7.0 phase separation is evident because of a high negative charge carried by the protein.

When the boundary between the serum and the cream layer is very sharp, the separation could be a consequence of flocculation. A more gradual transition between serum and cream phase would be observed due to creaming of the largest droplets in case of a broad particle size distribution.[16] In our case flocculation was evident, thus the phase separation was very fast. Xanthan gum can increase the rate of creaming due to the depletion flocculation (a volume restriction effect). At higher added hydrocolloid concentrations, the depletion interactions can be stronger. In this case creaming is inhibited due to the viscoelastic character of the interconnected regions of emulsion droplets that can become flocculated into a gel-like network.[17]

Table 4 *Kinetic constants of the destabilization process of pseudoemulsions containing different types of WPCs, olive oil and xanthan gum (0 or 0.1% w/w)*

	Thickness				Backscattering			
Samples	A_l	y_l	k_l (h^{-1})	R^2	A_0	y_0	k (h^{-1})	R^2
EWP1a	-4.80	37.32	0.474	0.986	15.11	26.21	0.480	0.938
EWP1	-0.45	38.78	0.014	0.999	12.69	40.56	0.024	0.995
EWP2	0.005	40.32	0.013	0.999	14.29	42.85	0.021	0.992
EWP3	-0.76	38.70	0.019	0.997	14.72	40.33	0.026	0.996
EWP4	-1.31	39.00	0.029	0.997	15.53	38.03	0.043	0.990
EWP5	-0.96	37.94	0.028	0.998	15.90	38.46	0.044	0.982

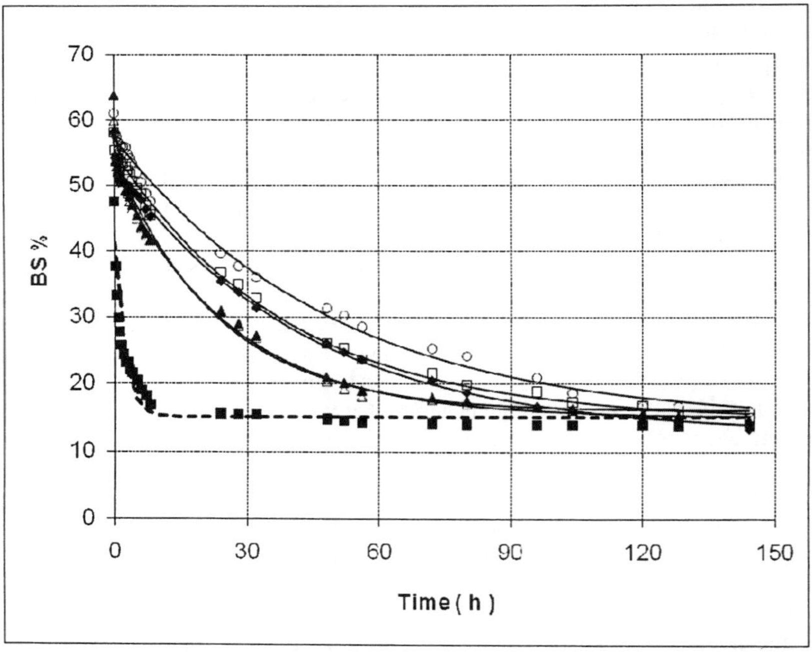

Figure 2 *Backscattering change kinetics of pseudoemulsions (EWPi) containing olive oil, xanthan gum (0.1% w/w) and different WPCs (● WP1, ○ WP2, □ WP3, Δ WP4, ▲ WP5, ■ WP1a)*

3.4 Gelation kinetics

Gelation is a result of protein denaturation and aggregation. The rapidly rising values of G′ in Figure 3 indicate a structural modification of WP-solutions, e.g. a transition from a sol to a gel structure. In this research, gelation occurs at a temperature range from 71-76°C, depending on the type of WPC used. Gelation temperature did not depend on WPC commercial source and initial protein content (i.e. 65 or 80%), but seems to depend on qualitative composition of samples and specifically a) the β-lactoglobulin ratio and b) the CMP %. Therefore, with regard to WPC with the same protein content (i.e. WP2, WP3 and WP4) the greater the CMP % values the higher the gel point (72.08, 75.20 and 75.06°C). The same was also true for the β-lg/α-la ratio.Gel point was found by extrapolating the G′ values to zero.

The greatest increase in G′ was observed in WP5 samples and the lowest in WP3. WP2 showed an increase in G′ initially and a reached plateau value thereafter. On the other hand WP1 presented a relatively fast G′ increase (at 72.5°C). In this sample, an initial peak was observed, then G' decreased before increasing again at 74°C (Fig. 3). The greatest, final G′ values at 85°C were observed in WP5 and the lowest in WP3. Similar final G′ values to WP3 were observed in WP4 and WP2. The final G′ value indicates the final elastic character of the gels and it was clearly related to the protein concentration of the samples. For a 15% increase in the total protein amount in the commercial sample an approximately 3 fold greater G′ value was observed (e.g. WP4 and WP5 values). The β-lg content of WPCs is linked to their gelling ability, since it is the most important protein for gelation, whereas CMP is detrimental to gel strength and water holding properties.[15,18,19]

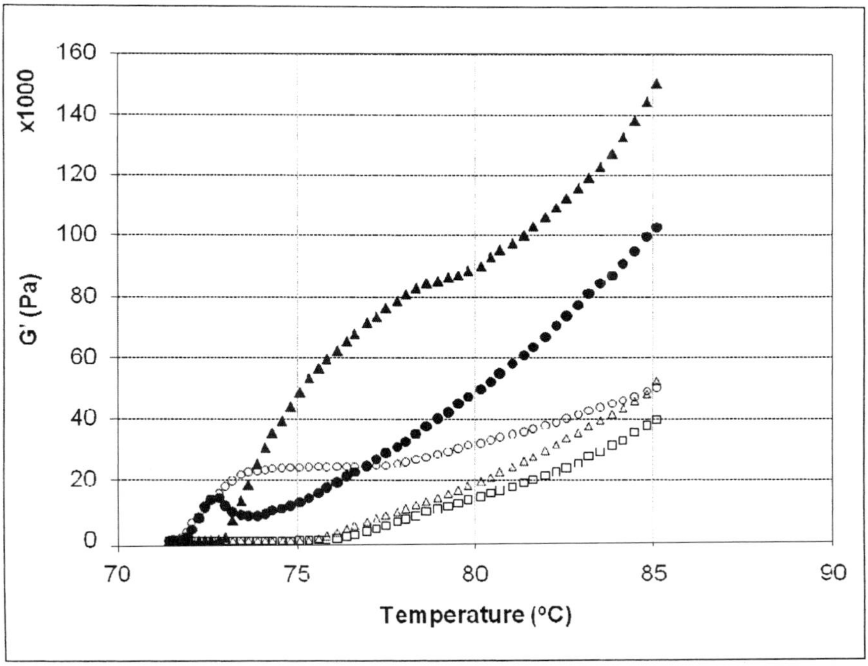

Figure 3 *Influence of temperature on G′ values of different WPCs solutions (20% w/w) (● WP1, ○ WP2, □ WP3, Δ WP4, ▲ WP5)*

Although the β-lg content of samples WP3 and WP4 was higher than that of WP2, their gelation temperature was higher. Solubility could be linked to the gel strength of samples investigated.[20] WP5 had the greatest solubility and the greatest gel strength values at heating. WP2 of low solubility index had a different behaviour compared to the rest of the samples as it was also observed in case of emulsion stability. However, differences among solubility values of all other samples were limited and general conclusions should be avoided.

4 CONCLUSIONS

From a technological perspective, the investigation of whey proteins of different protein profiles contributes to an understanding of their functional properties, i.e. differences in gelation kinetics or in their interactions with the gums. Whey protein concentrates, which had similar total protein content, had different emulsifying and rheological properties. The treatment used for their isolation seems to be quite important and determines the further applicability of the proteins. Further research is necessary for understanding the effect of treatment on the functional properties of whey protein concentrates.

REFERENCES

1. A. Tosi, L. Canna, H. Lucero and E. Re, *J. Food Chem.*, 2007, **100**, 794.
2. A.A. Perez, C.R. Carrara, C.C. Sánchez, L.G. Santiago and J.M.R. Patino, *Food Hydrocolloids*, 2009, **23**, 1253.
3. E. Dickinson, *Food Hydrocolloids*, 2003, **17**, 25-39.
4. A. Syrbe, W.J. Bauer and H. Klostermeyer, *Int. Dairy J.*, 1998, **8**, 179.
5. K.S. Kang and D.J. Petit in *Industrial gums. Polysaccharides and their derivatives*, eds. R.L. Whistler and J.N. BeMiller, Academic Press, New York and London, 1993, p 343.
6. A. Nussinovitch, *Hydrocolloid applications. Gum technology in the food and other industries*, Blackie, UK, 1997.
7. G. Moatsou, A. Hatzinaki, I. Kandarakis and E. Anifantakis, *Food Chem.*, 2003, **81**, 209.
8. T. Ji and Z.U. Haque, *Int. J. Food Sci.Technol.*, 2003, **38**, 453.
9. J. Li, M.M., Ould Eleya, and S. Gunasekaran, *Food Hydrocolloids*, 2006, **20**, 678.
10. C. Sun, S. Gunasekaran and M.P. Richards, *Food Hydrocolloids*, 2007, **21**, 555.
11. M.T. Patel and A. Kilara. *J. of Dairy Res.*, 1990, **73**, 2731.
12. M.A. De la Fuente, Y. Hemar, M. Tamehana, P.A. Munro and H. Singh, *Int. Dairy J.*, 2002, **12**, 361.
13. J.E. De Wit in *Developments in dairy chemistry*, ed. P.F. Fox, Elsevier, Essex, 1989, p 285
14. S.Y. Liao and M.E. Mangino, *J.Food Sci.*, 1987, **52**, 1033.
15. J.N. De Wit, G. Klarenbeek and M. Adams, *Netherlands Milk Dairy J.*, 1986, **40**, 41.
16. N. Neirynck, K. van Lent, K. Dewettinck and P. van der Meeren, *Food Hydrocolloids*, 2007, **21**, 862.
17. E. Dickinson, *Food Hydrocolloids*, 2009, **23**, 1473.
18. D.M. Mulvihill and J.E. Kinsella, *Food Technol.*, 1989, **41**, 102.
19. P.D. Velth and E.C. Reynolds, *J. Dairy Sci.*, 2004, **87**, 831-840.
20. R. Jost, *Trends Food Sci. Technol.*, 1993, **4**, 283.

PHASE EQULIBRIA AND RHEOLOGICAL BEHAVIOUR OF POLYSACCHARIDE-PROTEIN AQUEOUS SOLUTIONS WITH SUGAR

Asja Portsch, Fotios Spyropoulos and Ian T. Norton

Centre of Formulation Engineering, University of Birmingham, B15 2TT, Birmingham, UK

1 INTRODUCTION

Biopolymer model systems, such as those of proteins and polysaccharides, have been applied extensively to aid the understanding of structuring mechanisms in foodstuffs. The understanding of the structural behaviour of these simpler systems in relation to their processing (under high shear conditions), storage, handling (low shear) and consumption (medium shear) can help to create foods of a particular function and affect their marketability [1,2].

Polysaccharides, proteins, and sugars represent common structuring, nutritive and taste-imparting components of convenience dairy foods [2-5]. Their microstructure is formed using the intermolecular interactions between their components. One of the most important of these is the thermodynamic repulsion between proteins and polysaccharides. This incompatibility creates an emulsion-like microstructure and leads to segregation into two equilibrium phases which differ greatly in concentration [3,6,7]. Although added to enhance taste, sugars also affect the functionality of the macromolecules and can alter solvent quality [4-8]. It may therefore be expected that their presence affects the microstructure, and as such the phase equilibria and rheological behaviour of these foods.

To a large extent, it is the microstructure that governs the rheology, stability, mouthfeel properties and texture of the final product, and it may be altered by varying a range of factors, such as temperature, pressure, solvent type, pH, ionic strength, molecular weight, polydispersity, formulation, and shear [9,10]. However, it is the 'phase sense' of the system that is of the greatest importance in determining the overall structure. The phase sense, defined by which phase is continuous and which dispersed, can be influenced not only by the mass of each of the macromolecular components, but also by various other factors such as temperature, shear and gelation. In this way a variety of structures, i.e. droplets of various shapes and sizes may be obtained [1,8,10].

The aim of the study described herein was to investigate the influence of sugar (sucrose) concentration on the formation of the microstructure, the phase behaviour and the rheological behaviour of two model dairy aqueous-two-phase-systems (atps) at 20°C and pH 6. The atps under investigation were (i) an aqueous mixture of locust bean gum (LBG) and sodium caseinate (NaCAS), and (ii) an aqueous mixture of guar gum (GG) and sodium caseinate. These atps were first studied under quiescent conditions (using phase diagrams and microscopy) and subsequently analysed under shear (using viscometry). By investigating the structuring mechanisms in dairy aqueous emulsions, such as those of our two model systems, we herein complement the studies by Schorsch *et al.* of the effects of sugar on skimmed milk/LBG, native phosphocaseinate/LBG, and micellar casein/LBG systems at 5ºC and pH 6.8 [6,7]. In this study NaCAS was used in order to avoid the complexity of biological protein solutions, such as skimmed milk and whey, and LBG and GG were selected because of their widespread use as dairy thickeners and their tendency to separate out of protein solutions.

2 MATERIALS

Samples of GG (Sigma-Aldrich, Dorset, UK), LBG Grinsted® 246 (Danisco, Denmark) and NaCAS were kindly donated by Unilever Research Labs (Sharnbrook, Colworth, UK) and DMV International (Veghel, The Nederlands), respectively. To keep pH constant a citric buffer (Fisher Scientific, Loughborough, UK) was prepared. "Tate+Lyle" sugar (granulated sugar) was obtained from a local supermarket, and preservative potassium sorbate from Sigma-Aldrich (Dorset, UK).

3 METHODS

3.1 Phase diagrams

Bulk solutions of NaCAS (15% wt), LBG (1.5 % wt) and GG (1.5% wt) were prepared in a citrate buffer (pH 6). The bulk solutions, preserved using potassium sorbate (0.1% wt), were then mixed together with more buffer and sugar in order to produce binary solutions of various NaCAS/LBG (GG)/sugar partial concentrations, with constituents in the ranges (i) 0.0005-1.3% wt LBG (ii) 0.005-10% wt NaCAS, and (iii) 0-60% wt sugar. These concentrations were selected using a "trial and error" method in order that the threshold of miscibility could be determined as precisely as possible. The binary solutions were left at 20 ºC for 24 h and were then assessed for the incidence of phase separation.

For each sugar concentration, close pairs of data points (i.e. the bulk concentrations of the macromolecular components in the binary solutions) were selected from each side of the miscibility boundary. These sets of data points were selected in the immediate vicinity of the miscibility boundary and were used to calculate average data point values, which were then plotted in a two-axis phase diagram (data points not shown). The binodal curve, separating the immiscible and miscible regions, was then fitted through these points (Figure 1).

Various binary solutions from the immiscible regions were then assessed in terms of their phase-separated volume fractions, which were then used to draw tie-lines. In Figure 1, one tie-line is shown for each sugar concentration used. The slopes of the tie-lines (STL) were calculated in order to compare the asymmetries of the various systems. Their average values are given in the legends of Figure 1.

3.2 Confocal microscopy

A confocal scanning laser microscope (Leica DMIRE2, Heidelberg, Germany) with a He-Ne laser lamp of wavelength 633 nm was used to observe the quiescent microstructures of both systems for each sugar concentration and at 20 ºC. The images were obtained at 10x magnification at 400 scans/min, and were line-averaged over eight scans. The samples used were taken from the same tie-line in the phase diagram and are shown herein as a function of the volume fraction. An example showing microstructures of the LBG/NaCAS system with no sugar is given in Figure 2. Other sugar concentrations (0- 40% wt) were also examined. Before inspection, the samples were stained with Nile blue (0.01% wt, aq, Sigma-Aldrich, UK).

3.3 Viscometry

Steady-state flow measurements were carried out for each of the equilibrium phases (LBG/GG-rich top and NaCAS-rich bottom phases) and for five of their blends. The equilibrium phases were obtained after phase separation for 24 h at 20 °C for systems consisting of 0.6% wt LBG (GG), 4% wt NaCAS, and 0, 15, 20 and 40% wt sugar. The equilibrium phases were then mixed together to produce the blends (containing 5-95% of the polysaccharide-rich phase). In section 4.3 only the bottom NaCAS-rich equilibrium phase, and blends with 25, 50, 75 and 95% of the LBG-rich phase are shown, as an example. Other systems were also subsequently prepared using powdered ingredients. The equilibrium phases and polysaccharide solutions involved in their preparation were centrifuged (0.5-4 h, 5000 rpm, J2-21 High speed Beckman centrifuge).Viscometry tests were carried out for shear rates in the range 0.1-1000 s^{-1} at 20 °C, using cone and plate geometries (4°, 40 mm steel cone with a gap of 150 μm) and a rheometer (Bohlin Gemini 150 HR nano, Malvern, UK).

4 RESULTS AND DISCUSSION

4.1 Phase behaviour

The effect of sugar concentration on the phase behaviour of the two model atps was investigated at 20°C and pH 6. When the partial concentrations of the two components in either of the atps were sufficiently high, the mixtures became segregated into two coexisting equilibrium phases, namely a top (polysaccharide-rich), and bottom (protein-rich) phase.

Figure 1 shows the phase diagrams obtained for both systems and for up to 60% wt sugar. Each binodal corresponds to a system with a different sugar concentration. One tie-line is drawn for each of the sugar concentrations, and an average slope value is given. STL demonstrate the low compatibility between both sets of macromolecules and the high asymmetry of the phase diagrams (i.e. the existence of two phases having very different concentrations).

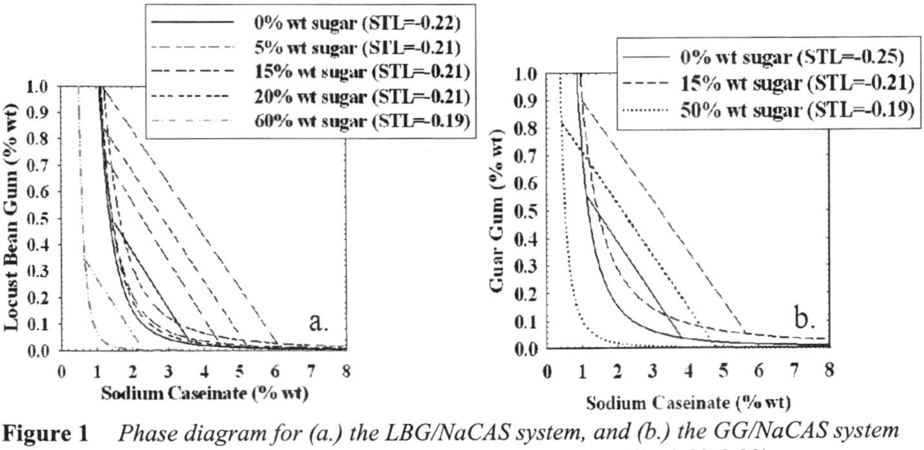

Figure 1 *Phase diagram for (a.) the LBG/NaCAS system, and (b.) the GG/NaCAS system at 20 °C, pH 6 using different sugar concentrations (R=0.89-0.99).*

Even though the tie-lines did not vary significantly with sugar concentration, clear shifts in binodals were nevertheless observed. The miscibility between the components increased when up to 15% wt sugar was present in the mixture. However, the components were less miscible when more than 15% wt sugar was added. For 50 and 60% wt of sugar, the miscibility was even less than that of the system without sugar.

The explanation of the observed effect of sugar on the phase behaviour of our systems may be found in studies of the effect of sugar on the individual macromolecular solutions [5,8]. In their study of LBG and GG, Richardson *et al.* [8] reported increased solubilities for solutions containing up to 20% wt of sugar. However, the addition of more than 20% sugar caused a reduction in the "solvent quality" and caused the molecules to compete for dissolution in the solvent. Similarly, Belyakova *et al.* [5] reported elevated hydrophilicity of NaCAS above the isoelectric point of the protein and for sugar concentrations of 20% wt. The authors proposed that the formation of hydrogen bonds between NaCAS and sugar is a probable explanation for the reduction in their molecular size.

Our findings related to the effect of sugar on the phase behaviour presented herein were supported by studies of other model systems by Schorsch *et al.* [6,7], who found that at 5°C and pH 6.8, the miscibility of systems containing 20% wt of sugar increases. The authors explained the phenomena as representing an improvement in the solvent quality, i.e. in terms of its change from a thermodynamically poor solvent to a thermodynamically good one. In contrast to our own findings, they found that the effect of the addition of sugar above 20% wt significantly changed STL, but not the positions of the binodals. The changes in STL point to the effect of the dilution/concentration taking place. This means that at higher sugar concentrations, the polysaccharide-rich phases become diluted and the protein-rich phases become more concentrated in protein. In our study, the addition of more than 20% wt sugar to either of the systems under investigation decreased the extent of the one-phase region of the phase diagram. This is a result of the system being too concentrated, which caused the polysaccharides and proteins to separate at lower concentrations. We also observed small changes in STL, but these were far less pronounced than those observed by Schorsch *et al.* [6,7]. The reason for this probably lies either in the differences in pH and/or T, or in the method used to obtain the phase diagrams.

4.2 Microstructure

The microstructures of the systems in the presence of the sugar concentrations of 0, 5 and 20% wt were all similarly droplet-like. Hence, in the following discussion only the system containing LBG and no sugar is used to illustrate the behaviour of most investigated systems (Figure 2). The only exception is the appearance of the samples with 40% wt sugar, which display non-droplet-like microstructures. We propose that these microstructures are products of a high viscosity environment and that droplets would form eventually.

The effect of sugar and of the volume fraction on the quiescent microstructure of our two model systems was tested by making a set of blends (mixtures of two equilibrium phases) from the same tie-line of the phase diagram (Figure 2). The confocal micrographs of such mixtures (Figure 2) showed droplet-like structures, which resembled traditional oil/water emulsions and may therefore be referred to as "water-in-water emulsions" [10].

By gradually increasing the volume fraction of the polysaccharide-rich phase from zero to 95%, we established that under static and low shear conditions, the phase sense depends only on the volume fractions of the phases. The phase representing the higher volume fraction assumes the role of the continuous phase, while the lower volume fraction phase becomes the dispersed phase. Figure 2 shows the microstructures of (a) pure protein-rich, (b, c) protein-continuous, (d) bi-continuous, and (e, f) polysaccharide-continuous systems. It may also be observed that under quiescent conditions, phase inversion occurs at equivolumetric conditions, although it does not occur instantly. Bi-continuous systems, where various parts of the system assume different roles (some parts are protein-continuous, others polysaccharide-continuous), form over a specific range of volume fractions.

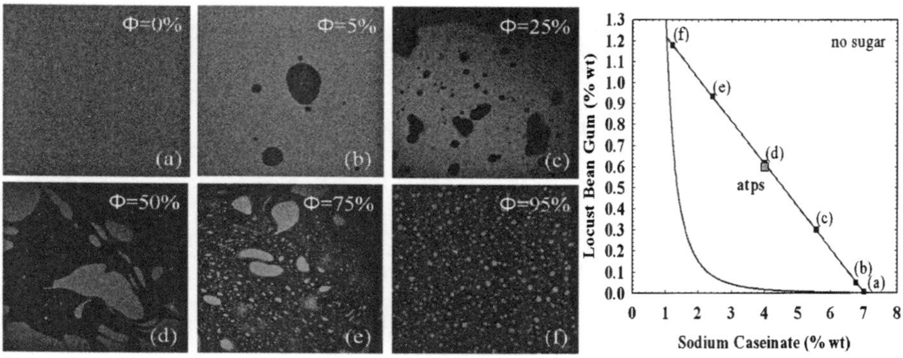

Figure 2 *Confocal micrographs (a)-(f) of LBG/NaCAS systems as a function of the volume fraction. The phase diagram shows their concentrations. F is the volume fraction of the LBG-rich phase (marked in black). Each micrograph is 1.5 mm wide.*

4.3 Flow behaviour

In the following section, the effect of the quiescent microstructure on flow characteristics is described, together with its behaviour under shear. The flow characteristics of three different microstructures (protein-continuous, bi-continuous, and polysaccharide-continuous) were investigated. The same approach to sample preparation was applied as in the previous section. The increase in the volume fraction of the more viscous polysaccharide-rich phase caused the overall viscosity of both systems to rise. While the protein-continuous samples exhibited normal shear thinning behaviour, complex flow behaviour was observed for the polysaccharide-continuous systems. For these samples shear thickening regions ("viscosity peaks") were observed at a shear rate of about 100 s^{-1} (Figure 3).

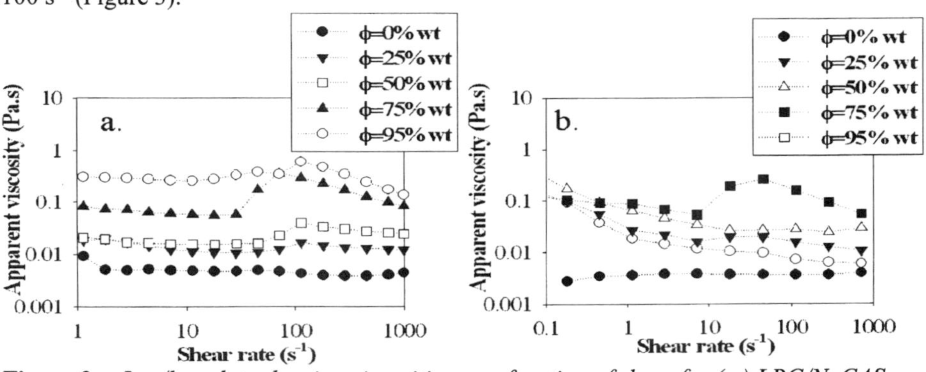

Figure 3 *Log/log plots showing viscosities as a function of shear for (a.) LBG/NaCAS and (b.) GG/NaCAS systems in the absence of sugar.*

The effect of sugar concentration on flow of the above microstructures was then investigated. In Figure 4, the flow behaviour of two different microstructures (25% and 75% LBG-rich atps) is shown. Regardless of their microstructure, the addition of sugar to these systems caused an increase in their average viscosities. For the polysaccharide–continuous samples, two distinct types of flow behaviour were observed, namely (i) complex flow behaviour for sugar concentrations below 20% wt and (ii) normal shear thinning flow behaviour for systems above 40% wt of sugar. The viscosities of protein-continuous systems with zero and 5% wt of sugar (Figure 4a.) were close to that of water.

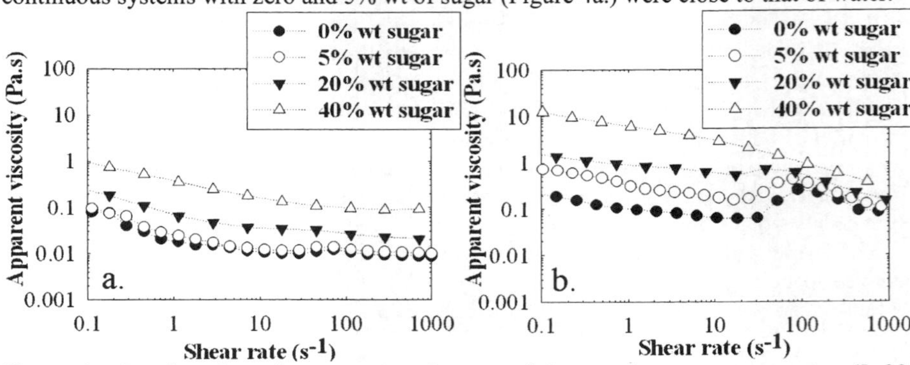

Figure 4 *Log/log plots of viscosity as a function of shear and sugar concentration (5, 20 and 40% wt) for the LBG/NaCAS system containing (a.) 25% of the LBG-rich phase, and (b.) 75% of the LBG-rich phase.*

Furthermore, a series of tests was designed to determine the cause for the existence of the complex flow behaviour in polysaccharide-continuous systems. First, the sample preparation was trialled by removal of the insoluble residues from LBG and GG solutions that were used in the preparation of our systems. Second, the sample preparation procedure was examined by comparing the initial preparation from equilibrium phases with instant preparation from powdered ingredients. Three distinct microstructures and both systems were tested. Third, the effect of pre-shear on flow behaviour of the samples was looked at (data not shown). Finally, flow behaviour of pure solutions and equilibrium phases was measured. GG, LBG, NaCAS solutions of various concentrations and centrifugation times (example of some in Figure 5), as well as centrifuged equilibrium phases were tested for the existence of the viscosity peaks.

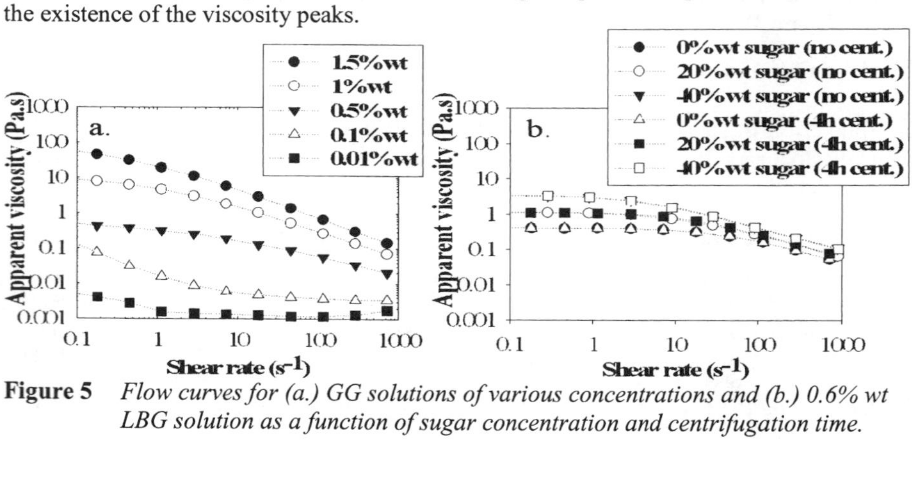

Figure 5 *Flow curves for (a.) GG solutions of various concentrations and (b.) 0.6% wt LBG solution as a function of sugar concentration and centrifugation time.*

The results of these tests showed that in systems that were initially polysaccharide-continuous, complex flow behaviour was observed, regardless of the method by which the samples were prepared and whether or not they were subjected to pre-shear. The same behaviour was absent in pure solutions (Figure 5) and in purified equilibrium phases. Hence, in relation to these findings, we believe that a two phase emulsion-like structure is required as a prerequisite for the occurrence of viscosity peaks [11].

After determining the underlying cause of the complex flow behaviour in polysaccharide-continuous samples, our attention was directed to finding an explanation for it. During the course of the experiments, we were able to determine that (i) initially polysaccharide-continuous systems differ from protein-continuous systems in their phase sense, (ii) polysaccharide-rich equilibrium phases are more viscous than protein phases regardless of the presence of sugar, (iii) the presence of sugar significantly increases the viscosities of both equilibrium phases. Because the dispersed phase determines the flow of the system and because the shear-induced rise in viscosity ("peaks") is measured for polysaccharide-continuous systems, the occurrence of a shear-induced phase inversion event seems a plausible explanation.

Shear-induced phase inversion is not a newly-discovered phenomenon [10,12]. It is believed to be triggered by the onset of the coalescence of the dispersed phase (i.e. the protein-rich phase in our case), which because of its low viscosity starts folding in the fractions of more viscous continuous phase (i.e. the polysaccharide-rich phase in our case). Eventually, through the formation of bi-continuous systems, phase inversion can occur. By taking the phase with the higher viscosity as the dispersed phase, the emulsion reduces its energy while under shear. In contrast, in protein-continuous samples the dispersed phase (i.e. the polysaccharide-rich phase in our case) is already the more viscous, and a phase inversion event does not therefore occur. At higher shear rates, the formation of smaller droplets takes place instead.

In a high sugar environment, the apparent viscosities of both equilibrium phases are elevated and shear-induced phase inversion does not occur. As a result of the high viscosity, medium coalescence (an important step that leads to the phase inversion event) of the dispersed phase is inhibited, and, as a consequence, complex flow behaviour is not observed. Furthermore, there are reports [13,14] of some systems under shear that no longer assume a droplet-like microstructure and form instead string-like structures to reduce the free energy of the system. However, since we were unable to observe the microstructure of our systems under shear, this latter explanation remains somewhat speculative and requires further verification.

5 CONCLUSIONS

The effect of sugar on the phase and flow behaviour of two model aqueous mixtures (LBG/NaCAS and GG/NaCAS) was investigated at 20°C and pH 6. The presence of sugar affected the behaviour of both substances. The addition of less then a 15% wt sugar increased the compatibility of these systems, but further increases in sugar concentration caused the miscibility to decrease. The presence of sugar affected the binodals more significantly than the tie-lines. The rheological behaviour of the systems was related to their quiescent microstructure. Depending on the phase sense, the systems exhibited either shear-thinning or complex flow behaviour. Shear-induced phase inversion occurred only in polysaccharide-continuous systems. The addition of up to 20% wt sugar did not affect phase inversion phenomena, however an additional increase of sugar concentration was found to repress it. The high viscosities of the equilibrium phases, caused by the high levels of sugar, are believed to be the reason for the observed inhibition of the phase inversion.

References

1 Norton, I.T. & Frith, W.J. (2001). Microstructure design in mixed biopolymer composites. Food Hydrocoll., 15, 543–553.

2 Hermansson, A.M., Langton, & Olsson, C. (2004). Designing structure into food, In: Gums and stabilisers for the food industry 12, Eds: P.A. Williams, G.O. Phillips, RSC, 3–12.

3 Vega, C. & Goff, H.D. (2005). Phase separation in soft-serve ice cream mixes: rheology and microstructure. Int. Dairy J., 15, 249–254.

4 Antipova, A.S., & Semenova M.G. (1995). Effect of sucrose on the thermodynamic incompatibility of different polymers. Carbohydr. Polym., 28, 359-365.

5 Belyakova, L.E., Antipova, A.S., Semenova, M.G., Dickinson, E., Matia Merino, L., & Tsapkina, E.N. (2003). Effect of sucrose on molecular and interaction parameters of sodium caseinate in aqueous solution: relationship to protein gelation. Colloids Surf., B, 31, 31–46.

6 Schorsch, C., Clark, A.H., Jones, M.G., & Norton, I.T. (1999). Behaviour of milk protein/polysaccharide systems in high sucrose. Colloids Surf. B, 12, 317-329.

7 Schorsch, C., Jones, M.G., & Norton, I.T. (1999). Thermodynamic incompatibility and microstructure of milk protein/locust bean gum/sucrose systems. Food Hydrocoll., 13, 89–99.

8 Richardson, P.H., Willmer, J., & Foster, T.J. (1998). Dilute solution properties of guar and locust bean gum in sucrose solutions. Food Hydrocoll., 12, 339–348.

9 Tolstoguzov, V.B. (1998). Functional properties of protein-polysaccharide mixtures. In: Functional properties of food molecules, pp. 252–277. Eds: S.E.Hill, D.A.Ledward, J.R.Mitchell. Aspen Publishers: Gaithersburg.

10 Foster, T.J., Brown C.R.T. & Norton I.T. (1996), Phase inversion of W/W emulsions, In: Gums and stabilizers for the food industry 8, Eds: G.O. Phillips, P.A. Williams and D.J. Wedlock, Oxford University Press, 288-306.

11 Efthimiadu, I. & Moore, I.P.T. (1994). Phase inversion of liquid-liquid dispersion produced between parallel shearing planes. Chem. Eng. Sci., 49, 1439–1449.

12 Brown, C.R.T., Foster, T.J., Norton, I.T., & Underdown J. (1995). Influence of shear on the microstructure of mixed biopolymer systems. In: Biopolymer mixtures, Eds.: S.E. Harding, S.E. Hill and J.R. Mitchell, Nottingham University press, Nottingham, 65–83.

13 Wolf, B., Scirocco, R., Frith, W. J., & Norton, I. T. (2000). Shear-induced anisotropic microstructure in phase-separated biopolymer mixtures. Food Hydrocoll., 14(3), 217–225.

14 Spyropoulos, F., Frith, W.J., Norton, I.T., Wolf, B., & Pacek, A.W. (2008). Sheared aq. two-phase biopolymer-surfactant mixtures. Food Hydrocoll., 22 (1), 121–129.

INFLUENCE OF PREPARATION METHOD ON THE QUALITY OF XANTHAN-LOCUST BEAN GUM MIXED GELS

G. Sworn and E. Kerdavid
Danisco France SAS, 20 rue Brunel, 75617, Paris, France

1 INTRODUCTION

One of the key functionalities of xanthan gum that set it apart from other hydrocolloid thickeners is its ability to interact synergistically with galactomannans such as guar gum, cassia gum, tara gum and locust bean gum and with structurally similar polysaccharides such as the glucomannan konjac. This interaction results in either a synergistic increase in viscosity in the case of guar gum or the formation of strong self supporting gels as seen with locust bean gum and konjac glucomannan. Initial studies proposed a model in which the unsubstituted, galactose free (smooth) regions of the galactomannans bind to xanthan in its ordered state.[1] This model has been used to explain the difference in degree of interaction between galactomannans of differing degrees of substitution. For example locust bean gum which has a galactose content of 17 – 26% forms self supporting gels with xanthan gum whereas guar gum which has a galactose content of 33 – 40% forms weak gel networks resulting in a synergistic increase in viscosity. Subsequent studies, which showed that interactions were enhanced after heating the mixture to temperatures above the coil-helix transition of the xanthan, were interpreted as evidence that the binding occurred with the cellulosic backbone of the xanthan gum in the ordered state.[2,3] An alternative explanation was that when the hydrocolloids are mixed at temperatures below the setting point of the gel, they form an inferior, disrupted network with melting and resetting giving a stronger, coherent gel.[4] More recent studies tend to support this interpretation which favours the original model of Morris et al.[5,6,7]
It has been reported that increasing the ionic strength of the solvent reduces the associations between xanthan and galactomannans/glucomannans.[8,9] For example increasing salt concentrations or lowering of pH results in weaker gels or lower viscosity in the mixtures. The effect of salt concentration, pH and preparation procedure on a commercial xanthan gum alone and in mixtures with locust bean gum (LBG) have been studied by flow rheology and large deformation gel tests. The results are discussed in terms of the order-disorder transition of the xanthan gum and the influence of NaCl and pH on the xanthan conformation and its relative affinity for the LBG molecules.

2 MATERIALS AND METHODS

The xanthan, GRINDSTED® Xanthan 80 lot 4450399909 and Locust bean gum GRINDSTED® LBG 175 lot 4761020463 (Danisco) were used without further purification.
Xanthan-LBG mixed systems were prepared by two different procedures.
Procedure A: The NaCl was dissolved in deionised water prior to the dispersion of xanthan and LBG mixture. The dispersion was then heated in a water bath to 87°C and mixed for 15 minutes. For the effect of pH, 0.5% NaCl and 0.08% sodium citrate was pre-dissolved in the water and the pH of the hot solution was adjusted as necessary, with 50% citric acid solution. The solution was then cooled undisturbed and stored at room temperature overnight before measurements were made.
Procedure B: The xanthan and LBG mixture was dispersed in deionised water and was then heated in a water bath to 87°C and mixed for 15 minutes. The NaCl was added to the hot gum solution. The solution was then cooled undisturbed and stored at room temperature overnight before measurements were made.
Rheological measurements were made using an AR-1000 controlled stress rheometer fitted with a 2°, 60 mm diameter cone. The temperature ramp measurements were made at a rate of 2°C s^{-1} with automatic gap adjustment. Viscoelastic measurements were made within the linear viscoelastic region of the sample.
Samples for gel strength were prepared in 80 mm diameter, 45 mm height containers and were measured after overnight storage at room temperature using a Stable Micro Systems TAXT+ fitted with a ½ inch diameter flat ended probe programmed to penetrate the gel at 1 $mm.s^{-1}$. The gel strength was taken as the rupture point of the gel.

3 RESULTS AND DISCUSSION

Effect of NaCl on gel strength

The results in Figure 1 show that the preparation method can have a significant influence on the final gel strength of the mixed xanthan-LBG gels. When added prior to hydration of the gums, (Procedure A), the NaCl concentration has a strong influence on the gel strength. In contrast, when the NaCl is added at high temperature, after hydration of the gums, (Procedure B) there is almost no influence on the gel strength. At salt concentrations above 0.2% the gel strength is higher when gums were hydrated in the presence of the NaCl.

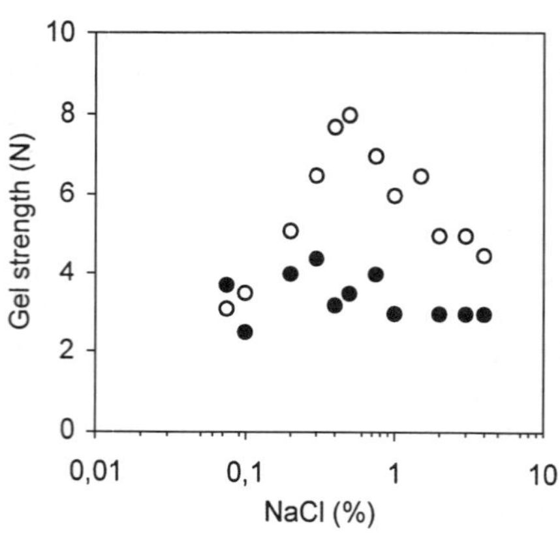

Figure 1. The influence of preparation procedure on the relationship between NaCl concentration and gel strength of xanthan-LBG mixed gels. (○) Procedure A hydration in NaCl (•) Procedure B hydration in deionised water (1% gum, 60/40 Xanthan/LBG).

It is interesting to discuss these results in terms of the thermal transition of the xanthan gum. Solutions of xanthan gum undergo a conformational transition during heating which is believed to be associated with the change from a rigid ordered state at low temperature to a more flexible, disordered state at high temperatures. This conformational change was first observed as a sigmoidal change in viscosity.[10] The conformational transition of 0.5% solutions of xanthan has been followed by measuring viscosity at $10s^{-1}$ as a function of temperature at different NaCl concentrations and an example of the results are shown in Figure 2.

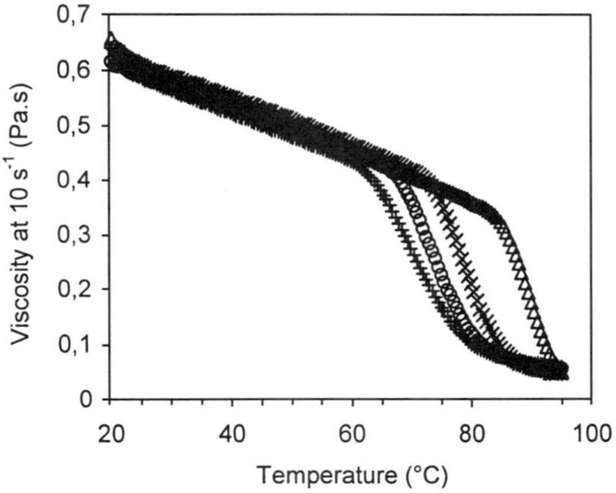

Figure 2 The conformational transition of 0.5% xanthan gum solutions measured by viscosity in (+) 0.05% (○) 0.10% (x) 0.15% (Δ) 0.275% NaCl.

The temperature at the onset and completion of the transition has been taken from Figure 2 and plotted as a function of salt concentration and the results are shown in Figure 3. As has been reported previously the temperature of the conformational transition increases with increasing ionic strength of the solution.[1,11]

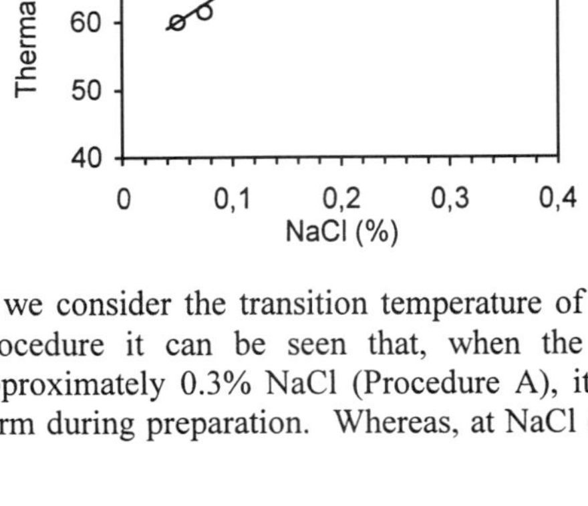

Figure 3 The temperature of the (○) onset and (●) completion of the thermal transition of 0.5% xanthan gum as a function of NaCl concentration.

If we consider the transition temperature of the xanthan with regard to the preparation procedure it can be seen that, when the xanthan is hydrated in the presence of approximately 0.3% NaCl (Procedure A), it will remain predominantly in the ordered form during preparation. Whereas, at NaCl concentrations below approximately 0.13%, the xanthan will, predominantly pass through the order-disordered transition during preparation of the gels. It is then evident that the highest gel strengths were obtained when the xanthan remained predominantly in the ordered form during the preparation of the gels (Procedure A above 0.3% NaCl). In procedure A below 0.13% NaCl and in procedure B the xanthan will predominantly pass through the transition during preparation and, although, after addition of NaCl, the ordered form will re-form prior to the onset of gelation of the xanthan-LBG gel, (Figure 4), the gel strength remains constant at approximately 3.5 N.

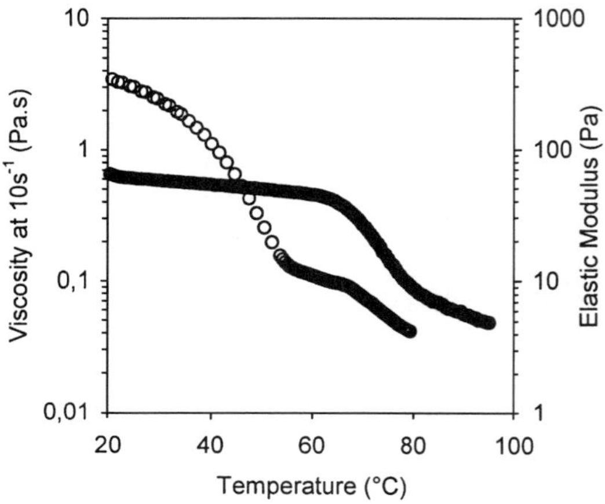

Figure 4. (○) Gelling profile of 1% 50/50 xanthan/LBG mixed gel in 0.075% NaCl measured by elastic modulus. (●) The conformational transition of 0.5% xanthan gum solution in 0.075% NaCl measured by viscosity.

Interestingly, a very similar effect was observed with the viscosity of 0.5% xanthan gum solutions (Figure 5). A solution heated to 87°C for 15 minutes in the presence of 0.5% NaCl retained virtually the same viscosity profile, when cooled, as an unheated solution, conversely, when the salt was added to the hot solution a significant drop in the viscosity at low shear rates was seen when the solution was cooled. In the first case the xanthan would be expected to remain predominantly in the ordered form throughout the heating whereas in the second case the xanthan undergoes an order-disorder transition.

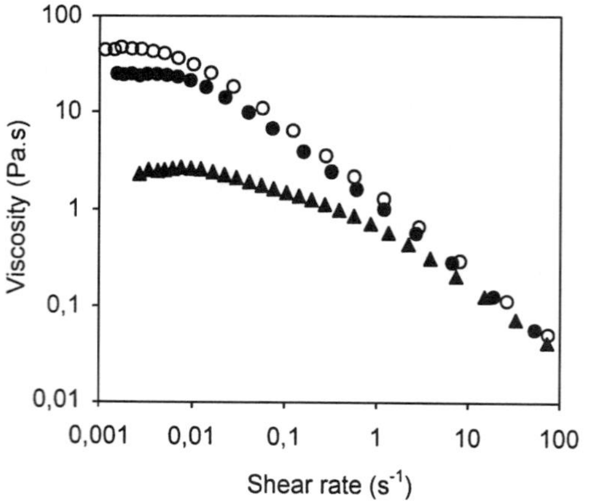

Figure 5 Flow curves of 0.3% xanthan gum solution hydrated in deionised water measured at 20°C after (○) addition of 0.5% NaCl, (●) addition of 0.5% NaCl, followed by 15 minutes at 87°C, (▲)15 minutes at 87°C, followed by addition of 0.5% NaCl at 87°C.

Xanthan has been described as having 3 conformational states.[12,13]

- I Native ordered form
  - never heated
- II Denatured ordered form
  - heated to transition temperature or above then cooled
- III Disorderd form
  - present at high temperature or low salt concentrations or both

The native ordered form (I) has been shown to have lower relative viscosity, and intrinsic viscosity compared to the denatured ordered form (II) but the same molecular weight suggesting a more compact conformation. It has also been demonstrated that addition of NaCl to the disordered form (III) at sufficient levels to induce the disorder-order transition to occur at 80°C results in a viscosity intermediate between the form I and II. This was interpreted as suggesting that form I is partially recovered on cooling.[14] The results in Figure 5 are consistent with this, since 0.5% NaCl is sufficient to induce the disorder-order transition at 87°C and therefore addition to the hot xanthan gum solution would result in a reduction of the viscosity upon cooling.

The effect of preparation procedure on the gel strength of Xanthan-LBG mixed sytems can be rationalised on the same basis. At low salt concentrations (< 0.1%) there is no difference in gel strength between the two preparation procedures since at these salt levels the completion of the transition temperature is below the preparation temperature and the xanthan will pass through the order-disorder transition before gelation with LBG even if the salt is present during preparation. In procedure B, (salt present during heating), as the NaCl level increases above 0.1% the gel strength increases as completion of the order-disorder transition increases above the preparation temperature. As a consequence less and less of the xanthan passes through the thermal transition until an optimum gel strength is reached at approximately 0.4% NaCl at which concentration the onset of the thermal transition would be expected to be above the preparation temperature (Figure 3). Increasing NaCl concentration above this optimum may continue to promote xanthan-xanthan interactions at the expense of the xanthan-LBG complex thereby reducing the gel strength.

In procedure A, the addition of the NaCl to the disordered xanthan above 0.1 % does not result in an increase in gel strength of the xanthan-LBG mixed system. In this case, the higher levels of NaCl will induce the xanthan disorder-order transition to occur at the preparation temperature and will result in partial recovery of form I on cooling. Since the reordering is complete before the gelation of the mixed system (Figure 4) it suggests that the more compact xanthan conformation of form I may result in poorer interactions with the LBG and could explain the lower gel strength seen with this procedure.

An alternative explanation for the lower gel strength seen with procedure A could be degradation of the xanthan molecule. The xanthan molecule has been shown to be degraded when in the disordered form, even at neutral pH.[15] In order to evaluate this a sample was prepared by heating the LBG to 87°C to hydrate then cooling to 65°C before adding the xanthan gum solution pre-hydrated in 0.075% NaCl at 65°C. It can be seen from Figure 4 that 65°C is above the gelling temperature of the mixed system but below the transition temperature of the xanthan gum. The gel strength of the gel was found to be virtually the same (3.5N) as when the xanthan had been heated above the transition temperature (3.7N) suggesting that degradation of the molecule was not a significant factor.

Effect of pH on gel strength

The gel strength of the xanthan-LBG mixed system remained relatively unchanged between pH 7 and 4. As the pH decreased below pH 4 a progressive drop in the gel strength was observed. Virtually identical behaviour was seen for the viscosity of a 0.3% xanthan gum solution in 1% NaCl (Figure 6).

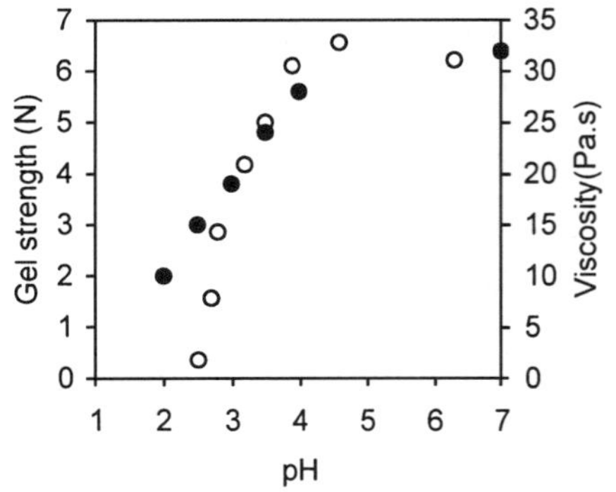

Figure 6 Effect of pH on (○) gel strength of 1% 60/40 xanthan/LBG gels in 0.5% NaCl and (•) Viscosity at 0.01 s^{-1} of 0.3% xanthan gum in 1% NaCl.

Xanthan is an anionic polymer and as such the pH sensitivity of the low shear viscosity is expected since changes in the pH will result in changes to the charge density of the xanthan which will in turn influence the molecular associations between xanthan molecules. It has been proposed that reducing pH progressively converts the carboxylate groups from the ionised to the un-ionised form ($COO^- + H^+ = COOH$), with consequent suppression of electrostatic repulsion between xanthan side chains.[16,17] This in turn could allow a more compact molecular shape which could explain the observed reduction in viscosity and synergy.

4 CONCLUSIONS

The quality of the mixed xanthan-LBG gels is greatest when the xanthan is maintained in the ordered state throughout the preparation process. Addition of the NaCl when the xanthan is in the disordered state has a negative impact on the quality of the gels. Addition of salt to the disorderd xanthan at high temperatures is believed to drive the xanthan molecules partially into the native ordered state. The viscosity of the xanthan gum and the quality of the mixed xanthan-LBG gels is reduced with reducing pH below pH 4.

These results show that the conformational state of the xanthan gum has a direct influence on the quality of the complexes formed with LBG.

References

1 Morris, E.R., Rees, D.A., Young, G., Walkinshaw, M.D. and Darke, A. (1977) Order-Disorder transition for a bacterial polysaccharide in solution. A role for polysaccharide conformation in recognition between *Xanthomonas* pathogen and its plant host. *J. Mol. Biol.*, **110**, 1-16.

2 Cairns, P., Miles, M.J. and Morris, V.J. (1986). Intermolecular binding of xanthan gum and carob gum. *Nature*, **322**, 89 - 90

3 Cairns, P., Miles, M.J., Morris, V.J. and Brownsey, G.J. (1987) X-Ray fibre-diffraction studies of synergistic binary polysaccharide gels. *Carbohydr. Res.*, **160**, 411 – 423.

4 Morris, E.R. (1990) Mixed Polymer Gels. In Harris, P. (Ed) *Food Gels*. Elsevier, London, UK, pp 291-359

5 Morris, E.R. and Foster, T.J. (1994) Role of conformation in the synergistic interactions of xanthan. *Carbohydr. Polym.* 23, 133 – 135.

6 Cronin, E.E., Giannouli, P., McCleary, B.V., Brooks, M. and Morris, E.R. (2002) Formation of strong gels by enzymic debranching of guar gum in the presence of ordered xanthan. In Williams, P.A. and Phillips, G.O. (Eds.) *Gums and Stabilisers for the Food Industry 11*, RSC, Cambridge, pp 289 – 296.

7 Fitzsimons, S.M., Tobin, J.T. and Morris, E.R. (2008) Synergistic binding of konjac glucomannan to xanthan on mixing at room temperature. *Food Hydrocolloids*, **22** 36 – 46.

8 Dalbe, B. (1992) Interactions between xanthan and konjac mannan. In Phillips, G.O., Williams, P.A. and Wedlock, D.J. (Eds) *Gums and Stabilisers for the Food Industry 6*, Oxford University Press, Oxford, pp 201 – 208.

9 Pettitt, D.J. (1982) Xanthan gum. In Glicksman, M. (Ed.) *Food Hydrocolloids Volume I*. CRC Press, Boca Raton, Florida pp 127 – 149.

10 Jeanes, A., Pittsley, J.E. and Senti, F.R. (1961) Polysaccharide B-1459: a new hydrocolloid polyelectrolyte produced from glucose from bacterial fermentation. *J. App. Polym. Sci.*, **5**, 519 – 526.

11 Sworn G. (2000) Xanthan gum. In Phillips, G.O. and Williams, P.A. (Eds) *Handbook of Hydrocolloids*. Woodhead Publishing Ltd, Cambridge, pp 103-116.

12 Holzwarth, G. and Prestridge, E.B. (1977) Multistranded helix in xanthan polysaccharide. *Science*, **197**, 757 – 759.

13 Milas, M. and Rinaudo, M. (1984) On the existence of two different secondary structures for the xanthan in aqueous solutions. *Polymer Bulletin*, **12**, 507 – 514.

14 Milas, M. and Rinaudo, M. (1986) Properties of xanthan gum in aqueous solutions: Role of the conformational transition. *Carbohydr. Res.* **158**, 191 – 204

15 Callet, F., Milas, M. and Rinaudo, M. (1989) On the role of thermal treatments on the properties of xanthan solutions. *Carbohydr. Polym.*, **11**, 127 – 137

16 Agoub, A.A., Smith, A.M., Giannouli, P., Richardson, R.K. and Morris, E.R. (2007) "Melt-in –the-mouth" gels from mixtures of xanthan and konjac glucomannan under acidic conditions: A rheological and calorimetric study of the mechanism of synergistic gelation. *Carbohydr. Polym.*, **69**, 713 – 724.

17 Rinaudo, M. and Moroni, A. (2009) Rheological behaviour of binary and ternary mixtures of polysaccharides in aqueous medium. *Food Hydrocolloids*, **23**, 1720 – 1728.

Food Applications

INDUSTRY REQUIREMENTS FOR HYDROCOLLOIDS IN BEVERAGE EMULSIONS

Matthias Schultz

Givaudan Schweiz AG, Science & Technology, Flavour Delivery Science, Überlandstrasse 138, CH-8600 Dübendorf, Switzerland

1 INTRODUCTION

Beverages represent one of the most important application areas for the flavour industry. An overview of the recent and projected development of the beverage market in different regions is given in Table 1. The figures are only related to the most important segment – carbonated soft drinks. Other categories such as juice drinks, fruit flavoured non-carbonates, functional (including sports and energy) drinks, flavoured alcoholic drinks and concentrates (syrups and squashes) also exist. Many flavours, citrus flavours in particular, however, can not just be blended in a beverage since they would not dissolve properly. Therefore, less soluble ingredients are often removed, solvents or solubilisers are added or, as usually done for cloudy beverages, the flavour is delivered in emulsified form as a so-called beverage emulsion.

Table 1 *Regional carbonates consumption – total volume 2002-2007 and projected until 2012 (in million litres). Source: Euromonitor International*

Total consumption	% Var.	Volume	% Var.	Volume
	2002-07	2007	2001-12	2012
North America	-0.6	58´535	-0.8	53´862
Latin America	+23.0	49´170	+13.8	55´943
Western Europe	+12.7	32´339	+9.8	35´504
Asia Pacific	+18.9	25´477	+18.0	30´067
Middle East & Africa	+21.5	20´305	+26.2	25´634
Eastern Europe	+28.0	12´627	+19.5	15´090
Australasia	+2.7	2´369	-1.1	2´342

Hydrocolloids may have quite a number of functionalities in beverages such as flavour modification and mouthfeel enhancement[1]. Most notably, however, they play an indispensable role in the preparation and stabilisation of beverage emulsions. This paper will focus on the use of hydrocolloids as emulsifiers in beverage emulsions.

2 FLAVOURS USED IN BEVERAGE EMULSIONS

Citrus flavours with Cola (containing Lime), Orange and Lemon being the most important, constitute the majority of flavours used in the carbonated soft drink segment. Grapefruit, Tangerine/Mandarin and Lime flavours also belong to this group. Citrus fruits are the main source of the flavour raw materials that accrue during the production and concentration of juice. The composition of peel oil, essence oil and aqueous essence, as to be expected for natural products, will considerably vary depending on climate, season and soil[2]. The characteristic flavour profiles are not caused by the monoterpenes, but by oxygenated terpenoids such as alcohols, aldehydes, ketones, acids and esters. The differences in their water solubility, partitioning coefficients and volatility have an important impact on their emulsification properties and, hence, on the formulation of beverage emulsions. Small content molecules can have a tremendous influence on the character of the flavour. Furthermore, the content of waxes in citrus oils will also influence their processing properties. Other important non-citrus flavours for cloudy soft drinks are Pineapple, Passionfruit, Peach, Mango and blends of flavours of tropical fruits. Their terpene content is comparatively low, and due to the better water solubility of the flavouring components emulsification would often not even be necessary or could be avoided by the addition of solvents such as ethanol or propylene glycol to the flavour. However, as long as the beverage should display cloudiness there are few alternatives to making these better soluble flavours part of an emulsion to produce sub-micrometer sized droplets which cause the scattering of light. Therefore, often limonene or vegetable oil are added to such flavours and then an emulsion similar to one containing citrus oils is created. Another way to achieve cloudiness in a beverage is the combination of flavours that are soluble in the final beverage with a so-called cloudifier or cloud emulsion. The oil droplets of such products consist of terpenes or vegetable oil without the addition of any flavour.

3 MAIN ASPECTS OF BEVERAGE EMULSIONS

Beverage emulsions are commonly used intermediates that often enable the beverage manufacturer to provide flavour, colour and cloudiness all at once by just blending the emulsion with an aqueous solution of sweetener and acid. A typical beverage emulsion contains 10-20% of a dispersed oil phase; the remaining part is the continuous water phase. Raw materials, processing and stability requirements of beverage emulsions have been reviewed recently[3,4,5].

The flavour and/or cloudifying oils make about half of the oil phase. Flavours display specific gravity values as low as e.g. 0.845 for a typical orange oil in contrast to a specific gravity of the surrounding aqueous phase of 1.00-1.05 depending on the type and content of sweetener. Proper weighting of the oil by adding so-called weighting agents constituting about the other half of the oil phase is therefore a delicate part of the art of beverage emulsion development. Weighting agents commonly used are glycerine esters of rosin ("ester gum"), damar gum, sucrose acetate isobutyrate and brominated vegetable oils. There are basically three sources of rosin for glycerine esters occurring on the oleoresin of pine trees used for ester gums – wood, gum and tall oil rosins – with major differences in their legislation. All three are similar in nature and composed of about 90% resin acids, mostly abietic acid, and 10% nonacidic neutral compounds[3]. Damar gum is collected from the trees of the *Dipterocarpaceae* family and used without further esterification, but has lost importance since it is no longer permitted in food and beverages in the European Union. Sucrose acetate isobutyrate is a mixture of sucrose esters, mostly sucrose diacetate hexaisobutyrate. In order to reduce its extremely high viscosity, it can be obtained on the market as a blend with fractions of orange terpenes, ethanol or medium chain triglycerides. The use of brominated vegetable oils has been much restricted due to bromine deposition in body lipids[6]; today they are rarely used.

If carotenoids are used as colorants in the beverage they need to be dissolved in the oil phase at temperatures up to 120°C for complete dissolution of the carotenoid crystals. This has consequences for the more volatile and easily oxidising flavour ingredients which must be kept away from such a high temperature dissolution step.

The main component of the water phase is the hydrocolloid emulsifier which will be looked at in more detail in Section 4. Other water phase ingredients include a preservative system usually consisting of sodium benzoate or potassium sorbate which are converted to the microbially active corresponding acids by the addition of citric acid. If synthetic water soluble dyes are used they will be dissolved in the water phase as well.

Usually, a two-step process is applied to manufacture a stable emulsion. First, oil and water phases are unified using high shear mixing. The resulting emulsion has a mean droplet diameter of about 5 μm. In

a subsequent treatment using a high pressure homogenizer the droplet sizes will be further reduced to below 1 µm (Fig. 1).
Beverage emulsions have typical dosages of 1-2 g / l beverage and must provide physical, chemical (sensorial) and microbiological stability over a shelf life of 12 months in concentrated (emulsion) and diluted (beverage) form.

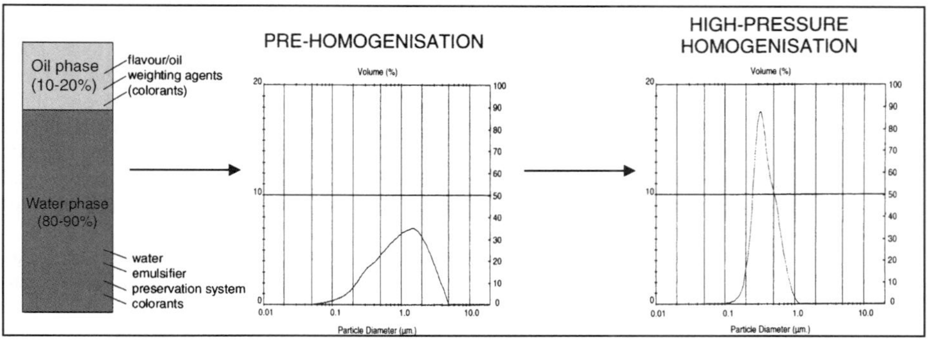

Figure 1 *Typical composition (left) and droplet size distributions after processing steps of beverage emulsions*

4 REQUIREMENTS FOR HYDROCOLLOIDS TO BE USED AS EMULSIFIERS AND STABILISERS IN BEVERAGE EMULSIONS

It has been pointed out that most hydrocolloids act as stabilisers of oil-in-water emulsions, but only a few can act as emulsifiers. The latter functionality requires substantial surface activity at the oil-water interface and the ability to facilitate the formation and stabilisation of fine droplets during and after emulsification. According to Dickinson[7] an effective emulsifier is therefore one that

(i) rapidly reduces the interfacial tension at the freshly formed oil-water interface,
(ii) binds strongly to the interface once adsorbed, and
(iii) protects the newly formed droplets against flocculation or coalescence.

Low molecular weight surfactants such as polysorbates would quite drastically reduce the interfacial tension (criteria (i)), but do not bind strongly enough and do not sufficiently protect the droplets from flocculation and coalescence. In contrast to surfactants, hydrocolloids have a comparatively high molecular weight that makes them slower to diffuse and to cover the oil droplet surface, but provides a strong, in part elastic interfacial film that strongly contributes to emulsion stabilisation by mostly steric repulsion between droplets. Hence, whereas surface active hydrocolloids usually fulfil criteria (ii) and (iii), high-pressure homogenisation is a necessary tool to sufficiently reduce the droplet size of flavour and other cloudifying oils, since higher energy input is required to achieve a sufficiently small droplet size at (compared with surfactants) higher interfacial tensions. Traditional and potential new emulsifiers have to be checked whether they fulfil the above mentioned technical criteria, but there is a number of additional requirements arising from quality assurance, economical, regulatory and consumer perception considerations. The more important of those are the following.

(iv) Constant quality and performance. As nicely described in a lecture at the previous Gums and Stabilisers Conference, "natural polymers are never uniform or simple. Their uniformity depends on more than one structural feature. Extraction processes alter the macro- and micro-structures and performance. Their most common feature is their variability which poses a big problem for both the producer and user of natural polysaccharides."[8]

(v) Global or at least regional regulatory acceptance. The more important flavour and fragrance companies as well as their major beverage customers operate globally. They market their brands in more than one region and, although the composition of beverages and their flavour profile traded under the same brand name might slightly differ in different markets, making use of the same list of ingredients is not only a huge benefit, but in some cases a must. Global regulatory admittance will also affect purchasing volumes, allow setting up long term contracts with suppliers and influence pricing.

(vi) Reasonable and constant price. Prices are driven by availability on the market which in turn depends on the way hydrocolloids are produced. Hydrocolloids derived from plant materials grown in regions subjected to climatic or political turbulences are at much higher risk than those produced by the action of enzymes or microbes. Hydrocolloid manufacturers can certainly reduce such risk by anticipatory stock taking and the choice of appropriate growing areas, but not all risks can be calculated. New and enhanced emulsifiers may often have comparable performance at significantly lower dosages and manufacturers are much encouraged to take care that there is a significant cost benefit for their customers.

(vii) Consumer acceptance. Consumers' liking of foods still mostly depends on aroma, taste, mouthfeel and other perception parameters. Hydrocolloids can much contribute to that. There is however much increased sensitivity of consumers towards health aspects. The term "naturalness" of a hydrocolloid is rather ill defined and there is much confusion about such terms. However, consumers would usually prefer plant derived materials that had not been subjected to further chemical treatment over chemically modified hydrocolloids. There is also a clear preference for materials derived from eatable parts of plants over tree exudates. Any additional proven health benefit (e.g. dietary fibres) will also be acknowledged.

Traditional emulsifiers and their potential substitutes have to be assessed how well they fulfil the criteria mentioned above. Some of the more common materials will be described below from this perspective. A summary for the more important emulsifiers is given in Table 2.

4.1. Gum arabic

Gum arabic, the exudate of *Acacia senegal* or *Acacia seyal* trees has a long tradition as an emulsifier in the food industry. It is not less than the gold standard of emulsifiers used in beverages. Nevertheless, its complex structure and the origin of its superb emulsifying and stabilising properties has long not been fully understood and is still a matter of thorough investigation. The actual state of the research is reflected by a number of recent publications[9,10,11,12].

For beverage emulsions, *A. senegal* derived gum is commonly used. It is now generally recognised that the emulsifying ability is due to the presence of an arabinogalactan protein (AGP) fraction accumulating at the interface between oil and water[13]. The protein backbone of that fraction is thought to constitute the hydrophobic part of the molecule with an affinity to the oil phase whereas the hydrophilic polysaccharide side chains remain in the aqueous phase. Other structural features of the gum such as molecular weight distribution also play a decisive role for its emulsifying and stabilising efficiency[14]. Recently, the existence of glycosyl-phosphatidylinositol lipids attached to the AGP fraction and their importance for the emulsifying efficiency of the gum has been reported[15].

Some attention has been paid on the interfacial behaviour of *A. senegal* derived gum[16]. Interfacial rheology experiments suggest that the mechanical stability of the adsorption layer formed by the gum around an emulsion droplet plays a significant role reflected by substantial shear elasticity of the interface[17]. It can be shown that the interfacial tension between a gum arabic (*A. senegal* type) solution and limonene is not further reduced when increasing the gum concentration in the solution to beyond 8%[18]. The droplet size in a model emulsion was found not to decrease any further when increasing the gum concentration to beyond 10%[19]. Nevertheless, typical gum concentrations in beverage emulsions with an oil phase content of up to 20 wt% are 15-20 wt%. The excess gum will be necessary to further stabilize the emulsion, i.e. to provide sufficient polymeric material in-between droplets and to increase the viscosity of the emulsion to effectively prevent collisions and coalescence. Indeed, electrokinetic measurements show that oil droplets in gum arabic stabilized emulsions bear a small but significant surface charge. Hence, the stabilisation mechanism is electrosteric, not purely steric, but the steric contribution is dominant[20].

When formulating emulsions using gum arabic as emulsifier its mineral content has to be taken into consideration. Demineralization has been found to be beneficial for emulsion stability[21]. Bivalent cations can react with certain synthetic dyes used in beverage emulsions and result in precipitation[22].

It is also known that gum arabic needs sufficient hydration time to unfold the polysaccharide chains enhancing the interfacial activity of the gum molecules.

The great variability within commercial gum arabic supplied to the market has been well illustrated[9]. This is a particular problem in times of inconsistent supply due to poor crops affected by climatic conditions or political instability in the growth regions. Quality control hence is demanding. Price is far from being constant, anyway the high dosage needed (usually a 1:1 emulsifier:oil ratio is necessary) causes relatively high cost. On the other side, gum arabic is more or less globally admitted. Consumer acceptance is good and might be even improved in the future since it can be claimed as dietary fibre with related health benefits[23].

4.2. Octenyl succinate modified starch

Once developed as alternative for gum arabic in times of supply shortages, octenyl succinate modified starch has long also found its place as high performance emulsifier in beverage emulsions[24]. When starch is esterified with n-octenyl succinic anhydride (OSA) the lipophilic n-octenyl chain is substituted onto the hydrophilic starch backbone resulting in the amphiphilicity necessary for interfacial activity. One of the main parameters for emulsification efficiency is the degree of substitution which in food is restricted to 0.03. It was found that increasing the degree of substitution further enhances the emulsifying ability of the starch, but that a degree of substitution of 0.07 is not necessarily beneficial over 0.03[25]. It was suggested that the OSA groups are probably distributed in the interior amorphous domains of amylopectin molecules as well as on the outside of the granule[26]. As for gum arabic the molecular weight distribution of modified starch is changed during high pressure homogenisation by degradation and "crumpling" of the polymer chains[27]. Although smaller, the dilatational response is reported to be similar to that of gum arabic on a qualitative basis. However, in contrast to gum arabic, modified starch can reduce the interfacial tension at the oil-water interface more effectively, but the formed layers lack the pronounced shear elasticity described for gum arabic. Its interfacial functionality is therefore said to resemble more that of traditional macromolecular surfactants[17]. Due to its higher emulsifying efficiency smaller amounts of OSA starch can be applied and lower pressures are necessary to produce the same small droplet size as with gum arabic. The need of a proper selection of homogenization parameters was emphasized in a study on the influence of emulsifier type and pressure on beverage emulsion stability[28]. Due to the tendency of starch to form inclusion complexes with flavours the flavour content may however need to be somewhat increased[29]. Furthermore, lower ethanol tolerance of starch based emulsions has been reported[3].

The variability of emulsifying starches is generally lower than that of gum arabic. Supply has never been at risk so far and prices have been fairly constant. Some threat to the constant price level may be posed by the increasing prices for starches due to the use of cultivated area for the production of biofuel, but so far this is not a serious problem. Regulatory acceptance is more or less global. As most of these starches are derived from waxy maize or tapioca, there is no major GMO issue. However, in the eyes of the beverage industry the chemical modification step to produce OSA starches makes the products appear less "natural", and this is not compensated by the fact that gum arabic is a tree exudate normally not consumed whereas starch is part of many foods.

4.3. Gum ghatti

Gum ghatti, the exudate of *Anogeissus latifolia* trees growing throughout the greater part of India, was described to be an excellent and cost efficient emulsifier in beverage emulsions. Gum ghatti has low viscosity but high solubility in water compared with many other polysaccharides and also has a similar viscosity profile to gum arabic. It has a sugar composition similar to that of gum arabic, but contains about 1.5 to twice as much protein[30]. When analyzing the oil adsorbed component it was found that it is widely distributed throughout the gum's molecular structure, from high molecular to low molecular weight. This component is a polysaccharide containing protein similar to gum arabic, but has larger molecular weight than gum arabic. The adsorbed component of gum ghatti has higher surface coverage and the amount of gum ghatti adsorbed is more than three times that of gum arabic[31]. It was demonstrated that emulsions containing gum ghatti showed higher emulsification producing capacity and better emulsion stability at lower concentrations than gum arabic[30,32]. Despite its large natural variability[32], its applicability for beverage emulsions has made progress. The main hindrance to more extensive use, however, is its limited regulatory acceptance in many parts of the world.

4.4. Sugar beet pectin

The origin of the emulsifying functionality of sugar beet pectin is still under study, but it is assumed that the content of proteinaceous material especially the fraction bound to polysaccharide in an AGP like manner is responsible for its surface activity[33]. It was demonstrated that proteinaceous material is adsorbed at the oil-water interface and that the emulsification efficiency is influenced by the accessibility of the protein and ferulic acid groups to the surface of the oil droplets, the proportion of the ester groups and the molecular mass distribution of the fractions[34]. A thorough study comparing the emulsifying properties of sugar beet pectin with those of gum arabic and soybean soluble polysaccharide found that sugar beet pectin required the smallest amount to cover the surface of the oil droplets and to activate the interface. Sugar beet pectin displayed the highest zeta potential indicating that electrostatic stabilisation plays a more important role when using sugar beet pectin. Long-term stability of emulsions however was lower than for gum arabic based emulsions suggesting that steric factors have an important effect on emulsion stability[19]. Hence, despite favourable attributes in terms of raw material abundance, dosage, cost, regulatory and consumer acceptance, sugar beet pectin has not yet found widespread use due to this limited ability to "stabilize emulsions with sufficient robustness in the hard world of beverage emulsions"[8].

4.5. Thermally maturated gum arabic

Gum arabic samples can be treated in the dry state in a range of physical forms to undergo a maturation process under strictly controlled conditions of temperature and humidity accelerating a process which would also proceed in nature during the aging of the trees or during storage of the crude harvested gum at slightly elevated temperatures in warehouses but over considerably longer time scales[35]. In the course of this process, gum arabic can be produced with precisely structured molecular dimensions. The controlling feature is the agglomeration of the proteinaceous components within the molecularly disperse system to increase the molecular weight and the amount of the emulsifying AGP fraction considerably improving the emulsifying ability of the gum[36]. The gum is hence recommended to be used in beverage emulsions at reduced levels compared with standard gum resulting in cost benefits and the elimination of the natural variability of standard gum arabic. Pressure homogenization has been shown to be necessary to unfold the AGP complex to fully explore these benefits[37]. Due to the fact that this approach overcomes some of the limitations of untreated gum arabic it certainly deserves further attention. It should be mentioned that thermal maturation has also been applied to enhance the properties of sugar beet pectin[38].

4.6. Other alternatives to gum arabic

Quite a number of other hydrocolloids has been proposed as possible replacements for gum arabic, in part with promising results, but have not yet been commercialized. Corn fibre gum[39], mesquite gum[40], fenugreek gum[41] and spruce galactoglucomannan[42] have been studied recently in emulsions with potential use in beverages.

While pure proteins such as whey protein isolate and sodium caseinate were found not to be suitable at the considerably acid conditions present in many beverages, the complexation of proteins with polysaccharides has been proposed to be studied as alternative emulsifiers in beverage emulsions[43]. Like the arabinogalactan-protein fraction in gum arabic such combinations should deliver the interfacial activity needed for efficient adsorption and film formation around the oil droplet. Complexes of whey protein and maltodextrin have been compared with gum arabic[44]. As a different type of protein-polysaccharide interaction, interfacial complexes were formed by electrostatic deposition of anionic polysaccharides onto cationic protein-coated droplets[45]. Although there are benefits such as much lower use levels and potential cost savings for these complexes compared with traditional emulsifiers, further research is necessary to overcome labelling, possible flavour-protein interaction and pH sensitivity issues.

Table 2 *Comparison how more important emulsifiers fulfil main requirements (for details see text). xx =fulfils criteria; x = partly fulfils criteria; o = does not fulfil criteria; *expected*

Criteria	Gum arabic	Modified starch	Sugar beet pectin	Gum ghatti
Technical applicability (reduction of interfacial tension, strong binding, droplet protection against coalescence)	xx	xx	x	xx
Constant quality and performance (sufficient volumes available, no batch-to-batch variability)	x	xx	x*	xx
Regulatory admittance (ideally admitted in all markets globally)	xx	xx	xx*	o
Reasonable and constant price (should reflect raw material price, no price fluctuations)	x	xx	xx*	xx
Consumer acceptance ("natural" perception, health claims, traditionally accepted)	xx	x	xx*	x*

5 REQUIREMENTS FOR HYDROCOLLOIDS PROVIDING OR SUPPORTING NEW FUNCTIONALITIES IN BEVERAGES

In general, currently available hydrocolloids serve the purpose of dispersing and stabilizing beverage emulsions and beverages containing them quite well. However, there is a growing need for a number of additional functionalities and solutions for problems not yet solved. Those include protection of flavours, controlled or targeted delivery of functional ingredients and masking of undesirable tastes from e.g. alternative sweeteners[5].

5.1. Flavour protection

The two main routes of citrus flavour deterioration in beverages are oxidation[46] and acid hydrolysis[47] which are common in low pH beverages, in particular under harsh storage or transport conditions. Oxidation is in part addressed by the addition of antioxidants such as butyl hydroxy anisol or tocopherol. Still, the effect of antioxidants is often not fully satisfying. Acid hydrolysis of sensitive flavour ingredients especially in citrus flavours can hardly be avoided. Flavours are commonly encapsulated in many foods[48] to retain them during processing, transport and storage, but the applicability of encapsulation systems in a water rich environment as represented by beverages is comparatively limited.

There have been few approaches based on hydrocolloids described to protect flavours or other ingredients prone to oxidation. Lipid oxidation rates of fish oil containing emulsions have been decreased by creating layer-by-layer membranes containing a cationic polymer (chitosan) suggested to repel ferrous ions which might serve as catalysts for lipid oxidation[49]. Unfortunately, there are few food grade cationic polymers available for such purpose which is one of the main gaps in the hydrocolloid field. Not only would such cationic hydrocolloids be useful for beverage applications, but for flavour encapsulation purposes in general.

5.2. Controlled or targeted delivery

Controlled release is not a big topic in beverages as the release usually has to be immediate in the mouth before the beverage is swallowed. Hence, there is little space for delayed or sequential release effects. In the field of functional drinks however there are situations where a pleasant flavour perception (immediate release) is combined with protection of functional ingredients which are not to be delivered before they reach the gastrointestinal tract. Appropriate encapsulation and dispersion of encapsulates might be a task for appropriate hydrocolloids. This is of particular importance if the ingredient is susceptible to the formation of unpleasant off-notes as it is the case for ω-3 fatty acids.

5.3. Bitterness and off-taste masking

The taste of beverages prepared with artificial sweeteners is often perceived to be unbalanced and to contain off-tastes not found when a beverage is sweetened with sucrose. With increasing concentration some high potency sweeteners develop a bitter or metallic aftertaste[50]. Finding new taste modifiers, particularly bitter-taste blockers, is an important goal for pharmaceutical and food industries. Hydrocolloids may play a role here. For example, cyclodextrins and their adducts linked with chitosan via succinyl or maleyl bridges were described to have bitter masking properties[51]. More research is needed in this direction to provide food and beverage grade solutions.

6 CONCLUSIONS

An industry person dealing with the development of beverage emulsions nowadays has quite a toolbox of hydrocolloid emulsifiers available. Besides the classical emulsifiers gum arabic and octenyl succinate modified starch there are however few other materials fully commercialized and ready to be used. Certainly, they are not just one by one replacements for gum arabic but require careful formulation work and stability testing on the basis of a thorough understanding of their structure-property relationships and their interaction with other beverage emulsion and beverage ingredients. Despite superior technical performance, even gum arabic and modified starch hardly fulfill all the criteria needed to make them universally applicable including constant performance and supply or overall consumer acceptance, although much progress has been made in performance enhancement, standardization and quality control of materials like gum arabic. Current trends in the food and beverage industry towards products related to health, wellness and lifestyle without compromising on the flavour profile require new and innovative hydrocolloids not only for emulsification and stabilization of flavours but for new functionalities together with a thorough understanding on how they work in a particular application.

References

[1] D'Angelo LL, Application of hydrocolloids in the beverage industry. In Williams PA, Phillips GO (Eds.), *Gums and Stabilisers for the Food Industry 13*, RSC Publishing, Cambridge 2006, pp. 421-430

[2] Staroscik JA, Wilson AA, Seasonal and regional variation in the quantitative composition of cold-pressed lemon oil from California and Arizona. *Journal of Agricultural and Food Chemistry*, **30** (1982) 835-837
[3] Tan C-T, Beverage emulsions. In Friberg SE, Larsson K, Sjoblom J (Eds.), *Food Emulsions*, Marcel Dekker, New York 2004, pp. 485-524
[4] Schultz M, Ringgenberg R, Emulsifying flavours into beverages. *Foods & Food Ingredients Journal of Japan* **213** (2008) 206-214
[5] Given Jr PS, Encapsulation of flavours in emulsions for beverages. *Current Opinion in Colloid & Interface Science* **14** (2009) 43-47
[6] Fourteenth report of the Joint FAO/WHO expert committee on Food Additives, *FAO Nutrition Meetings Report Series No. 48A* (1970)
[7] Dickinson E, Hydrocolloids as emulsifiers and emulsion stabilizers. *Food Hydrocolloids* **23** (2009) 1473-1482
[8] Phillips GO, Giving nature a helping hand. In Williams PA, Phillips GO (Eds.), *Gums and Stabilisers for the Food Industry 14*, RSC Publishing, Cambridge 2008, pp. 3-26
[9] Al-Assaf S, Phillips GO, Williams PA, Studies on acacia exudate gums. Part I: the molecular weight of *Acacia senegal* gum exudate. *Food Hydrocolloids* **19** (2005) 647-660
[10] Renard D, Lavenant-Gourgeon L, Ralet M-C, Sanchez C, *Acacia senegal* gum: Continuum of molecular species differing by their protein to sugar ratio, molecular weight, and charges. *Biomacromolecules* **7** (2006) 2637-2649
[11] Sanchez C, Schmitt C, Kolodziejczyk E, Lapp A, Gaillard C, Renard D, The acacia gum arabinogalactan fraction is a thin oblate ellipsoid: A new model based on small-angle neutron scattering and *ab initio* calculation. *Biophysical Journal* **94** (2008) 629-639
[12] Mahendran T, Williams PA, Phillips GO, Al-Assaf S, Baldwin TC, New insights into the structural characteristics of the arabinogalactan-protein (AGP) fraction of gum arabic. *Journal of Agricultural and Food Chemistry* **56** (2008) 9269-9276
[13] Randall RC, Phillips GO, Williams PA, The role of the proteinaceous component on the emulsifying properties of gum arabic. *Food Hydrocolloids* **2** (1988) 131-140
[14] Al-Assaf S, Katayama T, Phillips GO, Sasaki Y, Williams PA, Quality control of gum arabic, *Food & Food Ingredients Journal of Japan* **208** (2003) 771-780
[15] Yadav MP, Igartuburu JM, Yan Y, Nothnagel EA, Chemical investigation of the structural basis of the emulsifying activity of gum arabic. *Food Hydrocolloids* **21** (2007) 297-308
[16] Dickinson E, Hydrocolloids at interfaces and the influence on the properties of dispersed systems. *Food Hydrocolloids* **17** (2003) 25-39
[17] Erni P, Windhab EJ, Gunde R, Graber M, Pfister B, Parker A, Fischer P, Interfacial rheology of surface-active polymers: *Acacia senegal* gum versus hydrophobically modified starch. *Biomacromolecules* **8** (2007) 3458-3466
[18] Meier M, Schultz M, unpublished results
[19] Nakauma M, Funami T, Noda S, Ishikara S, Al-Assaf S, Nishinari K, Phillips GO, Comparison of sugar beet pectin, soybean soluble polysaccharide, and gum arabic as food emulsifiers. 1. Effect of concentration, pH, and salts on the emulsifying properties. *Food Hydrocolloids* **22** (2008) 1254-1267
[20] Jayme ML, Dunstan DE, Gee ML, Zeta potentials of gum arabic stabilised oil in water emulsions. *Food Hydrocolloids* **13** (1999) 459-465
[21] Buffo RA, Reineccius GA, Oehlert GW, Factors affecting the emulsifying and rheological properties of gum acacia in beverage emulsions. *Food Hydrocolloids* **15** (2001) 53-66
[22] Fang Y, Al-Assaf S, Sakata M, Phillips GO, Schultz M, Monnier V, Origin and thermodynamic properties of the instability of synthetic azo colorants in gum arabic solutions. *Journal of Agricultural and Food Chemistry* **55** (2007) 9274-9282
[23] Phillips A, Riley S, Hydrocolloids in health. In Williams PA, Phillips GO (Eds.), *Gums and Stabilisers for the Food Industry 14*, RSC Publishing, Cambridge 2008, pp. 306-312
[24] Trubiano PC, The role of specialty food starches in flavour emulsions. In Ho C-T, Tan C-T, Tong C-H (Eds.), *Flavor technology – Physical chemistry, modification, and process*, ACS Symposium Series 610, American Chemical Society, Washington 1995, p. 199-209
[25] Viswanathan A, Effect of degree of substitution of octenyl succinate starch on the emulsification activity on different oil phases. *Journal of Environmental Polymer Degradation* **7** (1999) 191-196
[26] Shogren RL, Viswanathan A, Felker F, Gross RA, Distribution of octenyl succinate groups in octenyl succinic anhydride modified waxy maize starch. *Starch/Stärke* **52** (2000) 196-204
[27] Nilsson L, Leeman M, Wahlund KG, Bergenståhl B, Mechanical degradation and changes in conformation of hydrophobically modified starch. *Biomacromolecules* **7** (2006) 2671-2679
[28] Dłużewska E, Leszczyński K, Effect of homogenisation pressure and some hydrocolloids on beverage emulsion stability. *Polish Journal of Food and Nutrition Sciences* **14** (2005) 293-298

[29] Meseck S, Bartsch A, Getränkeemulsionen: Inhaltsstoffe, Anforderungen und Anwendungsmöglichkeiten. *Flüssiges Obst* **69** (2002) 512-517
[30] Ido T, Ogasawara T, Katayama T, Sasaki Y, Al-Assaf S, Phillips GO, Emulsification properties of GATIFOLIA (gum ghatti) used for emulsions in food products. *Foods & Food Ingredients Journal of Japan* **213** (2008) 365-371
[31] Katayama T, Ido T, Sasaki Y, Ogasawara T, Al-Assaf S, Phillips GO, Characteristics of the adsorbed component of gum ghatti responsible for its oil-water interface advantages. *Foods & Food Ingredients Journal of Japan* **213** (2008) 372-376
[32] Al-Assaf S, Amar V, Phillips GO, Characerisation of gum ghatti and comparison with gum arabic. In Williams PA, Phillips GO (Eds.), *Gums and Stabilisers for the Food Industry 14*, RSC Publishing, Cambridge 2008, pp. 280-290
[33] Funami T, Zhang G, Noda S, Nakauma M, Asai I, Al-Assaf S, Phillips GO, Does pectin emulsification mechanism operate via an AGP (arabinogalactan protein) type fraction? *Foods & Food Ingredients Journal of Japan* **211** (2006) 255-263
[34] Williams PA, Sayers C, Viebke C, Senan C, Mazoyer J, Boulenguer P, Elucidation of the emulsification properties of sugar beet pectin. *Journal of Agricultural and Food Chemistry* **53** (2005) 3592-3597
[35] Al-Assaf S, Phillips GO, Aoki H, Sasaki Y, Characterization and properties of *Acacia senegal* (L.) Willd. var. *senegal* with enhanced properties (Acacia (*sen*) SUPER GUM™): Part 1 – Controlled maturation of *Acacia senegal* var. *senegal* to increase viscoelasticity, produce a hydrogel form and convert a poor into a good emulsifier. *Food Hydrocolloids* **21** (2007) 319-328
[36] Aoki H, Al-Assaf S, Katayama T, Phillips GO, Characterization and properties of *Acacia senegal* (L.) Willd. var. *senegal* with enhanced properties (Acacia (*sen*) SUPER GUM™): Part 2 – Mechanism of the maturation process. *Food Hydrocolloids* **21** (2007) 329-337
[37] Aoki H, Katayama T, Ogasawara T, Sasaki Y, Al-Assaf S, Phillips GO, Characterization and properties of *Acacia senegal* (L.) Willd. var. *senegal* with enhanced properties (Acacia (*sen*) SUPER GUM™): Part 5. Factors affecting the emulsification of *Acacia senegal* and Acacia (*sen*) SUPER GUM™. *Food Hydrocolloids* **21** (2007) 353-358
[38] Funami T, Nakauma M, Noda S, Ishihara S, Al-Assaf S, Phillips GO, Enhancement of the performance of sugar beet pectin as an emulsifier. *Foods & Food Ingredients Journal of Japan* **213** (2008) 347-356
[39] Yadav MP, Johnston DB, Hicks KB, Corn fiber gum: New structure/function relationships for this potential beverage flavour stabilizer. *Food Hydrocolloids* **23** (2009) 1488-1493
[40] Román-Guerrero A, Orozco-Villafuerte J, Pérez-Orozco JP, Cruz-Sosa F, Jiménez-Alvarado R, Vernon-Carter EJ, Application and evaluation of mesquite gum and its fractions as interfacial film formers and emulsifiers of orange peel-oil. *Food Hydrocolloids* **23** (2009) 708-713
[41] X. Huang, Y. Kakuda and W. Cui, Hydrocolloids in emulsions: particle size distribution and interfacial activity. *Food Hydrocolloids* **15** (2001) 533-542
[42] Mikkonen KS, Tenkanen M, Cooke P, Xu C, Rita H, Willför S, Holmbom B, Hicks KB, Yadav MP, Mannans as stabilizers of oil-in-water beverage emulsions. *LWT – Food Science and Technology* **42** (2009) 849-855
[43] Dickinson E, Towards more natural emulsifiers. *Trends in Food Science & Technology* **4** (1993) 330-334
[44] Akhtar M, Dickinson E, Whey-protein-maltodextrin conjugates as emulsifying agents: An alternative to gum arabic. *Food Hydrocolloids* **21** (2007) 607-616
[45] Harnsilawat T, Pongsawatmanit R, McClements DJ, Stabilization of model beverage cloud emulsions using protein-polysaccharide electrostatic complexes formed at the oil-water interface. *Journal of Agricultural and Food Science* **54** (2006) 5540-5547
[46] Nguyen H, Campi EM, Jackson WR, Patti AF, Effect of oxidative deterioration on flavour and aroma components of lemon oil. *Food Chemistry* **112** (2009) 388-393
[47] Schieberle P, Grosch W, Identification of potent flavour compounds formed in an aqueous lemon oil/citric acid emulsion. *Journal of Agricultural and Food Chemistry* **36** (1988) 797-800
[48] Madene A, Jacquot M, Scher J, Desobry S, Flavour encapsulation and controlled release – a review. *International Journal of Food Science and Technology* **41** (2006) 1-21
[49] Guzey D, McClements DJ, Formation, stability and properties of multilayer emulsions for application in the food industry. *Advances in Colloid and Interface Science* **128-130** (2006) 227-248
[50] Schiffman SS, Booth BJ, Losee ML, Pecore SD, Warwick ZS, Bitterness of sweeteners as a function of concentration. *Brain Research Bulletin* **36** (1995) 505-513
[51] Binello A, Cravotto G, Nano GM, Spagliardi P, Synthesis of chitosan-cyclodextrin adducts and evaluation of their bitter-masking properties. *Flavour and Fragrance Journal* **19** (2004) 394-400

DYNAMIC RHEOLOGICAL STUDY ON O/W EMULSIONS AS AFFECTED BY POLYSACCHARIDE INTERACTIONS USING A MIXTURE DESIGN APPROACH

I. Nor Hayati[1], Y.B. Che Man[2], I. Nor Aini[3] and C.P. Tan[2]

[1]Food Science Department, Faculty of Agrotechnology & Food Science, Universiti Malaysia Terengganu, 21030 Kuala Terengganu, Terengganu, Malaysia
[2]Food Technology Department, Faculty of Food Science & Technology, Universiti Putra Malaysia, 43400 Serdang, Selangor, Malaysia
[3]Product Development & Advisory Division, Malaysian Palm Oil Board, 43000 Kajang, Selangor, Malaysia

1 INTRODUCTION

Thickening polysaccharides such as xanthan gum (XG), carboxymethyl cellulose (CMC), and galactomannans (guar gum (GG) and locust bean gum (LBG)) are widely employed in the food industry because of their ability to improve the product's texture and rheology. Among them, XG and galactomannans are frequently used in combination, particularly in the emulsion-based food products due to their synergistic interaction that could provide better rheological properties of the products as well as being cost effective.[1] Furthermore, the effect of this synergistic interaction on the rheological behaviour of XG-galactomannan aqueous blends has incessantly been investigated.[2-6] In contrast, very little work has been carried out to investigate the effect of polysaccharide interactions (either synergism or antagonism) on emulsion rheological properties. Additionally, information on polysaccharide interactions other than XG-galactomannans is still limited.

It has been demonstrated previously that the use of XG, CMC, GG and LBG could generally improve the rheology and stability of the emulsions with partially crystallized fat droplets.[7] However, all of them showed a lack of functionality besides advantages on the overall emulsion characteristics. XG was recognized as the best polysaccharide to provide an emulsion with the best gel character, but considerably unstable due to depletion flocculation as well as partial coalescence. In a different way, CMC was able to produce an emulsion with excellent storage stability yet exhibited a very weak viscoelastic property. With the presence of GG and LBG, the emulsions demonstrated acceptable rheological properties and were found to be moderately stable upon storage. Based on these findings, binary or ternary blends of these polysaccharides in addition to XG-GG or XG-LBG could be exploited in order to produce emulsions with desirable rheological properties as well as storage stability.

A mixture experiment with an augmented simplex-centroid design has been widely used to obtain product formulations with optimized properties and to investigate the presence of either synergistic or antagonistic effects in a blend of two or more components. [8] By employing linear and non-linear regression models, the approach has been shown to be very useful in predicting the extent of the interaction effects on the properties of model or real food systems.[8-9] To date, no studies have been carried on the interaction effects of XG, CMC and LBG in blend formulations through the above experimental design and non-linear modelling.

2 METHOD AND RESULTS

2.1 Determination of Dynamic Rheological Properties of Emulsions

Blends of XG, CMC, and LBG containing pure XG, CMC, LBG and blends thereof were prepared according to an augmented simplex-centroid mixture design with 10 points. Binary blends of XG:CMC, XG:LBG and CMC:LBG were prepared at 1/2:1/2 ratio. Ternary blends of XG:CMC:LBG were prepared at the following ratios: 2/3:1/6:1/6, 1/6:2/3:1/6, 1/6:1/6:2/3 and 1/3:1/3:1/3. Blends of soybean oil and palm kernel olein (70:30) were used as a dispersed phase in a mayonnaise-like oil-in-water emulsion preparation. The emulsions were prepared using 40% oil, 10% water, 4% vinegar, 6% egg yolk and 40% polysaccharide blend solutions. The final percentage of polysaccharide in total emulsion formulation was 0.5%. Dynamic rheological measurements were performed in a controlled-stress rheometer using oscillatory tests (mechanical spectra) using a sand-blasted cone sensor. The experimental design and the mean values of the rheological responses of the emulsions, namely elastic modulus (G'), viscous modulus (G'') and loss factor ($\tan \delta = G'/G''$) at 1 Hz are presented in Table 1.

Table 1 *An augmented simplex centroid design with ten ratios of xanthan gum (X_1), carboxymethyl cellulose (X_2) and locust bean gum (X_3) blends and rheological responses of the respective emulsion*

Code	Component Proportion			Viscoelastic (at 1 Hz) Responses		
	X_1	X_2	X_3	G' (Pa)	G'' (Pa)	tan δ
A	1	0	0	84.03 ^{c}	16.78 ^{b}	0.198 ^{d}
B	1/2	1/2	0	37.02 ^{d}	8.76 ^{d}	0.237 ^{d}
C	0	1	0	3.70 ^{e}	4.27 ^{e}	1.153 ^{b}
D	0	1/2	1/2	1.58 ^{e}	4.34 ^{e}	2.755 ^{a}
E	0	0	1	29.46 ^{d}	22.78 ^{b}	0.774 ^{c}
F	1/2	0	1/2	171.43 ^{b}	23.83 ^{b}	0.139 ^{d}
G	2/3	1/6	1/6	238.00 ^{a}	32.07 ^{a}	0.136 ^{d}
H	1/6	2/3	1/6	2.41 ^{e}	2.93 ^{e}	1.227 ^{b}
I	1/6	1/6	2/3	80.89 ^{c}	17.82 ^{c}	0.220 ^{d}
J	1/3	1/3	1/3	157.07 ^{b}	28.86 ^{a}	0.186 ^{d}
Standard deviation range				0.12-14.51	0.12-3.33	0.001-0.135

$^{a-e}$ Means (n = 3) with the same superscript within the same column are not significantly different at $\alpha = 0.05$.

Combination of 50% XG and 50% LBG significantly ($p < 0.05$) gave a higher value of emulsion G' measured at 1 Hz (G' = 171.43 Pa), as compared to pure XG (G' = 84.03 Pa) (Table 1). However, both emulsions did not significantly differ in terms of their G'' and solid character ($\tan \delta < 0.3$). Conversely, combination of 50% XG and 50% CMC significantly reduced the elastic property of the emulsion. The viscoelastic behaviour of CMC:LBG emulsion almost resembled that of pure CMC. Thus, the emulsion behaved as a viscous flow rather than a solid body with $\tan \delta > 1$. For these emulsions, evolution of G' as a function of frequency applied in this study can be seen in Figure 1. Higher elastic property obtained for the XG:LBG emulsion compared to pure XG and LBG emulsions, could be related to a synergistic interaction between XG and LBG in the emulsion aqueous phase.[10] An intermolecular interaction may have occurred through the association of the locust bean's unsubstituted (i.e. galactose deficient) region, with the xanthan helix backbone (a cellulose backbone of β-[1⟶4]-*D*-glucopyranose glucan) forming junction zones [5,11] due to hydrogen bonding between water and hydroxyl groups of both polysaccharides.[12] Besides, regions of the LBG chain highly substituted by galactose residues could provide interruptions in this interaction, giving a matrix structure, and hence a gel results.[2] The gel formed thus allows a creation of complex network within the emulsion system, giving rise to a high elastic property.

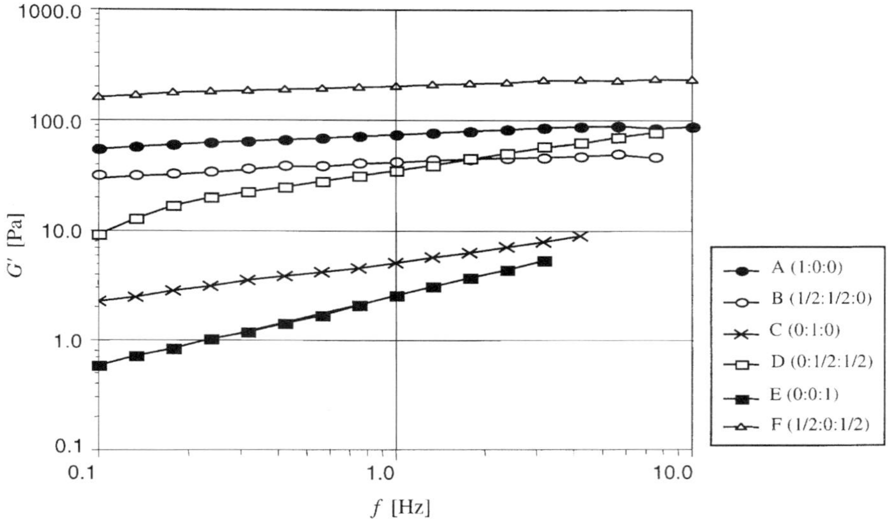

Figure 1 *Evolution of elastic modulus (G') as a function of frequency (f) (at strain of 0.5%) for emulsions containing pure and binary blends of polysaccharides. See Table 1 for sample codes.*

All emulsions with ternary blends demonstrated a gel-like character (*G'* > *G"*), except for the emulsion with a higher proportion of CMC (Table 1). The evolution of *G'* over the frequency range for these emulsions can be seen in Figure 2. Interestingly, the emulsions with higher proportion of XG and similar proportion of all polysaccharides demonstrated high *G'* values (at 1 Hz) of 238 Pa and 157 Pa, respectively. Particularly for the emulsion with a higher proportion of XG (~67%), its elastic nature was found to be significantly higher than that of pure XG emulsion as well as XG:LBG emulsion. The emulsion with an equivalent proportion of XG, CMC and LBG, was comparable with the XG:LBG emulsion in terms of their elastic modulus even though its XG level was lower (i.e. ~33% rather than 50%). Thus, it is reasonable to consider that the presence of CMC at ~33% in this ternary blend had contributed to another synergistic effect. It is likely that synergistic interaction might have been occurred between the unsubstituted region of LBG and the CMC cellulose backbone (β-(1⟶4)-*D*-glucopyranose polymer). The mechanism is believed to be similar to the one proposed for XG-LBG. Similar interaction in CMC-guar gum has been observed previously.[13] It is also believed that, the carboxymethyl deficient region of CMC could interact with the cellulose backbone of XG (since both polysaccharides contain hydroxyl groups to form junction zones). Consequently, a strong polymer complex might have occurred in the system that led to more complex network structure, reflecting a high elastic modulus. From these observations and explanations, we propose the occurrence of a ternary synergism of XG-CMC-LBG in addition to binary synergisms (XG-LBG and CMC-LBG) in our ternary blends. In a different view, CMC might have formed its own network interpenetrating with the network from XG-LBG synergism that finally resulted in more complex network structure as previously suggested to have occurred in a ternary blend of XG, LBG and methylcellulose. [6] However, this effect seemed not to be important when the level of CMC exceeded ~33% in the blend.

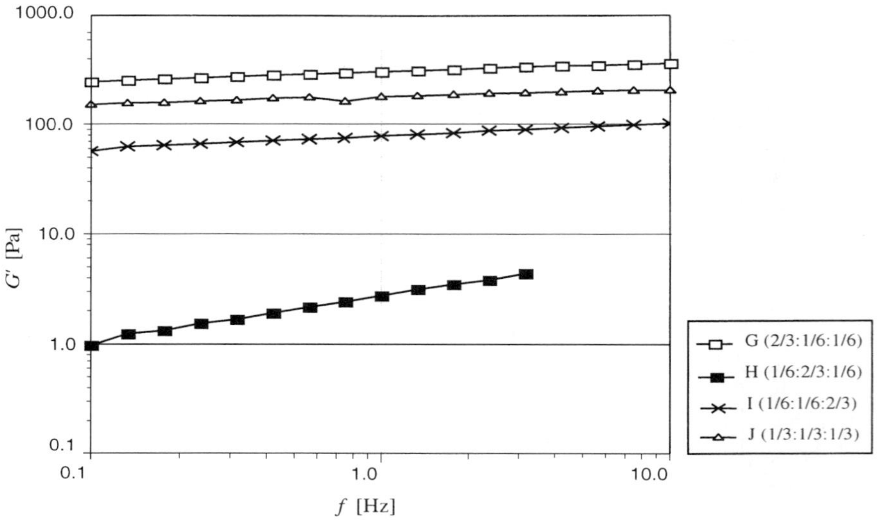

Figure 2 *Evolution of elastic modulus (G') as a function of frequency (f) (at strain of 0.5%) for emulsions containing ternary blends of polysaccharides. See Table 1 for sample codes.*

2.2 Determination of Fitted Regression Models, Contour and Surface Plots

The experimental domain consisted of different proportions of components of X_1 (XG), X_2 (LBG), and X_3 (CMC) between zero and one ($0 \leq X_i \leq 1$; $\sum X_i = 1$). The experimental domain was within an equilateral triangle (regular simplex). Mixture regression analysis was performed to determine estimated coefficients and significance of the model terms, the F-test and coefficient of determinations (R^2). The rheological responses were found to be best fitted with Scheffe's canonical special quartic (G' and tan δ) or full quartic (G'') models as shown in Table 2. It is clear that, both fittings finally provided no lack-of-fit (with p-value of 0.66-0.767) and high adjusted coefficient of determinations (with R^2 of > 0.9). The R^2 values were found to be essentially high and the variances found in all responses were explained well by the models. Once the estimated model equation for each response was obtained, contour and surface plots were generated. The plots show how a response relates to three components based on the model equation.

The highest positive coefficients of 12005.0 and 1408.9 found in the G' and G'' models, respectively showed that the strongest synergism could be obtained by using a ternary blend of XG:CMC:LBG at higher XG proportion. The effect can be clearly visualized in the pattern of the contour and surface plots of both responses, demonstrating a maximizing synergistic effect at increasing level of XG (Figures 3 and 4). The plots also exhibited a minimizing effect towards the point that corresponds to higher CMC. One can notice a limitation of both models in estimating the data for the points fall below the contour lines of $G' = G'' = 0$. This is due to the fact that the estimated value would be < 0, which is not true for the real elastic and viscous moduli. However, the equations are applicable to predict future data from the points of interest that are located beyond the constraint lines of $G' = G'' = 0$.

Table 2 *Fitted regression models for rheological response with lack-of-fit probability and adjusted coefficient of determination (R^2) of model fits*

Response	Fitted Regression Model	Lack-of-fit (p) [c]	Adjusted R^2
Elastic Modulus [a]	$G' = 82X_1 + 1X_2 + 29X_3 + 463X_1X_2 - 56X_2X_3 + 12005X_1^2X_2X_3 - 4433X_1X_2^2X_3 - 1639X_1X_2X_3^2$	0.306	0.9933
Viscous Modulus [b]	$G'' = 16.3X_1 + 3.4X_2 + 23.0X_3 + 17.6X_1X_3 - 36.3X_2X_3 + 67.6X_1X_3(X_1-X_3) + 1408.9X_1^2X_2X_3 - 215.8X_1X_2(X_1-X_2)^2$	0.336	0.9969
Loss Factor [a]	$\tan\delta = 0.19X_1 + 1.15X_2 + 0.77X_3 - 1.76X_1X_2 - 1.14X_1X_3 + 7.16X_2X_3 + 11.10X_1^2X_2X_3 - 91.19X_1X_2X_3^2$	0.066	0.9776

[a] Fitted with special quartic model. [b] Fitted with full quartic model. [c] Not significant at 95% confidence level (p < 0.05).

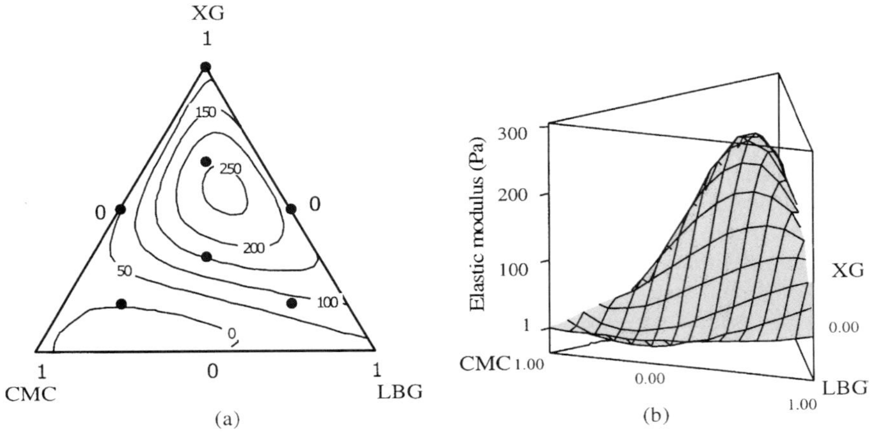

Figure 3 *Contour (a) and surface (b) plots for elastic modulus (G') (at 1 Hz) response fitted by a special quartic model. XG, Xanthan Gum; CMC, Carboxymethyl Cellulose; LBG, Locust Bean Gum*

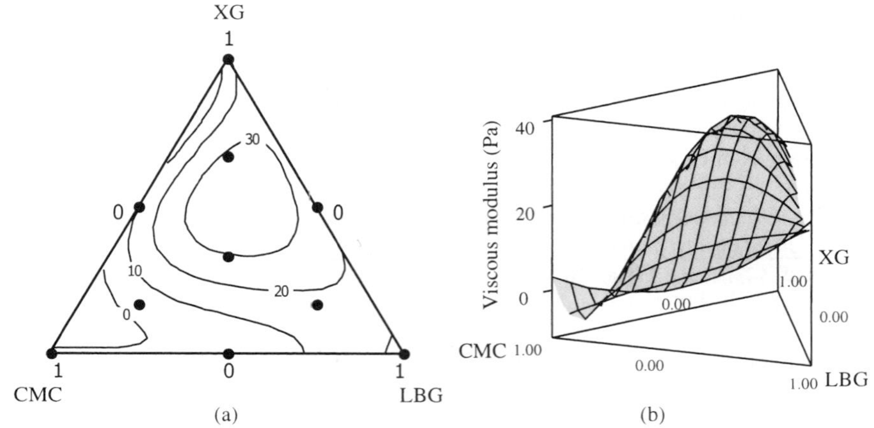

Figure 4 *Contour (a) and surface (b) plots for viscous modulus (G") (at 1 Hz) response fitted by a full quartic model. See Figure 3 for abbreviations*

The tan δ exhibited a different pattern of contour and surface plots (Figure 5). The response was predicted to be most affected by the antagonistic interaction among XG-CMC-LBG at higher LBG proportion with the highest negative coefficient (91.19). Moreover, it is obviously seen in the surface plot that the tan δ will gradually drop with a simultaneous substitution of CMC by XG and LBG. Thus, both XG and LBG will contribute to more elastic behaviour than viscous behaviour, causing the emulsion to behave more like a gel (tan $\delta < 1$). However, it is important to note that, a viscoelastic material always within the limit of $0 < \tan \delta < \infty$ and $\tan \delta = 0$ expresses an elastic ideal material.[14] Thus, any data from the points on or below the contour line of $\tan \delta = 0$ was not applicable to the real system studied. This reveals a limitation of the model to predict future data whereby the predicted data would be $\tan \delta \leq 0$.

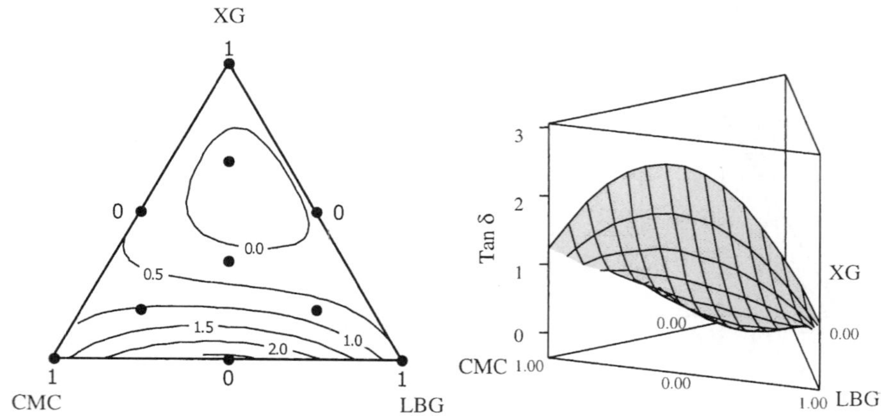

Figure 5 *Contour (a) and surface (b) plots for tan δ (G"/G') (at 1 Hz) response fitted by a special quartic model. See Figure 3 for abbreviations*

2.3 Model Validation

To validate the model equation of each response, two new ternary blends at ratios 1/2:1/5:3/10 (50% XG, 20% CMC and 30% LBG) and 3/10:1/5:1/2 (30% XG, 20% CMC and 50% LBG) (points within the interior of simplex) were chosen. Two new emulsions were prepared with these blends and their rheological responses were determined. An actual mean of each response was compared to the respective predicted value calculated using the respective equation model. The model is said to be valid when the actual mean response either matched well with the fitted value or fell within the range of the predicted value at 95% confidence interval.[15-16]

As shown in Table 3, the actual mean for each response fell within the range in which the estimated mean response for the new combinations are expected to fall. Thus, both special and full quartic models were experimentally checked as good and valid. However, this was an exception for the tan δ response, since the predicted value for each new combination was less than 0 (the value was not possible for the viscoelastic range). This observation provided evidence for the limitation of the model to predict future data for tan δ as previously discussed.

Table 3 *Predicted value for responses of emulsions with new component proportions xanthan gum (X_1), carboxymethyl cellulose (X_2) and locust bean gum (X_3) blends as compared to actual value*

Proportions			**Predicted values**			
X_1	X_2	X_3	**Fit**	**SE of Fit**	**95% CI**	**Actual values** [a]
			Predicted by a special quartic model			
Elastic Modulus (G')(Pa)						
1/2	1/5	3/10	254.632	3.565	(247.239, 262.024)	251.377 ± 3.213
3/10	1/5	1/2	160.085	3.566	(152.691, 167.480)	160.367 ± 2.739
Loss Factor (tan δ)						
1/2	1/5	3/10	-0.052	0.023	(-0.010, -0.004)	0.148 ± 0.003
3/10	1/5	1/2	-0.194	0.023	(-0.241, -0.146)	0.160 ± 0.004
			Predicted by a full quartic model			
Viscous Modulus (G") (Pa)						
1/2	1/5	3/10	37.418	0.845	(35.665, 39.171)	37.083 ± 1.055
3/10	1/5	1/2	26.604	0.730	(25.089, 28.119)	25.667 ± 0.345

[a] Data are presented as mean ± standard deviation, from three independent determinations (n = 3). SE, standard error; CI, confidence interval.

3 CONCLUSION

The synergistic interaction effect of ternary XG:CMC:LBG blends at approximately 33-67% XG levels appeared to be much stronger than that of the binary XG:LBG blend at 50% XG level. Nevertheless, an antagonistic interaction effect became important as CMC level in blends was more than 33%. The use of a mixture design with regression modelling was shown for the first time to be a valuable tool in better elucidating these interaction effects.

References

1 P.A. Williams and G.O. Phillips, G.O, 'Introduction to Food Hydrocolloids' in *Handbook of Hydrocolloids*, eds., G.O. Phillips, and P.A. Williams, Woodhead Publishing Limited, Cambridge, 2001, Chapter 1, pp. 1–19.
2 K.C. Symes, *Food Chem.*, 1980, **6**, 63.
3 C. Schorsch, C. Garnier, and J. Doublier, *Carbohydr Polym.*, 1997, **34**, 165.
4 J. Higiro, T.J. Herald, S. Alavi, and S. Bean, *Food Res Int.*, 2006, **39**, 163.
5 H.A. Khouryieh, T.J. Herald, F. Aramouni, and S. Alavi, *Food Res Int.*, 2006, **39**, 964.
6 Rinaudo, M. and Moroni, *Food Hydro*, in press.
7 I. Nor Hayati, Y.B. Che Man, C.P. Tan and I. Nor Aini, *Food Hydro.*, 2009, **23**, 233.
8 L.B. Karam, C. Ferrero, M.N. Martino, N.E. Zaritzky and M.V. Grossman, *Int J. Food Sci Technol.*, 2006, **41**, 805.
9 F. Depypere, D. Verbeken, O. Thas and K. Ewettinck, *Food Hydro.*, 2003, **17**, 311.
10 E.A. Makri and G.I Doxastakis, *Food Hydro* 2006, **20**, 1141.
11 I.C.M. Dea, A.H. Clark and B.V. McCleary, *Carbohydr Res* 1986, **147**, 275.
12 F. Franks, *Cyrobiol.*, 1983, **20**, 335.
13 L. Zhang and T. Kong, *Col Polym Sci.*, 2006, **285**, 145.
14 G. Tabilo-Munizaga and G.V. Barbosa-Canovas, G.V. *J. Food Engine.*, 2005, **67**, 147.
15 J. Cornell, *Experiment with Mixtures, Design, Models, and the Analysis of Mixture Data,* 3rd edn., John Wiley & Sons, Inc., New York, 2002, Chapter 2, p. 22.
16 S.P.J. Namal Senanayake and F. Shahidi, *Food Chem.*, 2002, **77**, 115.

IMPACT OF WATER ON FLAVOR DELIVERY SYSTEMS

P.E. Bouquerand[1] and V. Normand[2]

[1]Material Science Research Department, Firmenich S.A., 1217 Meyrin 2 Geneva, Switzerland
[2]Material Science Research Department, Firmenich Inc.,Princeton, 08536 N.J.,USA

1. INTRODUCTION

Granular Flavor Delivery Systems fulfill different purposes like flavor stabilization, easy handling, visual impact and especially controlled release. They are produced by broad range of technologies and processes, the most common being spray-drying, extrusion and fluid-coating [1, 2]. Despite a constant effort to investigate and develop new technologies and use new compounds, water soluble polysaccharides and more specifically starch hydrolysates, remain the main ingredients for flavour encapsulation. Therefore, the main release mechanism is observed through dissolution in water[3]. Among all advantages of using starch hydrolysates (price, sourcing, sustainability…), the most interesting feature is linked to its conventional plasticizer or solvent: water. Water is the most common and cheapest plasticizer on earth. Here is presented a review of the major models dealing with water effects on starch hydrolysates and their use to address production, stability and packaging issues.

2. THEORETICAL CONSIDERATIONS

Starch hydrolysates are polydisperse homopolymers characterized by a dextrose equivalent (DE) lower than 20 for maltodextrins and higher than 20 for glucose syrups. The impact of water on commercial starch hydrolysate based delivery systems depends strongly on the homopolymer molecular weight distribution. A quantification of the polydispersity of the material is obtained by size exclusion chromatography (SEC). From the molecular distribution profile, two useful average molecular weights can be calculated[4]. The viscosimetric molecular weight $\overline{M_v}$ (Equation 1) allows determination of the intrinsic viscosity $[\eta]_0$ using Mark-Houwink equation (Equation 2).
In Equations 1 and 2, k and α are constants related to the degree of molecular expansion depending of the local stiffness of the polymer backbone and polymer-solvent interaction (Equation 2).

$$\overline{M_v} = \left(\frac{\sum_i x_i M_i^{\alpha+1}}{\sum_i x_i M_i} \right)^{1/\alpha} \quad (1)$$

In Equation 1, x is the molar fraction, M the molecular weight and α is the Mark-Houwink exponent.

$$[\eta]_0 = k\overline{M_v}^{\alpha} \quad (2)$$

In Equation 2 (Mark Houwink model), $[\eta]_0$ is the intrinsic viscosity ($dl.g^{-1}$), k and α are constants specific of the molecular species.
From characterization of a starch hydrolysate by its intrinsic viscosity and its density, the impact of water on viscosity is predictable using Spurlin, Martin and Tennent's model[5,6] (Equation 3).

$$\eta_{solution} = \eta_{water}\left(\frac{d_{polymer} d_{water}(100 - c_{water})[\eta]_0}{d_{water}(100 - c_{water}) + d_{polymer} c_{water}} \exp\left(K \frac{d_{polymer} d_{water}(100 - c_{water})[\eta]_0}{d_{water}(100 - c_{water}) + d_{polymer} c_{water}} \right) + 1 \right) \quad (3)$$

In Equation 3, (extended Spurlin, Martin and Tennents's model), η is the viscosity, d is the density, c is the concentration in weight percent, K is 1.34.
Another useful average molecular weight is the effective molecular weight $\overline{M_E}$ characteristic of the dry glass transition temperature value (Tg_{dry}), the value of which is derived from Equation 4.

$$\overline{M_E} = \frac{\sum_i x_i \kappa_i M_i}{\sum_i x_i \kappa_i} \quad (4)$$

In Equation 4, the κ function takes into account the potential plasticization of the long homopolymer chains by the short oligomer chains.[4] From the $\overline{M_E}$ value, Tg_{dry} is estimated using Fox & Flory model (Equation 5) where (Tg_∞ is 520K, the Tg_{dry} of an infinite maltopolymer like starch[7]).

$$Tg_{dry} = Tg_\infty - \frac{a}{\overline{M_E}} \quad (5)$$

In Equation 5 (Fox and Flory model), a is equal to $53 kgKmol^{-1}$ for maltooligomers. The impact of water on the starch hydrolysate glass transition is then predictable using Couchman's model (Equation 6) because the specific heat of transition (ΔC_p) evolves linearly with starch hydrolysate average number molecular weight (M_n)[4].

$$Tg_{(blend)} = \frac{\phi_{water}\Delta C_{p(water)}Tg_{(water)} + (1-\phi_{water})\Delta C_{p(dry)}Tg_{(dry)}}{\phi_{water}\Delta C_{p(water)} + (1-\phi_{water})\Delta C_{p(dry)}} \tag{6}$$

In Equation 6 (Couchman's model), ϕ_{water} is the water weight fraction in w/w.

3. APPLICATION OF THE THEORIES

3.1 Water in processing maltopolymers for encapsulation

Spray-drying is a practical example where the effect of water predicted by Eq.3 is very useful. When a starch hydrolysate solution is to be spray-dried, there is an antagonism between the amount of water to be removed and the maximization of the surfaces[8] (particle size) to achieve the fastest mass transfer, fastest drying. A rough picture at the two ends is that a high dilution means time and energy consumption and a low dilution means slow drying kinetics and large particles sticking to the drying chamber wall. Therefore, an optimum viscosity is to be determined for each spray-drier configuration. Consequently, the amount of water necessary to prepare the feed solution can be adapted for each carrier maltopolymer according to a specific average molecular weight molecular characteristic of the molecular weight distribution ($\overline{M_v}$) (Figure 1).

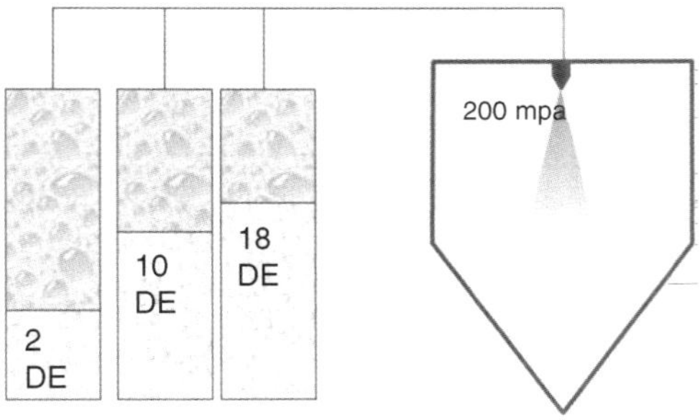

Figure 1: *Raw material dilution for optimum viscosity spray-drying following SMT 's model (Eq.3).*

3.2 Water in stability issues for maltopolymers based encapsulation products.

Modeling of the impact of water with the help of relevant physical parameters can solve very practical issues when dealing with flavor granular delivery systems. A first example is the moisture stability as evaluated by dynamic vapor sorption system (DVS).

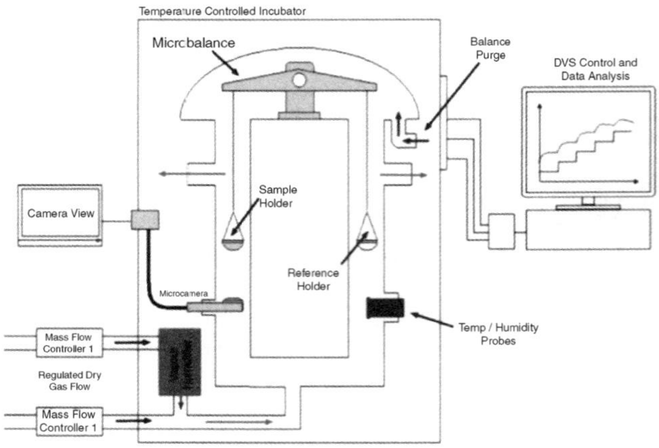

Figure 2 *Dynamic Vapour Sorption System (courtesy from SMS-UK)*

A DVS is a micro-balance where the sample is exposed to a controlled relative humidity under defined isothermal conditions (Figure 2). A classical experiment is to apply different relative humidity steps and wait for equilibrium with each step to evaluate the water sorption isotherm. The weight increase represents a water uptake quantity which is insufficient to characterize the product stability. However knowing the initial Tg and water content of the sample and using Couchman's model together with the weight increase, the Tg is predicted as a function of the moisture pick-up recorded by the DVS[9]. When the predicted Tg value equals the measurement temperature the sample becomes rubbery and loses the encapsulated volatiles. This application of the Tg prediction to encapsulation system moisture stability is not straightforward. A deconvolution is necessary between flavor losses (volatiles) and water uptake (Figure 3).

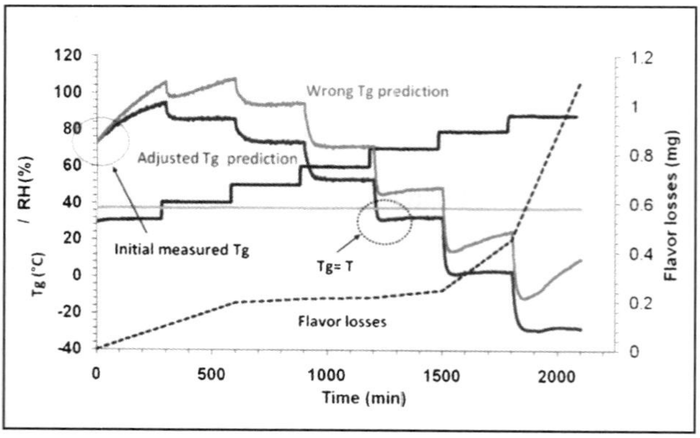

Figure 3 *Tg prediction including flavor losses*

Water sorption isotherms have also been used to normalize hygroscopicity[10]. From a process engineer point of view, lower molecular weight polysaccharides are more hygroscopic than higher molecular weight polysaccharides. Maltose is more hygroscopic than starch, more difficult to dry and more sensitive to moisture. Polysaccharides sigmoidal water sorption isotherms are well fitted by the Guggenheim, Anderson and de Boer's model (GAB) (Equation 7)[9-10].

$$X_W = \frac{X_m K C a_W}{(1 - K a_W)(1 - K a_W + K C a_W)} \quad (7)$$

In Equation 7 (GAB model), X_W is the total water weight fraction relative to the dry basis, X_m is the water fraction interacting directly with the polymer tested, K and C are two constants linked to water interactions with surrounding molecules, a_W is the water activity. From sorption isotherms at different temperatures an Arrhenius plot of C(T) and K(T) GAB parameters is possible. The parameters of the GAB equation represent one state of water toward the substrate (X_m monolayer) and two enthalpic contributions of state change, K from free water to polycondensed water (ΔH_{pl}) and C from polycondensed water to water on the substrate (ΔH_{mp}) (Figure 4).
The values ΔH_{mp} and ΔH_{pl} represent thermodynamic equilibrium between the three types of water interaction with the polymer moieties. The ratio of those two enthalpies, representing the different types of the water/substrate interaction, is a dimensionless number reflecting the affinity of the compound for water[10].

Table 1: *Selected hygroscopic number values.*

Compound	Hygroscopicity n°
Glucose (DP1)	97.5
Maltose (DP2)	52.7
18 DE Maltodextrin	19.7
10 DE Maltodextrin	16.6
2 DE maltodextrin	9.9
Starch (DP>60)	8.0
Yeast cell	6.46

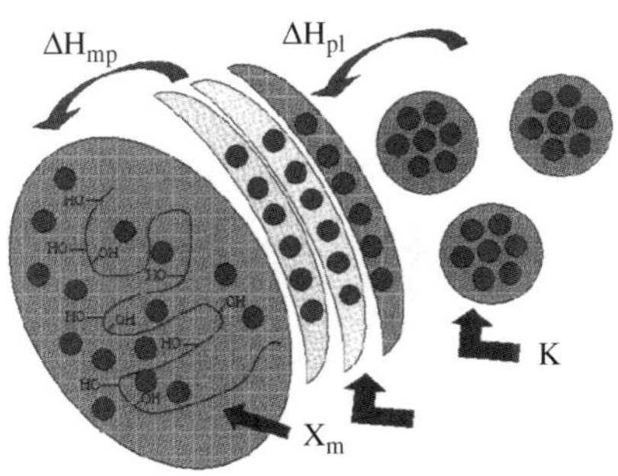

Figure 4: *GAB parameters and water/substrate interaction (the polymer is characterized with hydroxyl groups, black circles represent water molecules).*

$$Hygroscopic\ n° = \frac{Strong}{Weak} = \frac{\Delta H_{mp}}{\Delta H_{pl}} \geq 0 \quad (8)$$

The dimensionless hygroscopic number (Equation 8) is linearly related to the inverse of the degree of polymerization of maltopolymers, but is not limited to this specific polymer species. Any compound can be characterized by this dimensionless hygroscopic number.

Absolute comparison between very different compounds regarding their affinity with water is possible. This value is universal and does not depend on temperature, initial water content or morphology etc. Yeast cell is a good example (Table 1).

3.3 Water transfer through packaging material.

Modeling water impact can solve some important industrial issues such as the choice of a packaging material applied to a specific granular delivery system. A practical question asked by an industry supplying flavored granular delivery systems is: "Is this new packaging with supplier specification of 0.7g/m^2/day for water permeability suitable to this specific granular delivery system?" Considering a bag of 25 kg of flavor granules with standard carrier composition, according to Eq.6, 1 kg of water is necessary to decrease the Tg from 48°C to 24°C. In the absence of counter pressure of water in the bag, 4 years of diffusion is necessary when the bag is stored in liquid water at 24°C. But according to Equation 7, a counter pressure exists which is resolved by the mass transfer theory where the difference of water activity from outside to inside the bag is the driving force for the water penetration (Equation 9).

$$\frac{dm}{dt} = k\frac{S}{V}\left(a_{W\,inside} - a_{W\,outside}\right) \qquad (9)$$

In Equation 9, k is a kinetic constant in which the thickness, the density and the diffusion coefficient of water through the medium are hidden. S and V are the surface of the bag and the containing volume of the bag respectively.

To obtain the transfer constant k, small sealed bags containing relative humidity recorders were fully immersed in water and the data analyzed according to the resolution of Eq.9. Therefore, mass transfer of water through the bag material provides water gas inside the bag.

When the bag is filled with 25kg of maltopolymer based particulate material, the water gas equilibrates with the particles until equilibrium is reached. The counter pressure generated slows down the penetration of water as time increases but the condensed water on the hygroscopic medium lowers the Tg until it reaches the storage temperature (Figure 4).

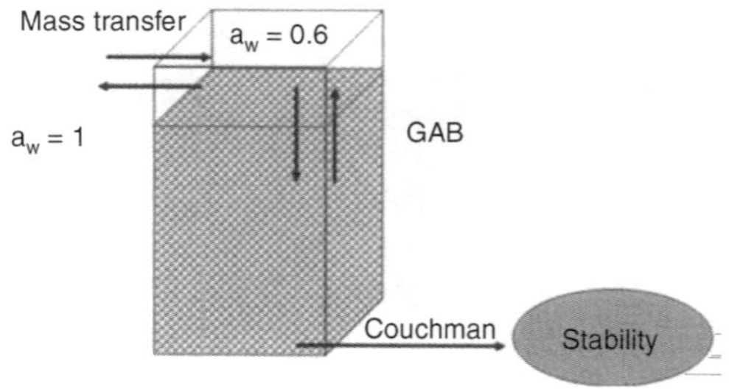

Figure 5: *Schematic representation of the product influence on water penetration through packaging*

Combining the different models presented above, a more precise evolution of Tg with time is calculated (Equation 10):

Using Eq.10, 40 years would be necessary to lower the Tg from 48 to 25°C when bags are immersed in water at 25°C. This result is 10 times higher than the initial rough estimation. Also, it is far above any industrial expectation.

$$\begin{cases} m_g(t+\delta t)=\dfrac{MV}{RT}\left[1-(1-a_W(t))\exp\left(-\dfrac{kS}{V}(\delta t)\right)\right] \\ m_p(t+\delta t)=m_p(t)+m_g(t+\delta t) \\ X_w(t+\delta t)=\dfrac{m_p(t+\delta t)}{m_0(1-\phi_{w0})} \\ \phi_w(t+\delta t)=\dfrac{X_w(t+\delta t)}{X_w(t+\delta t)+1} \\ Tg(t+\delta t)=\dfrac{\phi_w(t+\delta t)\Delta C_{pw}Tg_w+(1-\phi_w(t+\delta t)-\phi_{flav})\Delta C_{pcarrier}Tg_{carrier}}{\phi_w(t+\delta t)\Delta C_{pw}+(1-\phi_w(t+\delta t)-\phi_{flav})\Delta C_{pcarrier}} \\ a_W(t+\delta t)=\dfrac{-(KX_w(t+\delta t)(C-2)-KCX_m)-\sqrt{(KX_w(t+\delta t)(C-2)-KCX_m)^2-4((X_w(t+\delta t))^2K^2(1-C))}}{2X_w(t+\delta t)K^2(1-C)} \end{cases} \quad (10)$$

Equation 10 is used to assess the packaging performance of a mass m_0 of flavor granules containing ϕ_{W0} g of water and ϕ_{flav} g of flavor for 1g of granules. The water activity $a_W(t)$ in the first equation is calculated from the last equation knowing the water content at the origin (ϕ_{W0} in the third equation). In this simplified model, the mass of water penetrating the packaging m_g condenses entirely onto the particles the water mass of whichbecomes m_p before equilibrating with the headspace in the bag through the equilibrium driven by the sorption isotherm. The Tg is then calculated for all time increments.

4. Conclusion:

Knowing the chemical structure and the molecular weight distribution of the granular delivery systems, the impact of water is predictable in various aspects of the life of the delivery system. This understanding can help to resolve very practical issues and anticipate granular systems behavior and performance.

5. References

1 S. Gouin, *Trends in Food Science & Technology*, 2004, **15**, 330-347.
2 A. Madene, *Int. Journal of Food Science & Technology*, 2006, **41**, 1-21.
3 P.E. Bouquerand *et al.*, *AIChE Journal*, 2004, **50**, 3257-3270.
4 F. Avaltroni *et al.*, *Carbohydrate Polymers,* 2004, **58**, 323-334.
5 H.M. Spurlin *et al.*, *Journal of Polymer Science*, 1946, **1**, 63-74.
6 V. Normand *et al.*, *AgroFOOD industry hi-tech,* 2009, **20**, 14-17.
7 D. Benczédi *et al.*, *Macromolecules*, 1998, **31**, 3055-3061.
8 A. Gharsallaoui, *Food Research International*, 2007, **40**, 1107-1121.
9 P.-E. Bouquerand *et al.*, *Food Biophysics,* 2008, **3**, 182-185.
10 V. Normand *et al.*, *Starch/Stärke,* 2007, **59**, 100-102.

STABILITY OF LOW FAT STARCH-BASED DAIRY DESSERTS WITH INULIN ADDITION: VISCOELASTICITY AND PARTICLE SIZE DISTRIBUTION

A. Tarrega, J. D. Torres and E. Costell

Instituto de Agroquímica y Tecnología de Alimentos (CSIC). P.O. Box 73, 46100 Burjassot, Spain.

1 INTRODUCTION

Inulin is prebiotic dietary fiber composed of linear non-digestible polysaccharide of β-(2-1) linked fructose residues with a terminal glucose residue unit. Native inulin contains molecules of different polymerisation degree (from 2 to 60) and commercial products containing only short-chain inulin (2-7 units) or long-chain inulin (22-25 units) are obtained industrially. Inulin possesses the nutritional aspects of dietary fiber,[1] stimulates the selective growth of beneficial bifido bacteria,[2] increases the absorption of calcium by the body and reduces the triglyceride content of the serum and in the liver.[3]
Apart from its nutritional benefits, inulin is an interesting ingredient in the formulation of novel foods for different technological reasons that depend on the inulin average chain length. The short-chain inulin or oligofructose is much more soluble and sweeter than native inulin and it can be useful to partially replace sucrose in a formulation[2,4] or to replace it totally when combined with other non-caloric sweeteners. Long-chain inulin, with a high degree of polymerization, is more thermally stable, less soluble and more viscous than the native one[5] and has been used as a fat substitute in different dairy products.[6-9] In a recent study, the addition of long-chain inulin improved creaminess and consistency of low fat custards mimicking those of the full fat custard, but at the same time increased the sensation of roughness. It is probable that this last effect can be due to the presence of small crystals or crystal aggregates of inulin in the product. In concentrated aqueous solutions both native and long-chain inulin have been shown to develop a gel structure formed by a network of crystalline particles.[10,11] The characteristics of these gels have been shown to depend on multiple factors. In general, both crystallization rate and gel firmness increased with inulin concentration, size of inulin chain, shear treatment, and with the presence of seeding crystals after preparation, which depends on the thermal treatment.[12,13] As occurs in aqueous systems, when inulin is included in a product formulation the crystallization process could also be expected to occur, thereby affecting time-course rheological properties. The aim of this work was to study the variations in rheological properties and in particle size distribution of low fat custard desserts enriched with short-chain inulin, long-chain inulin and the mixture of both types of inulin during refrigerated storage.

2 MATERIALS AND METHODS

2.1 Materials

Two types of inulin were used in this study: inulin with high level of short-chain molecules (2-10 monomers, Frutafit CLR) and long-chain inulin ($\geq$ 23 monomers, Frutafit TEX!) from Sensus (Brenntag Química, Spain). The following ingredients were also used for preparing custard samples: hydroxypropylated tapioca di-starch phosphate (C* PolarTex 06741; Cerestar Ibérica, Spain), colorant (Vegex NC 2c; CHR Hansen S.A.), vanilla aroma (37548A; Lucta S.A.), skimmed milk powders (Central Lechera Asturiana, Spain), mineral water (Font Vella, Spain) and commercial sucrose.

2.2 Samples composition and preparation

Custard samples were prepared with 7.5% of short-chain inulin (CLR sample), long-chain inulin (TEX sample) or a mixture (50:50) of both inulin types (MIX sample). One custard without inulin (CONTROL sample) was prepared. All samples included fixed amounts of starch (3.75%, w/w), milk (80%, w/w), sucrose (6%, w/w), colorant (0.052%, w/w) and vanilla aroma (0.016, w/w). Milk was prepared 24h in advance by dissolving 13.5% (w/w) milk powder in mineral water and stored in refrigeration (4±1°C).
Samples were prepared in batches of 800 g. All ingredients were weighed in a flask and mixed under magnetic stirring during 10 min. The flask was placed in a water bath at 97±1°C and stirred constantly with a propeller stirrer. After 10 min the product temperature reached 85±1°C and heating was continued at this temperature for 15 min. After the heating process, the evaporated water was replaced gravimetrically. The sample was cooled in a water bath at 20°C until it reached temperatures of about 40°C and then the aroma was added. The sample was homogenised, transferred to a closed flask and stored in refrigeration (4±1°C). Particle size distribution and rheology were measured after 1, 2, 3, 6 and 7 days of storage.

2.3 Rheological measurements

Viscoelastic properties were measured in a controlled stress rheometer RS1 (Thermo Haake, Karlsruhe, Germany), with parallel plates geometry of 6cm diameter and 1mm gap. During the measurement a temperature of 10 ±1 °C was maintained using a Phoenix P1 Circulator device (Thermo Haake, Karlsruhe, Germany). Samples were allowed to rest for 10 minutes before measurement, two replicates were run for each sample and a fresh sample was loaded for each run.
First, stress sweeps were run at 1 Hz to determine the linear viscoelastic region. After that, frequency sweeps were performed over the range f = 0.01-10 Hz and the values of the storage modulus (G'), the loss modulus (G''), the loss tangent angle (tan δ), as a function of frequency, were calculated using the Rheowin Pro software (version 2.93, Haake).

2.4. Particle size distribution

Particle size distribution analysis was determined using a Laser Diffraction Particle Size Analyzer (MasterSizer 2000, Malvern Instrument Ltd., Worcestershire, England). A refractive index value of 1.53 was selected for measurements considering the starch refractive index as a reference. About 0.2 g of custard was dispersed in distilled water at room temperature (20 ± 2 °C) until an obscuration of 0.2 was obtained. The sample was placed under ultrasonic dispersion for 2 min to ensure particles were independently dispersed and thereafter maintained by stirring during measurement. Distributions were made in triplicate for each sample. Particle size calculations were based on the Mie-Scattering theory. Volume mean diameter values ($D_{4,3}$) and the percentage of volume corresponding to each observed population were obtained using the software provided with the equipment (Mastersizer 2000 V. 5.40).

2.5. Statistical analysis

The effects of sample type and the storage time on the viscoelastic parameters were studied using analysis of variance (ANOVA) of two factors with interaction. The Tukey test ($p=0.05$) was used to calculate the significant differences among mean values. Calculations were carried out with XLSTAT-Pro Version 2007 (Addinsoft, Paris, France).

3 RESULTS

3.1. Viscoelastic properties

Changes in the rheological behaviour of custard samples containing different types of inulin during storage time were studied. To measure the viscoelastic properties of the systems under study, mechanical spectra were obtained at a range of frequency values between 0.01 and 10 Hz. Figure 1 shows the mechanical spectra obtained for the four types of samples after one day and after seven days of storage. All the mechanical spectra obtained showed a viscoelastic response characteristic of gel systems with storage modulus values (G') higher than loss modulus values (G'') though some differences were observed depending on the type of inulin added and storage time. The day 1, all samples showed similar mechanical spectra with relatively low values of both moduli that correspond to the viscoelastic behaviour of weak gels. After seven days of storage, mechanical spectra for CONTROL and CLR samples were similar to those obtained the first day. However the samples containing long-chain inulin (TEX and MIX) showed after seven days of storage a different viscoelastic response corresponding to a stronger gel system with values of both G' and G" higher than those observed the first day.
In order to follow the changes occurring at different storage time and to compare among samples, the values of G', G'' and tan δ at a frequency of 1 Hz were taken (Table 1). For the three viscoelastic parameters, analysis of variance showed a significant interaction between the effects of the type of sample and storage time ($F=506.7$, $p< 0.001$; $F=254.6$, $p<0.001$ and $F=14.5$, $p< 0.001$ for G', G'' and tan δ, respectively) which confirms that variations in rheological properties of custards during storage depended on the type of sample. On one hand, for CONTROL and CLR samples, the G' and G''values remained without significant differences during storage time and tan δ values slightly increased. On the other hand, samples containing long-chain inulin (TEX and MIX) gradually increased

in the viscoelastic moduli (G' and G'') and decreased in tan δ values during storage time. The magnitude of these effects was higher for the sample enriched with long-chain inulin (TEX) than for the sample with the mixture of short and long-chain inulin (MIX). The beginning of these rheological changes took place sooner for TEX sample (after two days) than for MIX samples (after 3 days). An important change in the rheological behaviour was observed in the presence of long-chain inulin indicating that structural changes took place during storage. Accordingly to that observed in inulin concentrated aqueous systems[12], the formation of aggregates in the continuous phase of the of the custard system would explain the change in their viscoelastic properties.

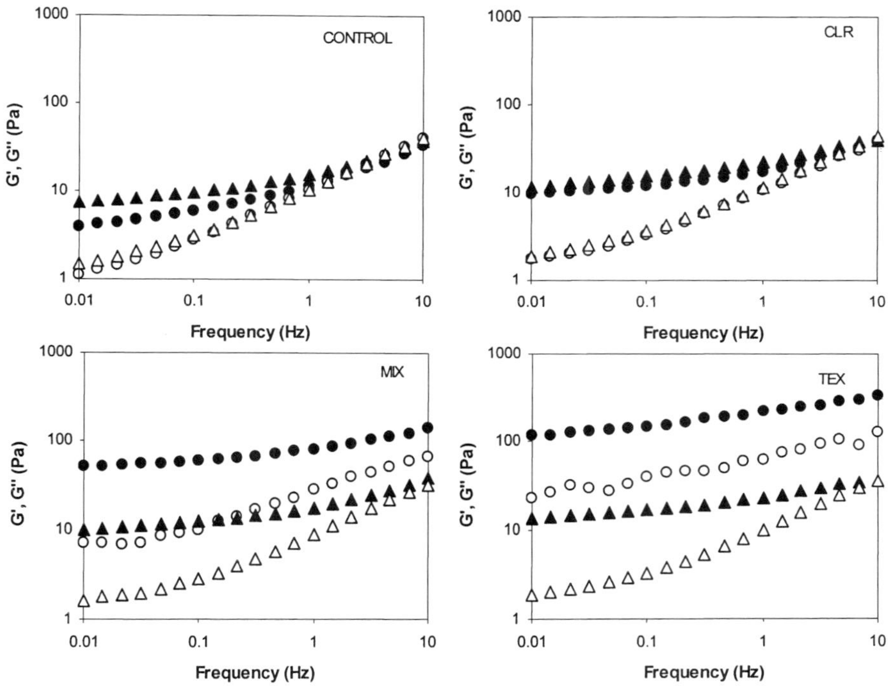

Figure 1 *Mechanical spectra obtained after 1 day (▲) and after 7 days (●) of storage for custard samples enriched with 7.5% of short-chain inulin (CLR), long-chain inulin (TEX), the mixture 50:50 of both inulins (MIX) and for the sample without inulin added (CONTROL). Storage modulus (filled symbols) and loss modulus (empty symbols).*

Table 1 *Storage modulus (G'), loss modulus (G'') and loss tangent (tan δ) values at 1 Hz for custard samples enriched with short-chain inulin (CLR), long-chain inulin (TEX), the mixture 50:50 of both inulins (MIX) and for the sample without inulin added (CONTROL).*

Sample	Time (Days)	G' (Pa)	G'' (Pa)	tan (δ)
Control	1	15.26^{ab}	11.05^{ab}	0.73^{g}
	2	19.33^{ab}	12.31^{ab}	0.64^{efg}
	3	17.74^{ab}	12.72^{ab}	0.72^{fg}
	6	11.56^{a}	11.21^{ab}	0.97^{h}
	7	11.89^{a}	10.60^{ab}	0.89^{h}
CLR	1	22.47^{abc}	11.38^{ab}	0.51^{bcd}
	2	23.80^{bc}	11.68^{ab}	0.49^{bcd}
	3	23.72^{bc}	12.27^{ab}	0.52^{cd}
	6	22.69^{abc}	13.54^{a}	0.60^{def}
	7	18.25^{bc}	10.37^{ab}	0.57^{cde}
TEX	1	23.71^{bc}	10.08^{ab}	0.43^{bc}
	2	59.54^{d}	18.50^{c}	0.31^{a}
	3	149.05^{f}	48.32^{e}	0.32^{a}
	6	242.65^{h}	75.33^{g}	0.31^{a}
	7	214.80^{g}	61.85^{f}	0.29^{a}
MIX	1	14.81^{ab}	8.29^{a}	0.57^{cde}
	2	18.17^{ab}	9.46^{ab}	0.52^{cd}
	3	31.84^{c}	13.64^{b}	0.43^{bc}
	6	62.89^{d}	23.80^{d}	0.38^{ab}
	7	81.93^{e}	28.17^{d}	0.34^{a}

3.2. Particle size distribution.

The changes in particle size distribution during storage time for custards containing different inulin type are shown in Figure 2. Custard containing short-chain inulin (CLR), showed a mono-modal distribution ranging from 10.0 μm to 91.2 μm with 34.5 μm of mean diameter ($D_{4,3}$). This distribution was the same to that obtained for the CONTROL sample corresponding to the population of particles of the basic system of custards that are mainly dispersed swollen starch granules. Regarding possible variations during storage, no changes in particle size distribution were observed neither for CONTROL sample nor for CLR sample. As expected, short-chain inulin addition did not affect the particle size distribution of this product because of the high solubility of this type of inulin.
However, the particle size distribution for samples containing long-chain inulin (TEX and MIX) varied depending on storage time. Particles below 10 μm were observed and attributed to aggregates of inulin crystals that as have been shown in inulin concentrated aqueous solutions [12] are formed from primary non-spherical crystals (0.5 by 3 μm).

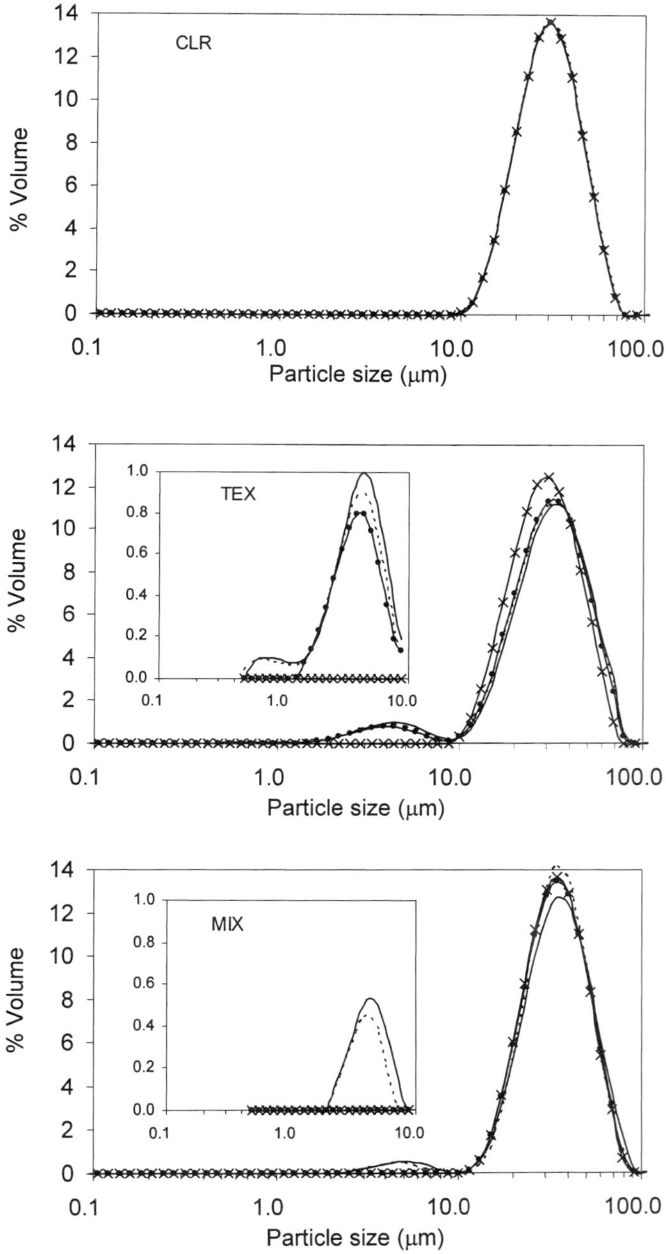

Figure 2 *Particle size distributions for custard samples enriched with short-chain inulin (CLR), long-chain inulin (TEX) and the mixture 50:50 of both inulins (MIX). Measurements were carried out at different time of storage: day 1 (-x-), day 2 (-•-) day 3 (- - -) and day 7(—).*

Table 2 *Variation during storage time of the percentage of volume occupied by the population of particles below 10 μm for custard samples enriched with short-chain inulin (CLR), long-chain inulin (TEX), the mixture 50:50 of both inulins (MIX) and for the sample without inulin added (CONTROL).*

		% Volume				
Sample	Range (μm)	Day 1	Day 2	Day 3	Day 6	Day 7
CONTROL	-	-	-	-	-	-
CLR	-	-	-	-	-	-
TEX	0.48 – 10.00	-	5.73	6.62	7.53	7.47
MIX	1.90 – 8.70	-	-	2.43	3.37	3.13

For both TEX and MIX the particle size distribution of day 1 was the same as that obtained for the CONTROL sample. After three days of storage, MIX sample showed that besides the peak corresponding to the basic custard system a new small population appeared located at lower particle size (1.9-10 μm) (Figure 2). The relative percentage of volume corresponding to this population initially was 2.43% and increased during storage time until reaching 3.13% of the volume (Table 2). The same population of small particles was observed in the TEX sample but it appeared sooner (after 2 days) and occupying a higher percentage of volume (from 5.73 to 7.47 %) than in the case of the MIX sample. Furthermore, after three days of storage, this distribution showed a shoulder from 0.48 to 1.45 μm indicating a new group of smaller particles that should correspond to the non aggregated inulin crystals.

Conclusions

In summary, results showed that custard enriched with 7.5% of short-chain inulin proved stable during the storage. However the viscoelastic properties of custard containing inulin (7.5% of long-chain inulin or 7.5% of the mixture of short and long-chain inulin) considerably changed during storage time leading to more elastic systems. In these samples, the formation of aggregates of inulin crystals was observed as a new population of small size particles (below 10 μm). The percentage of these particles increased during storage following the same trend as the changes in rheological behaviour. This confirms that the progressive aggregation of inulin crystals that took place in the continuous phase of the custard system, thus increasing the effective volume fraction, was responsible for the important rheological changes. An important reduction in the magnitude of these effects was observed using the 50:50 mixture of short and long-chain inulin instead of long-chain inulin.

ACKNOWLEDGEMENTS

To MICINN of Spain for financial support (Project AGL 2007-63444). To CHR Hansen S.A., Lucta S.A., Brentag Química and Central Lechera Asturiana for providing free samples of the ingredients.

REFERENCES

1. G. Flamm, W. Glinsmann, D. Kritchevsky, L. Prosky and M. Roberfroid, *Critical Reviews in Food Science and Nutrition*, 2001, **41**, 353-362.
2. M. B. Roberfroid, J. A. E. Van Loo and G. R. Gibson, *Journal of Nutrition*, 1998, **128**, 11-19.
3. N. Kaur and A. Gupta, *Journal of Biosciences*, 2002, **27**, 703-714.
4. F. P. De Castro, T. M. Cunha, P. L. M. Barreto, R. D. D. M. Amboni and E. S. Prudencio, *International Journal of Dairy Technology*, 2009, **62**, 68-74.
5. T. Wada, J. Sugatani, E. Terada, M. Ohguchi and M. Miwa, *Journal of Agricultural and Food Chemistry*, 2005, **53**, 1246-1253.
6. G. El-Nagar, G. Clowes, C. M. Tudoric, V. Kuri and C. S. Brennan, *International Journal of Dairy Technology*, 2002, **55**, 89-93.
7. P. Kip, D. Meyer and R. H. Jellema, *International Dairy Journal*, 2006, **16**, 1098-1103.
8. A. Tarrega and E. Costell, *International Dairy Journal*, 2006, **16**, 1104-1112.
9. T. Paseephol, D. M. Small and F. Sherkat, *Journal of Texture Studies*, 2008, **39**, 617-634.
10. C. L. M. Hebette, J. A. Delcour, M. H. J. Koch, K. Booten, R. Kleppinger, N. Mischenko and H. Reynaers, *Carbohydrate Research*, 1998, **310**, 65-75.
11. E. Chiavaro, E. Vittadini and C. Corradini, *European Food Research and Technology*, 2007, **225**, 85-94.
12. A. Bot, U. Erle, R. Vreeker and W. G. M. Agterof, *Food Hydrocolloids*, 2004, **18**, 547-556.
13. J. P. M. van Duynhoven, A. S. Kulik, H. R. A. Jonker and J. Haverkamp, *Carbohydrate Polymers*, 1999, **40**, 211-219.

PHYSICAL AND ANTIMICROBIAL PROPERTIES OF ENZYME-MODIFIED STARCH-BASED FILMS INCORPORATED WITH GARLIC OIL

M.H. Norziah[1*], L.L. Fang[1], and A.S. Ruri[1]

[1*]Food Technology Department, School of Industrial Technology,
Universiti Sains Malaysia, 11800, Penang, Malaysia.
E-mail: norziah@usm.my

1. INTRODUCTION

Edible and biodegradable films and coatings are an important area of food research with enormous commercial potential and great significance to the environment. Edible films can improve the appearance of food; protect its properties during their storage and handling. Generally, polysaccharides and proteins are well known for being good film formers[1]. Corn zein films were characterized to be able to form tough, glossy, hard, greaseproof coatings while soy protein isolated films were characterized by their functional properties of cohesiveness, adhesiveness, water and fat absorption, and emulsification[2]. Sometimes a composite film formulation, for example, caseinate-pullulan film with beewax lamination, can be tailor made to suit to the needs of a specific commodity due to the lack of characteristic functionality if used individually[3]. Edible films from amylose and collagen are probably the most effective forms of soluble packaging in terms of performance, adaptability to products, and production operations and costs[4]. The simultaneous utilization of starch and hydrocolloid in films can modify the material mechanical properties[5]. Starch-based films have poor mechanical properties and high moisture sensitivity. The film formed is brittle and easily cracks. The addition of hydrocolloid and plasticizers in starch-based film forming solution are usually done to improve mechanical properties of the edible films formed[6]. Plasticizers can decrease intermolecular attractions between adjacent polymeric chains, thus can increase film flexibility[7]. Incorporating antimicrobial agents into edible films or coatings provides a way to improve the safety and shelf-life of foods. The more commonly used antimicrobials include benzoic acid, lysozyme, lactoferrin, bacteriocins and plant-derived secondary metabolites, such as essential oils and phytoalexins[8], and sorbic acid[9]. The essential oil such as garlic oil are regarded as alternatives to chemical preservatives, and their use in foods meets the demands of consumers for minimally processed natural products, as reviewed by Burt[10]. The antimicrobial activity of spices such as garlic, onion, cinnamon, cloves, thymes and sage has been studied[6,8,10,11,12,13].

This paper presents work on the study of the development of enzyme-modified sago starch based edible films plasticized with sorbitol (as plasticizer) and garlic (as antimicrobial agent). In this study, high amylose content sago starch was used in the film formation. The effects of sorbitol and garlic oil on mechanical (tensile strength and elongation at break) and physical properties (water vapour permeability and moisture sorption isotherm) of the starch-based edible film were evaluated. The suitability and the effect of garlic oil as antimicrobial agent against Staphylococcus aureus was also evaluated.

2. MATERIAL AND METHOD

Sago starch (*Metroxylan sagu*) was obtained from Sim Company Sdn. Bhd. Penang, Malaysia. The other chemicals used in this analysis were sorbitol was supplied by Rhodia Malaysia, carrageenan (TIC Pretested Colloid) was kindly donated by TIC Gum Inc. US, pullulanase enzyme from Novo Nordisk (Bagsvaerd, Denmark), Nutrient broth used to grow the *Staphylococcus aureus* was obtained from Merck KGaA. The Muller-Hinton Agar and TPC-Agar (Total Plate Count Agar) were purchased from Oxoid, Basingstoke (Hampshire, England), Garlic oil was kindly donated by a local company. Chemicals used in saturated salt solutions: potassium carbonate, potassium acetate, magnesium chloride, strontium chloride, sodium bromide, sodium chloride, potassium chloride, lithium chloride and phosphorus pentoxide are of analytical grade.

2.1 Enzyme-modified sago starch

Enzyme-modified sago starch was prepared by enzymatic debranching procedure as described in our previous works[14]. Sago starch slurry (20 %, w/v) was prepared in sodium acetate buffer, pH 5.0. Pullulanase enzyme, 20 % (v/w of starch) was added and the starch suspensions were incubated at 58 °C in an orbital shaker (Certomat® SII; B. Braun Biotech Intl., Germany) for 20 h. The enzyme was deactivated at 75 °C for 15 min followed by centrifugation at 3000 × g for 15 min. The precipitate collected and dried in an oven at 40 °C is the debranched or enzyme-modified sago starch (EMS).

2.2 Organism and preparation of cultures

Staphylococcus aureus culture was obtained from the Microbiology Laboratory of the School of Industrial Technology (USM, Malaysia). The bacterial culture was grown on nutrient broth at pH 7.0 ± 0.2 at 25 °C.

2.3 Analysis of amylose content

The amylose content in the enzyme-modified sago starch was determined following the method described by Mc Grance[15] with slight modifications. Absorbance value was determined at 600 nm by using a UV-Visible spectrophotometer (UV-160A, Shimadzu, Tokyo, Japan). The percentage of amylose present was calculated.

2.4 Preparation of sago starch/carrageenan film

The film forming solution were prepared from a mixture of enzyme-modified sago starch (EMS) and carrageenan in ratios of 2:1 and 3:1 based on total weight of the two biopolymers (5 g) including sorbitol (20%) in 200 ml distilled water. Garlic oil was incorporated into the film forming solution at concentrations of 0.1 % and 0.3 % (v/w). Carrageenan powder was solubilized into distilled water by heating at 50 °C for 2 hr and added to the film forming solution. The mixture was further heated to 85 °C with continuous stirring for 45 min before it was cooled to room temperature. The solutions (80

g) were cast onto polyacrylic plates (16cm × 16cm × 3cm) followed by oven drying at 40 °C for 24 hr. The dry films obtained were peeled off and stored for further analysis. The thickness of films was measured at nine different positions by using the Dial Thickness Gauge (Mitutoyo, Japan).

2.5 Antimicrobial activity of films

The zone of inhibition assay on solid media was used for determination of the antimicrobial effects of films against *Staphylococcus aureus*. The edible films were cut into 6 mm diameter disks and then placed on Mueller Hinton agar plates, which had been previously seeded with 0.2 ml of inoculums containing indicator microorganisms in the range of 10^5-10^6 CFU/ml. The plates were then incubated at 37 °C for 24 hr observation. The plates were then examined for "zone of inhibition" of the film discs. The contact area was used to evaluate growth inhibition underneath the film disk in direct contact with target microorganisms in the agar. The area of the whole zone was calculated and then subtracted from the film disk area, and this difference area was reported as zone of inhibition. Experiments were carried out in duplicate.

2.6 Water vapor permeability

Water vapor permeability (WVP) tests were conducted following the ASTM[16] method E 96-00. The test film was sealed as a patch onto a glass permeation cell containing silica gel (RVP = 0) with 1.5 cm deep headspace. The glass permeation cells were 4.5 cm in diameter and 3.2 cm high. The cell was then placed in a desiccators maintained at 30 °C and RVP = 0.56 using a saturated solution of sodium bromide. The cell was weighed daily over 6 days. The samples were measured in triplicates.

2.7 Mechanical properties

Tensile strength and elongation at break of films were tested using a TA-TX2 texture analyzer (Stable Micro System, Surrey, U.K.) in accordance with ASTM[17] method D882-00. Five sample strips (14.0 × 2.0 cm) of each formulation were cut and clamped between tensile grips. Each film strip was placed in pneumatic grips (25 psi) and stretched at 1.0 mm/s and with an initial distance between the grips of 100 mm. The parameters determined were tensile stress (Mpa) and elongation at break (%E). Seven replicates were tested.

2.8 Water sorption isotherm

Sorption isotherms of starch films were determined at 30 °C according to the procedures described by Chang et al[18], with some modifications. The edible films were cut into small pieces (3 × 5 mm), and then placed in a desiccator maintained at 30 °C with P_2O_5 for 8 days. The samples were weighed to the nearest 0.0001 g into pre-weighed weighing bottles. The analysis was done in duplicate. The samples were equilibrated in air tight containers over saturated solutions of lithium chloride, potassium acetate, magnesium chloride, potassium carbonate, strontium chloride, sodium chloride and potassium chloride.

2.9 Statistical analysis

SPSS 12.0 was used in this study to analyze the data. One-way variance analysis was carried out using Duncan's test with confidence level as $p \leq 0.05$.

2.10 Scanning electron microscopy (SEM)

Films were observed using SEM (Leo Supra 50vp, Oberkochen, Germany). Film pieces (0.5 × 0.5 cm) were mounted on aluminium stubs using a double-sided tape and then coated with a layer of gold (100 to 200 mm thickness).

3. RESULTS AND DISCUSSION

The enzyme modified starch (EMS) prepared had a high content of amylose, 53.6% than the native starch which can then form stronger gels. All films prepared were easy to handle and transparent. They were also easily removed from the casting plate surface and had good flexibility and a smooth surface (based on visual observation). The average thickness of films was 0.62 ± 0.05 mm.

3.1 Antimicrobial activity

Inhibitory activity was measured based on the clear zone surrounding a circular film disk. If there is no clear zone, it is assumed that there is no inhibition. The results showed that there are significant differences ($p<0.05$) between the films with and without incorporation of garlic oil. From the result (Table 1), the films without added garlic oil did not show any inhibitory effect against the tested microorganism in terms of surrounding clearing zone. The EMS/Carrageenan (2:1) with no added sorbitol and 0.3% garlic oil showed highest inhibitory zone area (58.52±9.17 mm^2) compared to the other films. This result was similar with the study of Pranoto[5], who reported that *Staphylococcus aureus* was more sensitive to garlic oil incorporated edible films as compared to *E.coli* 0157:H7 and *Salmonella enteriritidis*. However, there are no significant differences between films with added garlic oil at both the concentrations of 0.1% and 0.3%.

3.2 Water vapor permeability

Films showed a markedly significant difference between the WVP for each film ratio after the addition of sorbitol. The presence of sorbitol has decreased the WVP. A similar trend was observed in both 2:1 amd 3:1 EMS/Carrageenan films (Figures 1 and 2). In addition, the results show that films without sorbitol and 0.3 % garlic oil had the highest WVP value compared to films with sorbitol for both film ratios. Talja[19] indicated that high WVP of fresh potato starch film without sorbitol could be caused by micro cracks in the film. Thus, the result for EMS film without sorbitol might be due to similar condition. Apart from the sorbitol effect, it was also observed that incorporation of garlic oil in films without sorbitol at each ratio, did show any significant effects on WVP. Films with sorbitol also showed the similar result. This might be due to the small amount of garlic oil added that could not give any significant effect to the film's WVP.

Table 1 *Antimicrobial activity of starch-based film incorporated with sorbitol and garlic oil against Staphylococcus aureus*

Films	EMS		EMS/Carrageenan, (2:1)		EMS/Carrageenan, (3:1)	
	Inhibitory Zone[A]	Contact Area[B]	Inhibitory Zone[A]	Contact Area[B]	Inhibitory Zone[A]	Contact Area[B]
0% Sorbitol & 0% Garlic oil	0.00 ± 0.00^a	-	0.00 ± 0.00^a	-	0.00 ± 0.00^a	-
0% Sorbitol & 0.3% Garlic oil			58.52 ± 9.17^b	+	48.42 ± 15.81^b	+
20% Sorbitol & 0% Garlic oil			0.00 ± 0.00^a	-	0.00 ± 0.00^a	-
20% Sorbitol & 0.1% Garlic oil			48.68 ± 10.46^b	+	48.27 ± 8.32^b	-
20% Sorbitol & 0.3% Garlic oil			57.33 ± 9.75^b	+	53.90 ± 13.51^b	-

Mean values (n=5) with different superscript within the column are significantly different ($p<0.05$);

+: represents an inhibitory effect and -: represents no inhibitory effect.

A are measurements of diameter of inhibitory zone (in cm); B is contact area in agar directly underneath film pieces.

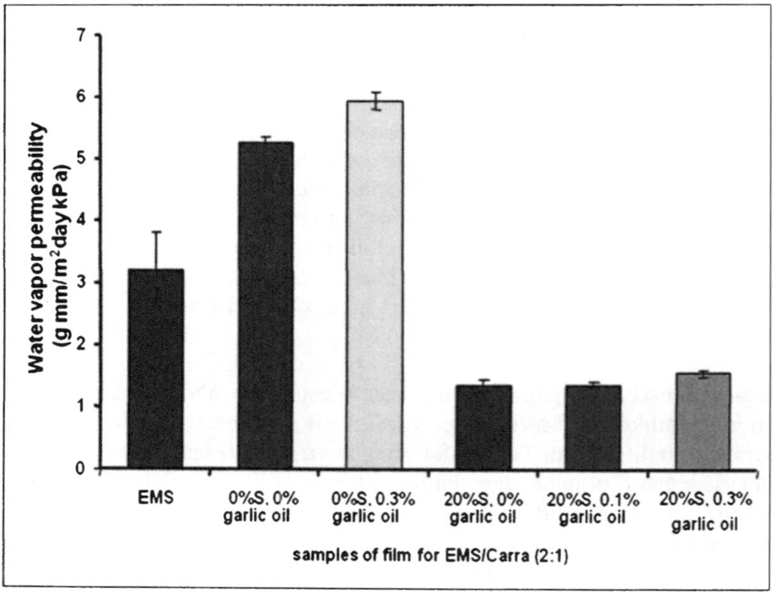

Figure 1 *Effect of sorbitol and garlic oil on water vapor permeability of films EMS/Carrageenan (2:1)*

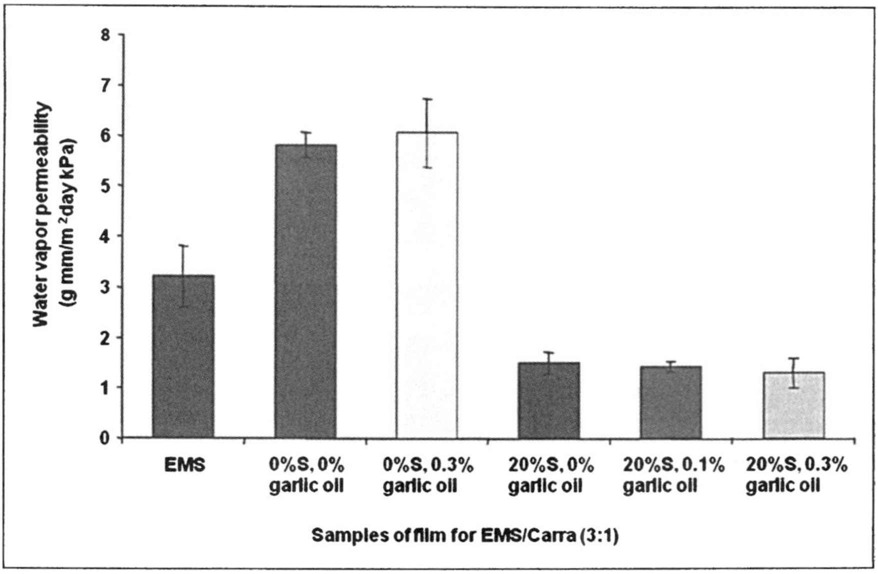

Figure 2 *Effect of sorbitol and garlic oil on water vapor permeability of films EMS/Carrageenan (3:1)*

3.3 Mechanical properties

The results given in Table 2 indicate that there are significant differences between the tensile strengths of films with and without sorbitol. The addition of sorbitol as plasticizer has reduced the tensile strength. In addition, a significant increase of elongation at break was also revealed by addition of sorbitol. Similar trends were observed for both ratios. Films without sorbitol also showed similar results. In general, EMS film without addition of carrageenan, sorbitol and garlic oil has shown the highest value of tensile strength and lowest value of elongation at break (Table 3). The EMS/Carrageenan (3:1) 20% Sorbitol, 0.3% Garlic oil was the lowest in tensile strength (0.21±0.01 MPa) and the highest in elongation at break (1.13 ± 0.10%) compared to the others. This might be due to more stress required to break down the tightly packed, ordered hydrogen-bonded network structure of the modified starch[20]. Moreover, the statistical analysis has shown that the tensile strength values of EMS/Carrageenan (2:1) films were significantly different from EMS/Carrageenan (3:1) films. The tensile strength values decreased with increased amount of starch and reduced amount of carrageenan.

3.4 Moisture sorption isotherm

Figures 3 and 4 show the results of moisture sorption isotherms for EMS/Carrageenan (2:1) and (3:1). The results show that the moisture content of films with and without sorbitol was increased as the relative humidity (RH) increased. Similar pattern was observed for both ratios. The results obtained just merely show that the moisture contents are increased gradually as the RH increased. In general, films without sorbitol were observed having higher water content than films with sorbitol at most RH or water activities for both ratios. Thus, this has indicated that the presence or absence of plasticizer will greatly affect the moisture sorption isotherm. The addition of sorbitol, which is hydrophilic plasticizer, will provide more active sites by exposing the hydrophilic hydroxyl groups to the film surface.

Table 2 *Tensile strength (MPa) for starch-based films incorporated with sorbitol and garlic oil*

Films	EMS	EMS/Carrageenan (3:1)	EMS/Carrageenan (2:1)
0% Sorbitol & 0% Garlic oil	0.37±0.01^{d}	0.23±0.01^{b}	0.30±0.02^{c}
0% Sorbitol & 0.3% Garlic oil		0.21±0.00^{b}	0.29±0.01^{c}
20% Sorbitol & 0% Garlic oil		0.18±.0.00^{a}	0.23±0.00^{b}
20% Sorbitol & 0.1% Garlic oil		0.18±0.00^{a}	0.22±0.01^{b}
20% Sorbitol & 0.3% Garlic oil		0.18±0.00^{a}	0.21±0.01^{b}

Means value (n=5) with different superscript within the column are significantly different ($p<0.05$).

Table 3 *Elongation at break (%E) for starch-based films incorporated with sorbitol and garlic oil.*

Films	EMS	EMS/Carrageenan (3:1)	EMS/Carrageenan (2:1)
0% Sorbitol & 0% Garlic oil	0.21±0.04^{a}	0.68±0.06^{b}	0.47±0.09^{b}
0% Sorbitol & 0.3% Garlic oil		0.70±0.09^{b}	0.54±0.06^{b}
20% Sorbitol & 0% Garlic oil		1.11±0.21^{c}	1.03±0.09^{c}
20% Sorbitol & 0.1% Garlic oil		1.11±0.16^{c}	1.06±0.07^{c}
20% Sorbitol & 0.3% Garlic oil		1.13±0.10^{c}	1.10±0.13^{c}

Means value (n=5) with different superscript within the column are significantly different ($p<0.05$).

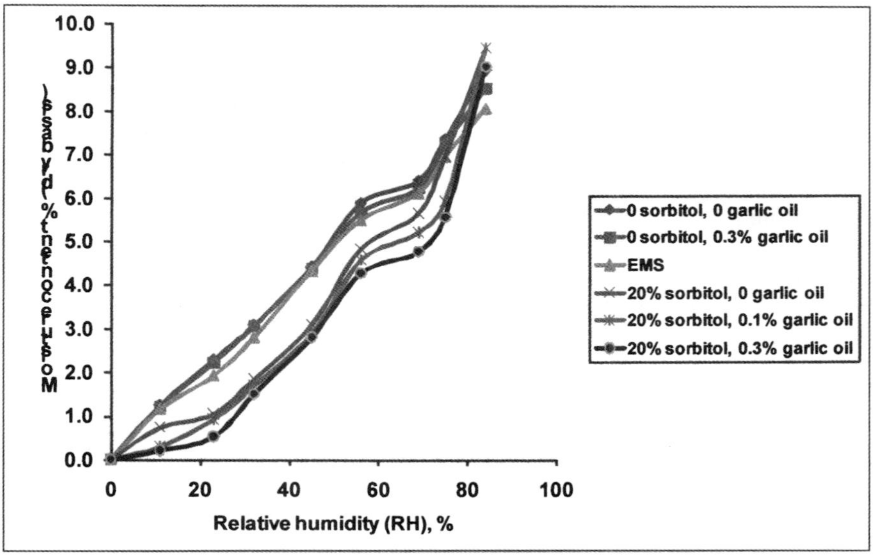

Figure 3 *Moisture sorption isotherm for EMS/Carrageenan (2:1) starch-based films incorporated with sorbitol and garlic oil.*

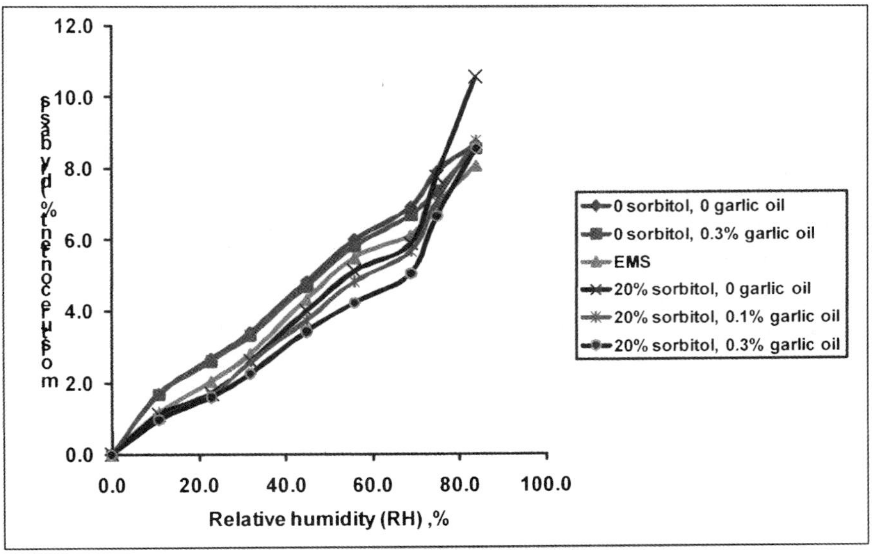

Figure 4 *Moisture sorption isotherm for EMS/Carrageenan (3:1) starch-based films incorporated with sorbitol and garlic oil*

3.5 Scanning electron microscopy (SEM)

From the SEM micrographs, EMS film was observed to be homogeneous and has pores (Figure 5). This is similar with the study of Garcia et. al.[7] which has shown that starch-based film without plasticizer formed brittle films with pores or cracks. The surface of EMS/Carrageenan film (Figure 6) was observed to be homogeneous and smooth without showing any pores or cracks. This might due to the fact that the hydrocolloid added can mix well with the starch slurry and thus, produce the homogeneous film forming solution. The result has shown that the surface of the EMS/Carrageenan/Sorbitol was observed to be homogeneous, rough and has pores. The primary role of a plasticizer is to enhance film flexibility and decrease brittleness[21]. Thus, there are no cracks observed in Figure 7.

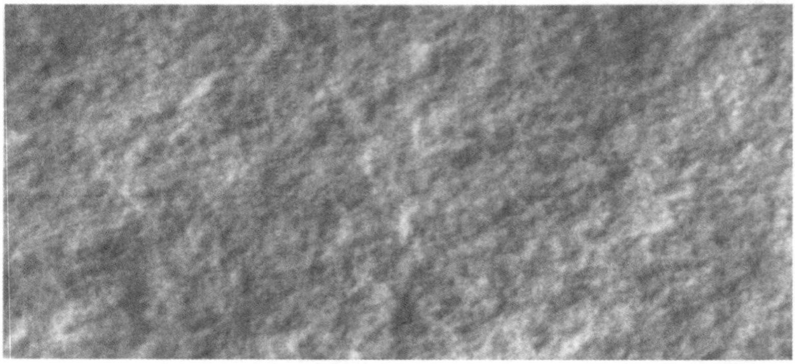

Figure 5 *Surface of enzyme modified sago starch film (EMS). Magnification 5000 X*

Figure 6 *Surface of EMS film with addition of carrageenan at ratio 3:1. Magnification 5000 X*

Figure 7 *Surface of EMS/Carrageenan (3:1) film with addition of sorbitol. Magnification 5000X*

4 CONCLUSION

The results obtained demonstrate that films containing garlic oil are effective in inhibiting the growth of *Staphylococcus aureus* based on the zone inhibition assay. The mechanical properties of films were greatly affected by the presence of plasticizer. The EMS/Carrageenan (3:1) films have lower tensile strength compared with films of EMS/Carrageenan (2:1). This indicates that films with better mechanical and barrier properties can be formed by increasing the amount of starch in the film forming formulation although there are no significant different between the WVP and %E for both ratios. The surface of EMS/Carrageenan/Sorbitol (3:1) was more homogeneous, rougher and has more pores compared to films without sorbitol.

Acknowledgement

The authors wish to express their thanks to Ministry of Technology and Science Malaysia for the grant provided for this study.

References

1. I. Arvanitoyannis, E. Psomiadou, A. Nakayama, S. Aiba and N. Yamamoto, *Food Chemistry*, 1998, **60**(4), 593.
2. T. Padgett, I.Y. Han and P.L. Dawson, *J. of Food Protection*, 1998, **61**, 1330.
3. R.N. Tharanathan, *Trends in Food Science*, 2003, **14**, 71.
4. R. Daniels, *Edible Coatings and Soluble Packaging*, In: Edible Coatings and Soluble Packaging, 1973, p.4.
5. P. Veiga-Santos, L.M. Oliveira, M.P. Cereda, A.J. Alves and A.R.P. Scamparin, *Food Hydrocolloid*, 2005, **19**, 341.
6. F. Debeaufort, and A. Voilley, *J. Agricultural Food Chemistry,* 1997, **45**, 685.

7. M.A. Garcia, M.N. Marthino, N.E. Zaritzky, and L. Plata, *Starch/Stàrke,* 2000, **52**, 118.
8. Y. Pranoto, S.K. Rakshit, and V.M. Salokhe, *Food Research International*, 2005, **38**, 267.
9. S. Guilbert, N. Gontard and L.G.M. Gorris, *Lebensm-Wiss.u.-Technology*, 1996, **29**, 10.
10. S. Burt, *Int. J. Food Microbiol*, 2004. **94**, 223.
11. Y. Pranoto, S.K. Rakshit, and V.M. Salokhe, *Lebensmittel-wissenschaft and Technologie*, 2005, **38**, 859.
12. A.C. Seydin, and G. Sarikus, *Food Research International*, 2006, **39**, 639.
13. A.R. Maria, M.R. Rosa, C.S. Robert, J.A. Roberto, H.M. Tara and M. Olga, *Postharvest Biology and Technology*, 2007, **45**, 254.
14. M. Maizurah, A. Fazilah, M.H. Norziah, and AA. Karim, Antibacterial Activity and Mechanical Properties of Partially Hydrolyzed Sago Starch–Alginate Edible Film Containing Lemongrass Oil, J. *of Food Science*, 2007, **72** (6), C324.
15. S.J. McGrance, H.J. Cornell, and C.J. Rix, *Starch/Stàrke*, 1998, **50**, 158.
16. ASTM. 2000. Standard test methods for water vapor transmission of materials, method E 96-00. Philadelphia, PA: American Society for Testing and Materials.
17. ASTM. 2000. Standard test methods for tensile properties of thin plastic sheeting, method D882-00. Philadelphia, PA: American Society for Testing and Materials.
18 Chang, Y.P., Cheah, P.B. and Seow, C.C. Plasticizing-antiplasticizing effects of water on physical properties of tapioca starch films in the glassy state. *Journal of Food Science*, 2000, **65**(3), 445.
19. R.T. Talja, H. Helén, Y.H. Ross, and K. Jouppilla, *Carbohydrate Polymers*, 2006.
20. S. Mali, L.S. Sakanaka, F. Ygamashita, and M.V.E. Grossman, *Carbohydrate Polymers*, 2005, **60,** 283.
21. L. Yang, and A.T. Paulson, *Food Research International*, 2000, **33**, 563.

EFFECTS OF ULTRAVIOLET TREATMENT ON THE MECHANICAL AND PHYSICAL PROPERTIES OF EDIBLE FILM FROM SAGO AND MUNG BEAN FLOUR

MS Ummi Shafiqah[1], A. Fazilah[1] and A.A. Karim[1]

[1]Food Technology Department, School of Industrial Technology, Universiti Sains Malaysia, 11800, Penang, Malaysia.

1 INTRODUCTION

Packaging is necessary for preserving the organoleptic, nutritional and hygienic characteristics of food during storage and marketing. The wide variety of packaging films can be divided into synthetic and edible or biodegradable. Constant progress in the technology of synthetic film preparation has expanded and supported their utilization in the food industry. However, most synthetic films are petrochemical-based and non-biodegradable, leading to environmental pollution and serious ecological problems[1]. The development of alternative edible and/or biodegradable packaging material to partially or totally replace synthetic polymers may offer opportunities to fulfill the consumers' demands and expectations of new food packaging materials that are convenient and environmentally friendly[2]. Edible films can be used as wraps or formed into pouches, bags, casings or sachets, resulting in reduced waste and improved recyalibility[3]. The use of biopolymers such as starch can be interesting because this polymer is quite cheap, biodegradable and edible[4]. Starch from different sources has been studied as a potential film-forming agent including from yam[5], wheat[6], tapioca[7], maize[8] and cassava[9].

Sago starch is extracted from the pith found inside the trunks of the sago palm *Metroxylan sagu* native to the East Indies. It is an important resource especially to the people in rural areas because it has various uses especially in the production of sago flour or sago pearl. Sago starch has been used for a long time in the food industry for the production of biscuits, noodles, bread, cake, fish crackers and many other traditional foods[10].Mung bean (*Vigna radiata (L.) Wilczek*) also known as green gram, golden gram and green soy is the seed of *Vigna radiata* which is native to India. Mung bean is similar in composition to other members of the legume family with 24% protein, 1% fat, 63% carbohydrate and 16% dietary fiber[11]. Mung bean is known as the best raw materials for starch noodles owing to its high amylose content and restricted swelling[12].

Starch films are excellent oxygen barriers due to their tightly packed, ordered hydrogen-bonded network structure and low solubility[13]. Unfortunately, there are some limitations for starch-based films due to their poor mechanical properties and high moisture sensitivity[14]. Therefore, improving properties of starch films through various treatments is desirable (enzyme, physical and chemical). [8] noted that thermal treatment enhanced the stability of maize starch films.

The modification techniques are often complex and time consuming; however irradiation can provide a simple and environmental-friendly alternative to change the physical, chemical and/ or biological characteristics of a product. This treatment requires limited sample preparation, is fast, does not require catalysts and does not induce a major temperature increase [15]. [16] reported that physical properties of a sago starch/PVA blend were improved by UV irradiation. According to [17], the tensile strength of starch-based plastic sheets was improved by electron beam radiation.

Ultraviolet (UV) irradiation as a physical, cost effective, non-thermal and environmental friendly technology has received increasing attention during recent years, which has been successfully applied for preservation and decontamination of food products[18]. In the USA, the code of Federal Regulations (Title 21, Part 179) permits use of UV radiation (wavelengths of 220 to 300 nm) on food products to control surface microorganism[19]. It also has been used in medical and pharmaceutical research to crosslink collagen and gelatin films[20, 21, 22]. According to [23], UV photodegradation of starch results in glycosidic bond cleavages, with a shortening of the amylose chain and a debranching of the amylopectin chain through free radical formation. [19]also stated that UV radiation showed potential for modifying properties of different proteins films such as solubility of sodium caseinate films, WVP of egg albumen films and TS of wheat gluten, corn zein and egg albumen.

In this work, edible films of sago and mung bean starch were produced by treating with ultraviolet (UV) radiation and the physical and mechanical properties were studied. The general aim of this work was to test the hypothesis that UV treatment could cross linked protein (from mung bean and sago) and starch and enhance the mechanical properties of the film.

2 MATERIALS AND METHOD

2.1 Materials

Sago starch (*Metroxylon sagu*) and mung bean starch (*Vigna radiata (L.) Wilczek*) were obtained from YKL Multi Enterprise, Penang, Malaysia. Glycerol from Qrec (New Zealand) was used in this study.

2.2 Preparation of sago and mungbean starch films

Starch films were prepared by adding 7 g to 200 mL distilled water followed by 2 g of glycerol. The solution was heated at 85 °C with continuous stirring for 1 h before it was cooled to room temperature. Films were cast by pouring 95 g of solution onto polyacrylic plates (16 x 16 cm) and then exposed to a UV light source (253.7 nm; 20 W) in a laminar flow cabinet at a distance 20 cm from the surface (Esco® PCR Vertical Laminar Flow Cabinet, Singapore). Films were exposed for 2 and 3 h. After exposure, films were oven dried at 40 °C for 24 h. Dried films were peeled and stored for further analysis.

2.3 Determination of amylose content in starch

Method used to determine the amylose content in starch is based on the procedure described by [24].

2.4 Film thickness

Thickness of films was determined using a micrometer (dial thickness gauge 7301; Mitutoyo Co., Tokyo, Japan) to the nearest 0.01 mm accuracy at 5 locations and the average value was calculated.

2.5 Mechanical properties

Tensile strength and percentage of elongation at break as evaluated in a tensile test was performed using a Texture Analyzer TA.TX2 (Stable Micro System, Goldaming, Surrey, UK) in accordance with[25]. The films were cut in strips (14.0 x 2.0 cm) and clamped between tensile grips. Force and distance were recorded during extension of the strips at 3.3 mm/s up to break with an initial distance between the grips of 100 mm. The parameters determined were tensile stress (MPa) and elongation at break (% E). 5 replicates were tested.

2.6 Water vapor permeability

WVP tests of the films were determined gravimetrically using a [26] procedure. The test film was sealed as a patch onto a glass permeation cell containing silica gel (RVP=0) with a 1.0 cm deep headspace. The cell was then placed in a desiccator maintained at 30 °C and RVP =0.52 using a saturated salt solution of magnesium nitrate. The cell was weighed daily over a 6-day period. Water vapor transport was determined from the weight gain of the cell. Changes in weight of the cell were plotted as a function of time. The constant rate of weight gain was obtained by linear regression with $R^2 \geq 0.99$. The water vapor transmission rate (WVTR) was calculated from the slope of the straight line divided by the test area. The samples were measured in three replicates. WVP ($g.m/m^2.s.Pa$) was calculated as $WVP = [WVTR / S (R_1 - R_2)] \times d$, where S= saturation vapor pressure (KPa) of water at test temperature, R_1 = RVP in the desiccator, R_2 = RVP in the permeation cell and d= film thickness (m).

2.7 Fourier Transform Infrared (FTIR)

The spectra of control and UV-treated (2 and 3 h) films were recorded by Fourier Transform Infrared (FTIR) spectrometry (System 2000, Perkin Elmer Wellesley, MD, USA) in the frequency range of 4000-650 cm^{-1}. The film samples were pressed against the objective to analyze them directly. The spectra were used to determine possible changes induced on UV treatment of the samples.

2.8 Statistical analysis

All the films were prepared in triplicate. SPSS software version 11.0 was used for all statistical analysis. All data were initially evaluated using analysis of variance (ANOVA). Differences in the properties of the film were determined by the Tukey test using $p<0.05$ as level of significance.

3 RESULTS AND DISCUSSIONS

In general all formulations produced easily handled and homogeneous films. The films formed were generally self-supporting, translucent, clear and colorless. The incorporation of glycerol increased the flexibility of the films. [27] reported that the addition of plasticizer leads to a decrease in intermolecular forces along polymer chains which improves the flexibility and facilitates the peeling process of the film from the casting plate. The average thickness of the films is 0.130 mm.

3.1 Mechanical properties

3.1.1 Tensile strength of starch films.

An edible film must withstand the normal stress during its application, subsequent shipping and handling of the food, in order to maintain its integrity and barrier properties. Tensile strength is the maximum tensile stress sustained by the sample during the tension test. Elongation at break is an indication of films flexibility and stretch ability (extensibility) which is determined at the point where the film breaks under tensile testing and is expressed as the percentage of change of the original length of the specimen between the grips of a film to stretch (extend) [28].

Fig.1 depicts the changes observed in tensile strength (TS) in sago and mung bean films after UV treatments. A significant increase in the TS was observed between control and films treated for 2 h (39% and 59% for sago and mung bean films, respectively). The increase in TS could be due to crosslinking within film structure by UV treatment. This observation is in agreement with 16 who reported that the TS of sago starch/ polyvinyl alcohol (PVA) blend films treated with UV radiation was increased up to 46% compared to non-treated films. They noted that the increase in TS is due to the increase in crosslinking density of the polymer chains. [29] and [30] provided further evidence of crosslinking of starch upon UV irradiation by electron spin resonance (ESR) measurements. It has been reported that the gamma irradiation induces free radicals at the C_1 position on the glucose molecule. Similar observation of starch crosslinking after UV irradiation has also been reported by [31].

The amylose content of starches appeared to have significant contribution to the TS of the films developed. Mung bean starch has higher amylose content (44.00 ± 2.2) than that of sago starch (25.40 ± 1.5), which resulted in films with a higher TS than sago films. This finding is also consistent with the results of [32] and [33] who reported that increasing amylose content leads to improved mechanical properties including TS and E which is probably due to the higher degree of crystallinity of amylose.

3.1.2 Elongation at break of starch based films

The effect of UV treatments on elongation at break (E) is shown in Fig. 2. In contrast to the TS, E of films decreased significantly with increasing UV radiation exposure time. Similarly, increased TS and decreased E have been reported for soy protein films crosslinked by heat-curing [34] and UV radiation [35]. Reduced E, accompanied by increased TS in physically or chemically crosslinked protein films, could be attributed to formation of a denser, tighter film structure and thus stronger but less extendable protein matrices [35].

3.2 Water vapor permeability (WVP)

Study of water vapor permeability of packaging films is extremely important as they are commonly used in product shelf-life predictions, as well as in tailoring film permeability properties for specific food applications. Knowledge of film thickness is essential when the water vapor permeability (WVP) of film is to be studied, as hydrophilic films often exhibit positive slope relationships between thickness and WVP[36]. Average film thickness obtained is about 0.130 mm. Results obtained (Table 1) show that WVP for all starch films are statistically insignificant although there was a marginal decrease for UV treated mung bean film. It indicates that UV irradiation treatment did not influence the WVP of the films. It is evidence that any variation in the WVP should not be attributed to the variation in thickness, as all the films having an equal thickness. Similar findings were reported by 36. They showed that UV treatment did affect the mechanical properties but did not influence the WVP for soy protein films. Ultrasound treatment has also been reported to increase mechanical strength of whey protein concentrate (WPC) and sodium caseinate films but did not affect film WVP[37].

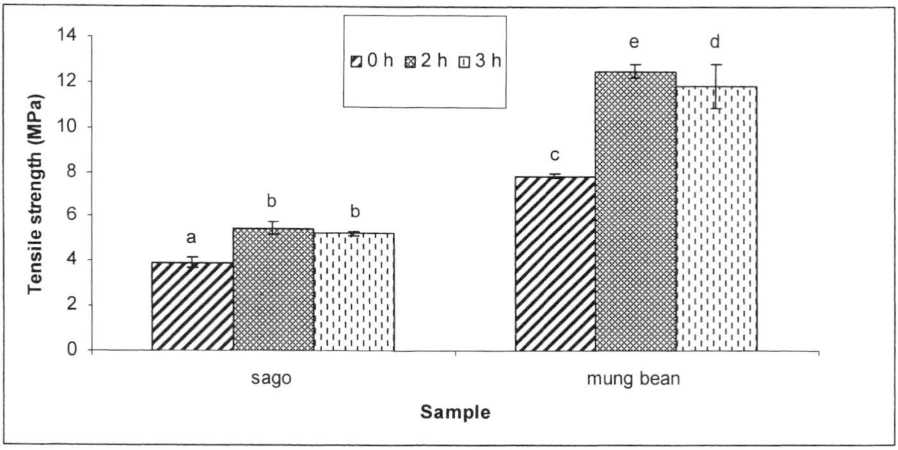

Figure 1 *Tensile strength of sago and mung bean films treated under UV light at 2 and 3 hours exposure time.* [a-e] *Mean value (n=5) with different superscript are significantly different (p<0.05)*

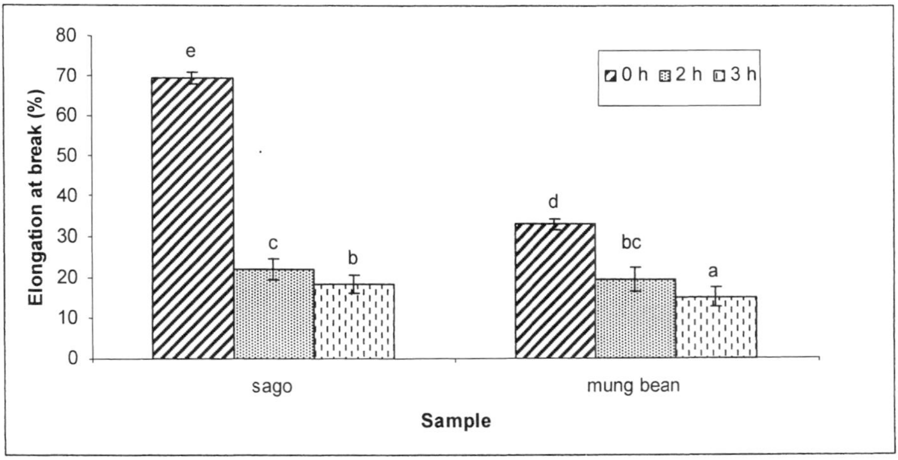

Figure 2 *Elongation at break of sago and mung bean films treated under UV light at 2 and 3 hours exposure time.* [a-e] *Mean value (n=5) with different superscript are significantly different (p<0.05)*

Table 1 *Thickness and water vapor permeability (WVP) of starch films treated at different exposure time*

Films type	Exposure time (hours)	Thickness (mm)	WVP (g.m/m^2.s.Pa) x 10^{-10}
Sago	0	0.130 ± 0.002^a	2.95 ± 0.14^a
	2	0.129 ± 0.001^a	2.94 ± 0.08^a
	3	0.131 ± 0.002^a	3.28 ± 0.24^b
Mung bean	0	0.130 ± 0.001^a	3.10 ± 0.11^{ab}
	2	0.130 ± 0.002^a	2.97 ± 0.04^a
	3	0.132 ± 0.004^a	3.09 ± 0.13^{ab}

Means ± standard deviation (n=3) with different superscript letters in each column are significantly different at ($p<0.05$).

3.3 Fourier Transform Infrared (FTIR)

The spectra of control and UV treated sago and mung bean starch films (Fig 3) show identical patterns, indicating insignificant changes among the functional groups. However, on closer examination, the absorption band at 2928.57 cm^{-1} (C-H stretching), 1647.13 cm^{-1} (δ-OH bend of water), 1417.18 cm^{-1} (C-H bending) and 1236.78 cm^{-1} (possibly C-O stretch) showed marked changes to have occurred. Slight shifts in the absorption bands at 2928.57 cm^{-1}, 1647.13 cm^{-1}, 1417.18 cm^{-1} and 1236.78 cm^{-1} were also observed between both films treated at 3 h exposure time. This might be possibly due to crosslinking that might have occurred in starch molecules after exposure to UV radiation.[38] noted that irradiation of natural polymers can result in degraded and cross linked polymers.

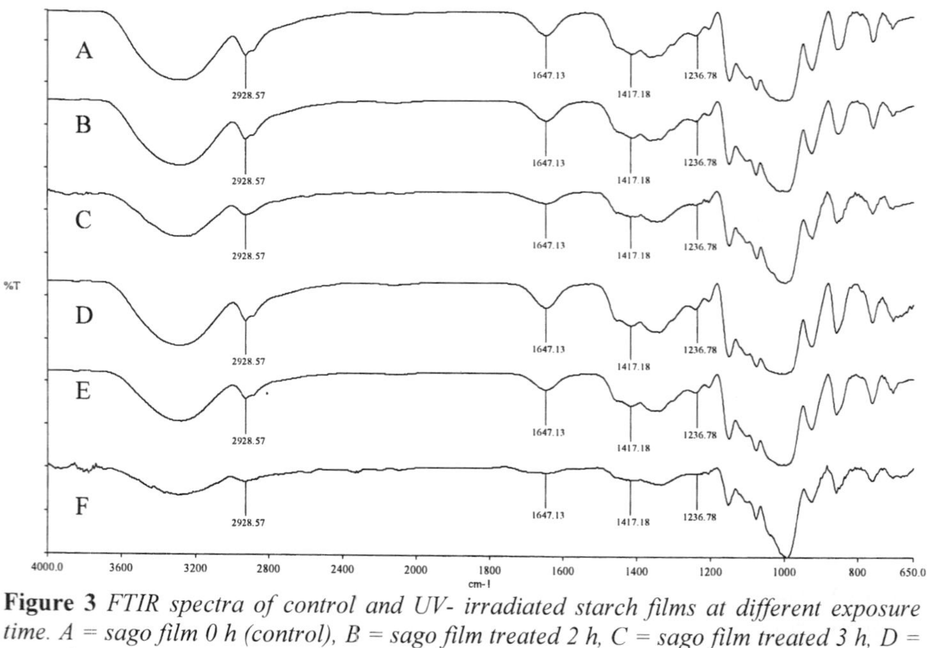

Figure 3 *FTIR spectra of control and UV- irradiated starch films at different exposure time. A = sago film 0 h (control), B = sago film treated 2 h, C = sago film treated 3 h, D = mung bean 0 h (control), E = mung bean film treated 2 h, F = mung bean film treated 3 h*

4 CONCLUSIONS

Sago and mung bean UV treated films (2 and 3 h exposure time) had higher tensile strength than non treated (control) films which suggests the possibility of using UV radiation for modifying the mechanical properties of starch films. Higher amylose content enhanced the mechanical properties of films as exhibited in higher tensile strength for mung bean films. On the other hand, UV irradiation did not affect the WVP of films developed. FTIR spectra of sago and mung bean films showed some changes in the absorption band at 2928.57 cm^{-1}, 1647.13 cm^{-1}, 1417.18 cm^{-1} and 1236.78 cm^{-1} for both films treated at 3 h exposure time.

5 REFERENCES

1 R.N. Tharanathan, *Trends in Food Science and Technology*, 2003, **14**, 71.
2 J.H. Han, *New technologies in food packaging: overview*, in Innovations in Food Packaging, ed. J.H. Han, Elsevier Academic Press, London, UK, 2005, p. 3
3 V.M. Hernandez-Izquierdo and J.M. Krochta, *J. of Food Science*, 2008, **73**(2), 30.
4 S. Mali and M.V.E. Grossmann, *J. of Agricultural and Food Chemistry*, 2003, **51**, 7005.
5 S. Mali, M.V.E. Grossmann, M.A. Garcia, M.N. Martino and N.E. Zaritzky, *Carbohydrate Polymers*, 2002, **50**, 379.
6 J, Delville, C. Joly, P. Dole and C. Bliard, *Carbohydrate Polymers*, 2003, **53**, 373.
7 P.T. Marques, A. Lima, G. Bianco, J.B. Laurindo, R. Borsali, J.F.L. Meins and V. Soldi, *Polymer Degradation Stability*, 2006, **91**, 726.
8 C.T. Vasques, S.C. Domenech, V.L.S. Severgnini, L.A.O. Belmonte, M.S. Soldi, P.L. Barreto and V. Soldi, *Starch/ Stärke*, 2007, **59**, 161.
9 C.M. Henrique, R.F. Teófilo, L. Sabino, M.M.C. Ferreira and M.P. Cereda, *J. of Food Science*, 2007, **72**(4), 184.
10 R.S. Singal, J.F. Kennedy, S.M. Gopalakrishnanm A. Kaczmarek, C.J. Knill and P.F. Akmar, *Carbohydrate Polymers*, 2008, **72**, 1.
11 W. Liu and Q. Shen, *J. of Food Engineering*, 2007, **79**, 358.
12 N. Ohwada, K. Ishibashi, Hironaka, K. and K. Yamamoto, *J. of Applied Glycoscience*, 2003, **50**, 481.
13 S. Mali, M.V.E. Grossmann, M.A. Garcia, M.N. Martino and N.E. Zaritzky, *Food Hydrocolloid*, 2005, **19**, 157.
14 P. Veiga-Santos, L.M. Oliveira, M.P. Cereda, A.J. Alves and A.R.P. Scamprin, *Food Hydrocolloid*, 2005, **19**, 341.
15 R.J. Woods and A.K. Pikaev, in *Applied Radiation Chemistry: Radiation Processing*, Wiley New York, 1994
16 M.A. Khan, S.K. Bhattacharia, M.A. Kader and K. Bahari, *Carbohydrate Polymers*, 2006, **63**, 500.
17 M. Zhai, F. Yoshii and T. Kume, *Carbohydrate Polymers*, 2003, **52**, 311.
18 T. Bintsis, E. Litopoulou-Tzanetaki, and R.K. Robinson, *J. of the Science of Food and Agriculture*, 2000, **80**, 637.
19 J.W. Rhim, A. Gennadios, D. Fu, C.L. Weller and M.L. Hanna, *Lebensmittel-Wissenschaft und -Technologie*, 1999, **32**, 129.
20 E. Fujimori, *Biopolymers*, 1965, **3**, 115.
21 K. Weadock, R.M. Olson and S.H. Silver, *Biomaterials for Medical Devices and Artificial Organs*, 1984, **11**, 293.

22 K. Tomihata, K. Burczak, K. Shiraki and Y. Ikada, *Polymer Preprints*, 1992, **33**, 534.
23 A. Merlin and J.P. Fouassier, *Makromolecular Chemistry: Macromolecular Symposium*, 1981,.**182**, 3053.
24 J.M. Scott, J.C. Hugh and J.R. Colin, *Starch/ Stärke*, 1998, **50**(4), 158.
25 ASTM (American Society for Testing and Materials). (1981a). *Standard test methods for tensile properties of thin plastic sheeting, Method D882-80a*. In 1981 Annual Book of ASTM Standards (Part 35), Philadelphia.
26 ASTM. (American Society for Testing and Materials) (1981b). *Standard test methods for water vapor transmission of materials. Method E96-80.* In 1981 Annual Book of ASTM Standards (Part 35), Philadelphia.
27 S. Guilbert, Technology and applications of edible protective films, in *Food Packaging and Preservation; Theory and Practic,* ed. M. Mathlouthi, Elsevier Applied Science Publishers, London, UK, 1986, p. 371.
28 T. Bourtoom, *International Food Research Journal*, 2008, **15**(2), 167.
29 J. Raffi and J. Agnel, *The Journal of Physical Chemistry*, 1983, **87**, 2369
30 J. Raffi, J.P. Agnel, C.C. Boizot, C. Theiry and P. Vincent, *Starch/ Stärke,* 1985, **37**, 228.
31 A.C. Bertolini, C. Mestres, P. Colonna and J.Raffi, *Carbohydrate Polymers*, 2001, **44**, 269.
32 D. Lourdin, G. Dela-Valle and P. Colonna, *Carbohydrate Polymers*, 1995, **27**, 261.
33 Palvianen, J. Heinämäki, P. Myllärinen, R. Lahtinen, J. Yliruusi and P. Forssell, *Pharmaceutical Development Technology*, 2001, **6**, 353.
34 A. Gennadios, V.M. Ghorpade, C.L. Weller and M.A. Hanna, *Trans. ASAE*, 1996, **39**, 575.
35 A. Gennadios, J.W. Rhim, A. Handa, C.L. Weller and M.A. Hanna, *J. of Food Science*, 1998, **63**, 225.
36 T.H. McHugh, R. Avena-Bustillos and J.M. Krochta, *J. of Food Science*, 1993, **58**(4), 899.
37 R. Banerjee, H.Chen and J.Wu, *J. of Food Science*, 1996, **61**, 824.
38 T. Kume, N. Nagasawa and F. Yoshii, *Radiation Physics and Chemistry*, 2002, **63**, 625

APPLICATION OF SUGAR BEET PULP AS DOUGH AND BREAD IMPROVER (BARBARI BREAD)

M. Majzoobi[1], A. Farahnaky[1], J. Jamalian[1], G. Mesbahi[1], F. Sariri[1]

[1]Department of Food Science and Technology, College of Agriculture, Shiraz University, Shiraz, Iran.

1 INTRODUCTION

Sugar beet pulp is the by-product of sugar factories. It is made up of hemicellulose (26%), pectin (24%), cellulose (23%), proteins (9%), sucrose (6%), lignin (4.5%), soluble minerals (4.0%), insoluble minerals (3%) and fat (0.5%).[1] It is mainly used as animal feed. Regarding the composition of sugar beet pulp, it may be possible to use it in human food mainly as a source of hydrocolloids particularly dietary fiber and proteins.
On the other hand, it has been reported that the quality of bread can be improved by addition of hydrocolloids to bread formulations.[2] Therefore sugar beet pulp may have the potential to be used in bread recipes as a source of hydrocolloids in order to improve bread quality. However the literature shows that not all the hydrocolloids can be used as dough and bread improvers.[3] The molecular structure of the hydrocolloids has a great impact on their functionality as bread improvers.
Barbari bread is a white flat bread which is common in many Asian and European countries. Unfortunately it loses its quality and goes stale quickly in a few days after baking causing a large economical damage.[2] Therefore improving the quality of Barbari bread and delaying its staling is of great importance. The main objectives of this study were to improve the quality of Barbari bread and also to postpone its staling by addition of sugar beet pulp powder (SBP).

2 METHOD AND RESULTS

Wheat flour was purchased from Khosheh milling factory in Shiraz, Iran. It had an extraction rate of 78% (weight of flour/100 kg clean wheat) according to the manufacturer. Sugar beet pulp was provided by Marvdasht sugar company, Marvdasht, Iran. Active dried yeast was obtained from the local market. All chemicals used in this study were purchased from Merck Company (Germany) and were of analytical grade.

2.1 Methods

The chemical composition of the flour including moisture, total protein, ash, fat and fiber content were determined according to American Association of Cereal Chemists (AACC) methods (AACC 1995).

To prepare SBP, the sugar beet pulp was dried in a cabinet drier at 50 °C for 8 hr to a moisture content of 3.5 %. It was then milled using a laboratory mill and sieved to obtain particle sizes in the range of 200-250μm. The SBP was packed in polyethylene bags and stored at 4 °C for further experiments.

The chemical composition of SBP including protein, fat, fiber, total and reducing sugar was determined using Association of Analytical Chemists (AOAC) methods (AOAC 2000).

2.2 Bleaching of sugar beet pulp using H_2O_2

Part of the sugar beet pulp was bleached using H_2O_2 according to a method described by Seres et al. 2005.[1] The bleached sugar beet pulp was washed and pressed to remove excess water. Then it was dried in a cabinet drier at 50 °C for 8 hr to a moisture content of 5.2 %. After that it was milled and sieved to obtain particle sizes in the range of 200-250μm. The bleached SBP was then packed in polyethylene bags and stored at 4 °C for further experiments.

2.3 Determination of rheological properties of the dough

A basic bread formula for making Barbari bread per 100g flour was used as follows: 2 g yeast (activated before being used according to the producer instruction), 2 g salt (NaCl) and 10 mg ascorbic acid. For this study different levels of SBP (0, 1, 3, 5, 7 and 10% w/w, flour basis) and bleached SBP (7 and 10% w/w, flour basis) were added to the dough ingredients at the mixing stage.

To study the rheological properties of the dough, all ingredients were placed in the mixing bowl of a Brabender Farinograph (Model FE022N, Germany) and the Farinograph parameters including % water absorption, dough arrival time, dough development time, dough stability time and dough softening after 12 min were determined from the Farinogram (AACC 1995).

2.4 Dough preparation and bread making

Bread dough was prepared by mixing all the ingredients with the appropriate amount of water (determined by Farinograph) in a laboratory dough mixer at 140 revolution/min (rpm) for 15 min. Then the dough was proofed, divided into portions of 400 g, rounded by hand and placed in the proofing cabinet once again for 45 min. The dough pieces were shaped to a thickness of 1 cm and diameter of 40 cm and were baked in an electrical baking oven (Model Karl Welkerkg, Germany) at 210 °C for 30 min. After that the breads were cooled down to room temperature. The samples at this stage were referred to as fresh bread.

To study the effects of SBP on bread staling, some of the fresh breads were packed in polyethylene bags. These samples were stored at 25 °C for 24, 48, 72 and 96 hours (mould growth was observed for the samples stored for 72 hr). Then the quality attributes of the breads were determined as follows:

2.5 Determination of quality of fresh and stored breads

2.5.1 Determination of the weight of the samples. Weight of bread samples containing different levels of SBP was determined using an accurate balance and it was repeated during storage time for 96 hr.

2.5.2 Textural characteristics of fresh and stored breads. A penetration test using a cylindrical probe with diameter of 10 mm was performed using a Texture Analyser (Model Stevens-LFRA). Probe speed and penetration depth were 1 mm/s and 10 mm, respectively. The experiment was performed at five different points of the bread crumb and the average value was reported.

2.5.3 Color determination of fresh and stored breads. The color of bread crust was evaluated using a modified method of Yam and Papadakis (2004)[4] and the color parameters; "L" (lightness), "a" (redness-greenness) "b" (blueness-yellowness) and were determined in the "Lab" mode of the software.

2.5.4 Density determination of fresh and stored breads. The volume of a piece of bread was determined by the rapeseed replacement method (AACC 1995).

2.6 Microscopic structure of fresh and stored breads

To study the microscopic structure of the bread crumb, scanning electron microscopy (SEM) (Model 5526, Mark Cambridge, UK) was used. To prepare each sample, thin layer of bread was first freeze dried and then a tiny part of the sample was sputter coated with gold. Finally the sample was transferred to the microscope where it was observed at 20 kV.

2.7 Sensory analysis

General acceptability, bread hardness, taste and color of the breads made with different levels of SBP were studied by sensory evaluation with 12 trained panelists using a ranking test and the test was performed in a standard booth. The highest score (5) was given to the best sample, while the lowest (1) was given to the worse samples. Since mould growth was observed on some the bread pieces stored for longer than 48 hr, these samples were not used for sensory evaluation.

2.8 Statistical analysis

The experiments were performed in a completely randomized design. All experiments were conducted in triplicate and the mean values and standard deviations were reported. Analysis of variance (ANOVA) was performed and results were separated using the Multiple Ranges Duncan's test ($\alpha < 0.05$) using statistical software of Statistical Package for the Social Sciences (SPSS) (SPSS, Inc. New Jersey, USA).

2.9 Results

2.9.1 Chemical composition of the raw materials. The chemical composition of the wheat flour, SBP and bleached SBP is given in Table 1. The results show that the SBP had significantly higher level of fiber compared to the wheat flour. The fiber content of bleached SBP was even higher than the SBP. Bleaching of the sugar beet pulp could reduce the protein and fat content of the pulp. Similar changes of the chemical composition of sugar beet pulp as a result of bleaching with H_2O_2 were reported by Seres et al. 2005 and Filipovic et al. 2007.[1]

Table 1. *Chemical composition of wheat flour, SBP, bleached SBP.*[*]

Chemical composition (% dwb)*	Moisture	Fiber	Protein	Fat	Ash	Total Sugar
Wheat flour	13.01	0.61	11.84	1.51	4.75	nd[1]
SBP**	3.50	61.88	7.02	1.01	5.96	20.01
Bleached SBP	5.20	75.09	2.18	0.73	6.61	14.07

* Dry weight basis, *Sugar beet pulp, [1]nd = Not determined

2.9.2 Rheological properties of dough. The results (Table 2) show that with increasing the level of SBP the water absorption of the dough increased significantly. Moreover addition of bleached SBP also increased the water absorption of the dough. This can be due to the presence of the some hydrocolloids in the SBP and bleached SBP. These hydrocolloids can absorb water and hence increase the water absorption of the dough. Comparison of the water absorption obtained for the same level of SBP and bleached SBP (e.g. 5%) shows that the addition of bleached SBP could increase the water absorption of the dough to a significantly higher level compared to the SBP (71.1% for bleached SBP and 69.2% for SBP). This can be due to the presence of higher amounts of fiber in the bleached SBP compared to SBP.

Dough development time (Table 2) increased with increasing the level of either SBP or bleached SBP. At the same level of SBP and bleached SBP, the development time of the dough containing bleached SBP was significantly higher than that made with SBP. This can be due to the longer time require for the hydrocolloids in the dough to absorb water and hence a longer time is required to form a dough with a constant consistency.

The results (Table 2) show that with increasing the powders in the formulation of the dough, the stability time increased significantly. At the same amount of SBP and bleached SBP, the latter caused longer stability time.

The results (Table 2) show that the dough samples containing the powders soften much less than the control. Furthermore dough samples containing bleached SBP had even less softening than those made with SBP.

Determination of the weight of the samples (data are not shown) showed that by increasing the level of SBP, the weight of the samples increased significantly. This may indicate that more water may remain in the bread samples containing SBP. Storage time, in contrast, had no effect on the weight of the breads since they were packed in polyethylene plastic bags and no moisture loss could be observed during storage time.

Figure 1 shows that with increasing the level of SBP, the density of the samples decreased significantly which can be related to the higher volume of the bread. Storage time, on the other hand caused a reduction in the density of the samples. This can be attributed to the shrinkage of the air bubbles, resulting in a reduction in bread volume. As the level of SBP increased, less reduction in bread density can be observed.

Figure 2 shows the effect of SBP and storage time on the bread crumb texture. The results show that with increasing the SBP in the recipe, the hardness of the bread crumb decreased. During storage time, the harness of the bread crumb increased significantly. On the other hand, bread samples made with SBP remained softer during the storage time.

□ **Table 2.** *Farinogarph parameters of dough samples prepared with different levels of SBP and bleached SBP*.*

Dough samples	Water absorption (%)	DDT[1] (sec)	DST[2] (sec)	DS[3] (BU[4])
Control	60.0 ± 0.0^{a}	180.0 ± 1.0^{a}	240.0 ± 0.0^{a}	$160.0\ 2.0^{a}$
SBP (1%)	63.5 ± 0.2^{b}	185.0 ± 1.1^{b}	$250.2\ 1.5^{b}$	$156.2\ 2.3^{a}$
SBP (3%)	65.1 ± 0.1^{c}	191.2 ± 1.2^{c}	$310.2\ 1.4^{c}$	$154.0\ 2.1^{ab}$
SBP (5%)	69.2 ± 0.1^{d}	198.3 ± 1.1^{d}	$325.3\ 2.3^{d}$	$152.1\ 2.3^{ab}$
SBP (7%)	73.0 ± 0.0^{e}	200.1 ± 1.3^{d}	$328.0\ 2.0^{d}$	$150.3\ 2.1^{ab}$
SBP (10%)	78.0 ± 0.1^{f}	203.2 ± 1.2^{e}	$330.0\ 2.5^{de}$	$146.2\ 1.1^{c}$
B-SBP[5] (5%)	71.1 ± 0.1^{g}	240.4 ± 2.5^{f}	$340.2\ 2.3^{f}$	$120.5\ 2.1^{d}$
B-SBP (7%)	75.2 ± 0.1^{h}	250.2 ± 2.5^{g}	$345.5\ 1.5^{g}$	$115.5\ 1.5^{e}$
B-SBP (10%)	78.3 ± 0.1^{i}	255.2 ± 2.1^{h}	$348.5\ 1.2^{h}$	$100.1\ 1.2^{f}$

D*Data are the averages of three replicated 1 ± 1 SD. Different letters in each column show the significant difference between the data ($p < 0.05$).[1] Dough development time, [2] Dough stability time, [3] Dough softening, [4] Brabender unit, [5] Sugar beet pulp powder.

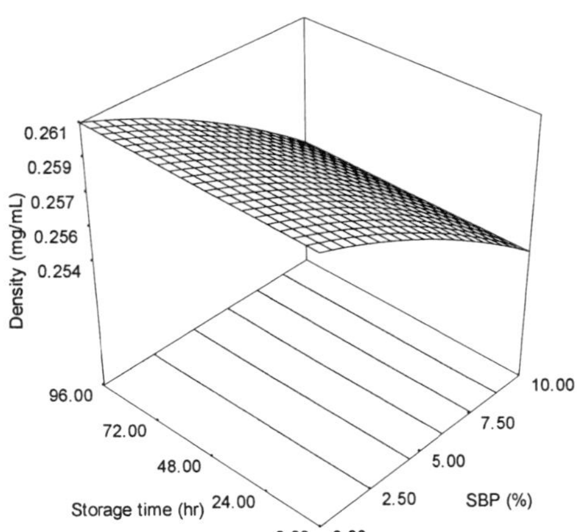

Figure 1. *Effect of storage time and the level of sugar beet pulp powder (SBP) on the density of the bread.*

Figure 3 shows that the crust lightness decreased with increasing the SBP level in bread recipe and storage time. Figure 4 shows that increasing the SBP in the bread recipe and storage time reduced the a-value of the bread crust. It seems that bread lightness was further affected by the SBP concentration rather than storage time. Determination of the b-value of the bread crust showed that this value decreased with increasing the SBP and storage time (data are not shown). Furthermore storage time had less effect on crust b-value compared to the level of SBP.

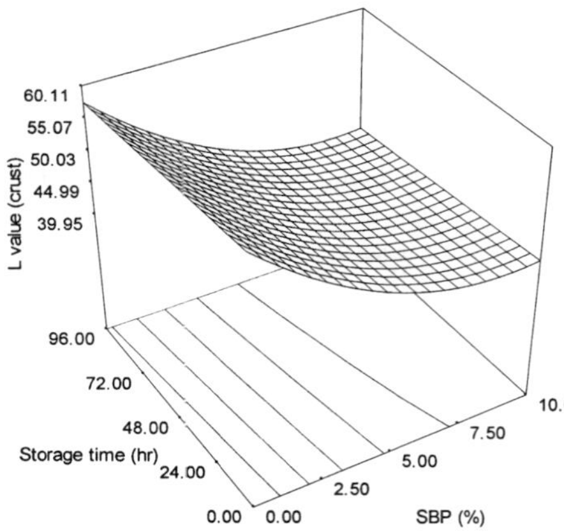

□

Figure 2. *Effect of storage time and the level of sugar beet pulp powder (SBP) on bread crumb texture.*

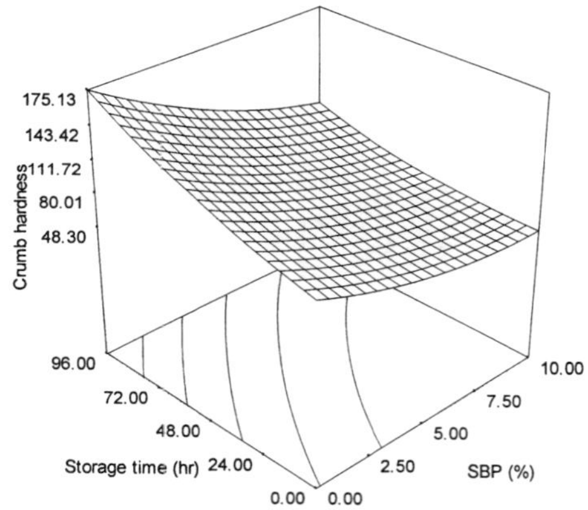

□**Figure 3.** *Effect of storage time and the level of sugar beet pulp powder (SBP) on bread crust L-value.*

2.9.3 Organoleptic evaluation of bread containing SBP. The sensory evaluation results (data not shown) indicated that increasing the level of SBP had a positive effect on bread texture since higher scores were given to the samples containing higher levels of SBP even when they were stored for 72 hr. In terms of bread color, the results show that the samples containing 5 and 7% SBP, received lower scores compared to other samples. It can be attributed to the reduction in bread lightness and increase in redness which can have negative effects on panelists' perception of color. Due to the staling, stored samples received lower scores for the color.

The results also show that addition of higher levels of SBP (5 and 7%) had a negative effect on taste and flavor of the bread, which was more pronounced during storage of the samples. The overall acceptance of the samples was lower especially when SBP was added at 5 and 7%. It seems that the overall acceptance was affected by the taste and color of the bread rather than the texture. The two former parameters were poor especially at high levels of SBP. The results showed that bleaching of SBP had positive effects on the organoleptic properties of the bread made with 5 and 7%.

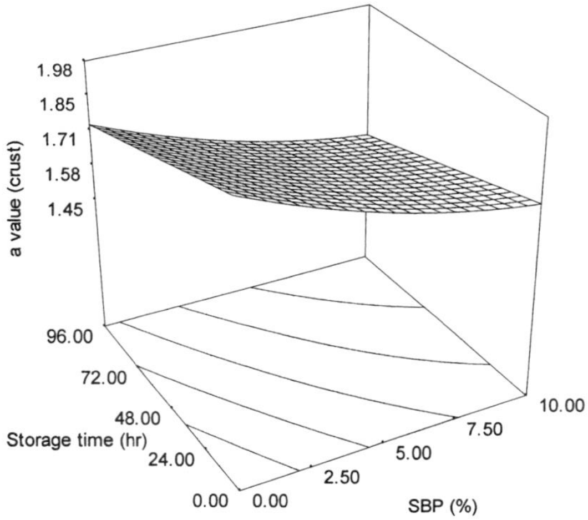

□ **Figure 4.** *Effect of storage time and the level of sugar beet pulp powder (SBP) on bread crust a-value.*

2.9.4 The microscopic structure of bread. The micrographs of the gas cell walls of the bread samples made with 0 and 5% SBP are presented in Figure 5. The gas cell wall of fresh bread made with no SBP (Figure 5-a) had a complex structure. Some swollen and deformed starch granules connected with other components of the bread with a number of cavities can be observed. The gas cell walls of the bread with no SBP after 96 hr (Figure 5-b), were similar to those of fresh bread, however it seems that the structure was shrunk during storage time.

Addition of SBP had a great impact on the gas cell walls of the bread (Figure 5-c). A continuous structure with a thicker appearance was observed for the gas cell wall of the fresh and stored bread for 96 hr. It seems that the components of bread were embraced by the hydrocolloids in the SBP. Therefore the underneath components were not clearly visible. Comparison of the micrographs obtained for the aged breads with 5% SBP and the one without SBP (Figure 5-b and -c) suggests that the hydrocolloids in SBP could make close interactions with other dough components and hence delay the interactions among them. Accordingly some of the processes involved on the staling of bread were retarded. Furthermore the hydrocolloids in SBP can strength the cell wall structure and hence reduce gas cell shrinkage due to the aging process. After storage of the bread containing 5% SBP (Figure 5-d), the structure was more compact and shrunk as the result of aging.

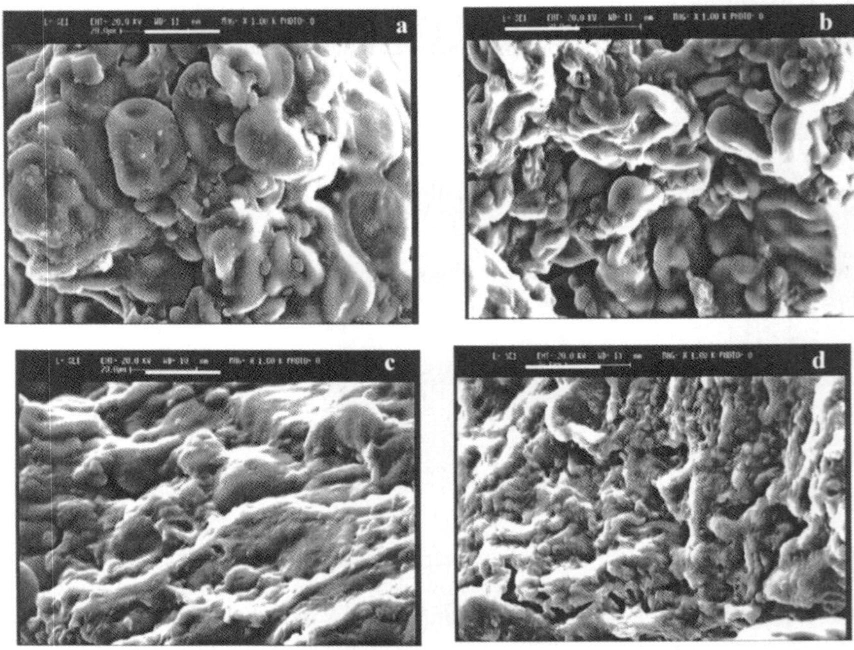

Figure5. *Scanning electron micrographs of fresh bread with 0% SBP (a), stored for 96 hr (b), fresh bread containing 5% SBP (c) and stored for 96 hr (d). Bars on the pictures are 20 μm.*

3 CONCLUSION

Sugar beet pulp can be used as a source of hydrocolloids to improve bread quality causing several changes to the dough and bread properties. Higher water absorption of the dough, higher dough consistency and less degree of softening were observed for the bread containing SBP compared to the ordinary bread. Therefore such dough can be used in long dough making processes. Furthermore SBP can increase the dough consistency and hence it is suitable to be used along with weak flours.

Breads made with SPB had lower density, softer texture and later staling compared to ordinary breads. The crust color of the bread at higher levels of SBP was slightly darker and the taste was to some extent undesirable compared to other samples. To avoid changes in bread color, taste and flavor, SBP should be used at levels < 5%. Otherwise SBP should be bleached before being used in bread formulation.

References

1 Z. Seres, J. Gyura, N. Filipovic and D. Soronja Simovic, *Eur. Food Res. Technol.,* 2005, 221, 54.
2 M. Majzoobi, A. Farahnaky and R. Ostovan, *Iran Agric. Res.,* in press.
3 N. Filipovic, M. Djuric and J. Gyura, *J. Food Eng.,* 2007, 78, 1047.
4 K. L. Yam and S. E. Papadakis, *J. Food Eng.,* 2004, 61, 137.

IMPROVING THE QUALITY OF PASTA USING HYDROXYPROPYL METHYL CELLULOSE, GLUTEN AND BLENDED WHEAT FLOUR

M Majzoobi[1*], R Ostovan[1], A Farahnaky[1], G Mesbahi[1], M Eskandari[1]

[1]Department of Food Science and Technology, College of Agriculture, Shiraz University, Shiraz, Iran.

1 INTRODUCTION

Pasta is defined as a product obtained by extrusion cooking of a dough made by semolina and water and then shaping the dough followed by drying to reach a moisture content of 12.5%. It is one of the most popular foods around the world which is produced in different shapes and sizes. The most common shape of pasta is in the form of spaghetti. Pasta is mainly used as an energy source due to its high content of carbohydrates. It is generally believed that the best quality of pasta is produced from semolina. However, recently other types of flours such as bread wheat and rice flours are used in pasta making. This may have diverse effects on the quality of the pasta. To compensate for undesirable effects of using non-semolina flour on pasta quality, some additives such as hydrocolloids, food colorants, etc. may be added in the pasta formulation.

Hydrocolloids are able to modify overall quality of the food products. They provide viscosity and texture, improve firmness, give body and mouth feel to the end product. In Iran the climate is perfect for durum wheat production. However a majority of farming land is specified to bread wheat production and hence there is insufficient durum wheat to produce semolina for all pasta factories. Therefore many pasta factories have to use normal or low quality bread wheat (with protein content < 10%) to make pasta. Unfortunately the resulting pasta has low quality. The main aim of this study was to improve the quality of pasta made with low quality bread wheat by addition of two hydrocolloids; namely hydroxypropyl methyl cellulose (HPC) and gluten.

2 METHODS AND RESULTS

2.1 Materials

Two types of wheat flour were purchased from Khosheh and Sepidan milling factories, Shiraz, Iran. The chemical composition of these two flours were determined according to the approved methods of AACC.

Gluten powder was obtained from Fars glucosin Company, Marvdasht, Shiraz, Iran. HPC was purchased from Stephan Co., India. Other chemicals were obtained from Merck Company, Germany.

2.2 Pasta Preparation

Pasta in the form of spaghetti was produced in Dena Macaron pasta factory. In this factory pasta is made by Sepidan flour (control sample) or by blending Sepidan and Khosheh flours (1:1 w/w) (blended flour). To make different samples, Sepidan flour was first mixed well with HPC (0.5% w/w, flour basis), gluten powder (3% w/w, flour basis), and Khosheh flour (1:1 w/w flour basis), separately. Then each mix was mixed with water to reach a moisture content of 31.5 %. The mixture was extruded through a single screw extruder (model Mircon, Iran) at 40 °C and 0.92 bars. The extruded dough was formed in the shape of spaghetti by passing through a circular die. The samples were then dried at 70 °C for 48 hr until they reached to the moisture content of 4-5 %. The samples were cut into pieces with the length of 30 cm and were packed in polypropylene bags with 700 g capacity.

2.3 Determination of Spaghetti Cooking Quality

2.3.1 Determination of water absorption.

10 g of the samples were placed into 300 g salted water (2% w/w NaCl in distilled water) and cooked in a cooking pan for 18 min. During cooking, every 30 sec, 5 g of the sample was removed from the cooking pan and drained for 2 min. The water absorption at each cooking time was determined according to Equation 1:

Water absorption (%) = [(weight of cooked spaghetti – weight of uncooked spaghetti) / weight of uncooked spaghetti] × 100 (1)

2.3.2. Determination of the optimum cooking time.

Optimum cooking time was determined by boiling the spaghetti in salted water (2% w/w, NaCl in distilled water) until the centre white core could no longer be seen when a small piece was crushed between two glass plates.

2.3.3. Determination of cooking loss.

A 10 g sample of spaghetti was placed into 300 g salted water and cooked to its optimum cooking time (as explained before). Then the sample was drained for 2 min and the cooking loss was determined according to the Equation below:

Cooking loss (%) = [(weight of drained residue in cooking water) / weight of uncooked spaghetti] × 100 (2)

2.4 Texture Analysis

The texture of the cooked spaghetti at optimum time was studied using a Texture Analyser (TA), model TA-TX, UK. In this experiment a metallic probe with circular die with a diameter of 0.5 cm was used. The probe was moving towards a piece of spaghetti with a constant speed of 1 mm/s and then entered the sample to the depth of 1 mm. After the probe returned back to its initial position, the maximum force for the compression of the sample was recorded by the instrument. This experiment was repeated at five different points of the sample and the average value was reported.

2.5 Color Determination

The color of spaghettis after cooking was evaluated using a modified method of Yam *et al.*[6] High resolution pictures of the samples were taken separately by a digital camera. Resolution, contrast and lightness of all images were set to 200 (dpi), 62 (%) and 62 (%), respectively. The pictures were saved in JPEG format and analyzed quantitatively using the Adobe Photoshop 8 software and the color parameters of L (lightness), a (redness-greenness) and b (blueness-yellowness) values were determined in the "Lab" mode of the software.

2.6 Sensory Evaluation

Spaghetti samples were first cooked for their optimum cooking time, drained and served to the 12 trained taste panelists. The samples were evaluated in terms of smell, taste, color, texture and general acceptability using a ranking test and were scored from 1 to 5 (higher score was given to the better sample). The analyses were performed in isolated booths in a standard taste panel area.

2.7 Statistical Analysis

The experiments were performed in a completely randomized design. All experiments were conducted in triplicate and the mean values and standard deviations were calculated. Analysis of variance (ANOVA) was performed and the results were separated using the Multiple Ranges Duncan's test ($\alpha < 0.05$) using statistical software of SPSS 13 (SPSS, Inc., New Jersey, USA).

2.6 Results and discussion

The chemical compositions of the flours used in this study are given in Table 1. The results indicate that the flour which was normally used in pasta making by the factory (Sepidan) had lower protein and gluten contents than the other flour; Khosheh. Both flours were commonly used for pasta production and were classified as weak bread flours.

Determination of the water absorption of the samples in given in Fig 1. The results show that the water absorption of the samples increased on increasing the cooking time. The sample containing HPC had the highest water absorption, while other samples had similar water absorption to that of control. The results may indicate that HPC may increase the water absorption of the spaghetti during cooking to a higher extend compared to the gluten and blended flours.

Table 1. *Chemical composition of the flours used in spaghetti production*[*].

Flour types	Moisture content (%)	Fat (%)	Ash (%)	Protein (%)	Dry gluten (%)
Sepidan	13.33 ± 0.12	1.14 ± 0.02	13.47 ± 0.25	0.26 ± 0.01	9.32 ± 0.03
Khosheh	13.35 ± 0.09	1.41 ± 0.40	14.12 ± 0.13	0.23 ± 0.01	10.22 ± 0.32

[*]Values in the table are the average of triplicates ± 1 SD.

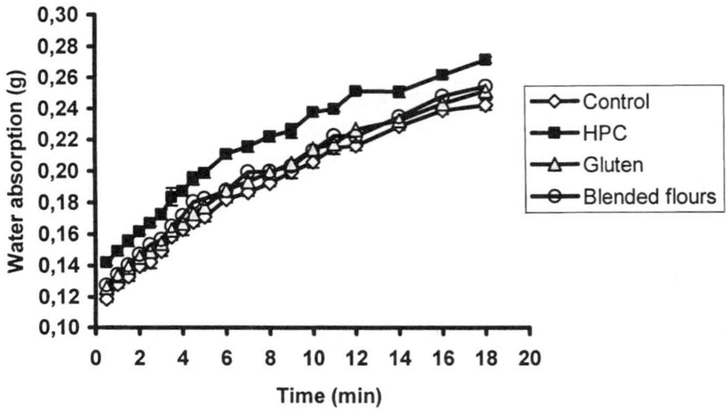

Figure 1. *The water absorption of the spaghetti samples during cooking for 18 min*

The results of determination of the optimum cooking time of the samples are given in Table 2. The results show that the optimum cooking time of the control and the sample produced by blending of the flours was the lowest, while it was the highest for the samples made with HPC and gluten. This may indicate that addition of HPC and gluten can increase the time required for fully hydration of the samples.

The control and the sample made with blended flours had the highest cooking loss while sample containing gluten had the lowest cooking loss followed by HPC sample. The firmness of the samples containing HPC and gluten was the highest followed by the sample made with blended flour and the control.

In pasta processing, gluten is mainly responsible for the formation of the structure and it is considered to be the most significant factor related to pasta cooking quality. Gluten consists of gliadin and glutenin and is responsible for elasticity and chewiness of pasta.

Table 2. *The quality of the spaghetti samples after cooking*[*].

Spaghetti types	Optimum cooking time (min)	Cooking loss (%)	Firmness (g-force)
Control	$11.33^{a} \pm 0.3$	$7.30^{a} \pm 0.26$	$80.00^{a} \pm 1.67$
HPC	$12.66^{b} \pm 0.3$	$5.90^{b} \pm 0.10$	$93.83^{b} \pm 2.13$
Gluten	$12.83^{b} \pm 0.3$	$5.77^{c} \pm 0.02$	$94.16^{b} \pm 1.95$
Blended flours	$11.67^{a} \pm 0.3$	$7.06^{a} \pm 0.15$	$84.50^{c} \pm 3.70$

[*]Values in the table are the average of triplicates ±1 SD. Different letters in each column indicates the significant difference ($\alpha < 0.05$).

Determination of the color of the samples (Table 3) showed that the lightness of the sample containing gluten powder was less than the others, while addition of HPC or blending of the flours had no significant effect on the lightness of the samples compared to the control. The a-value of the sample containing gluten was more than the other samples which had similar a-values. The b-value of the sample containing HPC was similar to that made with blended flours. These samples had higher b-value than the control. The sample made with gluten had the highest b-value compared to the others. Having slightly darker and yellowier color of spaghetti samples containing gluten can be due to the yellowish color of gluten powder which can change the color of the final product. HPC was a white powder which had no effect on the color of the samples.

Table 3. *The color parameters of spaghetti samples**.

Spaghetti types	L-value	a-value	b-value
Control	$82.9^{a} \pm 0.7$	$-4.8^{a} \pm 0.6$	$26.8^{a} \pm 0.8$
HPC	$82.2^{a} \pm 0.$	$-4.1^{a} \pm 0.7$	$28.1^{b} \pm 0.7$
Gluten	$79.1^{b} \pm 0.8$	$-1.1^{b} \pm 0.74$	$34.2^{c} \pm 0.8$
Blended flours	$82.1^{a} \pm 0.7$	$-3.8^{a} \pm 0.8$	$28.3^{b} \pm 0.7$

[*]Values in the table are the average of triplicates ±1 SD. Different letters in each column indicates the significant difference ($\alpha < 0.05$).

Table 4 shows the results of sensory analysis. In terms of smell, the spaghetti made with gluten received the highest score, while the control received the lowest. The HPC and blended flour samples received similar scores for their smell. The taste of the samples made with HPC and gluten was similar and significantly higher than that of blended flour and the control. The texture of the samples made with HPC, gluten and blended flours were all similar and better than the control. The color of the cooked spaghetti made with gluten was the best, followed by the sample made with HPC, blended flours and the control. The samples made with gluten received the highest score in terms of general acceptability, followed by the sample made with HPC, blended flours and the control.

Table 4. *Sensory evaluation of spaghetti samples as determined by taste panels**.

Spaghetti types	Smell	Taste	Texture	Color	General acceptance
Control	$2.58^a \pm 0.51$	$2.42^a \pm 0.51$	$1.92^a \pm 0.51$	$1.92^a \pm 0.51$	$2.41^a \pm 0.51$
HPC	$3.08^b \pm 0.67$	$4.00^b \pm 0.60$	$3.25^b \pm 0.45$	$4.08^b \pm 0.67$	$4.08^b \pm 0.66$
Gluten	$4.41^c \pm 0.51$	$4.16^b \pm 0.57$	$4.00^b \pm 0.60$	$4.25^b \pm 0.75$	$4.91^c \pm 0.29$
Blended flours	$3.00^b \pm 0.42$	$3.08^b \pm 0.66$	$3.25^b \pm 0.45$	$2.41^a \pm 0.51$	$3.23^d \pm 0.20$

* The values in are the averages of triplicates ±1 SD. Different letters in each column show the significant difference between the averages ($\alpha < 0.05$)

3 Conclusion

The results of this study indicated that it is possible to improve the quality of spaghetti made with low quality wheat flour using gluten powder and HPC. The following changes were observed when HPC, gluten and blended flours were used in spaghetti recipe:

Spaghetti made with HPC had significantly higher water absorption during cooking, while other samples had similar water absorption. This can be due the hydrophilic structure of HPC which can absorb a significant amount of water during cooking.

A major problem with control was its high cooking loss and soft texture after cooking. These properties were improved using gluten and HPC. Blending of the flours had no effect on cooking loss, while it could increase the firmness of the sample.

The control had a pale color which is not acceptable by the consumers. Using gluten could improve the color of the samples as determined by color analysis and taste panels. Using HPC and flour blending had no significant effect on color parameters of the samples as determined by color analysis. However spaghetti made with HPC received a higher score than the control and blended flours when it was tested for color by the taste panel.

The sensory attributes of the control in terms of taste, smell, color and general acceptability were low. All these parameters were improved by using gluten, HPC and blended flours. Gluten had greater effects on these parameters followed by HPC and blended flours.

References:

1. P. A. Williams and G. O. Phillips, Introduction to food hydrocolloids, In *Handbook of Food Hydrocolloids*, G. O. Phillips and P. A. Williams, Woodhead Publisher LTD. FL., 1982, P 1.
2. G. Owens, *Cereal Processing Technology*, CRC Press, London. 2001. p 112.
3. E. Carini, E. Vittadini, E. Curti and F. Antoniazzi, *J. Food Eng.*, 2009, 93, 400.
4. N. Sozer, *Food Hydrocol.*, 2009, 23, 849.
5. S. Chillo, M. Iannetti, V. Civica, N. Suriano, M. Mastromatteo and M.A. De Nobile, *J. Food Eng.*, 2009, In press.
6. A. Ocieczek, *Acta Agrophysica*, 2007, 9, 135.

AN INVESTIGATION ON THE FUNCTIONAL PROPERTIES OF CARBOXYMETHYL CELLULOSE (CMC) PRODUCED FROM SUGAR-BEET PULP

G. Mesbahi, M. Niakoosari, S. Savadkoohi, A. Farahnaky
Department of Food Science and Technology, Faculty of Agriculture, Shiraz University, Shiraz, Iran.

1. INTRODUCTION

Sugar-beet is one of the main sources of sugar production worldwide. Sugar-beet pulp is a by-product from the sugar industries. It has been reported to contain large amounts of cellulose amongst many other useful components. It is presently used for animal feed production. To improve pulp nutritional value as animal feed, molasses is added to the pulp. However, it might be converted to many useful derivatives with important applications in the food industry. There exists a potential to manufacture many added value products from this by-product of the sugar industry. Cellulose, pectin, hemi-cellulose and a number of minerals are amongst many products which could have a great potential in many industries particularly food and all could be derived from pulp.[1] The cellulose in pulp amounts to about 27 %.[2,3]
Hydrocolloids are complex compounds that have much functionality in food and related industries. They might be used for texture and consistency improvement, crystallization control, preventing syneresis, encapsulation, film formation, and physical stability of foods. Many of these functions are achieved in foods without the need to increase the energy content; hence they are very useful for diet foods.[4]
Mesbahi et al (2005) carried out a study to optimize the extraction of pectin from sugar-beet pulp and to improve the functional properties of this pectin for more potential uses. In comparison with citrus pectin, beet pectin did not have good gelling properties in the food products produced in this research, and could not produce a strong network to trap free water as much as citrus pectin does. However, beet pectin can be used as a thickening agent.[5] In another study Toğrul and Arslan (2003) extracted cellulose from beet pulp following elimination of fat, hemi-cellulose and pectin. They examined flow properties of sugar-beet pulp cellulose and its intrinsic viscosity–molecular weight relationship.[6] Adinugraha et al (2005) investigated synthesis and characterization of sodium carboxymethyl cellulose from Cavendish banana pseudo stem.[7] Pushpamalar et al (2006) examined the optimization of reaction conditions for preparing carboxymethyl cellulose from sago waste.[8] Gujral et al (2002) studied the effect of hydrocolloids, storage temperature and duration on the consistency of tomato ketchup.[9]
In another study by Sahin and Ozdemir (2004) five different hydrocolloids (tragacanth gum, guar gum, carboxymethyl cellulose, xanthan gum and locust bean gum) were added, at different levels to three different formulated ketchups and the effect of these hydrocolloids on the rheological properties of tomato ketchups was investigated. All

hydrocolloids increased the consistency of the tested samples; however, guar gum and locust bean gum caused the maximum increase, followed by xanthan gum, tragacanth gum and carboxymethyl cellulose (CMC).[10] Morris and Jacob (1965) stated that physicochemical properties of cellulose derivatives are influenced by various parameters including reaction type, degree of substitution, average molecular weight and molecular weight distribution.[4] Degree of substitution for a cellulose derivative compound is defined as the number of hydroxyl groups reacted with one unit of un-hydrated glucose.

CMC; E466, is a derivative of cellulose formed by its reaction with alkali and chloroacetic acid. It is used primarily because of its high viscosity and being non-toxic and non-allergenic. The objectives of this study were: to extract the cellulose from sugar-beet pulp, to manufacture CMC from the cellulose extracted from sugar beet pulp and to utilize the CMC in ketchup and investigate ketchup rheological properties.

2. MATERIALS AND METHODS

2.1 Sugar-beet pulp preparation and cellulose extraction

Sugar-beet pulp was obtained from a local sugar plant (Shiraz, Iran). The pulp used in the present work was collected from the diffusion stage of the sugar manufacturing process (no formalin added). The drying was carried out in a cabinet drier at 60 °C for 16 hours. It was then ground to pass a 60-mesh size screen. The ground pulp was heated to 97 °C in a water bath for about 10 min to inactivate pectic enzymes. Then, it was washed with water, filtered through a suction filter, dried at 50 °C and sieved. A 50-mesh fraction was used for further experiments. The dried and ground sugar-beet pulp (7 g) was defatted by extraction with chloroform: methanol (2:1, v/v) in a soxhlet for 6 h. The defatted sugar-beet pulp was mixed with 350 ml of 0.1 M Na_3PO_4 to bring to pH 7.5. Proteolysis was performed by the addition of protease (35 mg). After incubation overnight at 37 °C, the sample was filtered. The deproteinated sugar beet pulp was mixed with 700 ml of 0.25% ammonium oxalate (w/v) (pH 3.5). The mixture was shaken at 75 °C in a water bath for 60 min and filtered. 10% NaOH was added to the depectinated sugar-beet pulp. The mixture was shaken at 35 °C in a water bath for 22 h, filtered and washed with distilled water to remove the base. The hemicellulose-free sugar-beet pulp was mixed with 100 ml of distilled water, 5 ml of 10% acetic acid and 2 g NaCl. The mixture was shaken at 75 °C in a water bath for 60 min and filtered. The residue was washed with distilled water and ethanol to remove acid, and then dried at 50 °C in an oven for 16 h.[6]

2.2 Production of CMC from cellulose

Two grams of cellulose from the previous extraction process, 100 ml of solvent (isobutyl alcohol) and 20 ml of NaOH having concentrations of 30% were mechanically stirred for 90 min at 25 °C in a reaction vessel with the heat stabilization jacket. The mixture was filtered to a weight of about 6.5 g and shredded for 90 min in a shredder. Then 3.0 g of sodium chloroacetate was added and shredding continued at a temperature of 70 °C for 360 min. Water from a constant temperature bath was circulated through the jacket of the reaction vessel to maintain constant temperature during carboxymethylation. After the time of reaction was finished, the mixture was neutralized with acetic acid of 90% and filtered. The obtained cake was purified by washing with 70% methanol to remove undesired salts, filtered and dried at 70 °C.[8, 11-12]

2.3 Characterization of CMC

The moisture content, degree of substitution (DS), and purity of CMC were determined by the ASTMD 1439-94 standard method.[13]

2.4 Ketchup samples preparation

Ketchup was prepared based on the recipe provided by a major ketchup manufacturer.[14] Five formulations (with and without hydrocolloids) were prepared. Sugar-beet pulp CMC, commercial CMC (Merck), Sunrose (a commercial hydrocolloid mixture which is used extensively by Iranian ketchup manufacturers and contains high percentage of sodium carboxymethyl cellulose) and tragacanth gum were used as thickener.

2.5 Ketchup characterization

Ash content, pH, and Brix were determined based on the approved methods of ISIRI (2004).[14] Viscosity measurements were performed on a Deer rheometer (PDR) as instructed by Bourne at 27 °C.[15] Synersis was observed in production of ketchup and over three months storage period. Furthermore, sensory evaluation on texture, color, flavor and overall acceptance were determined based on the hedonic test using a group of 15 semitrained panelists.[16]

2.6 Statistical analysis

All tests were performed in triplicate. Analysis of variance (ANOVA) was performed using the Duncan's multiple-range test to compare treatment means. Significance was defined at $P<0.05$.[17]

3. RESULTS AND DISCUSSION

Table 1 summarizes chemical analysis of manufactured CMC and commercial CMC (Merck). The data indicate that degree of substitution (DS) of 0.52 and purity of just over 53% is much lower than that of Merck and even lower than those reported by Toğrul and Arslan (2003) which was 0.67. They stated that the DS is related to time-temperature combination.[18] In the present study, the reaction tank was made of plastic which probably requires a longer time to reach to reaction temperature.

Table 1 *Chemical analysis of manufactured CMC and commercial CMC*

Specifications	Amount (%)	
	Merck	Manufactured
Moisture	0.30 ± 0.00	7.00 ± 0.00
DS	0.91 ± 0.02	0.52 ± 0.02
CMC Purity	90.96 ± 3.48	53.94 ± 2.79

Table 2 shows the variation of pH, ash and brix for manufactured ketchup samples when supplemented with various hydrocolloids. The data show addition of manufactured CMC although it had a lower DS and purity (as illustrated in Table 1) did not impart any unwanted characteristics to the ketchup.

Table 2 *Chemical analysis of ketchup samples*

Ketchup Samples	pH	Ash	Brix
Control	3.83a* ± 0.01	4.50a ± 0.71	32.0a ± 0.5
Manufactured CMC	3.90a ± 0.02	4.50a ± 0.35	32.0a ± 0.1
Merck CMC	3.88a ± 0.03	4.25a ± 0.00	32.0a ± 0.1
Sunrose	3.87a ± 0.01	4.00a ± 0.35	32.0a ± 0.2
Tragacanth gum	3.80a ± 0.02	3.75a ± 1.41	32.5a ± 0.0

Table 3 shows the results of the rheological experiments. All data show a significant difference in apparent viscosity values ($p<0.05$). The viscosity of ketchup samples containing tragacanth gum was too high to be measured with the rheometer. After that, the ketchup supplemented with Sunrose had the highest apparent viscosity and the sample with no hydrocolloid showed an apparent viscosity which was almost 5 times lower. The data of the ketchup supplemented with manufactured CMC show a suitable viscosity although the CMC has a low DS. The viscosity of this sample is higher than that of ketchup with no hydrocolloid. Sahin and Ozdmir (2004) almost had the same results.[10]

Table 3 *Apparent viscosity of ketchup samples supplemented with hydrocolloids (Temp. = 27 °C and shear rate = 0.71 s⁻¹)*

Ketchup samples	Apparent viscosity (cP)
Tragacanth Gum	> 1000 ^{a*}
Sunrose	935.78 ^{b*} ± 63.01
Merck CMC	490.28 ^{c} ± 0.16
Manufactured CMC	289.71 ^{d} ± 32.42
Control	190.80 ^{e} ± 0.00

* *Data followed by a different letter in a column are significantly different (p<0.05)*

When analyzing the rheological data based on the power law equation, the flow behavior index, at 27 °C for ketchup containing Merck CMC and manufactured CMC were 0.34 and 0.41, respectively. This index for ketchup supplemented with Sunrose and for ketchup without any hydrocolloid were 0.33 and 0.45, respectively. The latter data clearly shows the shear thinning behavior of ketchup. The data is in agreement with those reported by Sahin and Ozdemir (2004).[10] In addition, the present data is a good comparison between CMC and other hydrocolloids. As the data shows even Merck CMC with high DS could not match in terms of viscosity with Sunrose or tragacanth gum. When analyzing the data for the syneresis phenomenon the results were interesting. No syneresis was observed for ketchup containing Merck CMC or tragacanth gum at any time during the storage of the ketchup. It however was observed for ketchup containing manufactured CMC and Sunrose and the control sample. The phenomenon was observed at the end of the first month of storage for the control sample and ketchup containing manufactured CMC. This for the ketchup supplemented with manufactured CMC was much less than the control sample. At the end of third month of storage some syneresis was observed in ketchup containing Sunrose but no increase was detected for ketchup containing manufactured CMC. All ketchup samples were examined for odor and flavor by experienced panelists. Table 4 summaries the results. The highest and lowest scores were awarded to control samples and ketchup containing tragacanth gum, respectively. The data shows no significant difference between ketchup containing tragacanth gum, Merck CMC and Sunrose ($p<0.05$). In terms of odor and flavor control samples and samples with manufactured CMC were amongst the favorites. This indicates some of the advantages of manufactured CMC when used as the hydrocolloid in the ketchup.

Table 4 *Sensory evaluation results for ketchup flavor and odor*

Ketchup samples	Score (1-7)	results
Control	$5.46^{a*} \pm 1.24$	Very good
Manufactured CMC	$5.26^{b} \pm 1.09$	Good
Merck CMC	$4.33^{c} \pm 1.17$	Medium
Tragacanth gum	$3.93^{c} \pm 1.57$	Medium
Sunrose	$4.46^{c} \pm 1.24$	Medium

* *Data followed by a different letter in a column are significantly different (p<0.05).*

Sensory evaluation results for texture of ketchup indicate the control sample produced the lowest consistency and ketchup containing tragacanth gum and Sunrose give rise to ketchup having a hard texture. The panelists gave "suitable" score to the ketchup containing manufactured CMC. Table 5 illustrates these findings.

Table 5 *Sensory evaluation results for ketchup texture*

Ketchup samples	Score (1-5)	results
Control	$1.86^{a*} \pm 0.63$	Soft
Manufactured CMC	$3.00^{b} \pm 0.53$	Suitable
Merck CMC	$3.53^{c} \pm 0.51$	Hard
Tragacanth gum	$4.46^{d} \pm 0.35$	Very Hard
Sunrose	$4.86^{d} \pm 0.51$	Very Hard

* *Data followed by a different letter in a column are significantly different (p<0.05).*

Color of the manufactured ketchup samples were also examined by the panelists (Table 6). In this assessment the samples containing Sunrose and tragacanth gum had the lowest score when compared with other samples. All other samples were considered having better color. The data in the latter group of samples were not significantly different ($p<0.05$). This is an indication that the manufactured CMC could be considered as a relatively suitable replacement for commercial hydrocolloids.

Table 6 *Sensory evaluation results for ketchup in terms of their colors*

Ketchup samples	Score (1-7)	results
Control	$5.26^{a*} \pm 1.16$	Good
Manufactured CMC	$5.26^{a} \pm 1.33$	Good
Merck CMC	$5.53^{a} \pm 1.18$	Good
Tragacanth gum	$4.20^{b} \pm 1.01$	Medium
Sunrose	$4.20^{b} \pm 1.30$	Medium

* *Data followed by a different letter in a column are significantly different (p<0.05).*

The panelists also scored the samples in terms of overall acceptance. They gave the lowest scores to ketchup supplemented with tragacanth gum and Sunrose, i.e. being the least acceptable and not being significantly different ($p<0.05$). However, the other three samples were given better marks and were considered as "relatively acceptable" by the panelists. These results were tabulated in Table 7.

Table 7 *Sensory evaluation results for ketchup in terms of their overall acceptance*

Ketchup samples	Score (1-7)	results
Control	4.20 [a*] ± 1.26	Relatively good
Manufactured CMC	5.73 [a] ± 0.59	Relatively good
Merck CMC	3.93 [a] ± 1.53	Relatively good
Tragacanth gum	2.80 [b] ± 0.86	Relatively bad
Sunrose	3.00 [b] ± 1.00	Relatively bad

* *Data followed by a different letter in a column are significantly different ($p<0.05$).*

The samples containing tragacanth gum and Sunrose were darker and appeared slightly more dehydrated which was unacceptable with examiners. This was probably due to higher water absorption capacity of these hydrocolloids when compared with CMC. Other samples including those supplemented with manufactured CMC seems to be more acceptable. The data for overall acceptance signified ketchups with tragacanth gum and Sunrose were unacceptable owing to having unacceptable color, texture, flavour and odour. It is noteworthy that all the results obtained were for 1% concentration of hydrocolloids. Thus reducing this concentration might improve the level of acceptance of the samples containing these hydrocolloids.

4. CONCLUSION

The study demonstrates the possibility of production of CMC and other useful value added products such as pectin and hemi-cellulose from sugar-beet pulp as a waste or by-product of the sugar industries which is presently used as animal feed. The most important issues in this process are the purity and degree of substitution of the manufactured CMC which are influenced by process conditions e.g. temperature, time and concentration of sodium hydroxide. Efficacy of the process for cellulose manufacturing from pulp was about 18%, which is fairly good. In the work carried out, the amount of CMC produced was also acceptable. The produced CMC has the DS of 0.52 which is useable for most foods, but it is possible to improve this value by simply enhancing the process conditions. By applying about 1% of this CMC to food products such as tomato ketchup, satisfactory results in terms of color, flavor, odour and texture and overall acceptance were achieved by the panelists. This is an indication that the manufactured CMC might be a relatively suitable replacement for other commercial hydrocolloids.

References:

1. www.persiansugar.com/modules.php?name=Content&pa=showpage&pid=243.
2. G. Mesbahi, 2009. Fundamentals of Sugar Processing Technology. 3rd ed., Agriculture Sciences Pub., Tehran, Iran, pp. 1-6, 54-57, 126-130.
3. G. Vaccari, E. Tamburini, G. Sgualdino, K. Urbaniec, and J. Klemes, J. Clean. Product., 2005, **13**, 499.
4. B. Morris, and S. Jacob, 1965. Chemical Analysis of Foods and Food Products. 3rd ed. D-van Nostrand Company Inc. New Jersey. pp: 30-31.
5. G. Mesbahi, J. Jamalian, and A. Farahnaky, Food Hydrocolloids, 2005, 19, 731.
6. H. Toğrul, and N. Arslan, Carbohydrate Polymers. 2003, **54**, 63.
7. M.P. Adinugraha, D.W. Marseno, and M. Haryadi. Carbohydrate Polymers, 2005, 1.

8. V. Pushpamalar, S.J. Langford, M. Ahmad, and Y.Y. Lim, 2006. Carbohydrate Polymers. 2006, **64**, 312.
9. S.H. Gujral, , A. Sharma, and N. Singh, International Journal of Food Properties, 2002, **5**, 179.
10. H. Sahin, and F. Ozdemir, Food Hydrocolloids. 2004, **18**, 1015.
11. J. Kötz, I. Bogen, Th. Heinze, U. and Heinze, W.M. Colloids and Surfaces.2001, **184**, 621.
12. H. Toğrul, and N. Arslan, Journal of Food Engineering, 2004, **62**, 271.
13. ASTM. 1973. Standard Methods of Testing Sodium Carboxymethyl Cellulose. D1439-72. pp. 104-113.
14. Institute of Standards and Industrial Research of Iran. 2004. Tomato Sauce. 1st ed., Standard No. 7963.
15. M.C. Bourne, 2002. Food Texture And Vicosity: Concept and Measurement. 2nd ed. Academic Press. New York. pp. 245-246.
16. B.M. Watts, G.L. Ylimaki, L.E. Jeffery, and L.G. Elias, 1989. Basic Sensory Methods for Food Evaluation. 1sted. The International Development Research Centre. Ottawa, Canada. pp. 66-79.
17. A. Basiri, 2005. Statistical Methods of Agriculture Sciences. 9th ed. Shiraz University Pub. , Iran, pp. 33-65.
18. H. Toğrul, and N. Arslan, 2003. Carbohydrate Polymers, 2003, **54,** 73.

Hydrocolloids and Health

DIETARY FIBRE AND SATIETY – NOT ALL FIBRE IS ALIKE

J.L. Slavin[1]

[1]Department of Food Science and Nutrition, University of Minnesota, 1334 Eckles Avenue, St. Paul, MN 55108 USA

ABSTRACT:

Diets low in energy and fat, such as those typically recommended for obese people are poorly satiating. Adding fibre to low-calorie/low-fat foods may enhance satiety. Consumption of high fibre diets is linked to lower body weight and body fat and less weight gain over time in epidemiological studies. Dietary fibre may impact body weight by many mechanisms including hormonal, intrinsic, and colonic effects. Adding bulk to the diet with fibre will also reduce the energy density of the diet. Satiety signals are generated both pre- and post-absorptively so different types of fibre may be effective by different mechanisms. Viscous fibres have been linked to improved satiety, but insoluble fibres that survive gut transit also are satiating. We conducted an acute, double-blind randomized study to compare the effects of four fibres on satiety. On five separate visits, healthy men and women (n=20) fasting subjects consumed either a low-fibre control muffin (1.6 g fibre) or a high-fibre muffin (8.0 - 9.6 g fibre) for breakfast. Subjects used 100mm visual analogue scales to rate hunger and appetite at baseline and at regular intervals for 180 minutes after muffin consumption. Responses were analyzed as change from baseline. Despite similar amounts of dietary fibre in the four high-fibre muffins, satiety responses varied among treatments. Subjects were significantly less hungry at 180 minutes after consuming either resistant starch or barley with oat fibres than after polydextrose; subjects also felt more satisfied after resistant starch and corn bran than after polydextrose. Additionally subjects were significantly more full after consuming resistant starch, barley with oat fibres, corn bran, and control muffins than after eating the polydextrose muffin. Results from this study support that not all fibres influence satiety equally. Effectiveness of different functional fibres on satiety must be balanced with gastrointestinal tolerance of these fibres. In general, resistant starches are well tolerated while oligosaccharides including fructo- and galacto- may cause gastrointestinal disturbances when consumed in quantities that impact satiety. Other factors to consider when evaluating satiety and fibre are dose of fibre and form of fibre. Generally small doses of any fibre are not effective in altering satiety. Our research with doses of fibre found that mixed fibres in a muffin were only effective at the highest amount of fibre fed, 12 gram dose. A dose response study with fenugreek fibre found that 8 grams of fenugreek fibre was effective in enhancing satiety. Food form studies suggest that whole foods containg fibre are more satiating than beverages, even when isolated fibres are added to the beverages. Thus, public health messages to increase consumption of dietary fibre are widely accepted, although scientific support for isolated fibres impacting body weight are lacking.

INTRODUCTION

Examining the relationship between dietary fibre intake and body weight is difficult because of the struggle to define dietary fibre and agree on recommended intake levels (1). In 2002, the Dietary Reference Intakes (DRIs) published definitions for fibre (2). *Dietary Fibre* consists of the nondigestible carbohydrates and lignin that are intrinsic and intact in plants. *Functional Fibre* consists of the isolated nondigestible carbohydrates that have beneficial physiological effects in humans. *Total Fibre* is the sum of *Dietary Fibre* and *Functional Fibre*. Non-digestible means not digested and absorbed in the human small intestine. Fibres can be fermented in the large intestine or can past through the digestive tract unfermented. Data were not available to determine an Estimated Average Requirement (EAR), and thus calculate a Recommended Dietary Allowance (RDA) for *Total Fibre*, so an Adequate Intake (AI) was developed instead. The AI for fibre is based on the median fibre intake level observed to achieve the lowest risk of coronary heart disease (CHD). A Tolerable Upper Intake Level (UL) was not set for *Dietary Fibre* or *Functional Fibre*. Nutrition Facts labels use 25 g dietary fibre per day for a 2,000 kcal/day diet or 30 g/day for a 2,500 kcal/day diet as goals for American intake.

Dietary fibre intake is less than recommended in the United States; usual intakes average only 15 g per day (2). Because of differences in analysis of fibre and fibre databases, it is difficult to compare fibre intakes across countries. Many popular foods contain little dietary fibre. Servings of commonly consumed grains, fruits, and vegetables contain only 1 to 3 g of dietary fibre. Legumes and high-fibre bread and cereal products supply more dietary fibre, but are not commonly consumed. Fibre intake is linked to calorie intake so as consumers are encouraged to eat less foods for weight control they usually consume less dietary fibre as a result.

Epidemiologic studies on dietary fibre and body weight

High intake of carbohydrates is linked to lower body weight. Saris (3) reported an inverse relationship between carbohydrate intake and body mass index (BMI) in 8 of 9 studies. The "fibre hypothesis" suggests consumption of unrefined, high-fibre, carbohydrate-based foods protects against industrialized diseases including diabetes, cancer, heart disease and obesity. However, this traditional diet is both high in dietary fibre and low in dietary fat, making it difficult to separate out the cause of these beneficial effects. In epidemiologic studies, usual intake of dietary fibre, soluble fibre, and insoluble fibre can be measured and then linked to body weight. Epidemiologic studies also associate body weight to intake of foods that are naturally high in dietary fibre, such as whole grains, fruits, and vegetables.

A multidisciplinary study of more than 5000 individuals found that obesity was associated with increased energy intakes and decreased consumption of fibre-rich foods like fruits and vegetables (4). In the Seven Countries Study, dietary fibre was significantly inversely associated with subscapular skinfold thickness (5). And a longitudinal study found fibre intake was inversely associated with body mass index at all levels of fat intake—even after adjustment for lifestyle factors and other confounding factors among young adults (6).

In the prospective Nurses Health Study, women who consumed more fibre consistently weighed less than women who consumed less fibre (7). Additionally, women in the highest quintile of dietary fibre intake had a 49% lower risk of major weight gain. More recently

Maskarinec et al (8) reported that plant-based foods and dietary fibre were most protective against excess body weight in a large ethnically diverse population. Howarth et al (9) examined the association of dietary composition variables with body mass index among US adults age 20 to 59 years in the Continuing Survey of Food Intakes by Individuals (CSFII) 1994-1996. For women, a low-fibre, high-fat diet was associated with the greatest increase in risk of overweight or obesity compared with a high-fibre, low-fat diet. Davis et al (10) matched 52 normal weight women to 52 overweight/obese women and found that the normal weight subjects had higher fibre and fruit intake than the obese subjects. Obese men and women have significantly lower dietary fibre intakes than lean men and women (11).

Intervention studies

Many intervention studies have been conducted to determine if dietary fibre intake assists in weight loss; the results have been inconsistent. In general, intervention studies are designed to monitor changes in body weight when subjects (either overweight or lean) are given low or high fibre diets. Birketvedt et al. (12) found the addition of dietary fibre to a low-calorie diet significantly improved weight loss. The fibre-supplemented group lost 8.0 kg while the placebo group lost only 5.8 kg. When postmenopausal women consumed higher fibre diets as part of a study on very low-fat diets and weight loss, the women who ate higher fibre diets lost significantly more weight (13). A 12-week, randomized, controlled study of 34 individuals found that a high-carbohydrate diet consumed ad lib (without energy restriction or change in energy intake) caused significant body weight and fat losses in older men and women (14).

Tucker and Thomas (15) conducted a prospective cohort design study with 252 women. The subjects completed baseline and follow-up assessments 20 months apart. Diet was measured with 7 day weighed food records. Fibre was expressed per 100 kcal. Across the 20 months, almost 50% of the women gained weight and fat. For each 1 g increase in total fibre consumed body weight decreased by 0.25 kg and body fat decreased by 0.25 percentage point. The authors concluded that increasing dietary fibre significantly reduces the risk of gaining body weight and body fat in women and that this related primarily to reducing energy intake over time.

Dietary fibre and weight management

Heaton (16) proposed that fibre acts as a physiological obstacle to energy intake by at least three mechanisms:

1) fibre displaces available calories and nutrients from the diet;
2) fibre increases chewing which limits intake by promoting the secretion of saliva and gastric juice, resulting in an expansion of the stomach and increased satiety; and
3) fibre reduces the absorption efficiency of the small intestine.

Intrinsic, hormonal, and colonic effects of dietary fibre decrease food intake by promoting satiation and/or satiety. Examples of these controls include, decreased gastric emptying and/or slowed energy and nutrient absorption. Gut hormone signaling also plays a significant role in satiety. Dietary fibre may also influence fat oxidation and fat storage. Satiation is defined as the satisfaction of appetite that develops during the course of eating and eventually results in the cessation of eating. Satiety refers to the state in which further eating is inhibited and occurs as a consequence of having eaten. Fibre plays a role in satiety, although not all fibres are equally effective (17)

Howarth and colleagues (18) summarized the effects of dietary fibre on hunger, satiety, energy intake, and body weight. Most studies with controlled energy intake reported an increase in post-meal satiety and a decrease in subsequent hunger with increased dietary fibre. With *ad libitum* energy intake, the average effect of increasing dietary fibre across all the studies indicated that an additional 14 g of fibre per day resulted in a 10% decrease in energy intake and a weight loss of over 1.9 kg through about 3.8 months of intervention. The effects of increasing dietary fibre were reported to be even more impressive in obese individuals. The results of increased dietary fibre from high-fibre foods or from fibre supplements on weight regulation were similar. Additionally, the beneficial effect of dietary fibre on weight regulation was seen for both soluble and insoluble dietary fibre. This group concluded that increasing the population mean dietary fibre intake from the current average of about 15 g/day to 25-30 g/day would be beneficial and may help reduce the prevalence of obesity.

Research on how different types of fibre affect appetite, energy and food intake have been inconsistent. Results differ according to the type of fibre and whether it is added as an isolated fibre supplement or naturally occurring in a food (table x. lists the main food sources for various fibres). Short-term studies that assess energy intake after subjects are fed fibre-containing meals suggest large amounts of total fibre are most successful for reducing subsequent energy intake. Additionally, higher viscosity fibres appear most successful in promoting satiety while new, soluble fibres appear to have minimal effects on satiety, even if consumed in very large doses. Longer-term studies that examine how both intrinsic and functional fibres impact satiety are required. Yet there is ample evidence to suggest that increasing consumption of high-fibre foods—especially viscous fibres—may decrease feelings of hunger by inducing satiation and satiety.

Few studies have compared the effects of different fibres on satiety in the same subjects. We hypothesized that certain types of fibre were more effective in enhancing satiety than others (19). Healthy men and women (n=20) participated in an acute, randomized double-blind, crossover study that compared the effects of four fibres and a low fibre treatment on satiety. On five separate visits fasting subjects consumed either a low-fibre muffin (1.6 g fibre) or one of four high-fibre muffins (8.0 – 9.6 g fibre) for breakfast. Subjects used four questions on 100 mm visual analog scales to rate satiety at baseline and at regular intervals for 180 minutes after muffin consumption. Responses were analyzed as area under the curve and significant differences from baseline. Satiety differed among treatments. Resistant starch and corn bran had the most impact on satiety, while polydextrose had little effect and behaved similar to the low-fibre treatment. The effective dose of fibre to enhance satiety is not known.

We measured the effect of 2 doses of fenugreek fibre on satiety in obese subjects (20). Most of the fibre in fenugreek is water-soluble galactomannans, similar to the fibre found in guar gum and locust bean gum. Subjects consumed either 0, 4, or 8 grams of fenugreek fibre mixed in the morning orange juice with a controlled breakfast. The 8 g dose of fenugreek fibre significantly increased mean ratings of satiety and fullness and reduced ratings of hunger and prospective food consumption ($P < 0.05$). Palatability was significantly reduced with increasing doses of fenugreek fibre. Satiety results were not related to postprandial blood glucose in this study.

In a follow-up study, a mixed fibre that included a range of viscous fibre known to alter satiety, guar gum, pectin, pea fibre, and barley beta-glucan was incorporated into a high calorie (500 kcals) muffin for breakfast (21). Healthy, normal weight men (n=10) and women (n=10) participated in a double-blind, crossover study. On four separate visits, fasting subjects consumed a muffin with 0, 4, 8, and 12 g of mixed fibres. Muffins had 500 calories and similar macronutrient content. Visual analogue scales rated hunger and satiety from 0 – 180 minutes. Food intake after muffin consumption was measured at an *ad lbitum* lunch buffet and for the rest of the day. In all satiety measures, the 12 g fibre muffin was signifcantly more satiating than the 0 g muffin. Muffins with 4 g and 8 g of fibre had intermediate effects. Despite significant differences in satiety measures after three hours, food intake did not differ among treatments.

The dose of fibre affects results. Mattes (22) compared a control breakfast bar to a breakfast bar containing alginate and guar gum (0.6 g fibre vs. 4.5 g fibre) in obese subjects. No significant treatment effects or cumulative effects of satiety were found with the fibre containing bar. Oligofructose (16 g/day) taken in 2 doses increased satiety during breakfast (23). Total energy intake was 5% lower during the oligofructose treatment. Higher doses of alpha-cyclodextrin (a cyclic oligosaccharide) resulted in greater satiety than a control, but caused increased gastrointestinal complaints and had reduced palatability (24). Overall, studies that report significant effects on satiety use larger doses of fibre.

Traditionally, high-fibre foods have been solid foods. Some of the newer functional fibres, such as resistant starches and oligosaccharides, are easily added to drinks and may not alter viscosity. Few studies on the satiating effects of drinks supplemented with these soluble, non-viscous fibres have been published, but we suspect they would have limited satiating effects. Moorhead et al (25) compared test lunches with carrots (200 g), as whole carrots, blended carrots, or carrot nutrients. Whole carrots and blended carrots resulted in significantly higher satiety. *Ad lib* food intake for the remainder of the day decreased in this order: carrot nutrients, blended carrots, whole carrots. They conclude that both fibre content and food structure are important determinants of satiety.

A similar study was conducted using apples, apple sauce, and apple juice (with added fibre) as a preload before a meal. The fibre supplement was a low-viscosity, apple-derived pectin supplement. Even though each of the three foods contained the same calories and fibre, subjects ate significantly less lunch when consuming the whole apple compared to either the applesauce, juice, or no preload. Again, this suggests that adding fibre to a beverage may not necessarily enhance satiety—and that solid foods may be more satiating than liquids (26).

Some studies support that fibre added to beverages can enhance satiety. Lyly et al (27) compared the satiety response in a beverage withou fibre, guar gum beverage, a wheat bran beverage, an oat beta-glucan beverage and wheat bread in 19 healthy volunteers. The guar gum beverage significantly decreased desire to eat in the subjects. The beverage with oat beta-glucan increased fullness. The authors conclude that dietary fibre in beverages can enhance their perceived satiety and decrease the desire to eat more than a beverage without fibre.

Many other studies support that increased dietary fibre intake promotes satiety, decreases hunger and thus helps provide a feeling of fullness (17). Fibre-rich foods tend to have a high volume and a low energy density and should promote satiation and satiety. Rolls (28) suggested humans may consume a constant weight of food. If this is true, then a constant weight of lower energy food (i.e. high fibre) may promote a reduction in body weight. High-fibre foods have much less energy density compared to high fat foods. Thus, high fibre foods can displace energy or calories. The bulking and viscosity properties of dietary fibre are predominantly responsible for influencing satiation and satiety.

The importance of the gut microbiota in energy regulation has recently been recognized (29). Ley et al (30) reported differences in the bacterial groups known as the Firmicutes and the Bacteroidetes between obese and lean human subjects. Additionally, when the obese individuals lost weight over a year, the proportion from Firmicutes became more like that of lean individuals. These are exciting findings that show that we depend on the bacteria in our gut and they may play a role in obesity prevention. And they suggest that we might manipulate the microbiotic environment to treat or prevent obesity (31).

Gut Hormones and Satiety

The organs of the gastrointestinal tract have an important role in sensing and signaling the regulation of energy homeostasis. The gut uses neural and endocrine pathways to communicate and control energy balance. Gut peptides play an important role in the regulation of food intake and energy homeostatis (32). Few gut hormones have been measured with a fibre question in mind. Gruendel et al (33) reported that acylated ghrelin was lowered with the addition of carob fibre to all liquid meal. The carob fibre, however, is also high in polyphenolic compounds which may be responsible for altered ghrelin levels. Erdmann et al (34) reported differences in ghrelin response between normal and obese subjects. Arabinoxylan, a soluble fibre, was fed to 11 subjects with normal glucose metabolism (35). Glucose response was not significantly different in subjects; however, plasma ghrelin was higher two hours after the fibre than after the control meal.

Oligofructose (21 g/day) was given to 48 overweight subjects for 12 weeks, compared to a placebo treatment (36). Supplementation with oligofructose significantly reduced body weight by 1.03 kg while the control group gained 0.45 kg during the 12 week study. The authors also reported a lower area under the curve (AUC) for ghrelin and a higher AUC for peptide YY.

CONCLUSIONS

Modest increases in consumption of fruits, vegetables, legumes, and whole and high-fibre grain products would bring adults close to achieving the dietary fibre recommendations of 14 g/1000 kcal. In addition, a higher fibre intake provided by foods is likely to be less calorically dense and lower in fat and added sugar. Many diseases of public health significance – obesity, cardiovascular disease, type 2 diabetes – as well as the less prevalent, but no less significant diseases of colonic diverticulosis and constipation, can be prevented or treated by increasing the amounts and varieties of fibre-containing foods. Promotion of a high-fibre diet by the dietetics professional cannot be overemphasized.

Not all fibres are euqally effective at promoting weight loss and satiety. Generally satiety is enhanced significantly by viscous fibre such as guar gum. Fibres that survive gut transit, such as corn bran and resistant starch, are also linked to improved satiety. Satiety is linked mostly to higher intakes of fibre, probably about 10 grams of fibre per serving is required to change satiety. Also, data on food form suggest that whole foods are more effective at enhancing satiety than beverages, even when fibres are added to beverages.

REFERENCES

1. J. L. Slavin,. *Nutrition*, 2005; **21**,411.
2. Institute of Medicine of the National Academies (2002) *Dietary Reference Intakes: Energy, Carbohydrates, Fiber, Fat, Fatty Acids, Cholesterol, Protein and Amino Acids*., The National Academies Press, Washington, DC, USA.
3. W.H.M. Saris, *Nutr. Rev.* 2003, **61**, S10.
4. L. Lissner L., A.K. Lindroos, and L. Sjostrom L, *Eur. J. Clin. Nutr.*, 1998, **52**; 316.
5. D. Kromhout, B. Bloemberg,, J. C. Seidell, A. Nissinen, and A. Menotti, *Int. J. Obesity*, 2001, **25**, 301.
6. D. S. Ludwig, M.A. Pereira, C. H. Kroenke, J.E. Hilner, L. Van Horn, M. L. Slattery, and D. R. Jacobs, *JAMA*, 1999, **282**, 1539.
7. S. Liu, W. C. Willett, J. E. Manson, F. B. Hu, B. Rosner, and G. Colditz, *Am. J. Clin. Nutr*, .2003, **87**, 920.
8. G. Maskarinec, T. Takata, I. Pagano, L. Carlin, M. T. Goodman, L. Le Marchand, A. M. Y. Nomura, L. R. Wilkens, and L. N. Kolonel, *Obesity*, 2006, **14**, 717.
9. N. C. Howarth, T. T. K. Huang, S. B. Roberts, and M.A. McCrory, *J. Am. Diet. Assoc.*, 2005, **105**, 1365.
10. J. N. Davis, V. A. Hodges, and M. B. Gillham, *J. Am. Diet. Assoc,.* 2006, **106**, 833.
11. W. C. Miller, M. G. Niederpruem, J. P. Wallace, and A. K. Lindeman, *J. Am. Diet. Assoc.*, 1994, **94**, 612.
12. G. S. Birketvedt, J. Aaseth, J. R. Florholmen, and K. Ryttig, *Acta. Medica.*, 2000, **43**, 129.
13. W. M. Mueller-Cunningham, R. Quintana, and S. E. Kasim-Karakas, *J. Am. Diet. Assoc.*, 2003, **103**, 1600.
14. N. P. Hays, R. D. Starling, X. Liu, D. H. Sullivan, T. A. Trappe, J. D. Fluckey, and W. J. Evans, *Arch. Intern. Med.*. 2004, **164**, 210.
15. L. A. Tucker, and K. S. Thomas, *J. Nutr.,*. 2009, **139**, 576.
16. K. W. Heaton, *Lancet*, 1973, **2**, 1418.
17. J. L. Slavin, and H. Green, *Nutrition Bulletin*, 2007, **32 (Suppl 1)**, 32
18. N. C. Howarth, E. Saltzman, and S. B. Roberts, *Nutr. Rev.*, 2001, **59**, 129.
19. H. J. Willis, A. L. Eldridge, J. Beiseigel, W. Thomas, and J. L. Slavin, *Nutr. Res.*, 2009, **29**, 100.
20. J. R. Mathern, S. K. Raatz, W. Thomas, and J. L. Slavin, *Phytother. Res.*, 2009 (in press).
21. H. J. Willis, W. Thomas, A. L. Eldridge, L. Harkness, H. Green, and J. L. Slavin, *Obesity Facts*, 2009, **2(suppl 2),** 142.
22. R. D. Mattes, *Phys. & Behaviour*, 2007, **90**, 705.
23. P. D. Cani, E. Joly, Y. Horsmans, and N. M. Delzenne, *Eur. J. Clin. Nutr.*, 2006, **60**, 567.
24. J. D. Buckley, A. A. Thorp, K. J. Murphy, and P.R. Howe, *Ann. Nutr. Metab.*, 2006, **50**, 108.
25. A. S. Moorhead, R. W. Welch, B. M. Livingstone, M. McCourt, A. A. Burns, and A. Dunne, *Br. J. Nutr.* 2006, **96**, 587.
26. J. E. Flood-Obbagy, and B. J. Rolls, *Appetite*, 2009, **52**, 416.
27. M. Lyly, K. H. Liukkonen, M. Salmenkallio-Marttila, L. Karhunen, K. Poutanen, and L. Lahteenmaki, *Eur. J. Nutr.*, 2009, (in press).
28. B. J. Rolls, *J. Nutr.*, 2000, 130 **(2 suppl),** 268S.
29. G. Wolf, *Nutr. Rev.*, 2006, **64**, 47.
30. R. E. Ley, P. J. Turnbaugh, S. Klein, and J. I. Gordon, *Nature*, 2006, **444**, 1022.

31. M. Bajzer, and R. J. Seeley, *Nature*, 2006, **444,** 1009.
32. K. G. Murphy, W. S. Dhillo, and S. R. Bloom, *Endoc. Rev.*, 2006, **27**, 719.
33. S. Gruendel, A. L. Garcia, B. Otto, C. Mueller, J. Steiniger, M. O. Weickert, M. Speth, N. Katz, and C. Koebnick, *J. Nutr.*, 2006, **136**, 1533.
34. J. Erdmann, M. Leibl, S. Wagenpfeil, F. Lippl, and V. Schusdziarra, *Regulatory Peptides,* 2006, **135**, 23.
35. M. Mohling, C. Koebnick, M. O. Weickert, W. Lueder, B. Otto, J. Steiniger, M. Twilfert, F. Meuser, A. F. H. Pfeiffer, and H. J. Zunft, *Hormone & Metabolic Res,* 2005, **37**, 303.
36. J. A. Parnell, and R. A. Reimer, *Am. J. Clin. Nutr.*, 2009, **89**, 1.

PHYSICO-CHEMICAL PROPERTIES OF HYDROCOLLOIDS DETERMINE THEIR APPETITE EFFECTS

A. Ström, H.M. Boers, R. Koppert, S.M. Melnikov, S. Wiseman and H.P.F Peters

Unilever R&D, Olivier van Noortlaan 120, 3130 AT Vlaardingen, The Netherlands

1 INTRODUCTION

The increasing prevalence of overweight and obese persons is a growing concern which reduces the quality of life for the affected people and increases the cost of health care for society as a whole[1]. There are a number of diet programs available aiming to reduce the intake of calories. However, many subjects fail to adhere to these programs and the reduction in weight is therefore not achieved or maintained[2,3,4]. Perceived hunger has been shown to be a significant predictor of failure to lose weight in clinical trials[5]. Targeting triggers in the body via gastrointestinal mechanisms has been shown to enhance satiety feelings[6] and may therefore help subjects to remain on a diet and reduce weight and/or maintain a healthy weight.

Populations with a high intake of fibres have been shown to have a lower percentage of overweight people[7,8,9]. Soluble fibres naturally present in food reduce the rate of gastric emptying[10] and increase feelings of satiety (increased fullness and reduced hunger)[11]. Furthermore, fibre-rich foods typically have a low energy density and may require more effort to ingest, which has been shown to increase satiation[1,8]. However, the effect of the addition of fibres to the diet, either by adding it back to the food or as a daily supplement has not always shown a clear benefit in terms of feelings of satiety[8,12,13]. Recent research has highlighted the link between the physico-chemical properties (e.g. viscosity, gel properties and solubility in case of dry food of certain polysaccharides) in food matrices and the effect on e.g. satiety[14,15,16]. Polysaccharides or fibre enriched food products, aimed for weight/hunger control applications, should therefore be analysed with respect to their physico-chemical properties under physiologically relevant chemical and mechanical conditions[17]. In addition, the interaction between the polysaccharides and the co-ingredients can influence their potential use in weight management products, both in terms of palatability and in-body functionality. To date, however, objective measurement of these properties has been lacking in many studies and this might also explain some of the discrepancies observed in the literature. Further advances in understanding the mechanical environment in the body would also be beneficial for better prediction of e.g. gastric viscosity and gelation.

Based on the literature, it appears that for polysaccharides the amount, viscosity and gastric gelation are all important factors for satiety[14,16,18]. It is not possible to clearly determine which one of these parameters is most important for satiety based on the literature. However, the viscosity of drinks needs to be high to increase satiety, making palatability low. Producing a palatable satiety drink based on gastric gelation might be a better option.

For gastric gelation it is believed that it is not only important to form a gel in the stomach, but the gel also requires certain strength in order to induce satiety[16,19]. Thus, studies have shown that the addition of a gelling polysaccharide (alginate) to chocolate-flavored meal replacement dairy drinks increases the feelings of satiety if the gel formed under acid conditions is strong enough[16]. However, the drinks used in these studies were physically unstable, would gel in the product before consumption and their viscosity upon consumption was unpleasantly high. To prevent premature gelation these gelling fibres were either added to the drink immediately prior to consumption or a two component system was used.

In this paper, we present a system that is physically stable over time, does not produce a gel in the product upon storage, and has a low viscosity at neutral pH, yet produces sufficient gelation and mechanical properties under acid conditions to enhance satiety[16,18]. The specific systems we focused on were either a mixture of alginate and high methoxy (HM) pectin[20,21] or a mixture of alginate together with $Ca_3(PO_4)_2$ This salt is insoluble at neutral pH, but releases calcium under stomach conditions, allowing the formation of a strong gel with the alginate[22]. These two systems were chosen for application in mildly acidic beverages and neutral product formulations, respectively. We present results on the efficiency of these systems in producing satiety, and also how they interact with other components in the formulation.

2 MATERIALS & METHODS

2.1 Materials

The sodium alginate was Manugel DMB from FMC (former ISP alginates) with a guluronate content of ~70%[16]. The HM pectin was JM 150 from CP Kelco with a methyl ester content of ~ 70% (data given by the supplier). Glucono-delta-lactone (GDL), MgCl, NaCl, KCl, NaOH and HCl were purchased from Sigma. $Ca_3(PO)_4$, was purchased from CFT Buddenheim. Orange juice (Brix number of 60-62, given by the supplier) and carrot juice (Brix number of 36-38, given by the supplier) were obtained from SVZ international BV, the Netherlands.

Test meals.
Two different test meals were used for three human studies: a protein drink in the first (325 ml) and third (125 ml) study and a fruit/vegetable juice flavoured drink without protein in the second study.
In the first clinical trial a meal replacement drink was used (325 ml, 190 kcal), where the regular skim milk powder and caseinate were replaced by 3% w/w whey protein isolate

(BiPro purchased from Davisco Food International Inc.). In the control drink the fibre source was carrageenan, cellulose (Avicel purchased from FMC Biopolymers) and Fibersol. In the two test drinks carrageenan and cellulose were replaced by 0.6 and 0.8% alginate, respectively. Total fibre content per drink was around 5 g. All drinks contained 0.26% of $Ca_3(PO)_4$, and the pH was 7. The drinks were sterilised, filled in aluminium cans and stored ambient. They were chilled (refrigerated) 24h before use in the human trial.

In the third human study the composition of the control was the same as in the first one, but instead of 325 ml only 125 ml was used (i.e. 60 kcal) and 1% alginate.

In the second human study a orange/carrot juice flavoured drink with HM pectin / alginate (1% total polymer in a ratio of 40:60) was used (100ml; 20 kcal). The control contained xanthan in a concentration to match the viscosity of the alginate / pectin drink. The pH was adjusted to 4.2.

Subjects. For all three studies volunteers were recruited from Vlaardingen and surroundings, the Netherlands. Selection criteria were age 18-60 years, slightly overweight, apparently healthy (measured by questionnaire) and not using medicines judged to influence the study results.

2.2 Methods

Rotational Viscometry. The viscosity of the solutions and drinks was measured using an AR 1000 from TA instruments. The geometry used was a cone (ϕ = 40mm, truncation = 71 μm, angle = 1.59°) and plate. The viscosity was recorded at T = 20°C and between shear rates of 0.03-1000 s^{-1}.

Large deformation rheology. Samples were prepared by mixing the appropriate amount of GDL (to obtain a final pH of ~2.6) in water. This solution was quickly added to the polymer solution, or the drinks, subsequently mixed and poured into Teflon moulds before onset of gelation. The moulds were cylinders, 12 mm in diameter and 12 mm in height. The samples were incubated at 37°C for 2 hours and the gels were than removed from the moulds. Flat plate compression tests were performed using a Texture Analyser (Stable Micro Systems, TA.XT Plus, UK). The crosshead speed was 5 mm/sec. and the measurement was performed at 20°C. 5 repetitions were performed on each sample.

pH meter. The pH of the solutions was measured using a Seven Multi meter (Mettler - Toledo) with a combined temperature and pH electrode Mettler inlab 413. The pH electrode was calibrated between pH 7 and 4.

Magnetic Resonance Imaging. In vivo images of the gastric content of two subjects after consumption of 250ml and 100ml of alginate-pectin fruit juice were carried out by a spin – echo EPI measurement on a whole body 0.5-T purpose built EPI scanner at Nottingham University in a similar manner as previously reported[16].

Study design:

All three human studies had a fully randomized, placebo-controlled cross-over design. In the first human study 24 volunteers were investigated during three test days. The subjects consumed 325 ml of the whey dairy drink with no, 0.6% or 0.8% alginate for breakfast. Subjects' self assessments of appetite (e.g. hunger, fullness) were measured every 30 minutes post-consumption over 5h. An electronic visual analogue scale (EVAS, iPAQ) was used for recording the volunteers' appetite feelings. The scale was anchored at the low end with the most negative or lowest intensity feelings (e.g. "not at all"), and with opposing terms at the high end (e.g. "very high").

In the second human study 22 volunteers consumed a 100 ml drink 15 minutes before lunch either with or without alginate/HM Pectin followed 15 minutes later by a fixed lunch amount (Slim.Fast), and 3 hours after that by an *ad libitum* afternoon meal (conditions 1 and 2). Alternatively, (conditions 3 and 4) the subjects received the same two drinks without/with the Alginate/HM Pectin, but now drinks were followed only by an *ad libitum* lunch 15 min. later. Appetite feelings were assessed for 3 hours post-consumption of the treatment. Food intake was measured during the *ad libitum* lunch or afternoon meal.

In the third human study 24 volunteers consumed a 125 ml whey protein drink with or without 1% alginate 30 minutes prior to an *ad libitum* lunch. Appetite feelings were assessed during 1 hour post-consumption of the treatment. Food intake was measured during the *ad libitum* lunch.

Statistics

In all three human studies the data were analyzed using analysis of covariance (ANCOVA) with baseline measurement as covariate. This model takes into account the possible baseline influence using correlation of the baseline with the score on each subsequent time point. Sources of variation taken into account were subject (as a random factor), treatment, period and period*treatment interaction. The difference of each treatment versus control was estimated using the Dunnett difference test (two-sided). The analysis was done per time point and also for Area Under the Curve (AUC), using SAS v9.3 PROC MIXED.

3 RESULTS AND DISCUSSION

3.1 Physico-chemical properties

3.1.1 Physico-chemical properties of alginate/protein mixtures.

In this system, the alginate forms a gel upon acidification, with calcium released from calcium diphosphate. As it is a meal replacer, the drink should contain protein and calcium to comply with nutritional guidelines. Therefore care must be taken to ensure that the drink does not contain free calcium in the formulation.

Alginates rich in G units form stronger acid and calcium gels than alginates rich in M units[22,23,24]. In order to have the strongest gels we have therefore used an alginate high in G content throughout this research. Alginate gels without calcium were acidified with glucono-delta-lactone[22] (GDL) and tested following the procedure described previously[16]. The force required to fracture (under compression) acidic alginate gels as a function of alginate is shown in Figure 1a. The target for this force to fracture is >11kPa and preferably >16kPa based on previous studies where alginate has shown to have an effect on satiety[16,18]. Thus, a gel with a force to fracture of at least 16kPa would require ~1.5% alginate, which exceeds the 1% addition of alginate allowable under current US regulations. Furthermore, solutions of high alginate concentrations are perceived as slimy, with bad flavor release properties[25]. Figure 1b shows the effect of the addition of an insoluble calcium source ($Ca_3(PO_4)_2$) to 1% alginate on the force required to fracture the gels as a function of calcium concentration. The insoluble calcium source becomes soluble as pH is lowered below 6-5 using GDL. A small amount of Ca^{2+} increases the gel strength at pH 3 in agreement with data obtained at neutral pH[22,23,24]. Shrinkage (and syneresis) is observed at higher Ca^{2+} concentration and it has been proposed that this occurs upon acidification due to exchange of calcium to protons at pH < pKa: the calcium induced gel changes with time into an acid gel which is denser than the calcium alginate network[26]. This implies that the gels will weaken with time when present in acid media (e.g. in stomach).

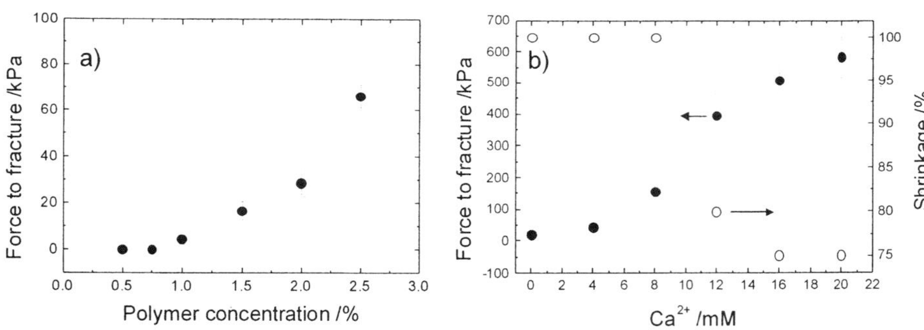

Figure 1: Force required to fracture acid alginate gels as a function of a) alginate concentration and b) of 1% alginate with theoretically calculated free calcium dissolved from the salt $Ca_3(PO_4)_2$ at pH 2.6 and T = 20°C.

The sensitivity of alginate to soluble calcium needs to be considered when formulating a protein-containing product (for example, a meal replacement drink). A suitable protein source for the alginate-containing beverage would be one which contains a low amount of multivalent ions to achieve a low product viscosity and a long shelf-life at neutral pH. Furthermore, as in the experiment above, an insoluble calcium salt should be added to the protein alginate solution to boost the gel strength under acidic conditions. This is illustrated in Figure 2, showing the viscosity of neutral solutions containing 3% protein from various sources, 0.8% alginate and 0.2 % of $Ca_3(PO)_4$ (~20mM Ca^{2+} under acid conditions) and the force required to fracture their respective calcium / acid gels. The viscosity of the solution containing skim milk powder (SMP) is considerably higher compared to the other solutions, and this sample forms a gel after a few hours. The SMP used in this study contained ~ 28mM Ca^{2+} (data given by the supplier) whereas the other protein sources contained less than 4mM Ca^{2+}. The force required to fracture the respective acid gels varied depending on protein source, but it is observed that the gel strength of all gels is lower than that of pure calcium alginate gels at pH 2.6. The presence of pure caseinate in the acidic calcium alginate gel made the gels brittle and grainy, and the gels fell apart upon handling.

There are limited data from the literature on alginate/caseinate/calcium systems. However, the strength of acid pectin/caseinate/calcium gels is known to be influenced by the ratio of pectin to caseinate in the presence of calcium. These systems go from typical caseinate gels to calcium pectin gels (weakened due to less developed calcium pectin crosslinks in the presence of caseinate) as pectin concentration is increased[27]. At intermediate pectin levels, the system shows considerable syneresis, similarly to the effect observed in this study for the calcium alginate - caseinate gels. As calcium alginate and calcium pectin form similar types of networks (the egg box model)[22,28,29], one could possibly expect to obtain coherent gels by changing the alginate-caseinate ratio.

Turning to the other proteins in Figure 2, the soy-containing samples showed a tendency towards phase separation whereas caseinate and whey containing samples appeared stable at neutral conditions.

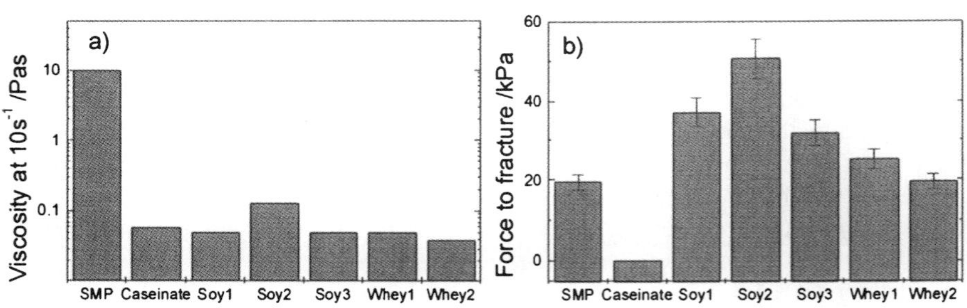

Figure 2: Viscosity (a) of 0.8% alginate + 0.2% $Ca_3(PO_4)_2$ + 3% protein from different sources at pH ~ 7 and b) force required to fracture their respective acid gels at T=20°C and pH ~ 2.6.

The sample with the lowest viscosity but still with an acceptable force required to fracture its acid gel (>16kPa) was a whey protein / alginate sample (whey2, BiPro from Davisco Int, US). Confocal scanning laser microscopy (CSLM) images of these samples at neutral conditions showed a polysaccharide continuous phase interspersed with protein aggregates, of which the largest ones appear to be around 25μm. Increasing Ca^{2+} concentration above 20mM further increased the force to fracture the gels.

Three chocolate flavoured drinks were prepared with the whey protein BiPro, one drink contained no alginate but cellulose and carrageenan, one contained 0.6% alginate and the third one - 0.8% alginate. The other polysaccharides were included in order to keep a similar fibre content in the drinks. In addition, all these samples contained fat, vitamin and mineral mix, extra calcium in the form of $Ca_3(PO_4)_2$, sucrose and sweetener, to represent a full product formulation. The viscosity of these drinks and the force to fracture their respective acidic gels are shown in Figure 3. Although the viscosity of the drinks is different, it can be assumed that the difference is not large enough to induce any satiety effect[30]. The force to fracture the gel at 0.8% alginate is actually higher than previously observed (Figure 2), which could possibly be related to altered ionic strength due to the presence of the mineral mix[31]. Further investigation is needed to better understand the influence of ionic strength on whey / calcium alginate gels at pH 2.6.

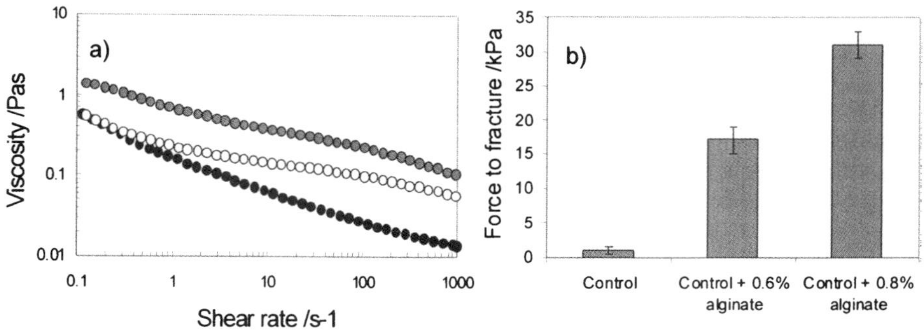

Figure 3: The a) viscosity of control (black symbol), control with 0.6% alginate (open symbol) and control with 0.8% alginate (grey symbol) as a function of shear rate and b) force to fracture their acid gels. All measurements done at T=20°C.

The viscosities of these drinks are considerably lower (4 and 2 times respectively), and also the alginate concentrations (0.6 and 0.8 versus 1 and 1.75%) are lower compared to the alginate drinks that have been tested previously and showed to enhance satiety[16,18]. Nevertheless, the acid gel strength is close to or greater than 16kPa (Figure 3b). These samples were used in a human intervention study in order to test whether a drink:

1) containing only whey protein (versus the previously used skim milk powder, which contains both casein and whey)
2) containing less alginate and
3) with lower viscosity upon consumption

is equally efficient in providing satiety compared to previous studies (see section 3.3 Human studies: satiety measurements).

3.1.2 Physico-chemical properties of alginate / HM pectin containing drinks.

Although a mixture of alginate and insoluble calcium works well in a neutral product, it would not work in a slightly acidified drink as the calcium salt would dissolve, leading to the gelation before consumption. Therefore, a system relying purely on acid gelation (at pH below product pH) would be required.

In this subsection, we look at synergistic gelation of alginate and HM pectin[20,21]. A study to find optimum ratio of alginate to HM pectin for acid gel strength showed that a ratio of 50:50 or 60:40 gives the strongest gel, in agreement with previous studies[20,21]. The ratio of 60:40 (alginate : HM pectin) was therefore used in what follows. The force required to fracture acid gels of the alginate / HM pectin mixtures compared to the alginate gels are shown in Figure 4 as a function of total polymer concentration.

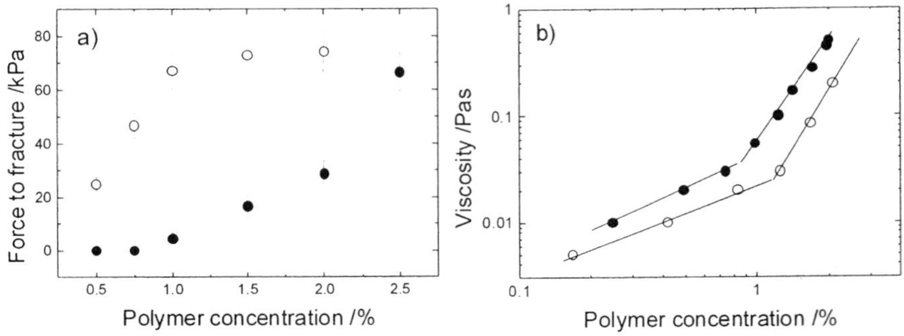

Figure 4: Comparison of a) force required to fracture acid alginate gels (solid symbols) and acid HM pectin alginate gels (open symbols) as a function of polymer concentration at pH ~ 2.6 and b) viscosity at pH ~ 7 and T=20°C.

It is shown that the gel strength of the HM pectin / alginate is larger for a given polymer concentration compared to alginate alone. The viscosity as a function of polymer concentration is lower for the mixed system, with the potential advantage that drinks of low viscosity could be made. Further investigation of this system showed that the pH of gelation (Figure 5) is close to the pKa of alginates in agreement with literature[20,20].

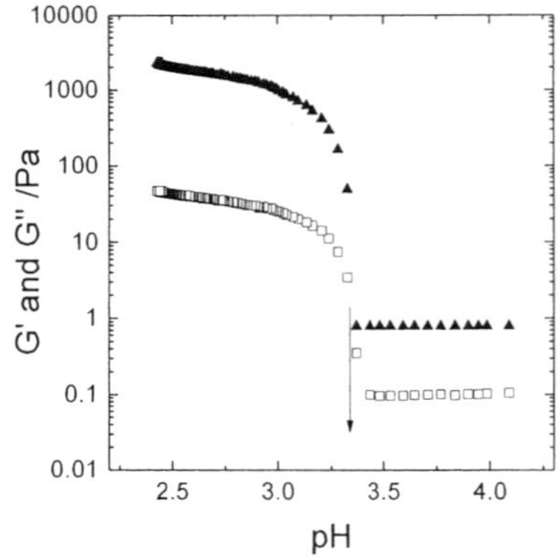

Figure 5: pH of gelation of 1% alginate / HM pectin mixture (in ratio 60:40) where filled triangles are G' and open squares G'' at f = 1Hz, strain = 0.5% and T= 37°C.

This is slightly lower compared to the pH of gelation of a calcium alginate gel, (which would start to gel upon solubilisation of the calcium salt – typically as pH goes below 6.0). This may be a disadvantage when applied as a gelation system to be activated in the stomach, where pH varies depending on whether or not the subject is in the fasting state.

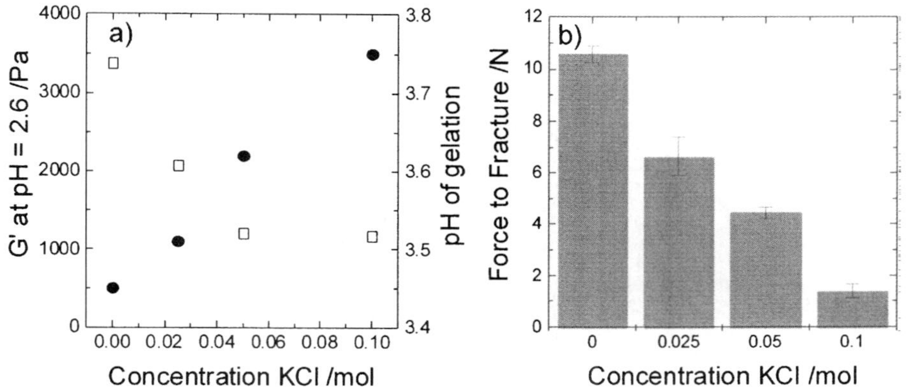

Figure 6: Influence of KCl concentration on a) the acid gel strength (open squares) of 1% of HM pectin / alginate gels and the pH of gelation (solid circles) measured at f=1Hz, strain = 0.5% and b) force to fracture 1% acid HM pectin / alginate gels at T=20°C

The presence of monovalent ions and Mg ions (the presence of Mg does not induce gelation of the alginate[22]) did not seem to be specific but related to ionic strength and screening of electrostatic interactions which could explain the increase in pH of gelation. However, to explain the reduction in gel strength is more difficult as the network mechanism for this system is not yet completely understood.

3.2 *In-vivo* study of gelation in stomach

In order to verify that the alginate / HM pectin system forms a gel in stomach, fruit and vegetable drinks with controlled mono- and di-valent ions, (containing 0.4% HM pectin and 0.6% alginate) were prepared for magnetic resonance imaging (MRI), *in – vivo*, in a similar fashion as previously shown for alginate in dairy drinks[16].

These drinks were consumed in a volume of 100ml and 250ml by two subjects after fasting. The viscosity of the fruit and vegetable drink is shown in Figure 8 and the force to fracture the drink under acid conditions was ~24kPa . It was observed that the drink formed gel lumps in the stomach within ~ 5 minutes (the time it took for the subject to consume the drink and be placed under the scanner) (Figure 7b) independent of the sample volume (100 and 250ml). The gel lumps remain after 15 min (Figure 7c), and they are clearly visible after 500ml of pure water was consumed (~30 min after initial drink, Figure 7d). The consumption of water increases contrast and facilitates visualization of the gels.

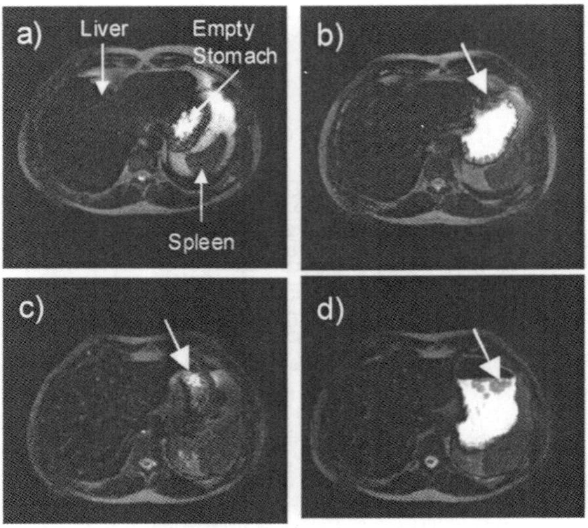

Figure 7: Magnetic resonance scan of a) an empty stomach, b) stomach upon consumption of 100ml of orange/carrot flavoured drink containing (in total) 1% alginate and HM pectin, ~ 5 min after consumption, c) ~15 minutes after consumption and d) after drinking an additional 500 ml of water, about 30 minutes after the consumption of the test sample. The white arrows indicate gel lumps.

This study was conducted on only two subjects, and the results are therefore only indicative. Nevertheless, it shows the ability of this system to form gels in the stomach despite the lower pH of gelation compared to the alginate / protein system, at least in a fasting stomach where pH is low. Furthermore, the gelation is rapid as gels are formed before a volume as low as 100 ml with low calorie content (20kcal) has time to empty from the stomach. Theoretically, these drinks should empty from the stomach with a rate of ~ 2-3kcal/min[32,33] giving a maximum residence time of the drinks of 7-10 minutes, whereas the MRI showed that the gels remained in the stomach for at least 15-20min (keeping in mind that only 2 subjects were tested). Another important point is that the size of the gelled particles being formed under gastric conditions is larger than that which could immediately be emptied from the stomach[34].

Given these initially positive results, the fruit alginate / HM pectin drinks were tested in a human intervention trial (see next section). A control, consisting of juice with xanthan was also used in the human study. The viscosity of the control (juice with xanthan) matches closely the drink containing the alginate / HM pectin (Figure 8). The force to fracture of the acid gel of the juice containing alginate and HM pectin is 24+/-3kPa whereas the xanthan containing drink did not form a gel.

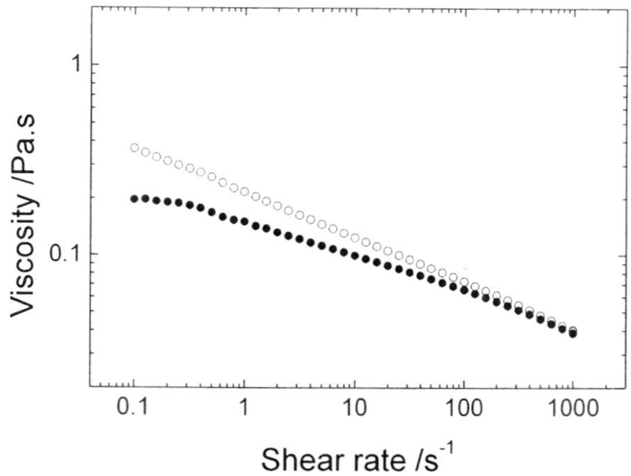

Figure 8: The viscosity as a function of shear rate for a juice containing xanthan (solid symbols) and alginate / HM pectin (open symbols) at T=20°C.

3.3 Human studies: satiety measurements

Three studies were performed. In the first one a 325 ml chocolate flavoured dairy drink with either 0, 0.6 and 0.8% alginate was tested. In the second and third one, 100 and 125 ml shots were tested with and without protein when taken shortly prior lunch.

Addition of alginate to a whey protein drink, consumed in a volume of 325 ml at breakfast gave a significant (more than 20%) enhancement of hunger control with 0.8% alginate, when represented as AUC (see Figure 9). Effects were also significant for most of the appetite scores with 0.6% alginate and a clear dose-response effect has been observed. This study shows the potential of using alginate as hunger control at concentrations lower than 1%, with whey as the protein source, and with reduced viscosity of the drinks upon consumption compared to previous studies[16,18]. Furthermore, this study was with a single component shelf-stable product containing the alginate in contrast to previous studies, where alginate has been either added to the protein drink prior consumption[16], or pre-gelation of the product was avoided by the use of a two component product[35,36], where the alginate and calcium sources have been separated and mixed prior to consumption, and where GDL has been added to boost acidification and gelation in stomach[36].

In contrast to the study where 325 ml was used, there was no effect on appetite feelings or *ad libitum* meal intake after the consumption of 100ml of low nutrient (20kcal) drink (juice containing alginate / HM pectin) compared to the juice without the fibers when consumed as a preload. Neither did the consumption of 125 ml of the whey dairy mini drink with added protein (3%) and 1% alginate (60kcal). These studies show that alginate-based drinks in lower volumes were not successful in enhancing the satiety or influencing *ad libitum* meal intake, despite the fact that the force to fracture of the acid gel (7N) is comparable to the alginate / whey protein system tested before.

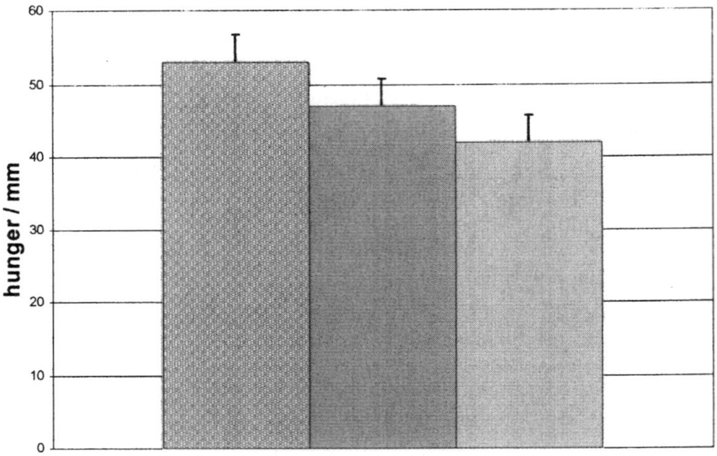

Figure 9: Area-under-curve of hunger (mean +/- SE) for 325 ml whey meal replacement drinks with 0, 0.6 and 0.8% alginate (** P< 0.001 vs. control).

Assuming that the change in study design (sample taken as preload or for breakfast) has not affected the study outcome, one can hypothesize that the lack of an effect on satiety for the two preload studies is related to either the low energy content/absence of protein in the juice drink or the low volume, at which they were consumed. The absence of protein and/or low energy content would reduce any effect related to the alteration of nutrient absorption due to the presence of gel (reduced gastric emptying rate and entrapment of nutrients in the gel) believed to be important for prolonged satiety[16]. However, the sample with the presence of protein did not show an effect either.

The use of an inflated gastric balloon has shown that volume is an important determination of satiation and satiety[37], but it has also shown that simultaneous nutrient delivery strengthens satiety synergistically[38]. The literature is more uncertain when it comes to studies carried out on gelling biopolymers taken as preload. A two compartment drink (237ml fruit flavored alginate pectin and 118ml fruit flavored calcium with low energy content (40kcal)) taken as a preload during one week showed no meaningful change in subjective measures of appetite, but a reduction in energy intake in a sub-population[35]. Unfortunately, no information is given on the viscosity of the samples used in this study or the strength of the acid gels. Another recent study showed reduction in energy intake upon consumption (during a week) of an alginate containing drink of 100ml (27 kcal). The drink was prepared from a dry powder prior to consumption. The subjects added 100ml of water to the powder, the mixture was shaken and consumed at once. GDL was present in the dry powder to facilitate the formation of a more homogeneous gastric gel compared to the design used in our studies, which implied gelation upon acidification by the gastric juice. This drink contained more alginate (1.5%) than tested in our study and the acid gel strength was considerably higher (30N versus ours of 2.7 and 7N). Overall this may indicate that drinks of low volumes, taken as a preload, may have an effect on meal intake if they form a gel with the sufficient strength under gastric conditions. If this is true, it indicates that the gel strength at lower volume should be higher than the gel strength at higher volumes, where, based on our work, we believe that systems forming gel strengths of >16kPa or >1.2N (measured as above) is sufficient to induce satiety upon consumption of a 325ml of protein-containing drink.

4 CONCLUSION

Our results show that gastric gelation can be a useful tool for appetite control. However, there is still a need to better understand the relationship between mechanical strength of the gel in the stomach and the consumed volume, and the impact of this relationship on the satiety enhancement. Once that is understood, further investigation is needed to understand the possible impact of nutrients in these drinks on their efficacy to provide satiety via gelling biopolymers. The study on the impact of nutrients is also necessary to ensure a shelf stable product with acceptable viscosity and sufficiently strong gel strength under gastric conditions.

It is difficult to compare outcomes of different clinical trials reported in the literature due to the differences in study design, but also due to insufficient data given on the material sourcing and mechanical properties of the samples used.

ACKNOWLEDGEMENTS

Ewoud Schuring is kindly acknowledged for statistical analysis of the data. Dave Mela is kindly acknowledged for input on the systems and design of human trials, Tim Foster, Robert Farr and Stephan Schumm for useful discussions on the physico-chemical properties of the systems (and Robert Farr also for proof-reading). We are thankful to Edward Haddeman and his team for helping to carry out the human trials. Yapeng Fang is acknowledged for excellent work on the HM pectin / alginate system, the preparation of study products by Khalid Farooq and Vanessa Bristow is highly appreciated. Caroline Hoad and Luca Marciani are acknowledged for carrying out the MRI scans.

References

1. B. Burton-Freeman, *J Nutr*, 2000, **130**, 272S
2. L.M. Morgan, B.A. Griffin, D.J. Millward, A. Delooy, K.R. Fox, S. Baic, M.P. Bonham, J.M.W. Wallace, I. MacDonald, M.A Taylor and H. Truby, *Public Health Nutr,* 2009, **12**, 799.
3. H. Truby, S. Bais, A. deLooy, K.R. Fox, M.B.E. Livingstone, C.M. Logan, I.A. Macdonald, L.M. Morgan, M.A. Taylor and D.J. Millward, *Brit Med J*, 2006, **332**, 1309.
4. H. Truby, *B Med J*, 2006, **332**, 1418.

5. L.G. Womble, D.A. Williamson, F.L. Greenway and S.M. Redmann, International Journal of Obesity, 2001, **25**,340.
6. H.P.F. Peters and D.J. Mela, Appetite and Food Intake: Behavioral and Physiological Considerations. Eds. R.B.S. Harris and R. Mattes, CRC Press, Taylor & Francis Group 2008.
7. D. S. Ludwig, M.A. Pereira, H. Candyce, J.E. Hilner, L. Horn, M.L. Slattery and D.R. Jacobs, *JAMA*, 1999, **282**, 1539.
8. J.L. Slavin, *Nutrition*, 2005, **21**, 411.
9. J.W. Anderson, P. Baird, R.H. Davis, S. Ferreri, M Knudtson, A. Koraym, V. waters and C.L. Williams, *Nutr Rev*, 2009, **67**, 188.
10. L. Benini, G. Castellani, F. Brighenti, K.W. Heaton, M.T. Brentegani, M. C. Casiraghi, C. Sembenini, N. Pellegrini, A. Fioretta, G. Minniti, M. Porrini, G. Testolin and I. Vantini, *Gut*, 1995, **36**, 825.
11. S.A. Moorhead, R.W. Welch, M. Barbara, E. Livingstone and A. Dunne, *Brit J Nutr* ,2006, **96**, 587.
12. N.C. Howarth, E. Saltzman and S.B. Roberts, *Nutr Rev*, 2001, **59**,129.
13. R.D. Mattes, *Physiol & Behav*, 2007, **90**, 705.
14. K.R. Juvonen, A.K. Purhonen, M. Salmenkallio-Marttila, L. Lahteenmaki, D.E. Laaksonen, K.H. Herzig, M.I.J Uusitupa, K.S. Poutanen and L.J. Karhunen, *J Nutr*, 2009, **139**, 461.
15. A.L. Jenkins, V. Vuksan, S. Panahi, P. Breitman , E. Jovanovski, M. Lyon, A.L. Rogovik, and L.A. Leiter, *Diabetica*, 2008, 51, S385.
16. C.L. Hoad, P. Rayment, R.C. Spiller, L. Marciani, C. Traynoer, D.J. Mela, H.P.F Peters and P.A. Gowland, *J Nutr*, 2004, **134,** 2293.
17. C.L. Dikeman, M.R. Murphy and G.C Fahey, *J Nutr*, 2006, **136**, 913.
18. K. Appleton, J. Hill, E. Haddeman, P.J. Rogers and D.J. Mela, *Proc. Nutr. Soc.*, 2004, **63,** 118A.
19. L. Marciani, P.A. Gowland, A.J. Fillery-Travis, P. Manoj, J. Wright, A. Smith, P. Young, R.J. Moore and R.C. Spiller, *Am J Physiol Gastrointest Liver Physiol,* 2001, **280,** 844.
20. V. Morris and G. Chilvers, *J Sci Food Agr,* 1984, **35,** 1370.
21. P. Walkenström, S. Kidman, A.-M. Hermansson, P.B. Rasmussen and L. Hoegh, *Food Hydrocolloid*, 2003, 17, 593.
22. K.I. Draget, Handbook of Hydrocolloids, Eds G.O. Phillips and P.A. Williams, CRC Press, Woodhead Publishing Limited, 2000, 379.
23. K.I. Draget, G. Skjåk-Bræk and O. Smidsrød, *Carbohydr Polym*, 1994, **25**, 31.
24. K.I. Draget, G. Skjåk-Bræk and O. Smidsrød, *Int J Biol Macromol*, 1997, **21**, 47.
25. M. Yanes, L. Durán and E. Costell, *Food Hydrocolloid*, 2002, **16**, 605.
26. B. Doumèche, M. Küppers, S. Stape, B. Blümich, W. Hartmeier and M.B. Ansorge-Schumacher, *J Microencapsulation*, 2004, **21**, 565.
27. L. Matia-Merino, K. Lau and E. Dickinson, *Food Hydrocolloid,* 2004, **18**, 271.
28. E. R. Morris, D.A. Rees, D. Thom and J. Boyd, *Carbohyd Res*, 1978, **66,** 145.
29. M.J. Gidley, E.R. Morris, D.A. Powell and D.A. Rees, *Int J Biol Macromol,* 1980, **2**, 332.
30. L. Marciani, P.A. gowland, R.C. Spiller, P. Manoj, R.J. Moore, P. Young and A.J. Fillery –Travis, *AJP-GI*, 2001, **280, 1227.**

31. S. Neiser and K.I. Draget, O. Smidsrød, *Food Hydrocolloid*, 1999, **13**, 445.
32. J.N. Hunt and M.T. Knox, *J Physiol* 1968, **194,** 327.
33. J.N. Hunt, J.L. Smith and C.L. Jiang, *Gastroenterology,* 1985, **89,** 1326.
34. P.O. Stotzer, G. Ringstrom, R. Saik and H. Abrahamsson, *Gastroenterology* 2002, **122**, A554.
35. C.L. Pelkman, J.L. Navia, A.E. Miller and R.J. Pohle, *Am J Clin Nutr,* 2007, **86**, 1595.
36. J.R. Paxman, J.C. Richardson, P.W. Dettmar and B. M. Corfe, *Appetite*, 2008, **51,** 713.
37. A. Geliebter, *Physiol & Behav,* 1988, **44**, 665.
38. S. Oesch, C. Ruegg, B. Fischer, L. Degen and C. Beglinger, *Physiol & Behav*, 2006, **87**, 903.

THE IMPACT OF PREBIOTIC FRUCTANS ON HEALTH WITH A FOCUS ON APPETITE AND BODY WEIGHT

H. Alexiou

BENEO-Orafti, Aandorenstraat 1, 3300 Tienen, Belgium

ABSTRACT

Inulin and oligofructose are the best- documented prebiotics. By interacting with the gut ecosystem, inulin and oligofructose influence a number of physiological functions, and the health benefits of these ingredients were demonstrated to extend outside the strict environment of the gut. In fact, much attention has recently been focused on their potential health benefits towards weight management. Several animal studies have shown that inulin and oligofructose are able to modulate the expression (and release) of gastrointestinal hormones involved in appetite regulation, with a concomitant lower body weight and adipose tissue gain. In parallel, since recent research has suggested a role for the gut microflora in energy homeostasis, experimental data have investigated the potential role of prebiotics in this context. Ingestion of a high-fat diet by mice was associated with changes in the microflora composition (e.g. decreased bifidobacteria), higher circulating LPS levels, and hence, was responsible for inducing an inflammatory state, as well as weight gain. In this model, the addition of oligofructose suppressed most of the metabolic alterations caused by the high-fat diet, and these effects were linked to a restoration of bifidobacteria levels. In humans, effects of prebiotics on appetite and energy intake were shown in healthy volunteers. Oligofructose supplementation resulted in lower daily energy intake together with various influences on appetite ratings. Adolescents receiving oligofructose-enriched inulin (Orafti®Synergy1) during 1 year had a lower body weight and body fat mass, as well as an appropriate body mass index (BMI) compared to the control group. Ultimately, data from a recent study demonstrate the ability of oligofructose supplementation to decrease body weight and energy intake in overweight and obese subjects. All together, these findings highlight exciting perspectives for the use of inulin and oligofructose as functional food ingredients in foods designed for weight management.

Key words: oligofructose, inulin, body weight, energy intake, microbiota.

1 THE CONTEXT: OBESITY AND DIETARY MODULATION OF ENERGY INTAKE

Obesity is today recognised as a disease, that is affecting an important and growing number of people worldwide.[1] Overweight and obesity are associated with a number of comorbidities, including metabolic disorders such as cardiovascular disease, type 2 diabetes or hypertension.[2] Causal factors of overweight and obesity are multiple and their interaction is complex. We can see those as the result of interactions between human biology and environmental factors, being involved in the complex process of regulation of energy balance in order to maintain body weight. Central to the problem is however today's society characterised by excessive energy intake and by minimal energy expenditure, the so-called "obesogenic environment". Rapid industrialization, increased living standards and enhanced availability of foods are amongst others, causal factors of this "obesogenic environment". In this context, multiple strategies are developed, notably by the food industry, to help attenuating energy intake through dietary means. These ways include (but are not restricted to) the formulation of low energy density foods or foods able to modulate appetite control.

In this respect, dietary fibres are undoubtedly essential compounds of our diet, contributing to our health by many aspects. Dietary fibres have a low caloric value, and a high intake of dietary fibres contributes to lower the energy density of foods.[3] Certain dietary fibres are able to increase bulk and viscosity, thus influencing satiation and satiety.[4] In observational studies, the consumption of dietary fibres is associated with a lower weight gain.[5] The physiological effects of several fibres are however dependent on their specific physical and chemical properties.[6]

The aim of the present review is therefore to focus on specific, prebiotic dietary fibres, i.e. inulin and oligofructose.

2 INTRODUCTION TO THE HEALTH BENEFITS OF INULIN AND OLIGOFRUCTOSE

Inulin and oligofructose are widely known and described for their prebiotic properties. Their ingestion indeed beneficially affects host health by selectively stimulating the growth and/or activity of one or a limited number of bacteria in the colon, mainly bifidobacteria and lactobacilli.[7] Inulin and oligofructose are naturally present in many plants and edible fruits and vegetables.[8] Native chicory inulin is produced on an industrial scale by extraction from the root of the chicory plant. It is a polydisperse ß(2-1) fructan composed of a mixture of oligo- and polysaccharides that are linear chains of fructose. The degree of polymerisation (DP) in native inulin ranges from 2 to about 65, with an average DP of 10. Inulin can be partially hydrolysed by means of an endoinulinase, to produce oligofructose in which the DP varies from 2 to 8 (average DP= 4).

Because of their specific configuration, made of β(2-1) bonds, inulin and oligofructose resist hydrolysis and absorption in the human upper gastrointestinal tract. Inulin and oligofructose are as a matter of fact classified as "non-digestible carbohydrates" and are recognised as dietary fibres in most countries worldwide. They are completely fermented in the lower part of the gut, mainly, if not exclusively in the colon. The colon harbours a large population of microorganisms, composed of over 1000 different species, which represent around 50% of the dry solids content of the large intestine. The gut microbiota is involved in several metabolic functions of the host, such as the metabolism of xenobiotics, but also of numerous non-digestible substrates in the colon, thus supporting the growth of colonic cells, through the production of short-chain fatty acids.[9,10] This microbial ecosystem thus plays a fundamental role in maintaining good health and imbalances in the composition of the gut microflora have been linked to the aetiology of various diseases. The selective fermentation of fructans by bifidobacteria (and possibly by a few other genera) is a central event in their health-benefits. In healthy volunteers, is has been shown that addition of inulin and oligofructose to their diet selectively stimulates bifidobacteria, while in parallel, a number of potentially harmful species are reduced. This has been repeatedly demonstrated in well-controlled intervention studies, all pointing out to a selective and significant increase in bifidobacteria after the intake of inulin and oligofructose.[11-17]

Inulin and oligofructose can thus be viewed as valuable colonic nutrients, of which the fermentation by the colonic microflora raises the production of metabolic compounds, i.e. short-chain fatty acids (SCFA, mainly propionate, butyrate and acetate), with a potential influence on metabolic events initiated in the gut, but also outside the gastrointestinal tract. Butyrate for instance, is largely accepted for its role in the growth and maintenance of healthy epithelial cells. Inulin-type fructans have hence been shown to modulate indirectly physiological effects such as mineral absorption, by enhancing calcium absorption.[18-21]

Recent research further highlights the role of prebiotic inulin and oligofructose in influencing enteroendocrine activity in the colon.

3 IMPACT ON GUT HORMONES, AND MODULATION OF WEIGHT IN ANIMAL MODELS

In a number of animal studies, the intake of fructans has been associated with modulation of the secretion of intestinal hormones, as a presumable consequence of the fermentation of these compounds in the colon. The hormones studied are secreted by enteroendocrine cells present in the gut, and are part of the very complex processes that regulate appetite and food intake. In these studies, the inclusion of inulin-type fructans (generally at a level of 10% w/w) in the diet of rats modified energy intake and body weight.

Wistar rats receiving a standard diet supplemented with fructans of different degrees of polymerisation (long-chain inulin, oligofructose or Synergy1, a patented oligofructose-enriched inulin) have a significantly lower energy intake compared to a control group. This resulted in a lower (epidydimal) fat mass accumulation (30% lower) in the rats whose diet was supplemented with oligofructose or oligofructose-enriched inulin. These effects occurred together with an increased concentration of glucagon-like peptide-1 (GLP-1) and proglucagon mRNA in the proximal colon, as well as raised GLP-1 in the portal vein. In parallel, the plasmatic levels of ghrelin, an orexigenic hormone that normally rises in states of food deprivation, remained significantly lower in the serum of rats treated with oligofructose or oligofructose-enriched inulin.[22] The concentration of another peptide, PYY (PP-fold peptide) was unchanged in the colon, but its caecal pool was increased, due to the enlargement of the caecal tissue.[23] GLP-1, as well as PP-fold peptide (PYY) are appetite-suppressing peptides.

These findings were confirmed in rat models of obesity. The addition of 10% (w/w) Synergy1 or oligofructose to the diet of Zucker <u>*fa/fa*</u> rat (genetic model of obesity) decreased their energy intake, and resulted in a lower body weight and lower (epididymal) fat mass development compared to the control group. The intake of fructans was associated with a decrease in the triglyceride content of the liver.[24,25] This effect is likely to result from a decrease in the hepatic lipogenesis, as the hypolipidemic effect of fructans was observed by others, too.[26-28] These effects were not observed when non-fermentable fibres (e.g. cellulose) were added to the diet, highlighting the importance of fermentation in the protective effects of fructans. Models of diet-induced obesity were also studied, in order to mimic the unbalanced, rich in fat-diets, characteristic of our Western society. Thus in high-fat fed rats, oligofructose intake decreased energy intake, and lowered body weight gain and fat mass development, compared to control rats.[29] Taken together, animal models consistently showed lower energy intake and lower body weight and fat mass gain compared to controls.

The observed effects on body weight in animals are persistent, as shown by a lifelong intervention study in rats where Synergy1 supplementation resulted in a lower body weight evolution during the whole life-span of healthy animals, and at the same time prolonged their survival rate at 24 months.[30]

4 MECHANISMS OF ACTION AND ROLE FOR GUT PEPTIDES?

Observations in animals were associated with modulations of gut peptides, mainly incretins, involved in appetite regulation. GLP-1 is a glucoincretin, i.e. a peptide secreted by the gut in response to nutrient intake. GLP-1 is produced in the intestinal epithelial endocrine L-cells, mainly present in the ileum and the colon, through a posttranslational processing of proglucagon, synthesized in these cells.[31] Higher circulating levels of this hormone are associated with lower subjective hunger feelings and lower food intake. Obese subjects have been reported to have reduced postprandial GLP-1 responses compared with lean subjects.[32]

Cani et al.[33] observed that oligofructose consumption by rats during 4 weeks resulted in an increased number of L-cells in the proximal colon, which was almost doubled. L-cells are produced from stem cells present in the intestine and that differentiate themselves into mature L-cells, under the action of specific differentiation factors. Oligofructose increased the expression of these differentiation factors (neurogenin 3 and NeuroD) in the proximal colon of rats. These data show that oligofructose is able to influence the pathway of L-cells differentiation, leading to an enhanced number of L-cells able to produce GLP-1.

In an attempt to explain the consistent link observed between the effects of fructans on gut hormones, and their effects on weight modulation and glucose metabolism, a study using GLP-1 receptor antagonist was setup. Mice receiving a high-fat diet exhibited lower food intake and body weight gain, as well as anti-diabetic effects. The treatment of the mice with the GLP-1 receptor antagonist, on the contrary, prevented the beneficial effects seen with oligofructose. The importance of GLP-1 in mediating the effects of oligofructose was also confirmed through the use of GLP-1 receptor knock-out mice (GLP-1R -/-) that were insensitive to the effects of oligofructose on body weight, food intake and parameters of glucose metabolism.[34]

5 OBESITY, MICROFLORA AND INFLAMMATION

There is today accumulating evidence showing that the endogenous gut microbial flora is involved in a number of physiological functions of the host, through host-microbe interactions.[35] The microflora has been described as an "organ" in itself, dedicated to the performance of metabolic functions the host is not able to carry.[36] Recent findings have also revealed that the microflora has an important role to play in several pathological conditions, such as inflammatory bowel diseases,[37] but also diabetes and obesity.[38] Obesity is a condition resulting from a persistent alteration in energy balance, and our microflora has recently been suggested to play a role in the host's energy metabolism. Bäckhed et al.[39] have shown that colonization of adult germ-free mice with a normal microflora (from distal gut) results in a dramatic increase of their adipose tissue (+ 60%) within 2 weeks, even though food intake was lower. Insulin resistance was also observed. Lean germ-free mice inoculated with the flora of genetically obese mice (*ob/ob*) had a greater tendency to gain total body fat in comparison with lean germ-free mice transferred with the flora of lean ones, thus suggesting that certain bacterial species are associated with a greater risk of adiposity accumulation.[40] In mice, as well as in humans, the qualitative composition of the gut microbiota has been shown to differ between lean and obese individuals.[40,41] When obese subjects lost weight after a diet, their flora resembled that of the lean individuals.[41] These findings unveil a potential cause-and-effect relationship between obesity and the gut microbiota, even though it still remains to be determined whether these compositional changes are a cause or a consequence in obesity.[42]

An important characteristic of obesity and insulin resistance is the subclinical low grade chronic systemic inflammation that has been described.[43,44] Recent studies have suggested an involvement of factors intervening in the composition of the gut microflora, and more specifically of bacterial lipopolysaccharide (LPS) (derived from gram-negative bacteria present in the gut), as trigger factor linking inflammation to the metabolic alterations provoked by obesity.[45] LPS- derived endotoxin is well-known for inducing the production of cytokines involved in the initiation of an inflammatory response by the host.

In a model of high-fat fed mice, a higher proportion of LPS-containing bacteria in the gut and a higher LPS concentration in the plasma were evidenced.[45] Since LPS is a component of gram-negative bacteria, it was explored whether modifications of the gut microbiota composition was able to modulate this state of metabolic endotoxaemia. Mice models of obesity (high-fat or *ob/ob*) receiving antibiotics were subjected to reduced metabolic endotoxaemia, reduced caecal LPS and inflammation as well as lower body and fat mass gain.[46] Recent human data have furthermore shown that type 2 diabetic subjects display higher circulating levels of LPS,[47] which partly mediates a state of sub-clinical inflammation.[48] These findings suggest that modulations of the gastrointestinal microbiota could be involved in obesity and chronic inflammation.

In another experience involving mice fed different types of diets (including standard, high-fat or high-fat supplemented with oligofructose), the high-fat diet increased endotoxaemia, but also modified the microbiota composition. The levels of *Bifidobacterium spp*. were considerably decreased, as well as those of other dominant groups, i.e. *E. rectale-C.coccoides* and *Bacteroides* MIB. Oligofructose supplementation increased bifidobacteria levels, and in addition, lowered the level of inflammatory markers, such as plasma endotoxin, IL-1 and IL-6. In mice, the oral intake of bifidobacteria was previously shown to decrease the rise in endotoxin levels, compared to a control group.[49] Consistent with previous findings, mice in the oligofructose group had a lower body weight gain and fat mass development.[50] These data suggest that the gut microbiota composition participates to the patho-physiological regulation of endotoxaemia, and sets the tone of inflammation and the occurrence of metabolic alterations consecutive to obesity.

To further understand the impact of LPS in high-fat diet-induced inflammatory events, the likelihood of altered intestinal permeability was tested. Indeed, genetically obese mice (*ob/ob* or *db/db*) have been shown to have an increased intestinal permeability, resulting in increased portal endotoxaemia.[51] In a study involving *ob/ob* mice, oligofructose supplementation decreased the plasmatic levels of LPS and cytokines, and in addition, significantly increased the mRNA level of proteins involved in gut permeability (ZO-1 and occludin) and decreased intestinal permeability. As previously shown, *ob/ob* mice receiving the prebiotic had a significantly lower energy intake, enhanced plasmatic GLP-1 and GLP-2 concentrations and reduced adiposity. GLP-2 is a gut peptide co-secreted with GLP-1 and is involved in the regulation of mucosal epithelial integrity.[52] The administration of a GLP-2 antagonist inhibited the beneficial effects of oligofructose towards inflammation, endotoxaemia and gut permeability, thus linking increased endogenous GLP-2 production following the prebiotic intake to its effects on reduced endotoxaemia and inflammation.[53]

These series of well designed experiments suggest a link between the bifidogenic effect of inulin and oligofructose, and the decreased gut permeability and inflammatory status, with potential implications on obesity and metabolic diseases.

6 HUMAN STUDIES

The effects of oligofructose on appetite and energy intake were initially investigated in a placebo-controlled intervention study (cross-over design) including 10 healthy volunteers. The normoweight volunteers received 8 grams of oligofructose twice daily (at breakfast and dinner) during 2 weeks. Appetite scores (satiety, hunger, fullness) were recorded after meals using VAS (visual analogue scale). These parameters, as well as food intake were assessed after ad-libitum free-choice buffets. Oligofructose intake decreased the energy intake at breakfast and lunch ($p<0.05$), as well as the total energy intake during the day by 5% ($p=0.05$), as compared to the placebo treatment. This was associated with various influences on appetite ratings.[54]

The ability to modulate weight itself can be seen as the ultimate aim of dietary interventions. In this context, first insights coming from a trial in adolescents have demonstrated the ability of fructans to contribute to weight modulation. Data of a 1-year randomised, double-blinded intervention trial in adolescents (n=100) showed that the administration of an oligofructose-enriched inulin (Synergy1) (8 g/day) resulted in a significantly lower increment in body mass index (BMI) compared to the control group (maltodextrin) (Figure 1)..In the prebiotic group, no significant modification in the BMI Z-score occurred, while the BMI Z-score of adolescents in the control group increased significantly. In addition, body weight and body fat mass were significantly lower ($p<0.05$) compared with the placebo group, thus supporting appropriate development of body weight during early adolescence with the prebiotic.[55]

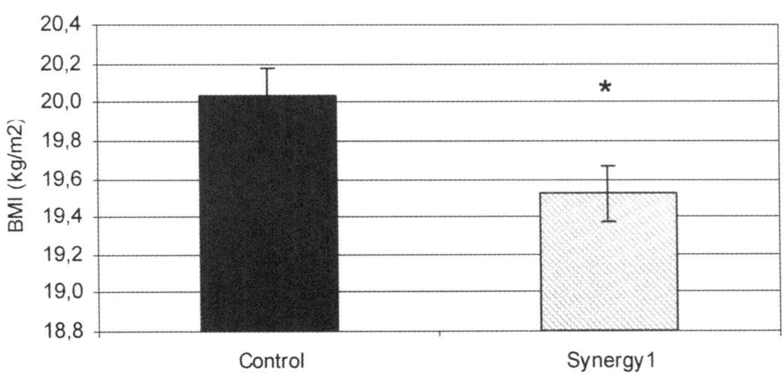

Figure 1 *Effect of 1-year supplementation with Synergy1 (8 g/day) or placebo on body mass index (BMI) (mean ± SEM) in adolescents. * p<0.05 (Adapted from 55).*

Confirmation of the ability of inulin-type fructans to modulate body weight recently came from a randomised, double-blind, placebo-controlled trial, involving 48 overweight or obese healthy adults. Subjects receiving oligofructose (21 g/day) experienced a significant reduction in body weight of 1.03 kg, while the control subjects gained 0.45 kg weight, after 12 weeks of intervention ($p<0.05$) (Figure 2). The weight loss affected mainly body fat mass, in particular trunk fat mass ($p<0.05$ versus control). The weight loss could be explained by the lower energy intake observed in subjects in the oligofructose group. The subjects did not modify their physical activity or dietary habits, thus attributing the observed results concerning body weight, fat mass and caloric intake to the oligofructose supplementation. An analysis of the blood levels of hormones involved in appetite regulation further revealed lower postprandial ghrelin and higher PYY responses in the subjects consuming oligofructose. Modulations in GLP-1 concentrations could not be detected.[56]

These findings highlight the important consequences food ingredients that influence the gut-brain axis of appetite control might have on weight modulation.

7 FORMULATION OF LOW ENERGY DENSITY FOODS

In addition to the aforementioned results, as low-digestible carbohydrates fermented in the colon, inulin and oligofructose display a low caloric value of 1.5 kcal/g,[57] which is lower than fully digestible carbohydrates or lipids. Due to their technological properties, inulin and oligofructose are easily used in food formulations to substitute sugar or fat. Inulin is able to substitute fat due to its ability to stabilise water into a particle gel network with a fat-like, creamy texture. Such inulin application gives the same mouthfeel as fat. Oligofructose is slightly sweet (about 35% sweetness compared to sucrose), which allows for natural sugar replacement.[58] Inulin and oligofructose thus allow formulating low energy dense foods, with a better-balanced nutritional composition.

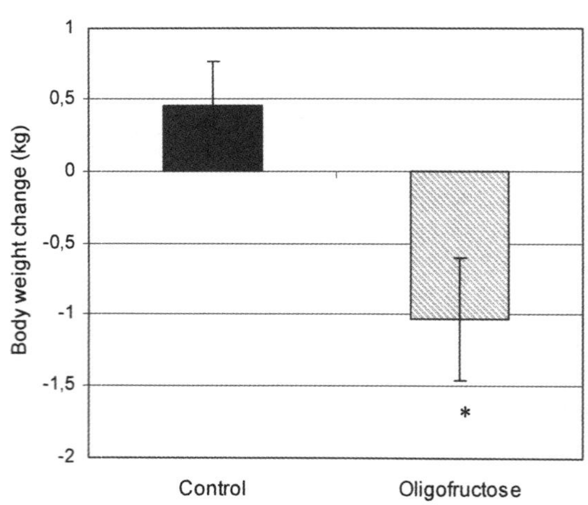

Figure 2 *Effect of 12 weeks supplementation with oligofrucose (21 g/day) or placebo on body weight modulation (mean ± SEM) in overweight and obese subjects. * p<0.05 (Adapted from 56).*

8 CONCLUSIONS AND PERSPECTIVES

Inulin and oligofructose are non-digestible carbohydrates, for which a sound research has today established a number of nutritional properties, such as their well-demonstrated prebiotic properties, overall contributing to improved gut health. Besides their bifidogenic effect, inulin and oligofructose play a major role in gastrointestinal physiology and evidence supporting the nutritional advantages of inulin-type fructans is rapidly accumulating. With obesity affecting an increasing proportion of the population, the recent involvement of inulin-type fructans on modulation of appetite and energy intake has unveiled new perspectives in weight management. Several animal studies have shown that fructans (and more specifically oligofructose and oligofructose-enriched inulin) are able to decrease body weight and body fat mass. In humans, first data support an effect on reduced energy intake, as well as on the prevention of undesirable weight gain during adolescence. Reduced body and fat mass weight, independent of any dietary intervention, was even lately shown in overweight and obese individuals. In addition, recent research highlights the importance of the gut microbiota composition as an under-appreciated target for the treatment of metabolic disorders, and thus its modulation through prebiotics appears as a potential new way to modulate weight and counteract diet-induced metabolic disease.
All together, these latest developments highlight the extraordinary potential of these prebiotic fibres as potential helping factors in the management of energy balance.

References

1 WHO. Obesity: preventing and managing the global epidemic: report of a WHO Consultation. Geneva, World Health Organization (WHO Technical Report Series No. 894), 2000.

2 WHO. Factsheet: Obesity and overweight, www.who.int/hpr/NPH/docs/gs_obesity.pdf, 2003.

3 J.L. Slavin, *Nutrition*, 2005, **21**, 411.

4 J.L. Slavin and H. Green, *Br. Nutr. Found.*, 2007, **32(Suppl I)**, 32.

5 R.M. Van Dam and J.C. Seidell, *Eur. J. Clin. Nutr.*, 2007, **61 (Suppl 1**), S75.

6 J.E. Blundell and V. J. Burley, *Int. J. Obes.*, 1987, 11 (Suppl 1), 9.

7 G.R. Gibson and M.B. Roberfroid, 1995, *J. Nutr.,* **125**, 1401.

8 J. Van Loo, P. Coussement, L. de Leenheer, H. Hoebregs and G. Smits, *Crit. Rev. Food Sci.,* 1995, **35**, 525.

9 J.H. Cummings and G.T. MacFarlane, *JPEN*, 1997, **21,** 357.

10 A. Belenguer, S.H. Duncan, G. Holtrop, H.J. Flint and and G.E. Lobley, *Curr. Nutr. Food. Sci.,* 2008, **4**, 109.

11 G.R. Gibson, E.R. Beatty, X. Wang and J.H. Cummings, *Gastroenterology*, 1995, **108,** 975.

12 E. Menne, N. Guggenbuhl and M. Roberfroid, *J. Nutr.*, 2000, **130**, 1197.

13 V.A. Rao, *Nutr. Res.,* 2001, **21,** 843.

14 K.M. Tuohy, R.K. Finlay, A.G. Wynne and G.R. Gibson, *Anaerobe*, 2001, **7,** 113.

15 H.J.M. Harmsen, G.C. Raangs, A.H. Franks, A.C.M. Wildeboer-Veloo and G.W. Welling, *Microb. Ecol. Health Dis.,* 2002, **14**, 211.

16 G.R. Gibson, H.M. Probert, J. Van Loo, R.A. Rastall and M.B. Roberfroid, *Nutr. Res. Rev.,* 2004**,** **17**, 259.

17 Y. Bouhnik, L. Raskine, K. Champion, C. Andrieux, S. Penven, H. Jacobs and G. Simoneau, *Nutr. Res.,* 2007**, 27,** 187.

18 I.J. Griffin, P.M. Davila and S.A. Abrams, *Br. J. Nutr.,* 2002, **87 (Suppl. 2)**, S187.

19 S.A. Abrams, I. J. Griffin, K.M. Hawthorne, L. Liang, S.K. Gunn, G. Darlington and K.J. Ellis, *Am. J. Clin. Nutr.*, 2005, **82**, 471.

20 K.E. Scholz-Ahrens and J. Schrezenmair, *J. Nutr.*, 2007, **137**, 2513S.

21 L. Holloway, S. Moynihan, S.A. Abrams, K. Kent, A.R. Hsu and A.L. Friedlander, *Br. J. Nutr.,* 2007, **97**, 365.

22 P.D. Cani, C. Dewever and N.M. Delzenne, *Br. J. Nutr.,* 2004, **92**, 521.

23 N.M. Delzenne, P.D. Cani, C. Daubioul and A.M. Neyrinck, *Br. J. Nutr.,* 2005, **93 (Suppl 1**), S157.

24 C.A. Daubioul, H.S. Taper, L.D. De Wispelaere and N.M. Delzenne, *J. Nutr.,* 2000, **130**, 1314.

25 C. Daubioul, N. Rousseau, R. Demeure, B. Gallez, H. Taper, B. Declerck and N.M. Delzenne, *J. Nutr.*, 2002, **132,** 967.

26 M. Fiordaliso, N. Kok, J.P. Desager, F. Goethals, D. Deboyser, M. Roberfroid and N. Delzenne, *Lipids*, 1995, **30**, 163.

27 N. Kok, M. Roberfroid and N. Delzenne, *Metabolism*, 1996, **45**, 1547.

28 E.A. Trautwein, D. Rieckhoff and H.F. Erbersdobler, *J. Nutr.*, 1998, **128**, 1937.

29 P.D. Cani, A.M. Neyrinck, N. Maton and N.M. Delzenne, *Obes. Res.,* 2005, **13**, 1000.

30 P. Rozan, A. Nejdi, S. Hidalgo, J-F. Bisson, D. Desor and M. Messaoudi, *Br. J. Nutr.,* 2008, **100**, 1192.

31 J.J. Holst, *Physiol. Rev.*, 2007, **87**, 1409.

32 C. Verdich, S. Toubro, B. Buemann, J.L. Madsen, J.J. Holst and A. Astrup, *Int. J. Obes.*, 2001, **25**, 1206.

33 P.D. Cani, S. Hoste, Y. Guiot and N.M. Delzenne, *Br. J. Nutr.*, 2007a, **98**, 32.

34 P.D. Cani, C. Knauf, M.A. Iglsesias, D.J. Drucker, N.M. Delzenne and R. Burcelin, *Diabetes*, 2006a, **55,** 1484.

35 L. Dethlefsen, M. McFall-Ngai and D.A. Relman, *Nature*, 2007**, 449,** 811.

36 P. Vaishampayan, M. Patole and Y.S. Shouche, *Curr. Sci.*, 2005, **88**, 25.

37 F. Guarner and J.R. Malagelada, *Lancet*, 2003, **361**, 512.

38 W. Jia, H. Li, L. Zhao and J.K. Nicholson, *Nat. Rev. Drug Discov.*, 2008, **7**,123.

39 F. Bäckhed, H. Ding, T. Wang, L.V. Hooper, G.Y. Koh, A. Nagy, C.F. Semenkovich and J.I. Gordon, *Proc. Natl. Acad. Sc.*, 2004, **101**, 15718.

40 P. Turnbaugh, R.E. Ley, M.A. Mahowald, V. Magrini, E.R. Mardis and J.I. Gordon, *Nature*, 2006, **444,** 21.

41 R.E. Ley, F. Bäckhed, P. Turnbaugh, C.A. Lozupone, R.D. Knight and J.I. Gordon, *Proc Natl Acad Sci,* 2005, **102**, 11070.

42 F. Bäckhed, *J. Pediatr. Gastroenterol. Nutr.*, 2009, **48 (Suppl 2),** S56.

43 A. Pradhan, *Nutr. Rev.*, 2007, **65,** S152.

44 K.E. Wellen and G.S. Hotamisligil, *J. Clin. Invest.*, 2005, **115,** 1111.

45 P.D. Cani, J. Amar, M.A. Iglsesias, M. Poggi, C. Knauf, D. Bastelica, A.M. Neyrinck, F. Fava, K.M. Tuohy, C. Chabo, A. Waget, E. Delmée, B. Cousin, T. Sulpice, B. Chamontin, J. Ferrières, J.F. Tanti, G.R. Gibson, L. Casteilla, N.M. Delzenne, M.C. Alessi and R. Burcelin, *Diabetes*, 2007b, **56,** 1761.

46 P.D. Cani, R. Bibiloni, C. Knauf, A. Waget, A.M. Neyrinck, N.M. Delzenne and R. Burcelin, *Diabetes*, 2008, **57,** 1470.

47 S.J. Creely, P.G. McTernan, C.M. Kusminski, M. Fisher, N.F. da Silva, M. Khanolkar, M. Evans, A.L. Harte and S. Kumar, *Am. J. Physiol. Endocrinol. Metab.*, 2007, **292**, E740.

48 O.S. Al-Attas, N.M. Al-Daghri, K. Al-Rubeaan, N.F. da Silva, S.L. Sabico, S. Kumar, P.G. McTernan and A.L. Harte, *Cardiovasc. Diabetol.,* 2009, **Apr 15,** 8:20, doi: 10.1186/1475-2840-8-20.

49 E.A. Griffiths, L.C. Duffy, F.L. Schanbacher, H. Qiao, D. Dryja, A. Leavens, J. Rossman, G. Rich, D. Dirienzo and P.L. Ogra, *Dig. Dis. Sci.,* 2004, **49**, 579.

50 P.D. Cani, A.M. Neyrinck, F. Fava, C. Knauf , R. Burcelin, K.M. Tuohy, G.R. Gibson and N.M. Delzenne, *Diabetologia*, 2007c, **50**, 2374.

51 P. Brun, I. Castagliuolo, V. Di Leo, A. Buda, M. Pinzani, G. Palù and D. Martines, *Am. J. Physiol. Gastrointest. Liver Physiol.,* 2007, **292,** G518.

52 J.L. Estall and D.J. Drucker, *Annu. Rev. Nutr.,* 2006, **26**, 391.

53 P.D. Cani, S. Possemiers, T. Van de Wiele, Y. Guiot, A. Everard, O. Rottier, L. Geurts, D. Naslain, A.M. Neyrinck, D.M. Lambert, G.G. Muccioli and N.M. Delzenne, *Gut*, 2009, **published online, Feb 24**, doi: 10.1136/gut.2008.165886.

54 P.D. Cani, E. Joly, Y. Horsmans and N.M. Delzenne, *Eur. J. Clin. Nutr.,* 2006b**, 60,** 567.

55 S.A. Abrams, I. J. Griffin, K.M. Hawthorne and K.J. Ellis, *J. Pediatr.,* 2007, **151**, 293.

56 J.A. Parnell and R.A. Reimer, *Am. J. Clin. Nutr.* , 2009, **89**, 1751.

57 M.B. Roberfroid, G.R. Gibson and N. Delzenne, *Nutr. Rev.,* 1993, **51**, 137.

58 A. Franck, *Br. J. Nutr.,* 2002, **87 (Suppl. 2)**, S287.

ENGINEERING INTERFACIAL STRUCTURES TO MODERATE SATIETY

N.C. Woodward, A.P. Gunning, P.J. Wilde, B-S. Chu & V.J. Morris

Institute of Food Research, Norwich Research Park, Colney, Norwich NR4 7UA, UK

1. INTRODUCTION

The prevalence of obesity has increased recently 3-fold. Europe-wide, obesity amongst the adult population is responsible for ~6% of health care costs, and this is likely to rise further given the trend of increasing obesity in children, and in adolescents as they enter adulthood. In the UK the estimated costs to the NHS of treating obesity-related conditions are £3.3-3.7 billion pa and this cost doubles when it includes the impact of treating patients who are simply overweight. A challenge is to moderate dietary intake of fats and a route to this goal is to manipulate the interfacial structure of emulsified fats in order to slow down the rate of fat digestion.

Within the GI tract there are a number of entero-endocrine feedback mechanisms that optimise the time required for digestion, absorption and systemic handling of nutrients. These feedback mechanisms act to prevent 'loss' of nutrients in the colon or wastage of energy in the extraction of nutrients from indigestible material. It is well-accepted that fat present in the small intestine can delay gastric emptying and increase the duodenal transit time. The sensations of lack of appetite and feelings of satiety ~4h after a meal, when the stomach has emptied, are mainly due to entero-endocrine responses to the presence of undigested food in the ileum. Infusion of undigested lipid in the jejunum has been found to delay chyme transfer through the ileocolonic junction[1]. Jejunal motility and transit time have been found to be inhibited by ileal lipid infusions[2]. There are a number of gut hormones that have been identified as playing a role in gastro-intestinal transit and brain responses. In particular the peptide YY (PYY) and cholecystokinin (CCK) have been identified with gastrointestinal handling of lipids. Secretion of CCK is stimulated by lipid present in the proximal ileum. It triggers contraction of the gall bladder thus providing the bile salts necessary for lipid digestion. PYY is mainly stimulated by the presence of lipid in the distal ileum[3]. The systemic infusion of PYY at physiological doses has been shown to suppress appetite markedly[4-5]. These data support the idea that moderating the rate of lipolysis in the small intestine provides a route to reduce appetite. The longer fat remains in the intestine the more satiety is promoted[6]: fat-induced satiety is intestinally mediated.

The majority of research on fat digestion is concerned with free fat which is emulsified in the stomach. However, for processed food emulsions it may be possible to control, in a rational way, the structure of protein-based interfacial structures, in order to moderate the adsorption of bile salts, thus controlling the level of adsorption of lipase-colipase complexes and the rate of lipolysis. The interfacial adsorption of bile salts in the duodenum facilitates lipolysis. The negative charge on the bile salts attracts colipase to the interface and the colipase anchors the lipase in a suitable configuration for lipolysis. The basis for controlling lipolysis through restricting colipase-lipase adsorption stems from novel observations on the nature of the surfactant displacement of proteins from interfacial structures. Visualisation by atomic force microscopy (AFM) of the structural changes occurring during displacement revealed a novel orogenic displacement mechanism[7-9]. The sampling of interfacial structures by Langmuir-Blodgett (LB) methods gave AFM images showing that, as a consequence of partial unfolding at interfaces, the proteins associate to form 'elastic' networks. Surfactants are generally more surface-active than proteins and, given sufficient time, at sufficiently high concentrations, they will eventually displace the protein. However, because the proteins are linked together it is not possible for the more surface-active surfactants to displace individual proteins. Thus it is necessary to explain how surfactants can displace proteins from interfaces. The AFM studies showed that the nature of the formation of the interfacial protein network results in defects which can be infiltrated by surfactants[7-8]. Having gained access to the interface the surfactants consolidate and expand their position generating domains of surfactant within the protein network. The growth of these domains compresses the protein network, causing it to fold and buckle, and eventually to break[7-8]. Once the protein network structure is broken it is possible for the surfactants to displace freed proteins, or protein aggregates from the interface into the bulk phase. The growth of the surfactant domains is opposed by the storage of elastic energy in the protein network. Recent studies have shown that the displacement of interfacial protein networks by bile salts under simulated *in vitro* duodenum conditions occurs through an orogenic process[10]. Thus, this article is concerned with the possibility of enhancing the mechanical strength of interfacial protein networks in order to limit the growth of bile salt domains, thus limiting the adsorption of colipase-lipase complexes at the interface, and hence reducing the rate of lipolysis in the duodenum: specifically the paper considers the effects of enzymatic crosslinking of interfacial proteins to limit surfactant adsorption at the interface and its effect on lipolysis.

2 METHODS

2.1 Proteins and surfactants

Sodium caseinate (DMV International) samples were supplied by Unilever, PLC. Samples of β-casein (C-6905, lot 12H9550), α-casein (C-6780, 041K7420) and κ-casein (C-0406, 051K7575), highly purified porcine pancreatic lipase (Type VI-S, activity ≥20,000 units/mg protein) and essentially salt-free porcine pancreas colipase were obtained from Sigma (Poole, UK). Protein solutions were prepared with surface pure water obtained from an Elga Elgastat UHQ water purification system (surface tension $\gamma = 72.6$ mN m^{-1} at 20°C, conductivity = 18.2 MΩ). Tween 20 (polyoxyethylene sorbitan monolaurate) was bought as a 10% solution from Pierce (Rockland Il, USA). Sodium glycodeoxycholate (NaGDC, 97% TLC), sodium taurocholate (NaTC, 95% TLC) and sodium dodecyl sulphate (SDS) from Sigma were stored at room temperature, and used without further purification.

2.2 Sampling of interfacial structures

Interfacial structures were prepared at an air-water interface using a PTFE Langmuir trough (surface area 0.05 m^2, volume 1 dm^3) equipped with a fixed and a movable barrier. The protein films were prepared by spreading them from a 2 mg ml^{-1} solution resulting in a surface coverage of 3 mg m^{-2} at the interface. Appropriate volumes of Tween 20 solutions, starting from an initial surfactant concentration of 0.5 μM, were added to the subphase to increase the surface pressure which was monitored by means of a ground glass Wilhelmy plate. At various, increasing surface pressures, the interfacial films were transferred onto freshly-cleaved mica using the LB technique. These samples were used to visualise the interfacial structures by AFM.

2.3 AFM imaging of interfacial structures

Imaging of transferred LB films was carried out using an East Coast Scientific AFM (ECS Ltd., Cambridge, UK). Images were collected in the dc (contact) mode, using Nanoprobe silicon nitride cantilevers (nominal force constant 0.38 N m^{-1}, Veeco Instruments Ltd). The measurements were made in a liquid cell, under redistilled *n*-butanol (Sigma).

2.4 Transglutaminase treatment of emulsions

Oil in water emulsions were prepared using olive oil (Sigma, Aldrich) as the oil phase. The olive oil was purified with Florisil resins (Sigma Aldrich): the oil was mixed with resin under gentle stirring for 3 hours and then separated by centrifugation at 12,000 rpm for 30 min. using a Beckman centrifuge, the supernatant was removed and filtered under vacuum. The purified oil was stored under argon, in the absence of light. Emulsion premixes were prepared by combining 1.25% w/w solution of sodium caseinate solution with olive oil in a blender cup in a 4:1 ratio. These mixtures were then pulsed in a Waring blender, 30 s low pulse, 30 s rest, 30 s high pulse, before passage 5x through an homogeniser at 20 psi. The final composition of the emulsion was 1% sodium caseinate, 20% olive, and the samples were allowed to stand for one hour. The size distribution and hence the surface area of the droplets was determined using a Coulter LS230.

Cross-linking of the sodium caseinate interfacial protein networks was achieved through the use of transglutaminase. The enzyme preparation used was a microbial transglutaminase ACTIVA-WM (Ajinomoto Europe) obtained from Vesantti Oy-Helsinki, Finland, which had been purified using the methods described by Lannto and co-workers[11]. The enzyme activity has been determined to be 11,000 nkat g^{-1} protein, using the method of Folk[12], and using 0.03M N-carbobenzoxy-L-glutaminylglysine as a substrate. Transglutaminase-treated emulsions were prepared by adding an appropriate concentration of the enzyme, mixing, placing the treated emulsion in an incubator at 40 °C, and shaking at ~150 rpm for ~20 hours. The transglutaminase was deactivated by immersion of the contained sample in a water bath at 90 °C for 5 min., followed by immersion in iced water for a further 5 min.

2.5 Protein displacement studies for emulsions

Displacement of sodium caseinate from emulsions by the neutral surfactant Tween 20 was monitored by measurements of surface charge/potential from laser electrophoresis studies and by protein loading assays using colourimetric methods. Emulsion-surfactant samples were prepared to cover a range of Tween 20 concentrations from 10^{-3} – 1% w/v. This was achieved by adding equal volumes of appropriate Tween 20 stock solutions to the emulsion and allowing the emulsion to stand for an hour before carrying out further analysis.

Surface potential measurements were made using a Nano DS equipped with a HeNe laser (633 nm) at 17° (Malvern Instruments): a concentration versus attenuation sweep was carried out on the emulsions to ensure an appropriate concentration was used for the measurements (0.005 % w/v). Measurements were made using a disposable folded capillary cell (DTS1060C-Malvern Instruments) in 10 mM phosphate buffer (pH 7). Typically 12 'sub runs' were carried out in triplicate. The electrophoretic mobility (U_E) was converted to a zeta potential (z) using Henry's equation (Equation 1) where ε and η, are the dielectric constant and viscosity of the medium respectively, and $f(Ka)$ is determined from the Smolochowski approximation.

$$U_E = \frac{2\varepsilon z f(Ka)}{3\eta} \quad (1)$$

The assay for the surface loading studies requires the preparation of the *o*-phthaldialdehyde (OPA) reagent on a daily basis. The OPA reagent was prepared using the method described by Cornec and co-workers[13]; 7.62 g sodium tetraborate was dissolved in 150 ml of ultrapure water, followed by the addition of 200 mg SDS and 176 mg dithiothreitol. 160 mg *o*-phthaldialdehyde was pre-dissolved in 2-4 ml ethanol then added to the mixture before adjusting the volume to 200 ml with water in a volumetric flask.

Surface loading studies were made following equilibration of the emulsions in the presence of Tween 20. The emulsions were centrifuged at ~13,000 rcf (relative centrifugal force) for 10 min., prior to removal of subnatant; this process was repeated twice more. In addition to analysing the subnatant from the displacement experiments it is also necessary to carry out a calibration curve for the protein over an appropriate concentration range. The resulting adduct absorbs at 340 nm, and a background correction is applied for the OPA reagent. 100 µl of the protein solution/subnatant is added to 3 ml OPA reagent, inverted to mix, then left for 5 min. to ensure complete reaction. In addition to measuring each sample at 340 nm, the absorbance was also measured at 600 nm to account for the contribution of residual emulsion to the absorbance obtained at 340 nm. The scattering factor was calculated to be 9.68, so the values obtained at 600 nm were multiplied by this value and the corrected absorbance subtracted from the absorbance at 340 nm. Any samples having an absorbance at 600 nm greater than 0.01 were not included in the final analysis. Protein concentrations were converted into mg m^{-2} using the specific surface area obtained from sizing measurements. All measurements were carried out in triplicate.

2.6 *In vitro* lipolysis studies

The reaction mixture employed for the *in vitro* lipolysis studies under simulated duodenum conditions consists of a mixture containing 9.7 mM bile salt (NaTC : NaGDC, molar ratio of 47.3:52.7), 0.15M NaCl, 0.01M CaCl2, 2 mM Bis-Tris buffer at pH 7 plus 2 nM of porcine pancreatic lipase and 10 nM of porcine pancreatic colipase. Low concentrations of lipase and colipase were used to slow down the lipolysis in order to make it possible to more easily detect changes in the rate of lipolysis. 0.5 ml of emulsion sample were treated in 15 ml of the reactant mixture at 37°C for 0.5 h. Lipolysis generates fatty acids which cause a lowering of the pH of the mixture. The level of lipolysis was monitored by titrating with 0.01N NaOH to neutral pH using a pHstat.

3 RESULTS AND DISCUSSION

In considering how to manipulate interfacial protein films to control lipolysis it is useful to investigate the behaviour of commercial protein isolates such as sodium caseinate and their susceptibility to displacement by surfactants. To investigate this susceptibility initial studies have been made on the behaviour at air-water interfaces. Studies at the air-water interface are easier to make than studies at oil-water interfaces. Previous research[8] has shown that the interfacial behaviour is generic and behaviour at the air-water interface can be used to infer behaviour at the oil-water interface.

Interfacial films of sodium caseinate were spread at an air-water interface on a Langmuir trough to a surface pressure of ~12 mN m^{-1}. These films were transferred onto freshly-cleaved mica using LB deposition and imaged by AFM to ensure uniform protein coverage. After addition of the surfactant Tween 20 to the subphase the interfacial films were sampled as a function of increasing surface pressure.

Figure 1 shows representative images of the interfaces at increasing surface pressure. The grey levels in the AFM images are a measure of the height of the samples and the light (grey) areas are regions occupied by protein. Tween 20 is removed by butanol and the dark areas in the images are representative of the regions within the interfacial film originally occupied by the surfactant. The nucleation and growth of surfactant domains within the protein network is characteristic of orogenic displacement.

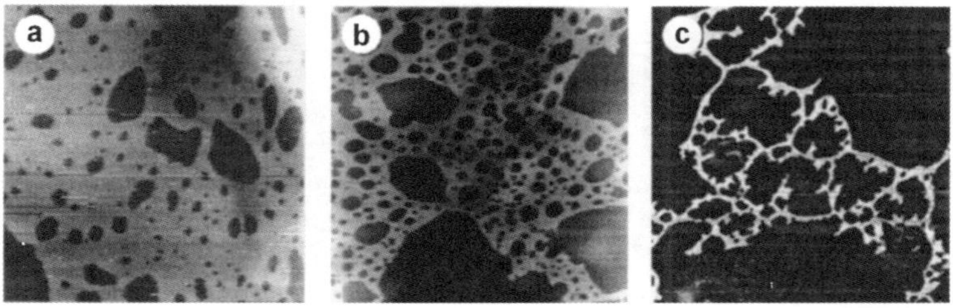

Figure 1 *AFM images showing displacement of sodium caseinate by Tween 20. (a) Surface pressure π = 20.3 mN m^{-1}, image size 12.5 x 12.5 μm. (b) π = 22.3 mN m^{-1}, image size 10 x 10 μm. (c) π = 23.7 mN m^{-1}, image size 15 x 15 μm.*

The images shown in figure 1 illustrate distinct stages in the development of the domain structures within the sodium caseinate network. At the lowest surface pressures in the presence of Tween 20 (π~20.3 mN m^{-1}, Figure 1a), the observation of numerous small holes within the sodium caseinate network is characteristic of the first stage (nucleation within defects) of the orogenic displacement mechanism. At higher, intermediate surface pressures (π~22.3 mN m^{-1}, Figure 1b), the surfactant domains have grown, reducing the area covered by protein. The domains are circular in shape with smooth boundaries. These types of domain structure are similar to those observed[7] in the displacement of β-casein, the major component of sodium caseinate, from air-water interfaces by Tween 20. At the highest surface pressures, approaching the collapse point of the film, the surfactant domains are now large and irregular in shape (π >23.7 mN m^{-1}, Figure 1c), leaving a fibrous-like network with a high degree of connectivity, not dissimilar to the types of networks observed previously for β-lactoglobulin[7] and whey protein isolate (WPI)[14]. It is noticeable that, despite the fact that the major component of sodium caseinate is β-casein, the sodium caseinate networks can resist breakage at surface pressures much higher than the collapse point of pure β-casein networks. The irregular, fibrous networks seen in sodium caseinate films at the highest surface pressure are unlike those observed for β-casein films. The circular domain structures seen in pure β-casein films suggest weak interactions between the proteins which are poor at resisting the growth of the surfactant domains[7]. The irregular domain shapes seen near the collapse point in sodium caseinate films are characteristic of strongly interacting protein structures which cause redistribution of stress around the domain boundaries allowing the domains to grow preferentially in the weakest regions of the network[7-8].

Sodium caseinate is a protein isolate which contains β-casein, α-caseins and κ-casein. The differences in behaviour between the displacement of sodium caseinate and β-casein could thus be due to the presence of the other minor components. Protein isolates can be modelled as simple mixtures of component proteins[8, 15]: these studies revealed that protein mixtures formed 'kinetically trapped' networks, showing little evidence of phase separation. The displacement mechanism was orogenic with the protein capable alone of forming the strongest network dominating the eventual failure (collapse) of the composite protein network. In view of these conclusions it is useful to contrast the displacement of sodium caseinate from air-water interfaces with the behaviour of the individual components; β-casein, α-caseins and κ-casein.

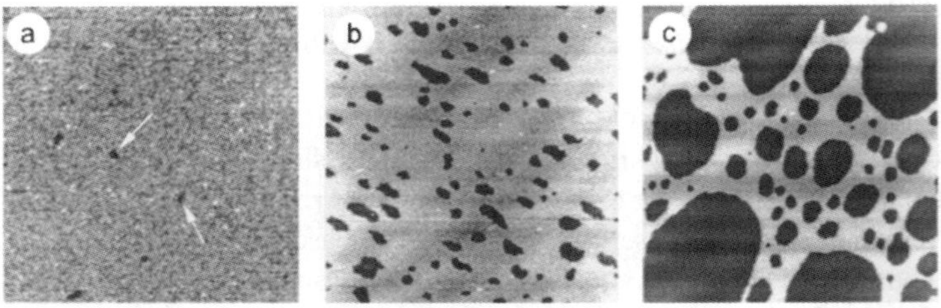

Figure 2 *AFM images showing displacement of β-casein with Tween 20. (a) Image size 1.6 x 1.6 μm, π = 15.9 mN m^{-1}. The white arrows indicate nucleating Tween 20 domains. (b) Image size 6.4 x 6.4 μm, π = 16.7 mN m^{-1}. (c) Image size 6.4 x 6.4 μm, π = 19.2 mN m^{-1}.*

Figure 2 shows AFM images representing the displacement of β-casein networks from air-water interfaces by Tween 20. As reported previously[7], the Tween 20 surfactant domains are essentially circular in shape, indicating uniform compression of a weak network as the surfactant domains grow in size. It is important to note that the growth of circular surfactant domains in β-casein occurs over a comparable range of surface pressures to the appearance of circular surfactant domains during the displacement of the sodium caseinate network by Tween 20 (Figure 1).

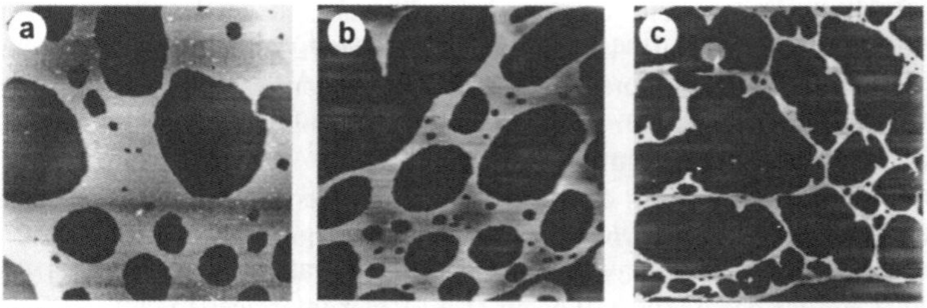

Figure 3 *AFM images showing displacement of α-casein with Tween 20. (a) Image size 10 x 10 μm, π = 21.1 mN m^{-1}. (b) Image size 15 x 15 μm, π = 21.5 mN m^{-1}. (c) Image size 20 x 20 μm, π = 22.3 mN m^{-1}.*

Figure 3 demonstrates that α-caseins also form weak protein networks at the air-water interface. The AFM images are similar to those seen previously[7] and in the present study (Figure 2) for β-casein, with a collapse point of π~23 mN m^{-1} similar to that seen for β-casein. The data is also consistent with that seen for sodium caseinate films over a similar range of surface pressures (Figure 1).

Figure 4 shows AFM data for the displacement of κ-casein from an air-water interface by Tween 20. The κ-casein was spread to a surface pressure of 6 mN m^{-1} and then compressed to a surface pressure of 14.4 mN m^{-1}. Tween 20 was added to the subphase to induce displacement of the film.

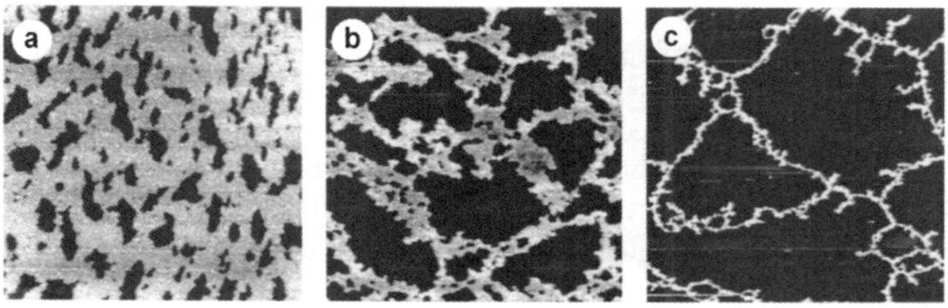

Figure 4 *AFM images showing displacement of κ-casein with Tween 20. (a) Image size 5 x 5 μm, π = 14.8 mN m^{-1}. (b) Image size 5 x 5 μm, π = 20.6 mN m^{-1}. (c) Image size 12.5 x 12.5 μm, π = 22.9 mN m^{-1}.*

For surface pressures $\pi<23$ mN m^{-1} the AFM images show that the displacement of κ-casein is unlike that observed for α-casein, β-casein and sodium caseinate films. Unlike the isotropic, circular domains observed for β- and α-caseins films (Figures 2 & 3) the surfactant domains within the κ-casein films are anisotropic (irregular) and resemble images seen previously for strongly-associated protein networks such as β-lactoglobulin and WPI[7-8, 14]. Furthermore, the behaviour of κ-casein networks, at surface pressures $\pi>23$ mN m^{-1} (Figure 4), resembles that seen for sodium caseinate networks (Figure 1).

These studies suggest that κ-casein is dominating the final failure of the sodium caseinate networks. This is consistent with previous studies on model mixed protein systems[8, 15] where the protein component, which alone forms the strongest network, dictates the failure of the mixed network. However, the result is surprising given that κ-casein is a minor component of sodium caseinate. A possible explanation is that in the early stages of displacement the weakly associating α-casein and β-caseins are preferentially displaced. The remaining κ-casein will be confined to a decreasing surface area as the surfactant domains expand, promoting κ-casein - κ-casein interactions and network formation.

These studies suggest that the use of weak network-forming proteins can be enhanced by incorporation of strong network-forming proteins. Although this improves the final failure characteristics of the network it does not alter the rate of growth of domains within the network, which will be determined by the weak network-forming proteins. In order to restrict domain growth to reduce adsorption of bile salts it is necessary to enhance the level of association of proteins throughout the network.

The studies shown above have been made at air-water interfaces. It is necessary to establish that similar behaviour will occur at oil-water interfaces in emulsions. Firstly it is necessary to demonstrate that the orogenic displacement observed in the interfacial studies occurs in emulsion systems. This can be achieved by surface loading studies which determine the protein concentration at the oil-water interface and can be used to investigate displacement by Tween 20. Figure 5 shows the displacement of sodium caseinate from tetradecane oil in water emulsions. The protein displacement is characteristic of that expected for orogenic displacement. There appear to be two stages of displacement, which could correspond to the initial removal of α- and β-caseins and then, at higher Tween 20 concentrations, removal of κ-caseins. Thus these surface loading studies demonstrate orogenic displacement and could be used to monitor the effects of enzymatic modification on displacement. The surface loading measurements are difficult to carry out and a simpler method of monitoring protein displacement would be useful.

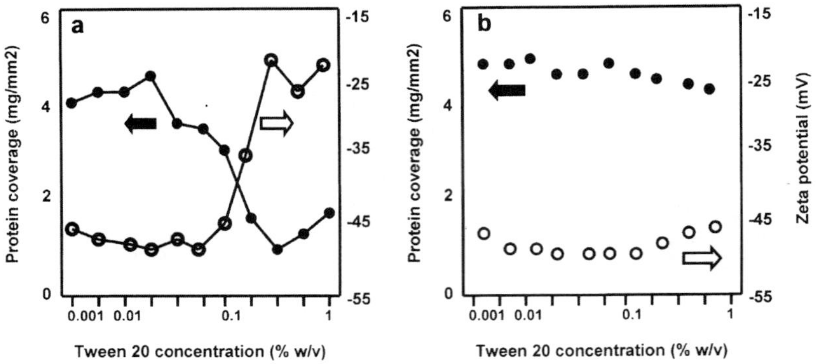

Figure 5 *Protein displacement data (•) and zeta potential data (o) showing (a) the displacement of sodium caseinate from olive oil emulsions with Tween 20 and (b) the inhibition of displacement by transglutaminase crosslinking of the interfacial structure.*

Displacement of protein with a neutral surfactant will lead to a decrease in surface charge and the surface charge or surface potential can be used to monitor adsorption of Tween 20 and displacement of protein. Figure 5a shows a comparison of the surface loading data and electrophoresis data showing displacement of sodium caseinate from tetradecane emulsions. The zeta potential measurements also demonstrate orogenic displacement and provide an easier method for assessing surfactant adsorption.

These methods can be used to investigate the effect of enzymatic crosslinking of interfacial sodium caseinate films on the ability to resist the growth of surfactant domains within the protein network. Previous studies have shown that the displacement of interfacial protein networks by bile salts occurs through an orogenic mechanism[10]. Thus, through the use of transglutaminase, it may be possible to increase the crosslinking within the protein network and control the area accessible to bile salts. Figure 5b shows that transglutaminase treatment of the interfacial sodium caseinate network can be used to inhibit the displacement of sodium caseinate with Tween 20: this is illustrated by both the zeta potential and surface loading data. The degree of displacement by Tween 20 provides a measure of the change in mechanical strength of the protein network on cross-linking with transglutaminase and can be used to assess the effects of cross-linking in samples prepared for *in vitro* lipolysis studies.

Figure 6a shows that crosslinking can be used to restrict the extent of adsorption of Tween 20 onto emulsion droplets stabilized with sodium caseinate. The black bars show the zeta potential for the emulsion droplets prior to addition of Tween 20. Data is shown for the uncrosslinked sample, the crosslinked sample and a control sample that has gone through all the preparative stages in the absence of transglutaminase. The grey bars show the changes in zeta potential on addition of a defined concentration of Tween 20. The uncrosslinked sample shows a large change in zeta potential due to protein displacement. At the Tween 20 concentrations used the protein should be removed from the interface (Figure 5a). The control sample shows similar results to the untreated sample but the crosslinked sample shows clearly evidence for restricted adsorption of surfactant onto the surface of the emulsion droplets.

Figure 6b shows the effects of the crosslinking with transglutaminase on *in vitro* lipolysis under simulated duodenum conditions. The action of the lipase generates fatty acids which lower the pH of the sample. Titration with 0.01N NaOH can be used to follow the extent of generation of fatty acids and hence to monitor the rate of lipolysis.

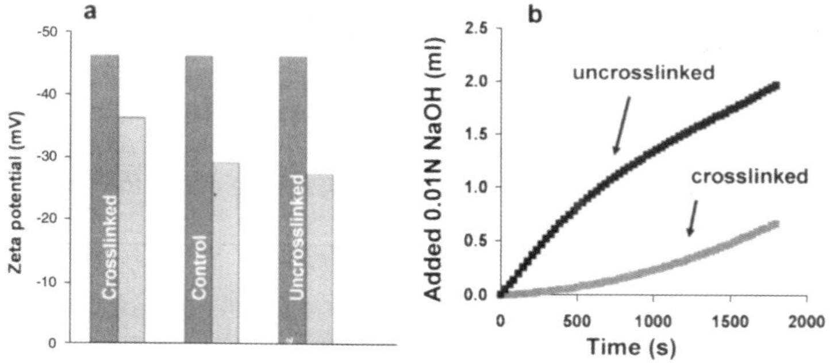

Figure 6 *Effects of transglutaminase crosslinking of interfacial sodium caseinate films on (a) the displacement of the protein from olive oil in water emulsions by Tween 20, monitored by zeta potential measurements, and (b) in vitro lipolysis under simulated duodenum conditions.*

The crosslinking clearly reduces the rate of lipolysis, suggesting that increasing the mechanical strength of the interfacial sodium caseinate films restricts growth of bile salt domains, reducing the adsorption of lipase-colipase complexes and hence the rate of lipolysis. These studies demonstrate the feasibility of modifying interfacial layers in processed emulsions to reduce the rate of lipolysis.

Emulsified processed foods have to pass through the mouth, throat and stomach before entering the small intestine. Any treatment of the interfacial structure has to survive this transit in order to exert its effect in the duodenum. Thus further generic information is required on the effects of pH, ionic strength, temperature and proteolytic enzymes on interfacial layers. A combination of these studies with the current research will provide a basis for attempts to engineer rationally interfacial structures to induce satiety and combat the growing health and social problems resulting from obesity.

References

1 J. Hammer, K. Hammer, K. Kletter, *Gut*, 1998, **43**, 111.
2 R.C. Spiller, I.F. Trotman, B.E. Higgens, M.A. Grimble, T.C. Lee, S.R. Bloom, J.J. Misiewicz and D.B.A. Silk, *Gut*, 1984, **25**, 365.
3 L. Pironi L , V. Stanghellini, M. Miglioli, R. Corinaldesi, R. Degiorgio, E, Ruggeri, C. Tosetti, G. Poggioli, A.M.M. Labate, N. Monetti, G. Gozzetti, L. Barbara, and V.L.W. Go, *Gastroenterology*, 1993, **105**, 733.
4 R.L. Batterham, M.A. Cohen, S.M. Ellis, C.W. Le Roux, D.J. Withers, G.S. Frost, M.A. Ghatei and S.R. Bloom, *New Engl. J. Med.*, 2003, **349**, 941.
5 R.L. Batterham, C.W. Le Roux, M.A. Cohen, A.J. Park, S.M. Ellis, M. Patterson, G.S. Frost, M.A. Ghatei and S.R. Bloom, *J. Clin. Endocrinol. & Metabol*, 2003, **88**, 3989.
6 R.C. Ritter, *Physiol. & Behaviour*, 2004, **81**, 249.
7 A.R. Mackie, A.P. Gunning, P.J. Wilde and V.J. Morris, *J. Colloid Interface Sci.*, 1999, **210**, 157.

8 V.J. Morris and A.P. Gunning, *Soft Matter*, 2008, **4**, 943.
9 A.R. Mackie, A.P. Gunning, P.J. Wilde and V.J. Morris, *Langmuir*, 2000, **16**, 2242.
10 J. Maldonado-Valderrama, N.C. Woodward, A.P. Gunning, M.J. Ridout, F.A. Husband, A.R. Mackie, V.J. Morris and P.J. Wilde, *Langmuir*, 2008, **24**, 6759.
11 R. Lantto, E. Puolanne, N. Kalkkinen, J. Buchert and K. Autio, *J. Agric. & Food Chem.*, 2005, **53**, 9231.
12 J.E. Folk, in *Metabolism of Amino Acids and Amines*, eds. H. Tabor and C.W. Tabor, Academic Press, New York, 1970, p. 889.
13 M. Cornec, P.J. Wilde, P.A. Gunning, A.R. Mackie, F.A. Husband, M.L. Parker and D.C. Clark, *J. Food Sci.*, 1998, **63**, 39.
14 N.C. Woodward, P.J. Wilde, A.R. Mackie, A.P. Gunning, P.A. Gunning and V.J. Morris, *J. Agric. Food Chem.*, 2004, **52**, 1287.
15 A.R. Mackie, A.P. Gunning, M.J. Ridout, P.J. Wilde and V.J. Morris, *Langmuir*, 2001, **17**, 6593.

MODELLING OF PHYSICAL PROCESSES IN THE SMALL INTESTINE

S. Bakalis, A. Tharakan, M.R. Jaime-Fonseca, P.J. Fryer and I.T. Norton

School of Chemical Engineering, University of Birmingham, Birmingham B15 2TT, UK

1 INTRODUCTION

A fundamental understanding of physical processes occurring in the gastrointestinal (GI) tract is crucial to develop and design new food and drug formulations. Formulation has a significant incidence in the rate of digestion and absorption of nutrients and bioactive components. For instance, the absorption rate of molecules such as glucose and cholesterol could be changed by adding biopolymers such as guar gum, pectin and fibres into a food system.[1-3] Furthermore, by changing the absorption rate of glucose in the blood it may be possible to reduce the risk of insulin intolerance and type II diabetes.[4] As such there is an need not only to provide consumers with functional foods but also to assess the behaviour of food and drug bioavailability under physiological conditions.

1.1. GI tract overview

The human GI Tract is a highly regulated system capable of processing various forms of food (solid, liquid or foams); different quantities (large meals, snacks); a range of constituents (carbohydrates, proteins, lipids) and with various physical and chemical properties (density, viscosity, solubility, diffusivity). The processing that the food undergoes in the human digestive system occurs in the mouth, stomach, small and large intestine. During eating, the mouth is used for the mechanical breakdown of food. Mastication reduces the particle size, mixing the food with saliva and enzymes. The stomach serves the function of mechanical processing (churning) mixing with hydrolytic enzymes, pH change and digestion. The small intestine is responsible for absorption of water and nutrients in which digestive reactions take place. The large intestine (colon) absorbs the remaining water from indigestible food matter, stores this unusable food matter (wastes) and then eliminates the wastes from the body.

Regardless of its importance relatively little work has been done to understand the physical processes in the upper gastrointestinal (GI) tract. In an effort to study the digestion and absorption of foods *in vivo* studies and *in vitro* models have been developed.

1.2 *In vivo* studies

There is a wide range of *in vivo* studies which have been conducted to monitor and analyse food digestion in the GI tract. Scintigraphy, ultrasonography and Magnetic Resonance Imaging (MRI) are some techniques developed for evaluating gastric motility, accommodation, emptying and intragastric processing of food in the upper GI tract.[5] Recent developments in functional MRI techniques have allowed *in vivo* visualisation of a range of formulations, for example. Images of alginate beads in the stomach have been presented by different authors.[6,7] They investigated the effect of model solid foods on satiety and reported that gastric emptying was delayed by the presence of gelled beads used as a model.

A commonly used method in studying the effect of formulation is measuring a response in the blood stream of a change induced e.g. by ingestion of a certain food product. Using a similar approach it has been long established in the literature that food formulation and more specifically increasing viscosity has reduced the rise in postprandial glucose and insulin levels in human subjects.[2,3]

A significant problem is that *in vivo* studies are expensive, time consuming and ethical issues can limit their use in human experimentation.

1.3 *In vitro* models

In order to design foods with specific functions, *in vitro* gut models have been developed to study how products perform in the digestion processes. The importance of this is demonstrated as a range of models that have been already developed and are briefly described in this section.

For instance, a series of five double-jacketed vessels was used as a simulator of the human Intestinal Microbial Ecosystem.[8] Similar studies have been done at the University of Reading (UK), in which chemostats were used to study bacteria inside the intestine.[9] The bioavailability of actives has been tested in the pharmaceutical area using similar methods.[10] TNO in the Netherlands has developed a Gastro Intestinal Tract Model (TIM) .[1] The model mimics interactions with gastric fluids and reactions while mixing is reproduced by altering external pressures. A range of physiological parameters (e.g. pH, temperature, movements, digestion enzymes) can be controlled. TIM has been used to validate the biodrug concept[11] and study the behaviour of orally administered drug dosage forms under physiological conditions.[12] Studies indicate that TIM shows promise as a useful tool for various research purposes dealing with; availability for absorption of mutagenic; as well as antimutagenic, components in food.

The Institute of Food Research (Norwich, UK) has recently developed a model of the stomach that simulates human digestion using a flexible walled vessel and a cylinder that processes the food at representative shear rates. During the digestion simulation acid and enzymes are added. The Model gut has been recently used to study the effect of phosphatidylcholine on the resistance of bovine lactoglobulin *in vitro* gastrointestinal proteolysis. The results indicated that biomolecular interactions between proteins and lipids altered patterns of proteolysis.[13] Phosphatidylcholine was shown to have a substantial protective effect during duodenal digestion indicating that lipid components in the gut allowed more protein to reach the mucosal immune system.[13]

The effect of wall motility on absorption profile was studied by an *in vitro* model of the small intestine at the University of Birmingham (UK).[14] The model consisted of a dialysis membrane and a mechanical system that simulated segmentation movements.

1.4 *In silica* models

Over the last decades effort has been devoted to develop mathematical models of various biological processes, in order to obtain a fundamental understanding of the effect of a range of physiological processes. A good example is the 3D model of a heart.[15] *In silica* models have been used to evaluate the residence time distribution, release and inactivation of bioactives such as probiotics, drugs and foods in the GI tract mimicking the intestinal fluid flow. They have proved to be useful techniques to predict the human fraction dose absorbed from orally administrated drugs.[16]

Computational models have been created to study the absorption process in the GI tract. Although these models were fitted to the experimental data their use is limited due to the complexity of the problem. As a first approach, an equation based model without considering complex fluid dynamics was developed.[17] The effect of peristalsis and segmentation on mixing has been also investigated[18] however absorption was not estimated in these models. Computational fluid dynamics (CFD) has been used to simulate segmentation and peristalsis in these *in vitro* and *in silica* models describing momentum transport during single phase flow which incorporates moving boundaries. Consideration was given to the design to ensure that the mechanical processing of the digesta in the human small intestine was being accurately represented.

The objective of this work was to develop a methodology that will allow a quantification of the effect of formulation/food properties on the absorption profiles in the GI tract. Emphasis was given to the effect of segmentation as means to enhance mixing and the mass transfer coefficient. All the experiments were performed in an *in vitro* small intestine model.[19]

2 METHODS

2.1 Small Intestine Model (SIM)

The SIM functions as a concentrically mass transfer exchanger composed of an inner dialysis membrane and an outer non-porous membrane (Figure 1). The inner membrane was a cellulose ester semi-permeable membrane (Spectra/Por 7, MWCO 8000 Daltons, size 8) and was purchased from Medicell International Ltd (London, UK). The non-active membrane was chosen for the controlled investigation into the effect of the biopolymer on the nutrient rate transfer across the membrane.

2.1.1 Intestinal Cell Set up Characteristics. The mass transfer cell (dialysis membrane) had a length of 0.5 m that could be adjusted as required; the cuff length was 0.12 m, the inner tube and the outer tube had a diameter of 0.03 m and 0.05 m, respectively.

2.1.2 Flow and Mixing in the SIM. The overall flow was constant for all the experiments in the system and was generated from a variable peristaltic pump outside of the membrane. The segmentation movement responsible for mixing of the intestinal content was reproduced by the concentric constriction of the tube with the use of inflatable bladders (Figure 1).

The mechanical squeezing mechanism and the delay times were controlled by a computer, allowing the opening or closing of four different solenoid valves.

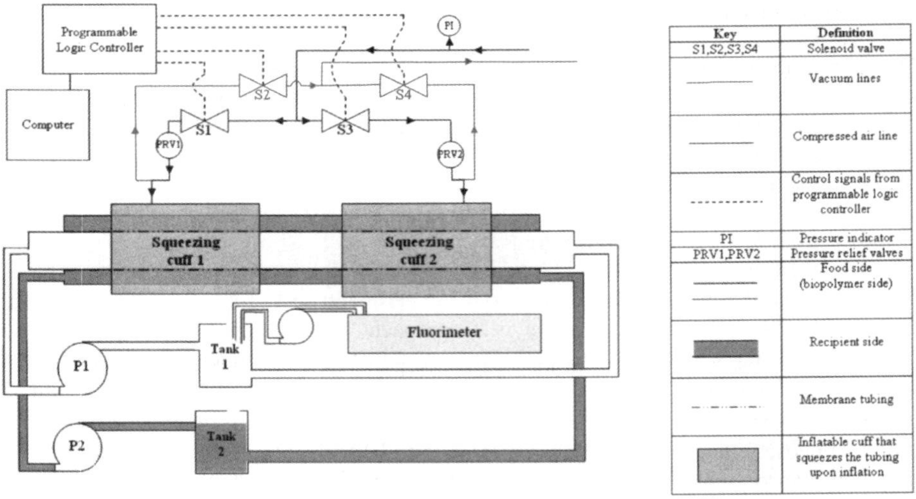

Figure 1 *A schematic of the intestinal cell set up*

2.2 Absorption of Glucose in the SIM

Effect of segmentation movements in the absorption of glucose was studied by two experimental conditions:

2.2.1 *Segmentation.* Net flow induced by peristaltic pump (P1, Figure 1) and local flow from segmentation by cuffs C1 and C2 (Figure 1) alternatively with a 2 second inflation time, 2 second deflation time, 2 second delay and 0.55 barg inlet air pressure.

2.2.2 No segmentation. Net flow induced by a peristaltic pump (P1) and no segmentation movements in the SIM.

The effect of formulation, specifically viscosity, in glucose absorption rates was also investigated. Experiments were performed using aqueous solutions as a control and two biopolymers, guar gum and carboxymethyl cellulose (CMC) when segmentation and no segmentation occur. All chemicals were supplied by Sigma-Aldrich Company Ltd. (UK). The biopolymer (1.0 and 5%, w/v) and glucose (1%, w/v) were homogeneously dispersed in distilled water using a magnetic stirrer for two hours. Each solution was added to the dialysis membrane in the SIM at room temperature. The zero shear viscosities of the biopolymers were determined using an AR1000 rheometer (TA instruments Newcastle, Delaware, USA) equipped with an acrylic plate geometry (0.06 m). Each experiment was repeated in triplicate at 25 °C. The apparent viscosity (μ) was calculated as a function of the applied shear rate ($\dot{\gamma}$) as $\mu = \tau / \dot{\gamma}$

Samples (1.0 mL) were taken from the recipient side every 10 minutes during one hour to determine glucose concentration outside of the membrane. The concentration of glucose was calculated following the method of 3,5-dinitrosalicylic acid (DNS) reagent for reducing sugars.[20] All experiments were performed at least in triplicate. Error bars on the graphs represent the standard deviations obtained from triplicates.

2.3 Mass Transfer Coefficients

The significance of the absorption mechanism and accurate estimates of the mass transfer coefficient were determined based on the experimental data. By plotting glucose increase (mM) against time (s), the rate of absorption of glucose was calculated. Then, overall mass transfer coefficients (K) were estimated by Equation 1.

$$N = K(\Delta c) \tag{1}$$

where N is the molar flux (mmol/m^2 s) and Δc (mol/m^3) is the concentration difference.

3 RESULTS

The glucose absorption rates (μmol/s) are given in Table 1. All measurements were obtained for at least three independent batches and mean ± standard deviation (std) is reported for each sample. Both segmentation motion and viscosity reduce the absorption rate of glucose. The increase in viscosity (from 1 to 200 mPa.s) caused a dramatic decrease in glucose absorption. Studies[3] showed that the presence of guar gum in glucose solutions fed to humans reduced the increase in blood glucose level, resulting in only 50% of that observed without guar. They proposed that the mechanism of reduction is by inhibiting the effects of intestinal motility on fluid convection. Possible mechanisms are that viscous polysaccharides act by delaying gastric emptying and thus slow delivery to the small intestine[3] and reduce the rate at which glucose is absorbed from the intestine (probably due to a reduced mixing efficiency).

Table 1 *Glucose absorption rates for experimental solutions*

Biopolymer	Concentration (%, w/v)	Viscosity ± std (mPa.s)	Absorption ± std (μmol/s)	
			Segmentation	*No segmentation*
Control	0.0	1.0 ± 0.2	1.41 ± 0.19	1.04 ± 0.33
Guar gum	0.1	2.0 ± 0.4	1.40 ± 1.02	1.03 ± 0.57
CMC	0.1	20.0 ± 0.2	1.20 ± 0.14	0.9 ± 0.98
CMC	0.5	200.0 ± 0.1	0.02 ± 0.26	-----

Figure 2 shows the increase in glucose versus time profile. The results showed that glucose absorption was accelerated when segmentation occurred in the SIM. After one hour glucose concentration outside of the membrane was around 6 mM when segmentation was induced in the SIM and 4 mM without segmentation for aqueous (control) solution. This behaviour was also observed for guar and CMC at higher viscosities as a consequence of the increasing mixing. Additionally, a reduction of the boundary layer on the dialysis membrane surface could occur.[1]

A good linear regression (R^2 = 0.97) was obtained when segmentation was applied giving a rate of absorption of 1.41 x 10^{-3} mmol/s from which N was determined (Equation 1). By dividing the molar flux (2.94 x 10^{-2} mmol/m^2.s) by the concentration (55 mmol) an overall mass transfer coefficient K, was obtained as 5.35 ± 0.07 x 10^{-7} m/s (water under segmentation) and 3.09 ± 0.12 x 10^{-7} m/s (water without segmentation). The membrane surface area was 0.05 m^2 and the recipient side total volume was 0.001 m^3. Similarly, K was estimated for the range of biopolymers used. Figure 3 shows the effect of viscosity in mass transfer of glucose under segmentation.

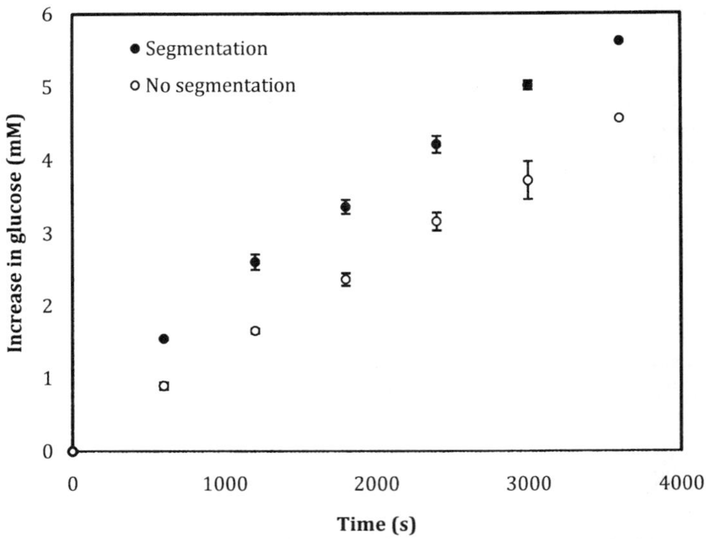

Figure 2 *Absorption of glucose in an aqueous solution (1.0 mPa.s) when segmentation and no segmentation motion take place in the SIM.*

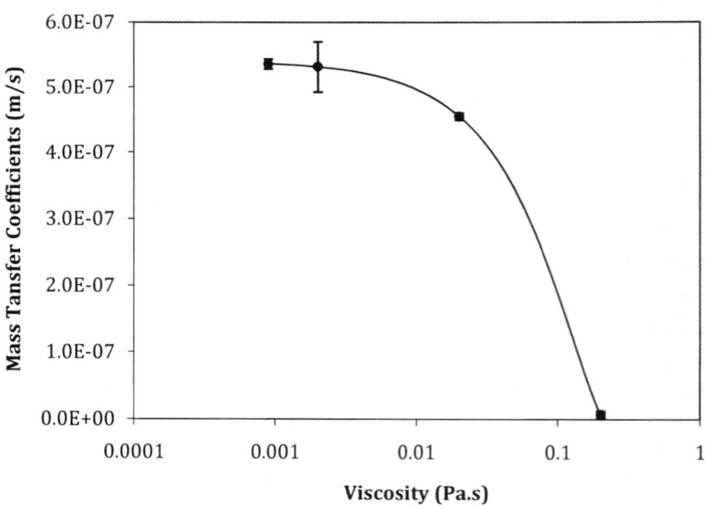

Figure 3 *Mass transfer coefficients as a function of the viscosity of the biopolymer under segmentation*

As the viscosity of the biopolymer was increased the absorption of glucose was reduced and the effect of segmentation motion was increased. These results showed the effect of changing the formulation of the system by increasing the viscosity.

4 CONCLUSIONS

Absorption of glucose through the semi-permeable membrane increased as segmentation was applied. An increase in viscosity resulted in a significant decrease of mass transfer in the SIM. This *in vitro* behaviour could be correlated within blood glucose levels in humans to understand the mechanisms by which biopolymers improve glucose tolerance.

References

1 M. Minekus, M. Jelier, J. Xiao, S. Kondo, K. Iwatsuki, S. Kokubo, M. Bos, B. Dunnewind, R. Havenaar, *Biosci. Biotechnolol. Biochem.*, 2005, **69**, 932.
2 P.R. Ellis, P. Rayment, Q. Wang, *Proc. Nutr. Soc.*,1996, **55,** 881.
3 N.A. Blackburn, A.M. Holgate, N.W. Read, *Br. J. Nutr.,* 1984, **52**,197.
4 I. Norton, P. Fryer, S. Moore, *AIChE J.,* 2006, **52**, 1632.
5 F. Kong, R.P. Singh, *J. Food Sci.*, 2008, **73**, R67.
6 C. Hoad, P. Rayment, E. Cox, P. Wright, M. Butler, R. Spiller, P. Gowland, *Food Hydroc.*, 2009, **23**, 833.
7 L. Marciani, P.A. Gowland, A. Fillery-Travis, P. Manoj, J. Wright, A. Smith, P. Young, R. Moore and R. C. Spiller, *Am. J. Physiol. Gastrointest. Liver Physiol.*, 2001, **280,** G844.
8 P. De Boever, B. Deplancke, W. Verstraete, *J. Nutr.*, 2000, **130**, 2599.
9 T. E. Steer, J.N. Gee, I.T. Johnson, G.R. Gibson, *Curr. Issues Intest. Microbiol.,* 2004, **5**, 23.
10 A. Dokoumetzidis, P. Macheras, *Inter. J. Pharm.*, 2006, ***321,*** *1.*
11 S. Blanquet, S. Marol-Bonnin, E. Beyssac, D. Pompon, M. Renaud and M. Alric, *Trends Biotechnol.*, 2001, **19,** 393.
12 S. Blanquet, E. Zeijdne, R.E. Beyssac, J.P. Meunier, S. Denis, R. Havenaar, M. Alric, *Pharm. Res.*, 2004, **21**, 585.
13 G. Mandalari, A.R. Mackie, N.M. Rigby, M.S.J. Wickham, E. N. C. Mills. *Mol. Nutr. Food Res.*, 2009, **53,** S131.
14 S. Bakalis, A. Tharakan, I. T. Norton, P. J. Fryer, *FASEB J.*, 2007, **21,** LB112.
15 D.M. McQueen, C.S. *Peskin Computer Graphics-Us,* 2000, **34**, 56.
16 F. Johansson, R. Paterson, in *Drug Absorption Studies: In Situ, In Vitro and In Silico Models*, ed. C. Ehrhardt, K-J. Kim, *Springer US*, 2008, ch. 21, pag. 504
17 B.R. Stoll, R.P. Batycky, H.R. Leipold, S. Milstein, D.A. Edwards, *Chem. Eng. Sci.* **55**,473.
18 B. Jeffrey, H. S. Udaykumar, K.S. Schulze, *Am. J. Physiol. Gastrointest. Liver Physiol.,* 2003, **285,** 907.
19 A. Tharakan, A. Rayment, P. J. Fryer, I.T. Norton, *ECCE-6,* 2007, 1.
20 G.L. Miller, *Anal. Chem.,* 1959, 426-428.

CHARACTERISING EFFECTS ON INTERFACIAL STRUCTURES DURING SIMULATED HUMAN DIGESTION OF EMULSIONS

J. Maldonado-Valderrama, A.P. Gunning, M.J. Ridout, N.C. Woodward, P.J. Wilde & V.J. Morris

Institute of Food Research, Norwich Research Park, Colney, Norwich NR4 7UA, UK

1. INTRODUCTION

The growing social and economic consequences of obesity are focussing attention on the need to control fat uptake in the diet. Controlling lipid digestion can improve health by modifying serum lipid levels and promoting satiety[1]. The rational design of processed food emulsions to control fat uptake represents a major challenge to the food industry. Rational manipulation of interfacial structures to modify dietary intake of fats requires a fundamental understanding of the effects of human digestion conditions on interfacial structures. During digestion pre-formed interfacial structures will be challenged by a variety of factors, which include changes in temperature, pH, ionic strength, and the effects of surfactants, biosurfactants and proteolytic enzymes. There are a growing number of papers dealing with the interfacial activity of lipases[2-5] but to date there are only a few generic studies on the effects of digestion conditions[6-7] on interfacial structures prior to lipase action in the duodenum[2], or on the modification of interfacial structures to influence lipase activity[8].

In the case of processed food emulsions it may be possible to modify the nature of protein-based interfacial structures to moderate the adsorption of bile salts, thus controlling the level of adsorption of lipase-colipase complexes, and the rate of lipolysis. Bile salts are important in controlling the interfacial activity of lipases: the negative charge on the bile salts attracts colipase to interfaces and the colipase secures the lipase in a suitable configuration for lipolysis. The idea for moderating lipolysis through restricting colipase-lipase adsorption stems from previous novel observations on the molecular mechanisms by which surfactants displace proteins from interfaces. Direct visualisation by AFM of the structural changes which occurred during displacement revealed a novel 'orogenic' displacement mechanism[9-10]. The AFM studies showed that, as a consequence of their partial unfolding at the interface, proteins associate to form 'elastic' networks[9-10]. Although surfactants are usually more surface-active than individual proteins and thus, given sufficient time, at sufficiently high concentrations, they will eventually displace the proteins, it is not immediately obvious how the surfactants displace proteins from the interface when they are linked together into a network. The AFM studies revealed that the formation of interfacial protein networks results in defects which allow infiltration by surfactants onto the interface[9-10]. The surfactants consolidate and expand these nucleated sites forming surfactant domains within the protein network. Growth of the domains compresses the protein network, causing it to fold and buckle, and then eventually to break[9-10]. Having broken the protein network it is then possible for the surfactants to displace proteins, or protein aggregates from the interface into the bulk phase. Recently it has been shown that the displacement of interfacial protein networks by bile salts under simulated *in vitro* duodenum conditions occurs by such an orogenic process[6]. There is thus interest in how digestion conditions will influence preformed interfacial networks, or modified networks, with a view to designing or modifying networks that can survive transit to the duodenum in order to control the level of bile salt adsorption.

Previous research[9-10] has shown that studies at air-water interfaces can be used to infer generic behaviour at oil-water interfaces. This article reports generic studies on the effects of digestion conditions in the stomach and small intestine on the structure of pre-formed interfacial protein (β-lactoglobulin) networks. A combination of interfacial and surface tension measurements, interfacial rheology, and atomic force microscopy (AFM), has been used to provide new insights into the changes in interfacial protein structures induced during digestion.

2 METHODS

2.1 Proteins and surfactants

Samples of crystallized and lyophilised β-lactoglobulin from bovine milk (90% by PAGE) and crystallized and lyophilised pepsin from porcine gastric mucosa (P6887: lot no. 056K767622, 3,260 units mg^{-1}) were obtained from Sigma, UK, stored at 4°C and -20°C respectively, and used without any further purification. The bile salts sodium glycodeoxycholate (NaGDC, 97% TLC) and sodium taurocholate (NaTC, 95% TLC) were also purchased from Sigma, UK, stored at room temperature, and used without further purification. Highly-refined olive oil (Fluka Chemie AG, Buchs, Switzerland) was used for the oil-water studies. The olive oil was purified with Florisil resins prior to use: the oil was mixed with the resins and agitated for 3 h, centrifuged at 12,000 rpm in a Beckman centrifuge for 30 min, filtered under vacuum and stored under inert gas in the dark. For the simulated duodenum studies all the samples were prepared in a 2 × 10-3 M BisTris buffer with 0.15 M NaCl and 0.01 M $CaCl_2$ (Fluka Chemie) at pH 7.0. Ultrapure water, cleaned using an Elga Elgastat UHQwater purification system (Elga Process Water, Marlow, U.K.), was used in the preparation of buffer solutions. All the glassware was cleaned with 10% Micro-90 cleaning solution and ethanol and repeatedly rinsed with distilled and ultrapure water. The interfacial tension of the clean interfaces was measured before every experiment to confirm the absence of surface-active contaminants, yielding acceptable values of 72.6±0.5 mN m^{-1} for the air-water interface at 20°C, and of 27.5±0.5 mN m^{-1} for the olive oil-water interface at 37°C. The mixed bile salt solutions were made by successive dilution from a concentrated stock containing 52.7% NaTC and 47.3% NaGDC and a total bulk concentration of 10^{-1} M.

2.2 Surface dilatational rheology

The surface dilatational modulus (E) of β-lactoglobulin films prior to and during displacement by bile salts at the air-water interface was measured by the ring-trough method[12-13]. Changes in surface area were kept below 5% of the total surface area in order to prevent disruption of the surface layer and the oscillating frequency was 0.13 Hz. A total volume of 250 ml of β-lactoglobulin solution 0.01 g l^{-1} was placed in a thermostatted glass dish (18 cm diameter, 3 cm deep). The bile salt concentration was increased sequentially by addition of bile salt solutions through the surface, outside the ring, from different stock NaTC/NaGDC solutions. Before the addition of bile salts the surface dilatational modulus of the β-lactoglobulin solution was measured for 1 h before the addition of bile salts in order to confirm the formation of a saturated surface layer of protein. After addition of each sample of bile salts the surface dilatational modulus of the system was recorded for 30 min. All measurements were made at 37 °C.

The interfacial dilatational modulus of β-lactoglobulin films prior to and during displacement by bile salts at the olive oil-water interface was measured following a similar protocol but by the oscillating drop technique. The amplitude of the applied oscillations in interfacial area was kept below 10% to avoid excessive perturbation of the interfacial layer, and the measurement frequency was 0.1 Hz. A drop of oil was formed in a protein solution (0.1 g l^{-1}). The saturation of the interface with protein was monitored by measuring the dilatational modulus as pure β-lactoglobulin adsorbed at the olive oil-water interface (~30 min). Bile salts were added sequentially from appropriate NaTC/NaGDC stock solutions and the surface dilatational modulus recorded for ~30 min after addition of bile salts. All measurements were made at 37 °C. Each measurement was carried out in triplicate and the reproducibility was within ±0.5 mN m^{-1}.

2.3 Surface shear rheology

Measurements of the surface shear modulus (G) were made with a TA Instruments AR2000 controlled stress rheometer (TA Instruments Ltd., Crawley, U.K.) operating in the controlled strain mode. Measurements on the protein solution (70 ml) were made with a polished aluminum bicone (6°, 60 mm diameter) located at the interface using a strain of 1% and a frequency of 0.5 Hz. The protein solution was poured into a glass dish at 20°C and the surface shear monitored for 1 h as the protein adsorbed onto the air-water interface. After one hour, the experimental conditions were changed by increasing the temperature of the thermostat to 37°C, by adding HCl to produce a final pH of 2.5, or by adding NaCl to produce a final ionic strength of 1.5×10^{-1}M. Additions were made via a j-shaped needle that was immersed in the subphase. After altering the conditions the cone was repositioned at the surface and the evolution of the surface shear rheology monitored for a further 1 h. In this manner the effect of the different conditions and any possible synergistic effects on the surface shear rheology of β-lactoglobulin adsorbed at the air-water interface were recorded in real time. The experiments were performed in triplicate to check the reproducibility of the data.

2.4 AFM of interfacial films.

A glass Petri dish (volume 500 ml) was used to prepare the interfacial films for imaging by AFM. Interfacial film structures were transferred onto freshly-cleaved mica using the Langmuir-Blodgett (LB) method. For the studies of displacement by bile salts the interfacial structures were monitored for the protein (β-lactoglobulin) film and then at increasing interfacial pressures after addition of bile salt mixtures. For studies under simulated stomach conditions the preformed β-lactoglobulin films were allowed to equilibrate for 1 h. After this period the subphase conditions were changed by increasing the temperature of the thermostat to 37°C or by adding HCl to produce a final pH of 2.5. The system was then allowed to equilibrate under the new conditions for 1 h. Finally studies of proteolysis were made by addition of pepsin (5 mg l^{-1}). The displacement of the interfacial films by Tween 20 was used as a means of assessing changes in the structure of the protein films due to changes in pH or temperature. Tween 20 was carefully added into the subphase from a stock solution of 10^{-3} M. All additions were made with a pipette into the aqueous phase behind the Teflon barrier so as not to disturb the surface. Successive addition of Tween 20 progressively increased the surface pressure of the system, and the displacement of the protein by Tween 20 was visualized at different levels of interfacial coverage, that is, at different surface pressures. Experiments using a preformed film prepared at 20°C were used to monitor Tween 20 displacement of the unmodified films. Samples collected at the air-water interface were allowed to dry in air before imaging, and then imaged in the dc contact mode under redistilled 1-butanol (Sigma, UK), using an East Coast Scientific AFM (ECS Ltd., Cambridge, U.K.) and Nanoprobe silicon nitride cantilevers (nominal force constant 0.38 N m^{-1}; Veeco Instruments Ltd).

3 RESULTS AND DISCUSSION

The milk protein β-lactoglobulin forms strong interfacial networks. It is useful, initially, to consider whether such networks will survive displacement by bile salts, should they remain intact on reaching the duodenum. Detailed studies[6] have shown recently that bile salts can displace completely β-lactoglobulin films from both air-water and oil-water interfaces. The main aspects of these studies are described below because, firstly they show that displacement by bile salts occurs through an orogenic mechanism, and secondly they demonstrate how, for complex scenarios such as digestion, studies at air-water interfaces provide useful generic information that can be translated into relevant generic behaviour at oil-water interfaces.

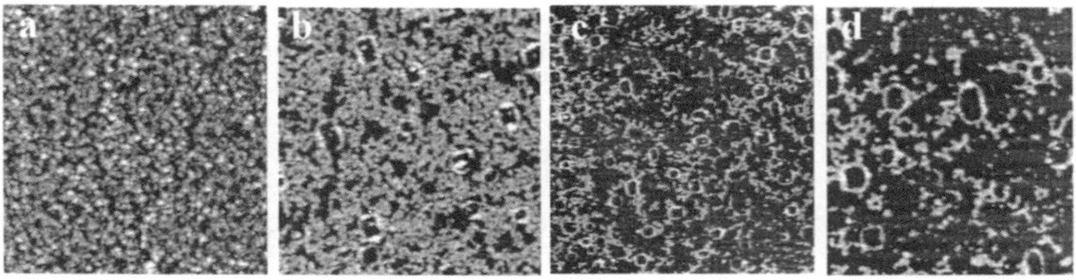

Figure 1 *AFM topography images of β-lactoglobulin films displaced by bile salts. (a) Image size 1 x 1 μm, surface pressure π = 25 mN m^{-1}. (b) Image size 2 x 2 μm, π = 26 mN m^{-1}. (c) Image size 4 x 4 μm, π = 26.9 mN m^{-1}. (d) Image size 2 x 2 μm, π = 26.9 mN m^{-1}.*

Previous studies[6] have shown that the behaviour of mixtures of the bile salts NaGDC and NaTC are additive and that they show no evidence of complex formation with β-lactoglobulin in solution prior to co-adsorption at air-water or oil-water interfaces. Thus for the present studies physiologically-relevant solutions of bile salts, composed of a mixture of 52.7%NaTC and 47.3% NaGDC, were used for the displacement studies. In accordance with reports in the literature[11] this mixture of bile salts showed no phase transitions over the range of concentrations and temperature appropriate to simulated duodenal conditions[6], allowing displacement studies by AFM to be modelled at room temperature. Figure 1 shows AFM images of the displacement of a pre-formed β-lactoglobulin network from an air-water interface by the bile salt mixture. It has been shown that LB transfer of protein networks, and partially displaced protein films, provides AFM images representative of the structures present at the air-water interface[9-10]. For example, comparative studies of the growth of surfactant domains in interfacial protein films using both Brewster angle microscopy (BAM) and AFM showed equivalent structures in the latter stages of domain growth for the interfacial films (BAM studies) and the samples LB transferred onto mica (AFM studies)[14]. Additional confirmation of the veracity of LB transfer comes from recent comparative studies of the structure of mobile phospholipid films at the air-water interface by BAM and, following LB transfer onto mica, by AFM[15]. Once again equivalent structures were observed by each technique suggesting LB transfer faithfully reproduces the interfacial structure. The images shown in figure 1 show the growth of bile salt domains within the β-lactoglobulin network indicating orogenic displacement by bile salts. The surface pressure range at which the displacement is detected by AFM is very high compared to other surfactants. However, bile salts completely displace the β-lactoglobulin at sufficiently high surface pressures[6] and at $\pi \sim 29.3$ mN m^{-1} (data not shown) the interface consists of islands of protein aggregates in a sea of surfactant (bile salts). The final surface pressure for the bile salts alone (26 mN m^{-1} at the air-water interface) is not particularly high for a surfactant: bile salts are facial amphiphiles, with a polar and a nonpolar face, and are believed to lie flat at the interface. Thus their ability to displace the proteins so effectively is quite surprising. Studies at oil-water interfaces showed similar generic behaviour[6], although subtle differences between the behaviour at olive oil-water and air-water interfaces were observed, which were attributed to the details of the interaction of the proteins with the olive oil: the domains formed in the β-lactoglobulin films at the olive oil-water interface were found to be more rounded than those seen at the air-water interface, suggesting that the β-lactoglobulin networks formed at olive oil-water interfaces are weaker than those formed at the air-water interface. This was attributed[6] to the influence of the nature of the oil phase on the unfolding and subsequent interaction of the proteins at the interface. The growth of surfactant domains within the protein network will weaken the network and this, and the collapse of the network, should be observable in surface rheology studies.

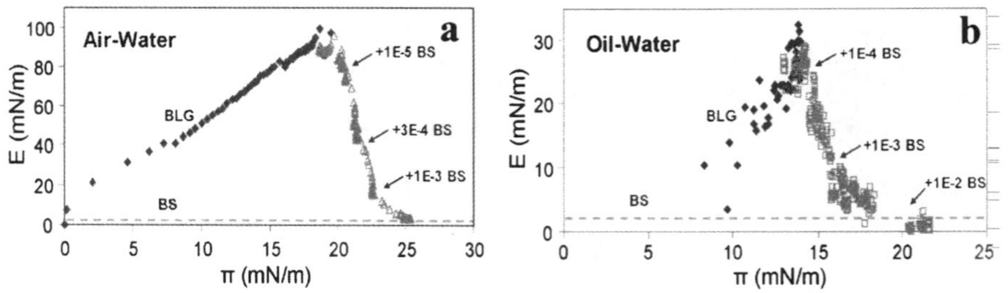

Figure 2 *Changes in surface dilatational modulus (E) at (a) air-water and (b) olive oil-water interfaces during adsorption of β-lactoglobulin (LBG) and its subsequent displacement by bile salts (BS).*

Figure 2 shows changes seen in the surface dilatational modulus at both the air-water (Figure 2a) and olive oil-water (Figure 2b) interfaces during the adsorption of β-lactoglobulin and its subsequent displacement by bile salts. These studies were made at 37°C. The oscillation frequency was sufficiently fast to emphasise the elastic component of the interfacial structure[13, 16-17] and the breakdown of the protein network caused by adsorption of the surfactant[13]. For a β-lactoglobulin film adsorbed at the air-water interface the dilatational modulus increases slowly to a value of ~100 mN m^{-1} in 1 h (Figure 2a), a value slightly lower than that reported previously for β-lactoglobulin in studies using the pendant drop technique[13] and with a ring-trough[16] at room temperature. Adsorption of β-lactoglobulin at the olive oil-water interface (Figure 2b) increases the dilatational modulus but the final value is significantly lower then that observed for β-lactoglobulin films adsorbed at the air-water interface. Oil-water interfaces are known to significantly affect the interfacial properties of β-lactoglobulin[16, 18-19]. The value of the dilatational modulus at the olive oil-water interface suggests the formation of a less cohesive network than that formed at the air-water interface. On addition of bile salts the protein network is severely disrupted, displaying a similar profile at both types of interface.

Comparisons of the dilatational experiments (Figure 2a) and the AFM images of the protein network at the air-water interface (Figure 1) show that the dilatational response to the presence of bile salts occurs before any release of protein occurs into the bulk. This was found to be true at both interfaces[6]. The growth of fluid domains within the protein network will reduce the dilatational modulus. At the highest surface pressures the amplitude of oscillations imposed on the network may become too large and cause breakdown of the network. This would explain why the modulus has vanished whilst an interconnected network is still observed by AFM (Figure 1 & 2a). The combined studies imply that bile salts will displace even intact, strongly-interacting protein networks. In order to design networks that can resist bile salt adsorption it is necessary to understand how transit through the stomach affects interfacial protein networks.

As a first step to understanding what happens to interfacial networks in the stomach preliminary studies have been made on the effects of ionic strength, temperature and pH on network structure using preformed β-lactoglobulin networks at air-water interfaces as a model system[20]. The results of this study are summarised in figure 3.

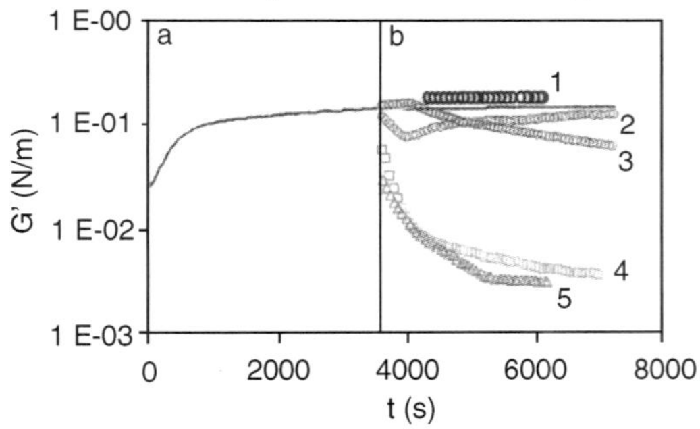

Figure 3 *The effect of physiological conditions in the stomach on the surface shear modulus (G') of a pre-formed β-lactoglobulin interfacial network as a function of elapsed time (t). (a) Adsorption of the protein at room temperature (20°C). (b) The effects of different conditions: 1-ionic strength (150M), 2-temperature (37°C), 3-pH (2.5), 4-all conditions, and 5-temperature and pH alone.*

Shear deformation is a direct measurement of the mechanical strength of the interface and should be sensitive to the effect of environmental conditions on the structure of the interfacial protein network. It is sensitive to intermolecular associations at the surface and therefore useful for characterising the strength of interfacial networks. Figure 3 (curve 4) shows that the combined effect of increasing the temperature, lowering the pH and raising the ionic strength is to reduce the surface shear modulus (G'). In order to identify the origins of this effect the influence of temperature, ionic strength and pH were investigated in isolation. Changes in ionic strength (curve 1) were found to have little effect on the modulus. Raising the temperature to 37°C (curve 2) led initially to a fall in G' but with time the modulus returned to its original value. Dropping the pH to 2.5 (curve 3) appeared

The effects seen in isolation did not appear to account for the changes seen in curve 4. However, raising the temperature and dropping the pH (curve 5) almost completely matched the behaviour shown in curve 4. This suggests an unexpected synergy between the effects of temperature and pH. The origins of this effect are discussed in more detail elsewhere[20] but it appears that at acid pH the changes observed on heating are no longer reversible and combine with the effects of pH to weaken the structure of the protein network. The effect on G' is quite dramatic. However, if the network is weakened then it becomes less able to sustain deformations than the pure protein network: deformations appropriate for studying the pure network may be too harsh for the weakened network, which may then suffer catastrophic failure on deformation. Thus it is useful to visualise the protein networks and the effects of temperature and pH.

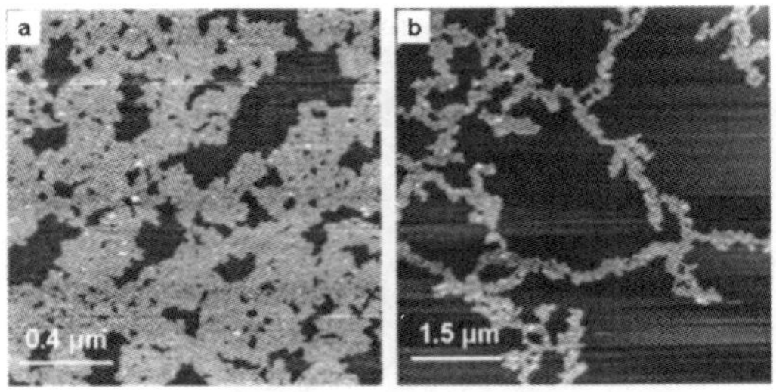

Figure 4 *AFM topography images of β-lactoglobulin displacement by Tween 20 (air-water interface, 20°C, pH 7): (a) β-lactoglobulin, π=20.2 mN m^{-1}; (b) β-lactoglobulin, π =24 mN m^{-1}.*

Visualisation of the displacement of modified β-lactoglobulin networks by Tween 20 provides information on the mechanical properties of the interfacial structures: the ease of growth and shape of the surfactant domains provides a monitor of the local elasticity of the network and the surface pressure at which the network fails provides a guide to the breaking strength of the network. Figure 4 shows the Tween 20-induced displacement of β-lactoglobulin from an air-water interface at representative surface pressures. Figure 5 shows equivalent images for pre-formed β-lactoglobulin networks 1 h after raising the temperature to 37°C. The AFM images (Figure 4 & 5) are similar showing nucleation and growth of large surfactant domains and a network which is clearly still intact at π=22 mN m^{-1}. This is consistent with the surface shear rheology data, suggesting that any initial temperature-induced changes in network structure recover with time, and that body temperature alone barely affects the eventual surface structure of the protein network.

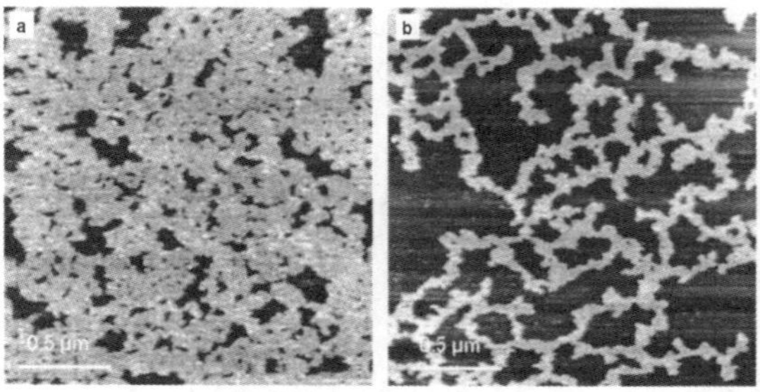

Figure 5 *AFM topography images of β-lactoglobulin displacement by Tween 20 (air-water interface, T=37°C, pH 7). (a) π=20 mN m^{-1}. (b) π = 22 mN m^{-1}.*

Figure 6 shows the effect of dropping the pH on the displacement of β-lactoglobulin networks by Tween 20. Dropping the pH produces marked changes in the surfactant domain structure during displacement. The images show a high level of small domains and, even at higher surface pressures, there is nucleation but little growth of the domains. The domain structure is similar to that seen in the displacement of β-lactoglobulin networks by charged surfactants[23], where the presence of charged surfactants favours uniform dispersal of the surfactant across the interface rather than a concentration in large domains. If the pH is raised to 7, prior to LB transfer, then the network structure re-arranges forming large surfactant domains. Comparison of the images in figures 4 & 6c suggests that at a surface pressure of 21 mN m^{-1} the surfactant domains are slightly rounder and larger after exposure to the acidic environment. However, again consistent with the surface shear rheology data, there is clearly a substantial interconnected protein network suggesting that the effect of pH is relatively mild.

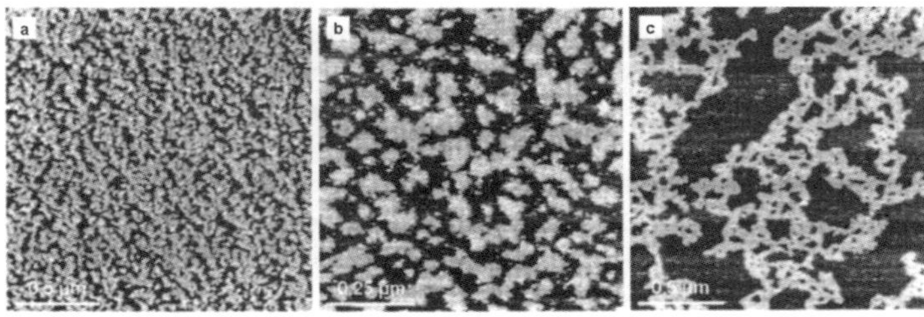

Figure 6 *AFM topography images of β-lactoglobulin displacement by Tween 20 (air-water interface). (a) π=20.5 mN m^{-1}, 20°C, pH 2.5 (b) π=20.5 mN m^{-1}, 20°C, pH 2.5. (c) π=21 mN m^{-1}, 20°C, pH 7.*

AFM images of β-lactoglobulin networks subjected to both a change in temperature and pH showed domains structures similar to those shown in figures 6 a & b but at a surface pressure of 22 mN m^{-1} the network structure appears to have collapsed showing a surfactant-dominated interface containing islands of aggregated β-lactoglobulin[20]. This is consistent with a synergistic effect of combined exposure to both body temperature and acid pH. These physiological conditions will weaken but not destroy the protein network.

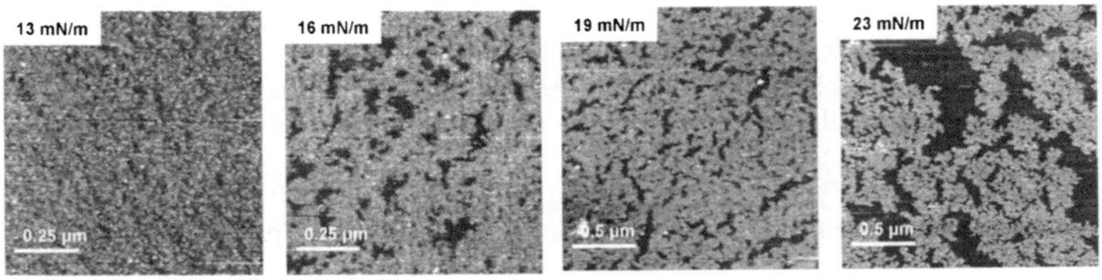

Figure 7 *AFM topography images of the displacement of preformed β-lactoglobulin networks by Tween 20 following subjection to simulated stomach conditions (temperature 37°C, pH 2.5) and the action of pepsin. The pH has been raised to pH 7 prior to LB transfer and imaging. Image sizes and surface pressures are indicated in each image.*

In addition to the effect of changes in physicochemical properties on entering the stomach the interfacial layers will be exposed to the effects of proteolytic enzymes. Thus another factor which needs to be considered is the extent of proteolysis and its effect on the integrity of interfacial protein layers. Figure 7 shows the effects of pepsin action on a preformed β-lactoglobulin interfacial network. The β-lactoglobulin was allowed to adsorb at 20°C, subjected to changes in temperature, pH, and compressed to a surface pressure of 15 mN m^{-1}, before being subjected to digestion by pepsin (5 mg l^{-1}). The samples were then re-adjusted to pH 7, perfused to remove pepsin and then displaced with Tween 20. The images at pH 7 can be compared with those shown in figure 4. It is clear that even after pepsin digestion an interconnected β-lactoglobulin network remains at the interface. The displacement by Tween 20, as seen by AFM, seems to start earlier for the pepsin-treated interface, showing holes at lower surface preseeures than the pure protein network (Figure 7). Furthermore, there appears to be no substantial weakening of the β-lactoglobulin network and the remaining interconnected network resists surfactant displacement to surface pressures similar to that needed for pure protein networks.

In the preparation of processed food emulsions the food product will contain not just protein, but also surfactants which may partially displace the protein network. To investigate the effects of surfactant displacement on proteolysis additional studies were made. The preformed β-lactoglobulin network was subjected to stomach conditions (37°C, pH 2.5) and partial displacement with Tween 20, used as a model surfactant, before pepsin treatment. The pH was then increased to pH 7, the sample perfused and then displaced with Tween 20. The images shown in figure 8 demonstrate that there is an unexpected synergism between the effect of the surfactant and pepsin. Partial displacement with surfactant prior to pepsin treatment leads to a substantial weakening of the β-lactoglobulin network, which is seen to have failed at 15.5 mN m^{-1}.

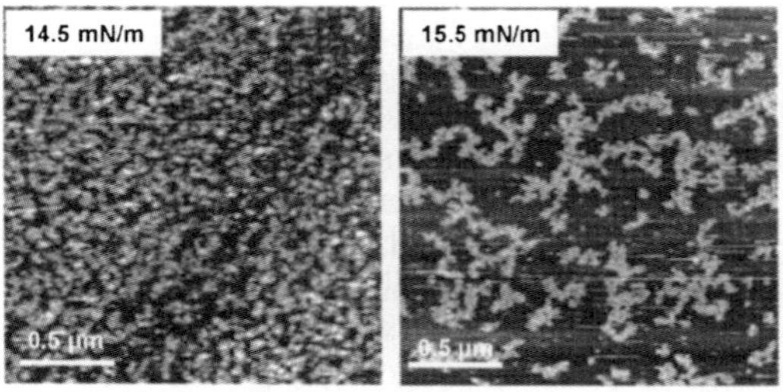

Figure 8 *AFM topography images of the displacement of preformed β-lactoglobulin networks by Tween 20 following subjection to simulated stomach conditions (temperature 37°C, pH 2.5), partial displacement with Tween 20 and then the action of pepsin. The pH has been raised to pH 7 prior to LB transfer and imaging. Image sizes and surface pressures are indicated in each image.*

These studies suggest that the folding of the protein network on partial displacement by surfactant exposes more of the protein to the action of pepsin. This is not inconsistent with recent studies on the degradation of interfacial β-lactoglobulin in emulsions by proteinase K and trypsin[22]: these authors found that the extent of proteolysis depended on the type of enzyme and the nature of the oil phase, suggesting that the extent of unfolding at the oil-water interface can determine exposure and the extent of enzymatic degradation.

The present studies have been made at air-water interfaces as a model system. The generic behaviour is expected to be the same at oil-water interfaces but the detailed behaviour may vary with the nature of the oil-phase. Thus the studies need to be validated for oil-water interfaces. Studies on the effect of pepsin on oil in water emulsions stabilised with β-lactoglobulin suggest that, unlike β-lactoglobulin in solution, which is resistant to proteolysis, β-lactoglobulin at oil-water interfaces is degraded by pepsin[7, 23]. The paper by Sarkar and coworkers[23] is particularly interesting. This paper shows a significant loss of the SDS-PAGE band for native β-lactoglobulin and a concomitant rise in a 15KD band, suggesting that, although all the adsorbed protein was degraded, the extent of degradation of individual proteins was small. This is consistent with the retention of an interconnected protein network as observed in figure 7. The observed synergism between surfactants and pepsin suggests that strategies that restrict the growth of surfactant domains within protein networks may also afford protection against proteolysis.

4 CONCLUSIONS

The present studies have demonstrated that interfacial structures formed by proteins used to stabilise emulsions are unlikely to survive displacement by bile salts in the duodenum, even if they were to survive transit through the stomach. Preliminary model studies using air-water interfaces suggest that the pepsin partially hydrolyses interfacially-adsorbed β-lactoglobulin, but does not break the interconnected interfacial protein network which is still able to resist displacement by surfactants. However, an unexpected synergism between surfactants and pepsin action has been revealed which can lead to rupture of the protein network. Generic aspects of proteolysis observed in these studies are likely to apply to the roles of proteolytic enzymes such as trypsin and chymotrypsin in the duodenum. The acid conditions in the stomach can also weaken but not break the protein network but this effect is heightened at body temperature. Altogether the studies suggest that interfacial protein networks are more resilient to stomach conditions than might have been expected, and that strategies designed to strengthen the mechanical properties of the networks may aid transit through the stomach, as well as controlling bile salt adsorption and rates of lipolysis in the duodenum.

Acknowledgements

The present studies were supported by the BBSRC through a core grant to the Institute, through the responsive mode grant BB/F020996/1 and by the EU, 7th Framework Programme, Marie Curie Intra-European Fellowship, proposal number 220570. JMV wishes to acknowledge additional support from project MAT2007-66662-C02-01 (Ministerio de Educacion y Ciencia (I+D+I), Spain).

References

1 L. Lundin, M. Golding and T.J. Wooster, *Nutrition & Dietetics*, 2008, **65**, S79.
2. P. Reis, K. Homberg, H. Watzke, M.E. Leser and R. Miller, *Adv. Colloid & Interface Sci.*, 2009, **147-8**, 237.
3. P. Reis, K. Holmberg, R. Miller, J. Kragel, D.O. Grigoriev, M.E. Leser and H.J. Watzke, *Langmuir*, 2008, **24**, 7400.
4 A. Aloulou, J.A. Rodrignez, S. Fernandez, D. van Oosterhout, D. Puccinelli and F. Carriere, *BBA-Mol. & Cell Biol. Lipids*, 2006, **1761**, 995.
5 M.E. Lowe, *J. Lipid Res.*, 2002, **43**, 2007.
6 J. Maldonado-Valderrama, N.C. Woodward, A.P. Gunning, M.J. Ridout, F.A. Husband, A.R. Mackie, V.J. Morris and P.J. Wilde, *Langmuir*, 2008, **24**, 6759.
7 A. Macierzanka, A. Sancho, E.N.C. Mills and A.R. Mackie, *Soft Matter*, 2009, **5**, 538.
8 S. Sandra, E.A. Decker and D.J. McClements, *J. Agric. & Food Chem.*, 2008, **56**, 7488.
9 A.R. Mackie, A.P. Gunning, P.J. Wilde and V.J. Morris, *J. Colloid & Interface Sci.*,

1999, **210**, 157.
10 V.J. Morris and A.P. Gunning, *Soft Matter*, 2008, **4**, 943.
11 N. Matubayasi, S. Sugiyama, M. Kanzaki and A. Matuzawa, *J. Colloid & Interface Sci.*, 1997, **196,** 123.
12 J.J. Kokelaar, A. Prins and M. De Gee, *J. Colloid & Interface Sci.*, 1991, **146**, 507.
13 B. Rippner Blomqvist, M.J. Ridout, A.R. Mackie, T. Wårnheim, P.M. Claesson and P.J. Wilde, *Langmuir*, 2004, **20**, 10150.
14 A.R. Mackie, A.P. Gunning, M.J. Ridout, P.J. Wilde and J. M. Rodriguez Patino, *Biomacromolecules*, 2001, **2**, 1001.
15 J.M. Rodriguez Patino, A.L. Caro, M.R. Rodriguez Ninö, A.R. Mackie, A.P. Gunning and V.J. Morris, *Food Chem.*, 2007, **102**, 532.
16 J. Benjamins, J. Lyklema and E.H. Lucassen-Reynders, *Langmuir*, 2006, **22**, 6181.
17 J. Krågel, M. O'Neill, A.V. Makievski, M. Michel, M.E. Leser and R. Miller, *Colloids & Surf. B*, 2003, **31**, 107.
18 E.M. Freer, K.S. Yim, G.G. Fuller and C.J. Radke, *Langmuir*, 2004, **20**, 10159.
19 S. Roth, B.S. Murray and E. Dickinson, *J. Agric. & Food Chem.*, 2000, **48**, 1491.
20 J. Maldonado-Valderrama, A.P. Gunning, M.J. Ridout, P.J. Wilde and V.J. Morris, *Eur. J. Phys. E-Soft Matter*, DOI: 10.1140/epje/i2008-10426-0
21 P.A. Gunning, A.R. Mackie, A.P. Gunning, N.C. Woodward, P.J. Wilde and V.J. Morris, *Biomacromol.*, 2004, **5**, 984.
22 S.S. Sakuno, S. Matsumoto, S. Kawai, K. Taihei and Y. Matsumura, *Langmuir*, 2008, **24**, 11483.
23 A. Sarkar, K.K.T. Goh, R.P. Singh and H. Singh, *Food Hydrocolloids*, 2009, **23**, 1563.

PECTIN–GALECTIN 3 INTERACTIONS: MOLECULAR ORIGINS OF BIOACTIVITY

V.J. Morris, R.J.M. Bongaerts and A.P. Gunning

Institute of Food Research, Norwich Research Park, Colney, Norwich NR4 7UA, UK

1. INTRODUCTION

Pectin is an integral component of non-graminaceous plant cell walls. The pectin cell wall network in fruits and vegetables is a source of both soluble and insoluble fibre. Pectin is also present in processed foods where it functions as a gelling agent or as a stabiliser in acidic milk drinks. In addition to the health benefits associated with its role as dietary fibre new aspects of pectin bioactivity are emerging. There is a growing interest in the role of natural polysaccharides present in the diet in moderating cancer progression and metastasis[1] and one of the most characterised bioactive roles for pectin is the potential use of a modified form of citrus pectin (MCP) as an anti-cancer agent[1]. An interesting facet of these studies is that the accumulated evidence in the literature would appear to suggest that MCP is effective for a range of cancers at all stages of development.

Such 'broad spectrum' activity would seem to imply that the effects are due to a simple generic mode of action. A molecular model has been proposed as a hypothesis for explaining pectin bioactivity[1]: it has been suggested that the treatments used to modify pectin cause the release, or result in the release during digestion, of fragments of pectin which can bind to, and inhibit the roles of a mammalian lectin called galectin 3 (Gal3). Gal3 is a biomarker for cancer and is a recognised target for therapy[2-4]. The ability to bind to and inhibit Gal3 is important because Gal3 regulates key stages in the development and spread of cancers[2-5]: these steps include the adhesion of cancer cells to tissue, the growth of blood vessels to tumours (angiogenesis) and factors which enhance the survival of cancer cells, such as self-association (emboli formation) and resistance to programmed cell death (apoptosis).

Gal3 is a member of a growing family of evolutionary-conserved lectins which are found in a range of species from lower invertebrates to mammals. The mammalian lectin Gal3 is literally a galactose-binding lectin which contains a ~14 KD carbohydrate recognition domain (CRD) with an affinity for β-galactoside residues[6-7]. Gal3 is a chimeric protein consisting of a globular CRD attached to a long thin 'collagen-like' tail, resulting in a 'tadpole-like' structure as revealed by electron microscopy[8]. Through the use of solid-phase binding and inhibition assays a number of natural ligands have been identified that will bind to galectins[9]. These molecules include the carbohydrates galactose, lactose, polylactosamine and N-acetyllactosamine (LacNAc). There are published x-ray studies on Gal3-lactose, Gal3-LacNAc and Gal3-LacNAc complexes[10-12] which have revealed the nature of the binding of the terminal galactose residues within the primary binding site. It has been established for lactose that hydrogen bonding to O4 and O6 of the galactose and O3 of the glucose is important in stabilising the binding of the ligand within the CRD. Galactose, lactose, polylactosamine and LacNAc are natural carbohydrate inhibitors for Gal3 and high-affinity carbohydrate inhibitors have been developed through derivatisation of LacNAc[12] and galactose[13]. Flexibility in the substitution at C3 of the galactose residue has allowed the development of higher-affinity ligands[13] for Gal3.

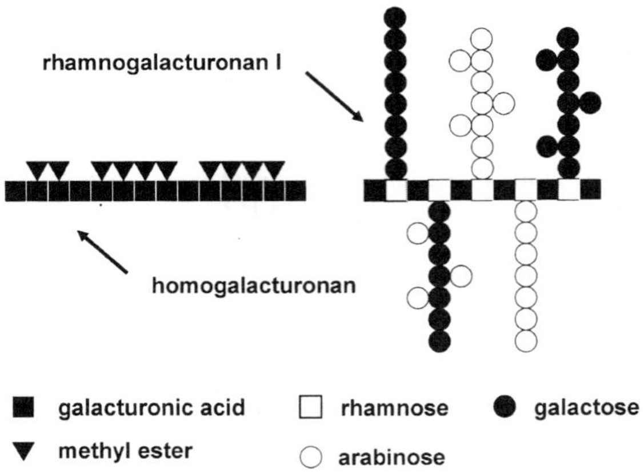

Figure 1 *Schematic diagram showing the key structural features of pectin extracts.*

Pectins are extremely complex structures. Pectin extracts have been shown to contain two main types of structure relevant to their functional properties: linear homogalacturonan backbone (smooth) regions, largely responsible for their gelling ability, and branched (hairy) rhamnogalacturonan 1 (RGI) regions[14], which are the likely origins of their bioactivity. Figure 1 shows a simple pictorial illustration of these components. Homogalacturonan backbones are composed of (1→4) linked α-D-galacturonic acid residues (Gal*p*A) in which some of the carboxyl groups are methyl-esterified. RGI units consist of a repeating disaccharide [→4)-α-D-Gal*p*A-(1→2)-α-L-Rha*p*-(1-)] of galacturonic acid and rhamnose (Rha*p*) residues. A proportion of the Rha*p* residues contain neutral sugar sidechains linked at C4 consisting mainly of α-L-arabinose (Ara*f*) and/or β-D-galactose (Gal*p*) residues. Pertinent features of the RGI regions relevant to this study are the arabinogalactan/galactan sidechains. There are two basic types of arabinogalactans, essentially based on linear (1→4) β-D-galactans and branched (1→3,6)-β-D-galactans linked to C4 of the backbone rhamnose residues. The (1→4) β-D-galactans may be linear unsubstituted chains or may contain predominately (1→3)-α-L-Ara*f* sidechains. All of the arabinogalactans contain terminal galactose residues and the simplest candidates for binding to Gal3 are linear (1→4) β-D-galactans: the simple linear structures might be expected to better facilitate binding within the Gal3 CRD.

The studies on MCP as an anti-cancer agent have used a modified form of citrus pectin which has been subjected to an alkaline treatment followed by an acid treatment[15-16]. The alkali treatment of pectin will lead to β-elimination reactions which will cause a depolymerisation of the polysaccharide backbone[17]. Alkali treatment will also de-esterify the homogalacturan regions generating lower molecular weight oligomers of polygalacturonic acid (PGA), plus RGI regions attached to polygalacturonic acid oligomers. This treatment should increase the solubility of the pectin. The acid treatment will preferentially cleave neutral sugars and is thus expected to lead to the release of modified hairy regions of the pectin chain[18]. An acid treatment would be expected to generate RGI regions lacking arabinan sidechains and with reduced substitution of arabinogalactan sidechains, plus free arbinogalactan and galactan chains. The potential importance of arabinogalactans or galactans is supported by published evidence for the anti-cancer activity of arabinogalactans[19-21]. Recent comparative studies of pectins from different dietary sources have shown that the galectin inhibitory behaviour is enhanced in pectic polysaccharides with higher arabinose and galactose contents, indicating the importance of arabinogalactans in the inhibition of Gal3-mediated cellular interactions[22].

Recent studies on RGI, which has been enzymatically cleaved from citrus pectin[23], have shown that a combination of endo-1,5-α-L-arabinanase, endo-1,4-β-D-galactanase, α-L-arabinofuranosidase and β-D-galactosidase will debranch the RGI sample, suggesting the presence of β (1→4)-D-arabinogalactan and/or galactan sidechains. Citrus pectin RGI, or modified RGI regions, are not available commercially. However, potato RGI regions are available commercially in reasonable quantities for experimental studies. Enzymatic analysis[24] of the structure of potato RGI regions suggests the presence of α-(1→5) linked arabinans and an abundance of β (1→4)-D-galactans poorly substituted with arabinose. Thus these samples provide a route to examining the potential importance of the β (1→4)-D-arabinogalactan/galactans as candidates for binding to Gal3. Treatment of potato RGI with α-L-arabinofuranosidase removes the arabinan sidechains and most of the arabinose sidechains on the arabingalactans leaving a modified RGI (mRGI) consisting of a rhamnogalacturonan backbone substituted with essentially linear (1-4) linked β-D-galactans.

This article reports studies aimed at investigating the potential binding of these mRGI regions to Gal3. For completeness studies have also been made of the possible interactions of PGA and RGI with Gal3.

2 METHODS

2.1 Pectin fragments and proteins

Potato RGI, modified potato mRGI and citrus pectin PGA samples were purchased from Megazyme International Ireland Ltd (Ireland). The RGI sample has been prepared by enzymatic treatment to remove the homogalacturonan backbone region. mRGI has been further treated with α-L-arabinofuranosidase to remove arabinose residues. The quoted sugar composition for the RGI extract is 62% galacturonic acid, 20% rhamnose, 12% galactose, 3.3% arabinose, and 1% xylose, and the quoted sugar composition for the mRGI sample is 88% galactose, 6% galacturonic acid, 3% arabinose, and 3% rhamnose. The PGA samples were prepared by a partial depolymerization of citrus pectin with polygalacturonase: the quoted sugar composition is 96% galacturonic acid with detectable amounts of galactose (1%), arabinose (0.2%), rhamnose (1.2%), xylose (1.0%), and glucose (0.3%). All carbohydrate samples were dialysed prior to coupling them to substrates. Recombinant human Gal3 was purchased as a Gal3-lactose complex from Sigma Chemicals Ltd (UK). The BSA used was fraction V. All the other materials were purchased from Sigma Chemicals Ltd (UK) unless otherwise stated.

2.2 Fluorescent labelling of carbohydrates

PGA, RGI, and mRGI samples were fluorescently-labelled with 5-(4,6-dichlorotriazinyl) aminofluorescein (5-DTAF, Invitrogen, UK) as described elsewhere[25]. Binding of DTAF-labelled carbohydrate to derivatised beads or slides was imaged in both bright-field and fluorescence modes on an inverted microscope fitted with an UPlanAPO x40, 0.85 NA objective lens (IX-70; Olympus, Japan). Epifluorescence micrographs were obtained with a dichroic filter block in the optical path (UM41001 HQ: FITC; Chroma Technology Corp., USA).

2.3 Preparation of activated substrates

Two types of substrates were used: silica beads and glass slides. Silica beads (mean diameter 6.8 μm) were purchased from Bangs Laboratories (Fishers, IN, USA). The procedures used to functionalise the beads are described in detail elsewhere[25-26]. Basically the beads were acid washed to regenerate surface hydroxyl groups and then coated with a layer of 3-mercaptopropyltrimethoxy silane (MTS). For flow cytometry studies the silanised beads were functionalised with a protein coupling agent *N*-γ-maleimidobutyryloxy succinimide ester (GMBS, Pierce, USA) allowing subsequent attachment of Gal3 or BSA to the silica beads (incubation with 0.1 mg ml^{-1} solutions, 22°C, PBS (phosphate buffered saline), 1 h). Functionalised beads were incubated with a 10 mg ml^{-1} solution of glycine to cap any unreacted succinimide groups, and then rinsed in PBS. Gal3 was coupled to the beads as a Gal3-lactose complex in order to avoid blocking the CRD. Prior to binding studies the lactose was released by washing with PBS.

For force spectroscopy studies the silanised beads were then functionalised with a carbohydrate coupling agent 4-(4-*N* maleimidophenyl) butyric acid hydrazide hydrochloride (MPBH, Pierce, USA) and then glued onto the end of the AFM cantilevers (long thin NP; Veeco Inc., USA) using a 2-part epoxy resin (40 min cure time; Permatex Inc., USA): the tip is dipped into the liquid resin and then pressed down onto the bead. Once the epoxy resin had hardened the cantilever-bead assemblies were incubated with mRGI solutions (0.1 mg ml^{-1}, 22°C, PBS, 1 h), rinsed in PBS and then incubated for 30 min in a 10 mg ml^{-1} glucose solution in PBS to cap any unreacted hydrazide sites on the bead, followed by rinsing in PBS. Fully functionalised tips were inserted into the liquid cell of the AFM and maintained under PBS solution. Covalent coupling of mRGI to silica beads was checked by visualising the attachment of fluorescently-labelled mRGI to the beads: the subsequent force spectroscopy studies used 'native' unlabelled mRGI.

Glass slides to be used for fluorescence microscopy and force spectroscopy studies were functionalised with Gal3, Gal3-lactose complexes, or BSA using procedures outlined above for silica beads. GMBS was used as the heterobifunctional crosslinking agent. Selected regions of the functionalised slides were incubated with the relevant protein solutions, washed in PBS, glycine-treated to cap any unreacted succinimide groups, rinsed in PBS and inserted into the AFM liquid cell.

2.4 Flow cytometry

For flow cytometry measurements Gal3-derivatised, BSA-derivatised, or glycine-capped control beads were incubated (1 h, 22°C, pH 7.4 – (PBS buffer) or 1h, 22°C, pH 12.0 – (PBS + NaOH)) with fluorescently-labelled carbohydrate (PGA, RGI or mRGI). Samples were then washed with buffer before re-suspension in PBS at pH 7.4 for flow cytometry. Data was collected using an inFlux flow cytometer (Cytopeia, Seattle, WA, USA – now Becton Dickinson). A minimum of 5000 events were acquired. Median fluorescence intensities of beads, selected on their forward and side-scatter characteristics, was measured with a photomultiplier tube using a 528/38 nm band pass filter. The binding of fluorescent DTAF-labelled mRGI, RGI, or PGA was measured from the fluorescence signal from individual beads. All data analysis was performed with FlowJo software (Tree Star, Ashland, OH, USA).

2.5 Force spectroscopy

Force spectroscopy studies were made using a TM Lumina (Veeco, Santa Barbara, CA. USA) combined AFM-inverted optical microscope. All binding measurements on functionalised glass surfaces were carried out under PBS. For inhibition studies the lactose solutions were added into the liquid cell to a concentration of 3.3 mg ml^{-1}. All the experimental data was captured in the layered imaging mode (at a rate of 10 μm s^{-1} in the Z direction and at a scan rate of 1 Hz). Force-distance curves were obtained by ramping the Z piezoelectric element of the tube scanner by a predetermined amount at each sample point over a selected scan area, and recording the cantilever deflection as it is advanced into and then retracted away from the sample surface.

2.6 Analysis of force spectroscopy data

The raw data obtained on each scan consists of 625 cantilever deflection-piezo displacement curves. These data were converted into force-distance curves using experimentally determined values for INVOLS (which quantifies the optical response of the cantilever when pressed against a hard surface) and the spring constant, based on thermal noise analysis[27-28]. These force-distance curves were analysed using 'in-house software' designed to identify adhesive events. Adhesive events were monitored by analysing the hysteresis region observed between the 'advance' and 'retract' portions of the force-distance curves, at the pull-off point. The detachment forces were calculated by fitting a straight line to the 'off' region of the retract curve and measuring the maximum rupture force in the hysteresis region of the curve. The data obtained on rupture forces was then displayed as histograms showing the measured distribution of detected rupture forces.

3 RESULTS AND DISCUSSION

Initial studies to assess the specific binding of mRGI to Gal3 were made using fluorescence microscopy (data not shown). The fluorescently-labelled mRGI was found to label regions of glass slides functionalised with Gal3 but not those regions functionalised with BSA or controls containing the capped coupling agent. Furthermore, the binding to Gal3 was inhibited on addition of lactose. These feasibility studies prompted further characterisation of the binding by flow cytometry.

Flow cytometry measurements showed binding of fluorescently-labelled mRGI to Gal3 immobilised onto silica beads at pH 7.4 (Figure 2a). The binding was found to be inhibited by lactose[25] suggesting a specific interaction. However, for completeness, further studies were also made with fluorescently-labelled PGA and RGI to determine whether these compounds would bind to Gal3: these compounds would not be expected to bind specifically to Gal3. Surprisingly, it was observed that both RGI and PGA (at pH 7.4) also showed evidence for binding to Gal3 (Figure 2a), although, in these cases, the binding of PGA and RGI was respectively not *or* only partially inhibited by lactose. This implied that the binding of PGA, and a substantial portion of the RGI binding, might not be specific, and might instead be due to non-specific charge-charge interactions. In order to test for the existence of non-specific binding the incubations with the labelled carbohydrates were carried out at pH 12 where no such non-specific charge-charge interaction should occur.

Following incubation at pH 12 the samples were washed and resuspended in PBS, at pH 7.4 for the flow cytometry measurements. These studies showed the binding of mRGI to Gal3, only a limited binding of RGI to Gal3, and no binding of PGA to Gal3 (Figure 2b). It was also demonstrated[25] that the binding of both mRGI and RGI to Gal3 was inhibited by lactose, suggesting specific binding in the carbohydrate recognition domain. The flow cytometry data therefore suggests specific binding of pectin-derived galactans to Gal3: this was then confirmed by force spectroscopy studies.

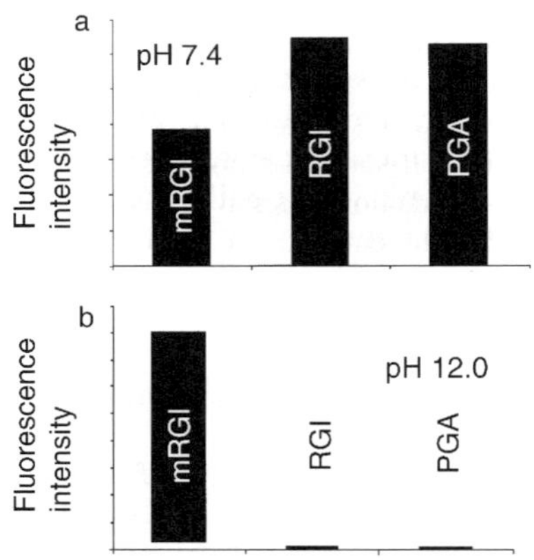

Figure 2 *Summary of flow cytometry data on the binding of mRGI, RGI and PGA to Gal3. (a) Incubation of the carbohydrates with Gal3 at pH=7.4. (b) Incubation of the carbohydrates with Gal3 at pH=12.*

Figure 3a shows an example of typical advance and retract force–distance curves which were obtained for the studies on potential interactions between mRGI and Gal3. For the curves shown the hysteresis between the advance and retract curves shows evidence of a binding event on retraction. The calculated detachment force will be a measure of the interaction between Gal3 and mRGI. Figure 3b shows a histogram of the observed binding events. The calculated modal value of the detachment force was found to be 276 pN.

The present experiments monitor interactions between Gal3 and mRGI. mRGI consists of a fragment of the pectin molecule in which linear galactan chains are attached to the RGI backbone. Previously we have measured the molecular weight of the mRGI fragment[25] by comparison with pullulan standards: the calculated molecular mass of mRGI was found to be 272 kDa. This value will be an overestimate of the molecular mass because of the complex shape of mRGI and the stiffness of the (1-4)-linked β-D-galactan sidechains, relative to the highly flexible pullulan standards. However, the value is consistent with values of 50–500 kDa reported for chemically-extracted potato RGI[29]. The length of the neutral sugar side chains in the RGI regions will vary with the source of the pectin, but chain lengths as long as 20 sugar residues have been reported[30]. High resolution NMR studies were used to estimate a mean chain length for galactan chains in potato mRGI[25]: this data showed that mRGI sidechains are linear (1-4)-β-D-galactans with an estimated average chain length of ~22 residues. Taken together this data indicates that mRGI contains several galactan chains attached to the RGI backbone. Thus the mRGI molecules may bind to several Gal3 molecules, as shown in the insert to figure 3b.

It is desirable to limit the interactions to that of single galactan chains in the mRGI with single Gal3 molecules. One way of achieving this would be to dilute the Gal3 with BSA and to thus increase the average separation between Gal3 molecules. An alternative, used in the present studies, is to add the inhibitor lactose to inactivate a high proportion of the immobilised Gal3 molecules. This also effectively increases the separation between active Gal3 molecules. Figure 3c shows the effect of such inhibition on the histogram of measured detachment force. Fewer binding events were observed during the scan and the modal detachment force was reduced to 79 pN. This suggests that the binding between mRGI and Gal3 has been largely restricted to the binding of just one galactan chain per mRGI region. The modal value of the measured detachment force is typical of that measured for other carbohydrate–lectin interactions[31-33], suggesting that there is a specific interaction between mRGI and Gal3 which involves binding into the CRD. Touhami and co-workers[32-33] report a value of 96±55 pN for single carbohydrate interactions between glucose and Concanavalin A and a similar value (121±53 pN) for single interactions between glucose and individual cell-surface-expressed lectins on live yeast cells. Reported studies on lactose–galectin 1 interactions[31] showed that the detachment force varied with loading rate and a measured detachment force of 34±6 pN was reported at a loading rate of 3 nN s^{-1}.

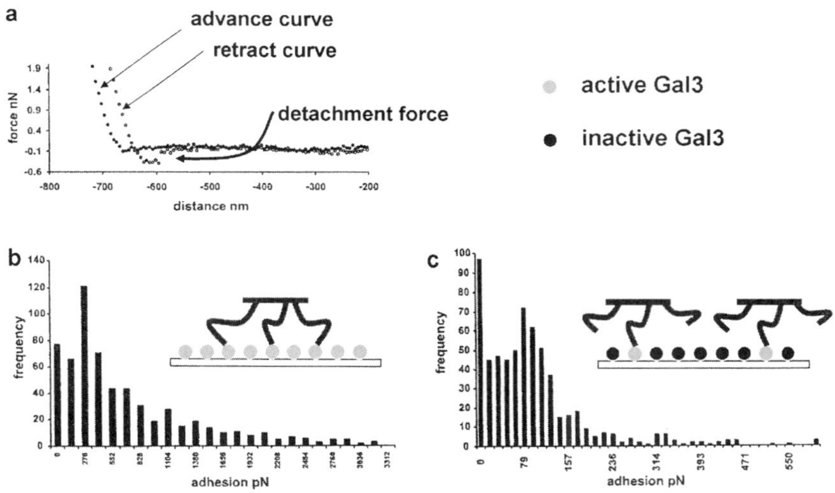

Figure 3 *Force spectroscopy data for mRGI–Gal3 interactions. (a) Advance–retract curves showing hysteresis and a measurable detachment force on retraction. (b) Histogram of detachment forces for Gal3-mRGI binding–insert shows suggested multiple binding of galactans to active Gal3 molecules. (c) Histogram of detachment forces after partial inactivation of Gal3 molecules by addition of lactose-insert shows suggested effects of inhibition on binding.*

4 CONCLUSIONS

The present studies show that there is specific binding between pectin-based galactan chains and Gal3. The studies indicate that binding occurs within the CRD and that it would be expected to inhibit functions of Gal3 involving binding to the CRD. The results provide support for the molecular hypothesis for the bioactivity (anti-cancer activity) of modified pectin and would account for the 'broad spectrum' activity reported from animal and human studies[1]. The present data suggests that this bioactivity results from modification to pectin extracts which facilitate access and binding of Gal3 to arabinogalactan or galactan sidechains from the rhamnogalacturonan I (RGI) regions. This is consistent with the expected results from the modifications used to prepare modified pectin extracts for clinical, animal and cellular studies.

Acknowledgement

The present studies were supported by the BBSRC through the core grant to the Institute.

References

1. P. Nangia-Makker, J. Conklin, V. Hogan and A. Raz, *Trends Mol. Med.*, 2002, **8,** 187.
2. A. Krzelśka and A. Lipińska, *Cellular & Mol. Biol. Letts.*, 2004, **9**, 305.
3. P. Demetter, N. Nagy, B. Martin, A. Mathieu, P. Dumont, C. Decaestecker and I. Salmon, *J. Pathol.*, 2008, **215**, 1.
4. A. Nakahara and A. Raz, *Anti-Cancer Agents in Med. Chem.*, 2008, **8**, 22.
5. Y. Takenaka, T. Fukamori and A. Raz, *Glycoconj. J.*, 2004, **19**, 543.
6. S.H. Barondes, V. Castronovo, D.N.W. Cooper, R.D. Cummings, K. Drickmer, T. Feizi, M.A. Gitt, J. Hirabayashi, C. Hughes, K. Kasai, H. Leffler, F.T. Lui, R. Lotan, A.M. Mercurio, M. Monsigny, S. Pillai, F. Poirer, A. Raz, P. W.J. Rigby, J.M. Rini and J.L. Wang, *Cell,* 1994, **76**, 597.
7. S.H. Barondes, D.N.W. Cooper, M.A. Gitt and H. Leffler, *J. Biol. Chem.*, 1994, **269**, 20807.
8. B. Birdsall, J. Feeney, I.D.J. Burdett, S. Bawumia, E.A.M. Barboni and R.C. Hughes, *Biochem.*, 2001, **40**, 4859.
9. P. Sörme, B. Kahl-Knutsson, U. Wellmar, B.G. Magnusson, H. Leffler and U.J. Nilsson, *Methods Enzymol.*, 2003, **363**, 157.
10. J.Seetharaman, A. Kanigsberg, R. Slaaby, H. Leffler, S.H. Barondes and J.M. Rini, *J. Biol. Chem.*, 1998, **273**, 13047.
11. P. Collins, K.I.P.J. Hidari and H. Blanchard, *Acta Cryst.*, 2007, **D63**, 415.
12. P. Sörme, P. Arnoux, B. Kahl-Knutsson, H. Leffler, J.M. Rini and L.J. Nilsson, *J. Amer. Chem. Soc.*, 2005, **127**, 1737.
13. P. Sörme, B. Kahl-Knutsson, S. Carlsson, U.J. Nilsson, P. Arnoux, J.M. Rini and H. Leffler, *FASEB J.* 2003, **17**, A162.
14. B.L. Ridley, M.A. O'Neil and D. Mohnen, *Phytochem.*, 2001, **57**, 929.
15. K.J. Pienta, H. Naik, A. Akhtar, K. Yamazaki, T.S. Replogle, J. Lehr, T.L. Donat, L. Tait, V. Hogan and A. Raz, *J. National Cancer Institute*, 1995, **87**, 348.
16. D. Platt and A. Raz, *J. National Cancer Institute*, 1992, **84**, 438.
17. C.M.G.C. Renard and J-F. Thibault, *Carbohydr. Res.*, 1996, **286**, 139.
18. J-F. Thibault, C.M.G.C. Renard, M. Axelos, P. Roger and M.J. Crepeau, (1993) *Carbohydr. Res.*, 1993, **238**, 271.
19. J. Beuth, H.L. Ko, V. Schirrmacher, G. Uhlenbruck and G. Pulverer, *Clin. Exper. Metastasis*, 1998, **6**, 115.
20. G. Uhlenbruk, J. Beuth, K. Oette, W. Roszkowski, H.L. Ko and G. Pulverer, *Naturwissenschaften*, 1986, **73**, 626.
21. B. Hagmar, W. Ryd and H. Skowmedal, *Invasion & Metastasis*, 1991, **11**, 348.
22. U.V. Sathisha, S. Jayaṛam, M.A.H. Hayaka and S.M. Dharmash, *Glycoconj. J.*, 2007, **24**, 497.
23. B.M. Yapo, P. Lerouge, J-F. Thibault and M.C. Ralet, *Carbohydr. Polym.*, 2007, **69**, 426.

24. J. Obro, J. Harholt, H.V. Scheller and C. Orfila, *Phytochem.* 2004, **65**, 1429.
25. A.P. Gunning, R.J.M. Bongaerts and V.J. Morris, *FASEB J.*, 2009, **23**, 415.
26. S.K. Bhatia, L.C. Shriverlake, K.J. Prior, J.H. George, J.M. Calvert, R. Bredhorst and F.S. Ligler, *Anal. Biochem.*, 1989, **178,** 408.
27. J.L. Hutter and J. Bechhoefer, *Rev. Sci. Instrum.*, 1993, **64**, 1868.
28. G. Meyer and N.M. Ámer, *Appl. Phys. Letts.*, 1988, **53**, 1045.
29. J. Obro, J. Harholt, H.V. Scheller and C. Orfila, *Phytochem.*, 2004, **65**, 1429.
30. P. Lerouge, M.A. O'Neill, A.G. Darvill and P. Albersheim, *Carbohydr. Res.*, 1993, **243**, 359.
31. W. Dettmann, M. Grandbois, S Andre´, M. Benoit, K. Angelika, A.K. Wehle, H. Kaltner, H-J. Gabius, and H.E. Gaub, *Archiv. Biochem. & Biophys.*, 2000, **383**, 157.
32. A. Touhami, B. Hoffmann, A. Vasella, F. Denis and Y.F. Dufrene, *Langmuir*, 2003, **19**, 1745.
33. A. Touhami, B. Hoffmann, A. Vasella, F. Denis and Y.F. Dufrene, *Microbiol.*, 2003, **149**, 2873.

Hydrogels for Medical Applications

A COMPARATIVE STUDY OF IOTA CARRAGEENAN, KAPPA CARRAGEENAN AND ALGINATE HYDROGELS AS TISSUE ENGINEERING SCAFFOLDS

N Mehrban, N Hunt, AM Smith and LM Grover

School of Chemical Engineering, University of Birmingham, Edgbaston, Birmingham, UK

1 INTRODUCTION

The increasingly aging population and lack of donor tissue has pushed tissue regeneration to the forefront of modern biomedical research and biopolymers traditionally found in the food and pharmaceutical industry are increasingly being used in tissue engineering. The development of tissue engineering provides the opportunity to take a sample of cells from a patient which are then cultured *in vitro* to organise into functional tissue that can then be implanted back into the patient, ultimately, overcoming problems of tissue rejection and the need for donor tissue. To grow replacement tissue requires a suitable substrate or scaffold for cells to attach, proliferate and organise into viable tissue that can be safely implanted into the body. Biopolymers and hydrogel forming biopolymers in particular, have been shown to be a promising scaffold material due having properties that resemble the environment of the mammalian extracellular matrix (ECM). In addition properties such as mild gelation conditions, potential for good mass transport of nutrients and waste molecules, nontoxic nature and biological compatibility provide an advantage over synthetic polymers and ceramic and metallic materials which have also been used as tissue engineering scaffolds.

As well as supporting cell growth and proliferation, the scaffold must have the mechanical characteristics needed to retain its structure for as long as it is needed [1]. Ideally gels used as scaffolds for tissue engineering would have a mild sol-gel reaction that can be controlled at 37°C and at physiological pH, have rapid gelation kinetics and be robust enough for 3D culture and implantation. This explains why much of the research on biopolymers for tissue engineering has been performed on materials such as alginate[2] and chitosan[3] collagen[4] agarose[5] and fibrin[6]. The relative success of the cell culture of the mentioned biopolymers has generated interest in other gel forming biopolymers such as gellan gum[7] and chemically modified gels such as RGD alginate[8]. Performance of the cells when encapsulated in these biopolymers varies due to the cellular interaction with the polymer chemistry and gel network structure.

In this study we have investigated the use of carrageenan as a 3D tissue engineering scaffold for mammalian cells and looked at the impact of structural differences between kappa and iota carrageenan has on the attachment, survival and proliferation of fibroblasts and in comparison with alginate which, is widely used in tissue regeneration. Carrageenan has been used in bacterial culture[9] and has been evaluated as a delivery vehicle for growth factors for mammalian tissue engineering applications[10] however there have been no reports of carrageenan being used exclusively for the 3D culture of mammalian cells.

2 METHODS AND RESULTS

Unless otherwise specified all reagents were obtained form Sigma Aldrich (Poole, UK).

2.1. Formation of hydrogel scaffolds for encapsulation of fibroblasts

NIH 3T3 murine fibroblasts were added to 2% (w/v) Na-alginate at a density of 0.75×10^6 cells/ml. The cell-gel suspension was then added drop wise into a bath of 100mM $CaCl_2$ and left to incubate at 37°C for 2h to form cross-linked spheres of 3.0±0.2 mm diameter.

1% *kappa* and 2% *Iota* carrageenan discs were formed using 3ml autoclaved gel which was cooled to 40°C before a 0.75×10^6 NIH 3T3 murine fibroblast cell suspension was added and pippetted repeatedly to ensure homogenisation. No extra cations were added to the gel solution as there were sufficient salts present within the supplemented media added to cell suspensions to cause the gel to cross-link.

All samples were cultured in high glucose Dulbecco's Modified Eagle's Media (DMEM) supplemented with 10% v/v foetal bovine serum (FBS) (PAA, Somerset, UK), 1% v/v penicillin-streptomycin (P/S), 2.25% v/v HEPES and 2% v/v L-glutamine and maintained at 37°C with 5% CO_2 and 100% relative humidity and media was changed three times weekly. Control samples without cell seeding were also prepared for degradation studies

.

2.2 Formation of hydrogel scaffolds for surface seeding

Alginate discs were prepared using sterile 2% low viscosity Na-alginate. The discs were formed by covering 3ml Na-alginate with 100 mM $CaCl_2$ for 2h in 6-well tissue culture plates. Following crosslinking, NIH 3T3 murine fibroblasts were seeded on the surface of the alginate at a density of 0.44×10^6 cells/ml

Iota and *Kappa* carrageenan gel cylinders were formed using sterile 2% (w/v) *iota*-carrageenan and 1% (w/v) *kappa* carrageenan in 12-well culture plates containing 3ml of the gel solution per well before addition of 1 ml 200 mM KCl. The cross-linked *kappa* and *iota*-carrageenan discs were then seeded at a cell density of 0.75×10^6. The gel discs were topped with supplemented DMEM and maintained at 37°C with 5% CO_2 and 100% relative humidity.

2.3 Analysis of cell morphology and survival

Following 10 days in culture, samples of the gels were removed and the cells were analysed using a Live/Dead® Viability/Cytotoxicity Kit. This procedure, based on the uptake of calcein AM and conversion to fluorescent calcein (green) by intracellular esterases in live cells and the uptake of propidium iodide (red) by cells with damaged plasma membranes, was used in conjunction with fluorescence microscopy to assess cellular responses to the alginate and carrageenan scaffolds. The results for the encapsulated samples showed that >95% of cells remained viable in alginate. The rounded morphology, however, also indicate that they were not well attached to the gel (Fig. 1A). Similarly cells encapsulated in *kappa*-carrageenan are rounded in appearance on day 10 (Fig 1B) with > 95% of the cells remaining live. Samples of *iota*-carrageenan (Fig. 1C) also had a >95% survival, however, the appearance of thin projections and overall elongation of the cells suggests attachment of the cells to the hydrogel. It is hypothesised that the lack of ionically bound junction zones within the structure of iota-carrageenan gels provides spaces between the polymer chains that are sufficient in allowing the cells to grow and proliferate similar to the findings of Jeon et. al. (2007)[11]. For surface seeded samples only a few cells were seen to be attached to alginate and *iota*-carrageenan after 10 days (Fig. 2A and 2B). These cells were shown to be alive, but with a rounded morphology without cytoplasmic projections indicating that the cells were poorly attached to the surface of the gel. Although the cells in alginate exhibit similar behaviour to kappa carrageenan when they are encapsulated, there is a marked reduction in cell number (<50%) and overall lack of attachment to the surface of the alginate samples. This poor attachment was similar to reported found by Harris et al (2007) who used gellan gum for surface seeding. The authors hypothesised that poor attachment was due to poor cell traction which prevented the cells from adhering to the surface of the material and exerting forces sufficient to allow migration, which may be the case for alginate and kappa carrageenan. Indeed, it has been shown that modifying alginate with the tri-peptide RGD dramatically improves cell attachment[8]. For *iota*-carrageenan, however, the cells appear to be attached to the surface of the gel although this cannot be truly confirmed without performing assaying for the presence of focal contacts or adhesion proteins.

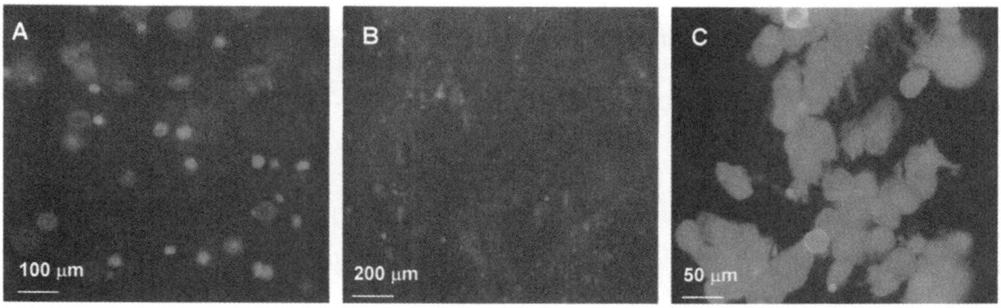

Fig. 1. 3T3 Fibroblasts stained with calcein-AM and PI 10 days post encapsulation in alginate (A), *kappa*-carrageenan (B) and *iota*-carrageenan (C).

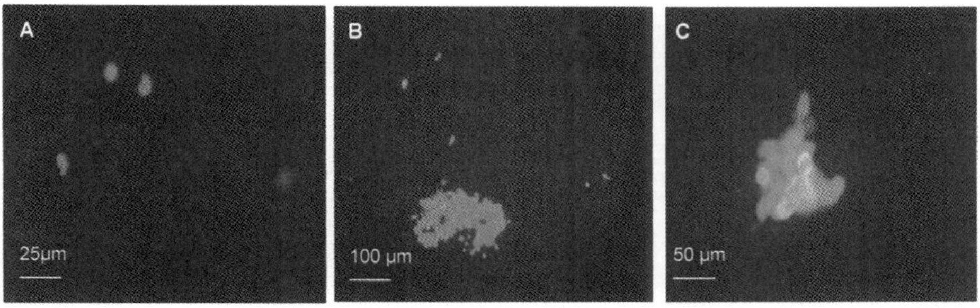

Fig. 2. 3T3 Fibroblasts stained with calcein-AM and PI 10 days after application to the surface of alginate (A), *kappa* -carrageenan (B) and *iota*-carrageenan (C).

2.2 MTT (3-(4,5-Dimethylthiazol-2-yl)-2,5-diphenyltetrazolium bromide) assay

To identify and quantify any cell proliferation within the gels an MTT assay was performed, whereby the tetrazolium ring of the MTT is cleaved by the mitochondrial dehydrogenases in living cells, producing purple formazan crystals. Dissolving these crystals and measuring the absorbance is an indication of the cell populations metabolic activity. This assay was applied to the cultures by adding 200 µl MTT to a 2 ml gelled sample. The absorbance readings (at 620 nm) obtained from the MTT assay over 12 days were used to determine the number of viable cells present in each daily sample. Figure 3 shows a comparison of the calculated cell number for all three gels. It can be seen that whilst there was a significant difference between the cell counts of all three gels ($P < 0.05$) there appeared to be no proliferation in the number of cells encapsulated in alginate. Cells encapsulated in *iota*-carrageenan show a gradual increase in the cell number over 5 days. This may be explained by the fact that iota-carrageenen gels have thé ability to spontaneously reform below their melting point following mechanical disruption[12] allowing cells to migrate throughout the gel without disrupting the network, unlike in kappa-carrageenan and alginate where the cells are confined within a more tightly bound gel network. After 5 days in culture, this number reduced (from 4.9×10^6 to 1.1×10^6 by day 12). Proliferation was also seen in *kappa*-carrageenan, however, not at the same rate as in *iota*-carrageenan and it was not until day 10 that the cells in kappa carrageenan were of comparable number to those in the iota-carrageenan.

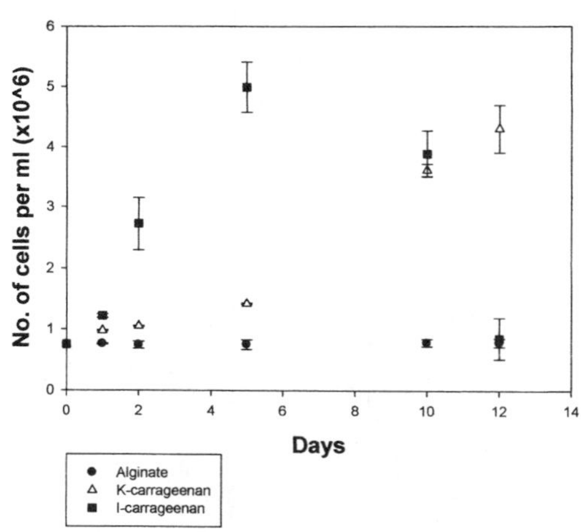

Fig. 3: Number of viable cells calculated from the absorbance of formazan of the encapsulated cells. Vertical error bars represent the standard deviation of the reported mean values (n=3).

2.4 Degradation

The changes in mass of gels with and without cells was determined by measuring the dry mass of the samples (by slow vacuum drying) on a daily basis in a vacuum freeze-drier (Edwards, EF03, Sussex, UK) for 16h. This was to quantify any fluctuations in mass within the samples which may occur as a consequence of polymer degradation, protein absorption from the culture media or formation of extracellular matrix.

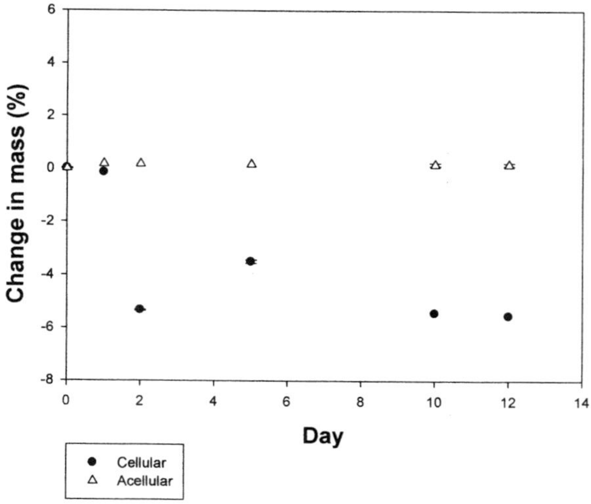

Fig. 4: Changes in the mass for *Kappa*-carrageenan with and without cells. Vertical error bars represent the standard deviation of the reported mean values (n=3).

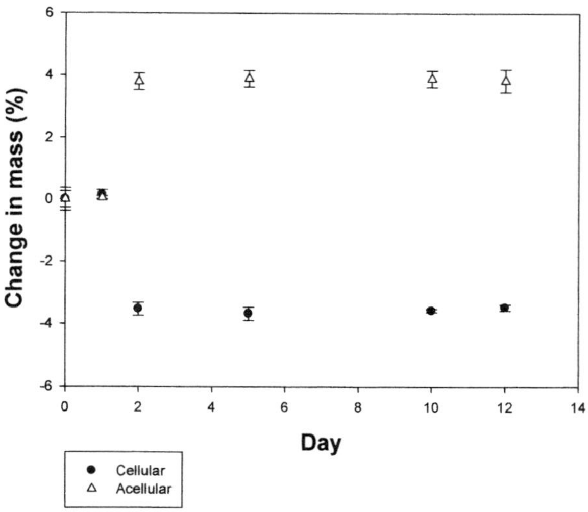

Fig. 5: Changes in the mass for *Iota*-carrageenan with and without cells. Vertical error bars represent the standard deviation of the reported mean values (n=3).

The degradation rates of the three gels were found to vary considerably, with no statistical difference seen between the cellular and acellular samples of alginate ($P > 0.05$) (results not shown). The acellular samples of *kappa*-carrageenan remained unchanged for the duration of the study, however, the addition of cells within the gel caused a gradual loss in total mass suggesting degradation of the polymer. The gel mass of acellular samples of *iota*-carrageenan increased by a total of 3.9% 2 days post encapsulation indicating the possibility of protein absorption from the culture media to the carrageenan, enabling cell attachment. This supports the findings from the proliferation assay and the observations of cell attachment. The *iota*-carrageenan containing cells showed a 3.5% reduction in mass which, is thought to be due to degradation of the gel which was visually apparent when samples were being transferred to the freeze dryer for measurements. The kappa carrageenan samples where observed to be more robust than the iota-carrageenan however the gel integrity of both types did reduce through the duration of the study. Kappa-carrageenan was initially observed to be a stronger gel in comparison with iota-carrageenan and interestingly, these physical properties may be the cause of initial cell proliferation observed in iota-carrageenan which were not seen in the kappa form. Eventual dissolution of iota-carrageenan lead to difficulty in quantifying the proliferation of encapsulated cells within the polymer at day 12 which accounts for the reduction in cellular proliferation measured using the MTT assay. The gel network of the kappa-carrageenan samples also began to lose its integrity, however, this was not apparent until day 10 of the study which correlates with a dramatic increase in the proliferation of encapsulated cells. This is in contrast to the alginate samples which appeared unchanged throughout. This demonstrates the necessity of investigating the mechanical properties of hydrogel scaffolds in relation to cellular behaviour.

3.0 CONCLUSION

In conclusion we have shown that cells can be encapsulated and remain viable in both kappa and iota carrageenan for up to 12 days which is comparable with alginate which is a widely used tissue engineering substrate. There was no evidence of attachment of cells to the alginate or the kappa carrageenan however some attachment was evident when using iota carrageenan as the scaffold for both encapsulated cells and surface seeded cells. It was also shown that proliferation of the cells within the scaffolds occurs at different rates possibly due to the mechanical properties and network architecture of the gels. Fluctuations in mass were also measured in the scaffolds which correlated with changes in cellular proliferation.

Subsequently, this study has highlighted the need for investigations into the mechanical properties of hydrogels when applied as cell scaffolds. The potential for utilising biopolymers and mixtures of biopolymers traditionally used in the food and pharmaceutical industry for this purpose is apparent as the subtle differences in gel architecture can have a dramatic effect on cell culture. This preliminary study suggests that both kappa and iota-carrageenan may be suitable hydrogels for use as a scaffold for tissue engineering purposes and as a tool for examining cells as proliferation occurs. Iota-carrageenan in particular, may have applications as a cell delivery vehicle due to the initial proliferation and subsequent dissolution. Furthermore, chemical modifications, and the development of polymer blends to change the structure and gelation behaviour may be of use to tailor the scaffolds further.

Acknowledgements

We would like to acknowledge the EU for the provision of a studentship under FP6 (N Hunt, NanoBioTact) and the BBSRC for the provision of a studentship (N Mehrban).

REFERENCES

1 J.J. Harris, A.M. Smith, S. Campbell-Lynch & R.M. Shelton (2008) *In Gums and stabilisers for the food industry 14.* 79-86

2 JL. Drury & DJ. Mooney (2003) *Biomaterials.* 24, 4337-4351

3 B Baroli (2007) *J. Pharm. Sci.* 96, 2197-2223

4 Z. Feng, M. Yamato, T. Akutsu, T. Nakamura, T. Okano, and M. Umezu (2003) *Artificial Organs* 27, 84 – 91

5 M.A. DiMicco, J.D. Kisiday, H. Gong and A.J. Grodzinsky (2007) *Osteoarthr. Cartilage* 15 1207–1216

6 K. Baar. (2007). Uniaxial stretch increases collagen synthesis in fibrin-based engineered tendons. *Tissue Engineering* 13, 1773.

7 A.M Smith, R.M Shelton, Y. Perrie, & JJ.Harris (2007) *J. Biomater. Appl.* 3, 241-54.

8 J T. Connelly, A J. García & M E. Levenston (2007) *Biomaterials.* 28, 1071-1083

9 J. Woodward (1998). *Journal of microbiological Methods,* **8**, 91-102.

10 V. Espírito Santo, A.M. Frias, M. Carida, R. Cancedda, M.E. Gomes, J.F Mano & R.L. Reis (2009) Biomacromolecules

11 O. Jeona, S.J Songa, K. Leea, M.H. Parkb, S. Leed, S.K. Hahne, S. Kimf & B.S. Kimc (2007) *Carbohydr. Polym.* 70, 251-257

12 H.J. Bixler, K. Johndro & R. Falshaw (2001) *Food Hydrocolloids* 15, 619-630

STRUCTURAL PROPERTIES OF NON AQUEOUS ETHYL CELLULOSE SYSTEMS USED IN TOPICAL DRUG DELIVERY

Lilia Bruno[1], Jiang Bin[1], Keat Theng Chow[1], Stefan Kasapis[2] and P.W.S Heng[1]

[1] National University of Singapore, Science Drive 4, Singapore 117543
[2] School of Applied Sciences, RMIT University, City Campus, Melbourne, Victoria 3001, Australia

1 INTRODUCTION

Aqueous gels are much more studied than non aqueous gels because hydrophilic gelling agents are readily available. However water poses a risk of hydrolytic reactions and the alternative formulations of hydro alcoholic compounds (e.g. ethanol) can neither eliminate the possibility of drug hydrolysis nor prevent undesirable variation in formulation owing to high evaporative potential[1]. A non-aqueous gel system consisting of ethyl cellulose and propylene glycol dicaprylate was reported to be a good delivery system for water-sensitive drugs showing good rheological and mechanical properties for different formulations and different storage time[2]. Among the many variants of cellulose, ethyl cellulose (EC) has garnered considerable attention since it can be used as a binder, dispersing agent, stabilizer, water conserving agent, and a slow releasing agent in many kinds of medicinal applications[3, 4, 5]. The EC repeat group can form inter- and intra molecular hydrogen bonds, which restricts the motion of the polymer backbone resulting in a material that is highly crystalline and oriented. EC does not dissolve in water but does dissolve in organic solvents. This makes it an ideal polymer for use in a non aqueous drug delivery system. Although EC is a popular polymer in film forming applications, its use in the formulation of non aqueous gels remains largely unexplored. Non-aqueous solvents are of increasing interest since they can be used as vehicles of topical drug delivery.[6] A topical dosage form that ensures clinical efficacy and patient acceptability must satisfy certain physicochemical and aesthetic criteria, but fundamental understanding of EC properties in non-aqueous systems is scarce in the literature. Formation of clear and thermoreversible gels has been reported upon cooling of EC/diester phthalate solutions.[7] Thixotropic gels with an extended linear viscoelastic region that exhibit desirable wetting behaviour in their function as a moisture barrier were produced with 11 to 16% EC in the formulation at various particle grades.[2,8] Blends of ethyl cellulose with poly(propylene carbonate) exhibited a range of viscoelastic consistencies from predominantly amorphous regions to cholesteric liquid crystalline structures in the rich EC composition range.[9] We felt that a study on the molecular characteristics of the polymer in a non-aqueous solvent will go a long way to fill in gaps associated with the functionality of these systems in industrial applications relevant to topical drug delivery.

2 METHODS AND RESULTS

Homogeneous gels of 12%w/w of ethyl cellulose in non aqueous solvent propylene glycol dicaprylate (PGD) were prepared and analysed for their rheological properties with respect to temperature and time. In addition, other techniques like micro and modulated differential scanning calorimetry and optical profile analysis were used to further elucidate the structural characteristics of the gel system.

2.1 Effect of temperature on the mechanical properties of EC / PGD system

The structural properties of the system were monitored using small-deformation dynamic oscillation since this approach possesses the advantage of preserving the delicate organization of the network as it forms. The effect of temperature on the gel system was determined by dynamic oscillation testing comprising of heating followed by a cooling cycle of 25°C to 90°C to 25°C. The heating profile showed a gradual reduction in G' and the cooling profile shows that the material restructures but it requires an isothermal run at 25°C for 48 hours for a complete recovery which indicates thermal hysteresis in the molecular network dynamics. This heating, cooling and isothermal run was repeated twice and in doing so specific features of tan δ and G' as seen in figure 1a and 1b show the involvement of two distinct molecular processes. The origin of this bimodal behaviour will be accounted for later, in discussions of the time course of structure formation in the EC/PGD gel.

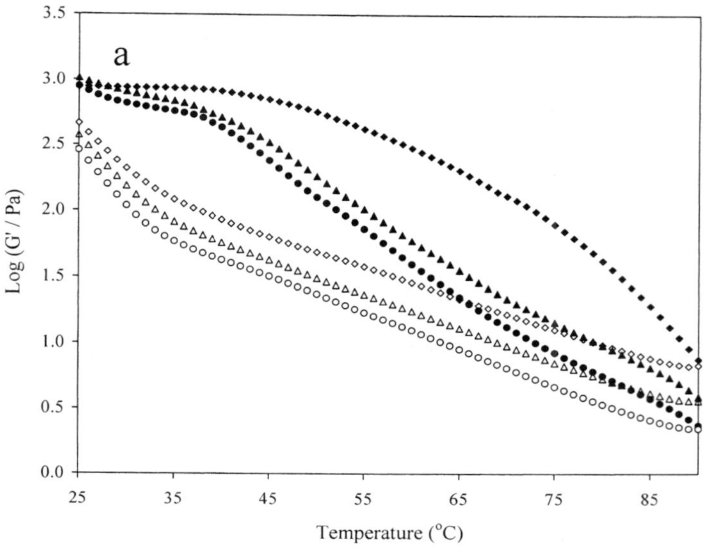

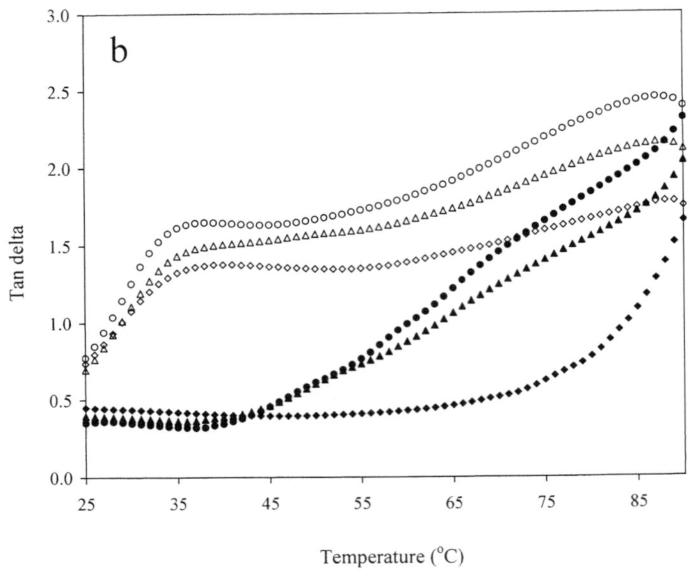

Figure 1 *Controlled heating (closed symbols) and cooling (open symbols) at 1°C/min of the shear storage modulus (a) and tan delta (b) for 12% EC freshly made in PGD - first temperature sweep (◆; ◇), second temperature sweep following aging at 25°C for 48 hours (▲; △), and a third temperature sweep following another aging period at 25°C for 48 hours (●; ○) (frequency: 1 rad/s; strain: 0.5%).*

2.2 Effect of time on the mechanical properties of EC / PGD system

To compare the effect of temperature and time an optical profile analyser was used to record the gel roughness at ambient temperature. Figure 2 shows the changes in particle size observed over a prolonged period of observation for untreated and thermally treated samples. As seen in the figure, the arithmetic mean roughness of polymer particles (R_a) is substantially lower for the heated samples as compared to the unheated ones. This shows that the structural characteristics of the gel once heated is distinct from the unheated sample. In the case of the untreated fresh gel the roughness values appreciably decreases after reaching a peak value at 3 days of observation. The magnitude of these changes shows that a time temperature equivalence exists for the system. Hence under long periods of observation, both time and temperature treatments should yield comparable morphologies for the ethyl cellulose gel.

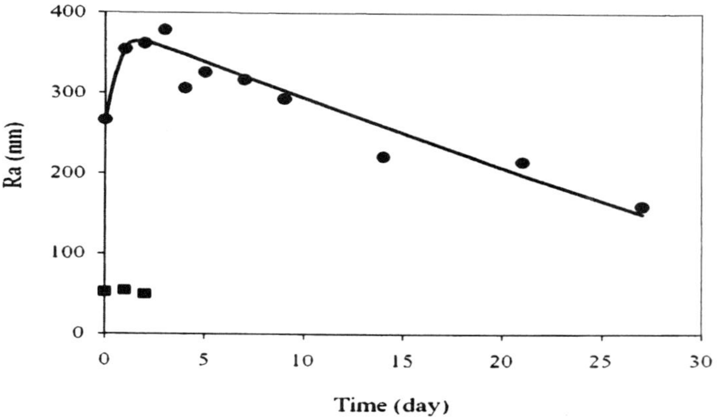

Figure 2 *Arithmetic mean roughness (R_a) of the fresh (●) and heated (■) gel of 12% EC in PGD with respect to time of observation.*

To further understand the effect of time on the properties of the gel system, a time sweep was carried out for a period of 6 days. As seen in figure 3 a considerable increase in the values of the G' and tan δ were noticed in the initial stages. Overall this investigation showed that the gelation process has reached a dynamic "pseudoequilibrium" and is predominantly time dependent.

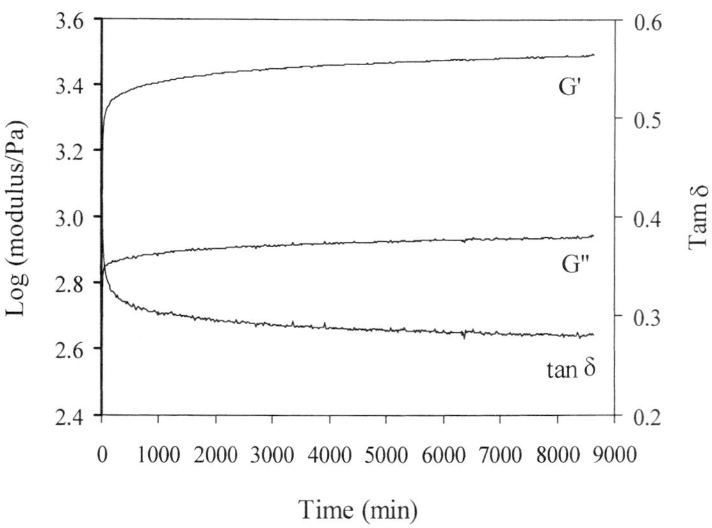

Figure 3 *Storage modulus, loss modulus and tan δ variation as a function of time at 25°C for the 12% EC/PGD gel.*

From figure 2 and 3 we can thus conclude that there are dynamic structures present in the system that are continuously evolving as a function of time (and temperature) of observation.

In order to better understand the nature of molecular processes involved in the structural disintegration seen in figure 2, the gel system was analysed using micro DSC. The structural disintegration seen can be due to either topological disentanglements, or helix to coil transitions of a low energetic content[10]. In the case of κ-carrageenan, for example, thermodynamic transitions recorded on cooling are likely to produce well resolved peaks, which are sharp at the beginning and quite broad towards the end. The former development is due to a co-operative process of structure formation with the latter being attributed to polydispersity and the formation of a 'heterogeneous' network encompassing a range of temperature-induced relaxation processes.[11] Figure 4a reproduces the heating and cooling thermograms of the gel system. The observed flat baselines on heating and cooling support the argument that the molecular interactions are similar to topological entanglements. The temperature range of observation was then further extended to subzero temperatures and the thermograms obtained are shown in figure 4b. The solvent and the gel system were both subjected to the same experimental conditions and it was seen that both scans were virtually identical and devoid of any heat flow pattern to suggest glass transition phenomenon. The absence of vitrification phenomenon argues that the gel system has a soft rubbery consistency even at very low temperatures. However, the lower range of experimental temperatures unveils two cooling exotherms that show similarity in shape and position hence denoting the freezing point of propylene glycol dicaprylate at -59°C.

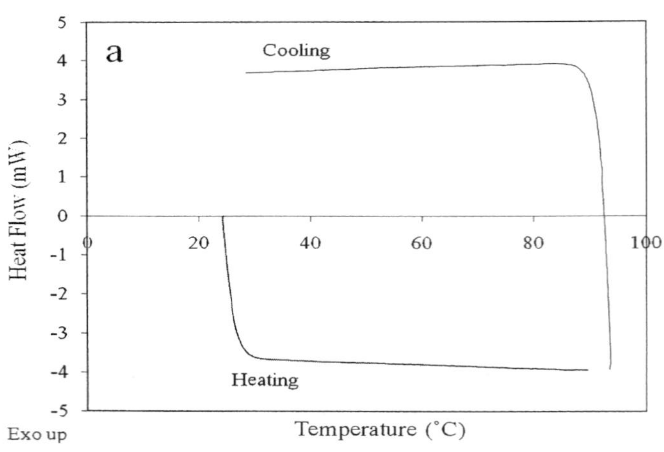

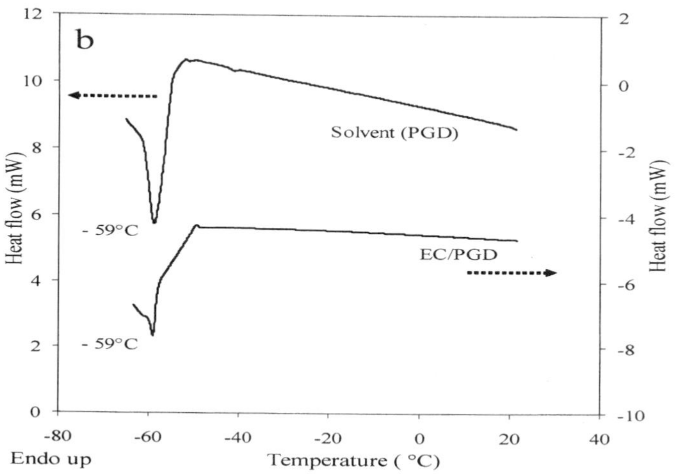

Figure 4 *(a) Micro DSC thermograms taken at a scan rate of 1°C/min for the 12% EC/PGD gel, and b) Modulated DSC thermograms taken at a scan rate of 5°C/min for the 12% EC/PGD gel and the PGD solvent.*

3 CONCLUSION

From the experimental observations so far, it can be concluded that solvent diffusion and consequent particle swelling is followed by particle erosion as a function of aging at ambient temperature or thermal processing, resulting in the release of polymeric strands that are able to form a three dimensional structure. The outcome of these two molecular mechanisms is a composite mixture of individual ethyl cellulose species supporting particle remnants, which is based on a balance of topological particle disentanglements and polymeric entanglements in solution.

References

1. Agrawal, A.M.; Manek, R.V.; Kolling, W.M.; Neau, S.H. *AAPS PharmSciTech* 2003, **4**, 1.
2. Heng, P.W,S.; Chan, L.W.; Chow, K.T. *Pharmaceutical Research* 2005, **22**, 676.
3. P. Sakellariou, R.C. Rowe, *Prog. Polym. Sci.*,1995, **20** , 889.
4. J. Hjartstam, T. Hjertberg, *Int. J. Pharm.*, 1998,**161**, 23.
5. C. Remunan-Lopez, M.L. Lorenzo-Lanosa, J.L. Vila-Jato, M.J. Alonso, *Eur. J. Pharm.Biopharm.*, 1998, **45**, 49
6. Qiu, L.Y.; Bae, Y.H. *Pharmaceutical Research* 2006, **23**, 1.
7. Lizaso, E.; Munoz, M.E.; Santamaria, A. *Macromolecules* 1999, **32**, 1883.
8. Chan, L.W.; Chow, K.T.; Heng, P.W.S. *Pharmaceutical Research* 2006, **23**, 408.
9. Zhang, Z.; Zhang, H.; Zhang, Q; Zhou, Q.; Zhang, H.; Mo, Z.; Zhao, X.; Wang, X. *Journal of Applied Polymer Science* 2006, **100**, 584.
10. Van de Velde, F.; Antipova, A.S.; Rollema, H.S. et al. *Carbohydrate Research* 2005, **340**, 1113.
11. Kasapis, S.; Al-Marhoobi, I.M. *Biomacromolecules* 2005, **6**, 14.

Physically cross-linked hydrogel from carboxymethylated curdlan prepared by freeze-thaw techniques

Juan Wu, Hongbin Zhang*

Advanced Rheology Institute, Department of Polymer Science and Engineering, School of Chemistry and Chemical Technology, Shanghai Jiao Tong University, Shanghai 200240, People's Republic of China

Abstract:
This paper describes a novel hydrogel prepared from carboxymethylated curdlan (CMc) using freeze-thaw techniques. CMc aqueous solutions can form physically crosslinked gels at low pH values by freezing without any chemical crosslinking agents. The formation and mechanical properties of CMc hydrogels are controllable and dependent on concentration, pH, number of freeze-thaw cycles, temperature and time for both freezing and thawing. A 1wt% CMc aqueous solution can form a gel at a pH below 2.40 by one freeze-thaw cycle, and after five freeze-thaw cycles, the resulting gel can persist even when the heating temperature keeps at 70°C for 20h. The equilibrium swelling ratios of CMc gels decrease with increasing the concentration or the number of freeze-thaw cycles, indicating that high concentration or the cycle procedure leads to significant densification of the gel structures. A preliminary understanding of the gelation is proposed based on FTIR spectra. The gelation of CMc may be attributed to the association of hydrogen bonds among carboxylic and hydroxyl groups on the backbone of the CMc chains as junction sites of the network structures. This novel physical hydrogel consisting only of CMc has potential for biomedical and drug delivery application due to its inherently desirable biocompatibility and bioactivity.

1. Introduction

Hydrogel is always defined as a three-dimensional macromolecular network, which can imbibe and retain large amounts of water or biological fluids[1]. Due to its swollen network, it has a widespread application in different technological areas, e.g. as materials for contact lenses and devices for drug delivery. Hydrogel can be prepared by chemical and physical methods. For chemically crosslinked hydrogel, it may have some disadvantages, for example, the existence of organic crosslinking reagents and solvents that are usually toxic, and generally it is hard to remove traces of these agents completely[2]. In addition, industrially the purification process increases the cost of the products. However, physical crosslinking without any chemical agents can avoid these disadvantages. In recent years, there has been increasing interest in physically crosslinked hydrogels. A physical hydrogel can be prepared by crosslinking through ionic interactions, crystallization, hydrogen bonds and so on[3].

Curdlan, a naturally occurring linear polysaccharide composed of 1, 3-beta-D-glucan,

*Corresponding author. Tel.: +86 21 54745005; fax: +86 21 54741297.
E-mail address: hbzhang@sjtu.edu.cn (H. Zhang).

can form physical hydrogels by various methods, some of which are heating a curdlan aqueous suspension, neutralizing or dialyzing an alkaline solution of curdlan against water[4,5]. But curdlan is insoluble in water, which limits its biological applications. Carboxymethyl substitution is considered as an effective method to improve the functional properties for curdlan. After carboxymethylation, curdlan has good water solubility as well as good bioactivity, and it has been added successfully to cosmetics due to its immunocompetence[6]. Recently, we found that, although carboxymethylated curdlan (CMc) can not form hydrogels by the above mentioned methods, it can form hydrogels by using another easy and convenient method, i.e., freeze-thaw technique. This method is similar to that applied to poly(vinyl alcohol) (PVA), which forms a tough and swollen gel after several freeze-thaw cycles[7-9]. This physical CMc hydrogel may have potential for biomedical and drug delivery applications due to its desirable biocompatibility and bioactivity.

In this paper, we report that CMc forms a physical hydrogel by a freeze-thaw procedure. The formation conditions and characteristics of hydrogels were investigated and the gelation mechanism of CMc was also discussed.

2. Materials and Methods

2.1 Materials

Curdlan (Mw 1.96×10^6) was a gift from Takeda-Kirin Foods Corporation (Tokyo, Japan) and used without further purification. All chemical reagents used are analytical grade.

2.2 Synthesis of carboxymethylated curdlan (CMc)

Curdlan was carboxymethylated by the method described by Sasaki[11]. The procedure is briefly as follows: a suspension of 3.0 g of curdlan in 80 mL 2-propanol was stirred for 30 min at room temperature. Then, 8.0 mL of a 30 wt% sodium hydroxide solution was added dropwise to the suspension and vigorous stirring was continued at room temperature for 90 min. Next, 3.0 g of monochloroacetic acid (MCA) was added over three separate intervals of 10 min each and the mixture was stirred at 55°C for 4h. The suspended product was recovered by filtration and washed successively with methanol/water (4:1 v/v), methanol and acetone, respectively. The precipitate was dissolved in water and dialyzed against distilled water at 4°C for 4 days. The retentate was frozen and lyophilized to give the final product. The degree of substitution (DS) of CMc was 0.83 units of carboxymethyl group per one glucose unit in curdlan, determined by conductometric back titration in accordance with a previous report[12]. The products obtained were characterized by FTIR (Paragon 1000, Perkin Elmer, Inc., USA) and 1H NMR (Mercuryplus 400 MHz, Varian, Inc., USA) as described in our previous work[6] to confirm the successful carboxymethylation.

2.3 CMc hydrogel preparation and characterization

CMc aqueous solutions at different concentrations (0.5wt%, 1wt%, 3wt%) were prepared using distilled water and adjusted to different pH values with 1M HCl. Typically freeze-thaw procedures are as follows. The samples were frozen at -18°C for 48h, and allowed to thaw at 25°C for 2h. This is the first freeze-thaw cycle, but the following cycles are -18°C for 24h, thawing at 25°C for 2h. One, three and five such freeze-thaw cycles were performed for each sample. In addition, to understand the gelation of CMc, 2M urea, a destructor of hydrogen bonds, was added into the initial

CMc solutions, then the solutions were subjected to the same treatments.
To measure the swelling properties of the obtained hydrogels, the gels were immersed in phosphate buffer-physiological salt (PBS, pH=7.4) solutions. The equilibrium swelling ratio was calculated via $(W_s-W_d)/W_d$, where W_s is the weight of the sample after equilibrium swelling and W_d is the weight of the dried sample. In order to probe into the gelation mechanism of CMc further, the structure features of native CMc powders and CMc dried gel were also investigated by FTIR.

3. Results and discussion

We prepared physical hydrogels of CMc for the first time, which can only form in certain conditions, e.g. at low pH values via freeze-thaw treatments. The appearance of hydrogels obtained can range from weak and soft to tough and rubbery with increasing the concentration and the number of freeze-thaw cycles, or decreasing the freezing temperature and pH as shown in Fig.1 and Fig. 2. It was found that pH has a significant influence on the gelation of CMc. 1% CMc solutions at pH below 2.40 can form gels whereas solutions at pH of 2.94 cannot form gels, and the densification and subsequent stiffness of the hydrogels became strong with increasing the number of freeze-thaw cycles. Generally speaking, the formation and mechanical properties of CMc hydrogels are closely related to CMc concentration, number of freeze-thaw cycles, temperature and time for both freezing and thawing process. In addition, the syneresis of hydrogels increased with the number of freeze-thaw cycles, but it can be reduced by increasing the CMc concentration.
For hydrogels prepared from CMc solutions at different concentrations and pH values after one, three and five freeze-thaw cycles, their equilibrium swelling ratios are shown in Table 1. The equilibrium swelling ratios decreased with increasing the number of freeze-thaw cycles, indicating that these cycle procedures led to significant densification of the gel structures. It was found that high concentration samples had a lower swelling ratio than low concentration ones.

Fig. 1 The appearance of hydrogels formed from 1wt% CMc solutions at pH=1.50 and with one (a) and three (b) freeze-thaw cycles

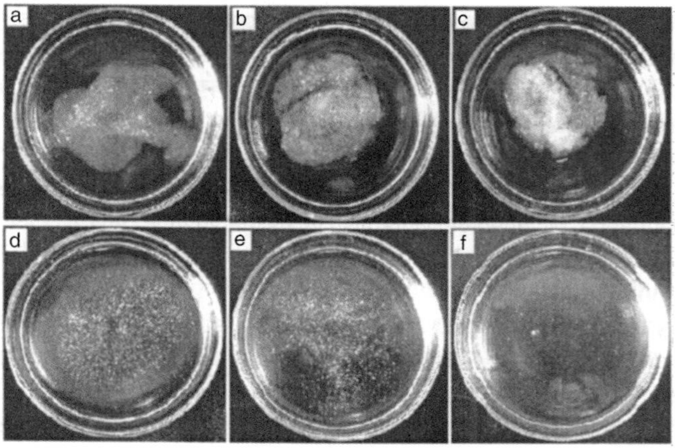

Fig. 2 Hydrogels formed from 1wt% CMc solutions at different pH and with different freeze-thaw cycles (a: 1 cycle, b: 3 cycles, c: 5 cycles at pH=1.50; d: pH=1.97, e: pH=2.40, f: pH=2.94 with 1 cycle)

Table 1 Equilibrium swelling ratios of CMc hydrogels prepared at different conditions

CMc gels C(wt%)	pH	Cycles	Equilibrant swelling ratio
0.5	1.50	5	18.01
1	1.50	5	14.74
3	1.50	5	10.48
1	0.76	5	9.88
1	2.40	5	17.39
1	1.50	1	57.30
1	1.50	3	16.55

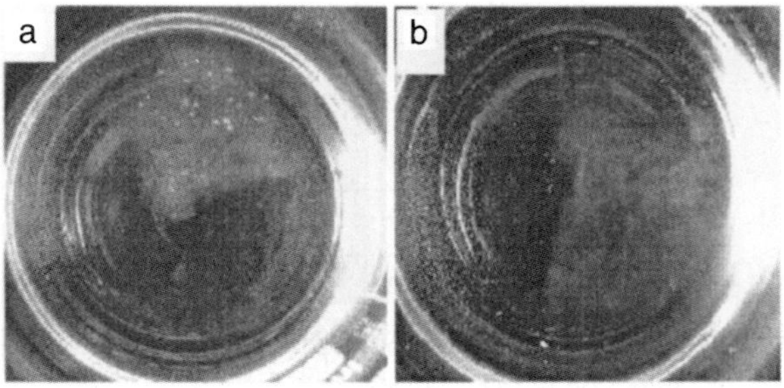

Fig. 3 1wt% CMc hydrogels immersed in PBS at room temperature (a: fresh gel; b: gel in twenty days)

Additionally, a CMc gel immersed in PBS at room temperature was not dissolved but only slowly diffused, as shown in Fig. 3 observed after twenty days. We also found that even when heated up to 70°C for twenty hours, the gel still did not melt. At about

90°C, obvious melting was observed. As for the influence of pH on the melting of the gel, even though the pH of the soaking solution was adjusted to 11.48, the gel was not dissolved. The interior structure of the hydrogel seems to be of considerable steadiness to prevent the penetration of the hydroxyl ions. However, the CMc gel was readily dissolved in 2M NaOH. In addition, we found that 1wt% CMc solution containing 2M urea could not form a gel. This fact, along with the fact that CMc gel can melt in 2M NaOH solution or by heating at elevated temperatures, strongly implies the occurrence of hydrogen bonding in CMc hydrogels. For CMc, based on its chemical structure in aqueous solution at low pH, hydrogen bonding might be the only driving force for the gelation. This supposal is confirmed further by the following FTIR measurements.

Fig. 4 shows the FTIR spectra of both native CMc powders and CMc dried gel. Quite different features between the two samples are observed though their chemical structures are identical. In the IR spectrum of CMc powders, the peaks at 1606 and 1430 cm^{-1} are assigned to asymmetrical and symmetrical C=O stretching of COO^-, respectively, while for CMc dried gel, the strong broad peak at 1738 cm^{-1} assigned to C=O stretching of COOH may indicate the occurrence of interaction between COOH groups. As for the shoulder peak at about 1640 cm^{-1}, it should be due to the in-plane deformation of the water molecule. This strongly bound water cannot be completely eliminated from the sample. This similar phenomenon in FTIR spectra can be seen for many polysaccharides, for example, native curdlan[6] and konjac glucomannan[13]. The peak at 1243 cm^{-1} for CMc dried gel, which did not appear in the spectrum of native CMc power, is assigned to C-O stretching coupled with O-H in-plane bending[14].

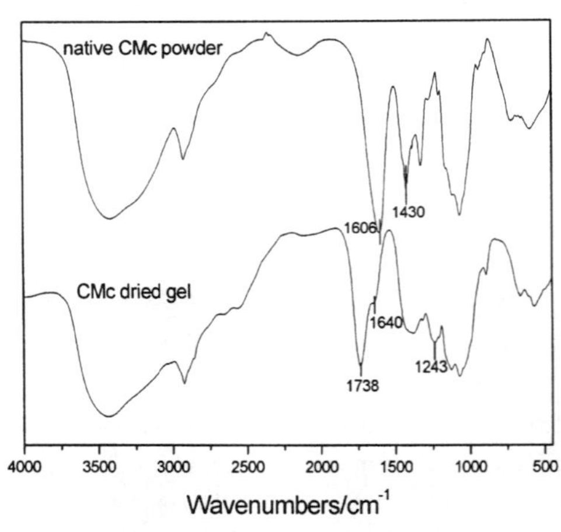

Fig. 4 FTIR spectra of native CMc power, and CMc dried gel formed from 1wt% CMc solutions at pH 1.50 and with 1 freeze-thaw cycle

The hydrogen bonds between carboxylic groups of carboxylic acids[15] or polymers like poly(acrylic acid)[14] have long been investigated. For CMc, the existence of hydrogen bonds may be in several associated forms as shown in Fig. 5. The formation of hydrogen bonding may result from interactions between carboxylic groups, between carboxylic and hydroxyl groups, and between hydroxyl groups. Taking into account that at about 3400 cm^{-1}, both the spectra are dominated by a broad band assigned to the stretching vibration modes of OH groups and that no obvious shift to low wavenumber was found in this band of CMc dried gel, the hydrogen bonding induced from OH groups in gels may be weak in comparison with those from COOH groups. Interactions between COOH groups seem to play a predominant role in the formation of hydrogels. Polysaccharides containing COOH groups seem to be of such gelling ability under low pH and freeze-thawing. It has been reported that hyaluronan is able to form such a physical gel[16]. In addition, interestingly, we also found that carboxylmethylated cellulose, gellan gum and xanthan gum that have carboxylic groups can also form hydrogels at low pH by freezing.

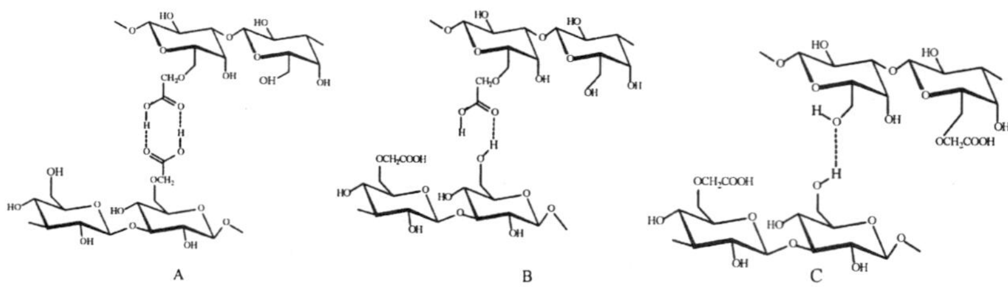

Fig. 5 Hydrogen bonding between COOH and OH groups in CMc hydrogels

In the solution state, CMC chains form temporarily entangled network structures. The decreases of the temperature and pH of the solution gives rise to suppression of both the electrostatic repulsion between CMc chains and the movement of the macromolecules, which make CMC chains pack close to each other to facilitate the formation of hydrogen bonding, consequently to form gel network structures. As for the effect of freeze-thawing cycles, the porous structure of hydrogels is imprinted during the first freeze-thaw cycle. Upon successive cycles, the structure of CMc progressively acquires a higher stability by the further formation of hydrogen bonding.

ACKNOWLEDGEMENTS

The authors are thankful for the financial support from the Shanghai Leading Academic Discipline Project (No. B202).

Reference:

1 Coviello, T., Matricardi, P., Marianecci, C. and Alhaique, F. (2007). Polysaccharide hydrogels for modified release formulations. *Journal of Controlled Release*, 119, 5–24.

2 Janes, K.A., Calvo, P. and Alonso, M.J. (2001). Polysaccharide colloidal particles as delivery systems for Macromolecules. *Advanced Drug Delivery Reviews*, 47, 83–97.

3 Hennink, W.E., Van Nostrum, C.F. (2002). Novel crosslinking methods to design hydrogels. *Advanced Drug Delivery Reviews*, 54, 13–36.

4 Harada, T., M. Terasaki, A. Harada. (1993). Curdlan, In *Industrial Gums*, 3rd Edn, Whistler, R. L., BeMiller, J. N., Eds., Academic Press, New York, pp. 427-445.

5 Nishinari, K. Zhang, H. and Funami, T. (2009). Curdlan, In *Handbook of Hydrocolloids*, G. O Phillips, P. A. Williams, Eds., Woodhead Publishing Ltd., England, Chapter 20, pp. 567-591.

6 Jin, Y., Zhang, H., Yin, Y. and Nishinari, K. (2006). Comparison of curdlan and its carboxymethylated derivative by means of Rheology, DSC, and AFM. *Carbohydrate Research*, 341, 90–99.

7 Ricciardi, R., Auriemma, F., Gaillet, C., De Rosa, C. and Laupretre, F.. (2004). Investigation of the Crystallinity of Freeze/Thaw Poly(vinyl alcohol) Hydrogels by Different Ttechniques. *Macromolecules*, 37, 9510-9516.

8 Ricciardi, R., Gaillet, C., Ducouret, G., Lafuma, F. and Laupretre, F. (2003). Investigation of the relationships between the chain organization and rheological properties of atactic poly(vinyl alcohol) hydrogels. *Polymer*, 44, 3375–3380.

9 Peppas, N.A. and Scott, J.E. (1992). Controlled release from poly(vinyl alcohol) gels prepared by freezing-thawing processes. *Journal of Controlled Release*, 18, 95-100.

10 Ricciardi, R., D'Errico, G., Auriemma, F., Ducouret, G., Tedeschi, A.M., De Rosa, C., Laupretre, F. and Lafuma, F. (2005). Short Time Dynamics of Solvent Molecules and Supramolecular Organization of Poly (vinyl alcohol) Hydrogels Obtained by Freeze/Thaw Techniques. *Macromolecules*, 38, 6629-6639.

11 Saasaki, T., Abiko, N., Sugino, Y. and Nitta, K. (1978). Dependent on chair length of antitumor activity of (1-3)-β-D-glucan from *Alcaligenes var faecalis myxogenes* IFO 13140, and its aciddegraded productions. *Cancer Research*, 38, 379-384.

12 Eyler, R. W., Klug, E. D. and Floyd, D. (1947). Determination of Degree of Substitution of Sodium Carboxymethylcellulose. *Analytical Chemistry*, 19, 24-27.

13 Zhang, H., Nishinari, K., Williams, M., Foster, T. and Norton I. (2001). Gelation Behaviour of Konjac Glucommanan with Different Molecular Weights. Biopolymers, 59, 38-50.

14 Dong, J., Ozaki, Y. and Nakashima, K. (1997). Infrared, Raman, and Near-Infrared Spectroscopic Evidence for the Coexistence of Various Hydrogen-Bond Forms in Poly(acrylic acid). *Macromolecules*, 30, 1111-1117.

15 Marechel, Y. (1987). Infrared Spectra of Cyclic Dimers of Carboxylic Acids: the Mechanics of Hydrogen Bonds and Related Problems. In *Vibrational Spectra and Structure*, Durig, J. R., Ed., Elsevier, Amsterdam, Chapter 5.

16 Okamoto, A. and Miyoshi, T. (2002). A biocompatible gel of hyaluronan, in *Hyaluronan 2000*, J.F. Kennedy, et al., Eds., Woodhead Publishing, Cambridge, p. 285-292.

PREPARATION OF CMC-ACID GEL AND ITS APPLICATIONS

M. Takigami[1], N. Nagasawa[2], Y. Maehara[3], S. Takigami[4] and M. Tamada[2]

[1]Gunma Industry Support Organization, c/o Japan Atomic Energy Agency, Takasaki, Gunma 370-1292, Japan
[2] Japan Atomic Energy Agency, Takasaki, Gunma 370-1292, Japan
[3] Tomioka General Hospital, Tomioka, Gunma 370-2393, Japan
[4] Center for the Materials Research by Instrumental Analysis, Gunma University, Kiryu, Gunma 376-8515, Japan

1 INTRODUCTION

Carboxymethylcellulose (CMC) is one of the chemically modified carbohydrates used in the food industry as a thickener, a stabilizer and a dispersing agent. As CMC is prepared from cellulose and chloroacetic acid under concentrated alkaline conditions, carboxymethyl groups are in the form of the sodium salt. The applicable field can be extended to other fields by gelling CMC, using ionizing radiation[1,2], crosslinking agents[3,4], or multivalent metal ions. The former two provide gels crosslinked by covalent bonds and the latter by ionic bonds. The gels prepared by ionizing irradiation are transparent and absorb a considerable amount of water, and are free of contamination. However, crosslinking agents and multivalent metal ions remain in the gels prepared by the latter two methods. Therefore, these methods are not suitable for the production of foods.

We reported that CMC paste was gelled after immersion in hydrochloric acid solution[5]. The paste was prepared by mixing CMC and water at a CMC concentration of 20%. The resulting gel was elastic and seemed to be different from the gels prepared by the previous methods. Combination of ionizing radiation and immersing the CMC paste in acid gave stronger gels. In the present research, CMC gels were prepared by mixing CMC and acid solution and the properties of the gels were measured.

2 METHOD

2.1. Preparation of Gels

CMC 1380 (sodium salt, degree of substitution (DS):1.34, viscosity of 1% solution at 25°C:1640mPa·s, Daicel Chemical Industries, Ltd.) was used throughout the experiment unless otherwise described. CMCs with various DS (0.65 – 2.2) and viscosities from the same manufacturer were also used. The acids were reagent grade and used after dilution with purified water.

CMC was mixed with acid solution using a mixer (Hybrid defoaming mixer ARE-250, Thinky Co., Japan). By mixing, CMC turned into a soft and elastic gel as shown in Figure 1. The mixture was moulded into a cylinder with a diameter of 20 mm and thickness of 10 mm

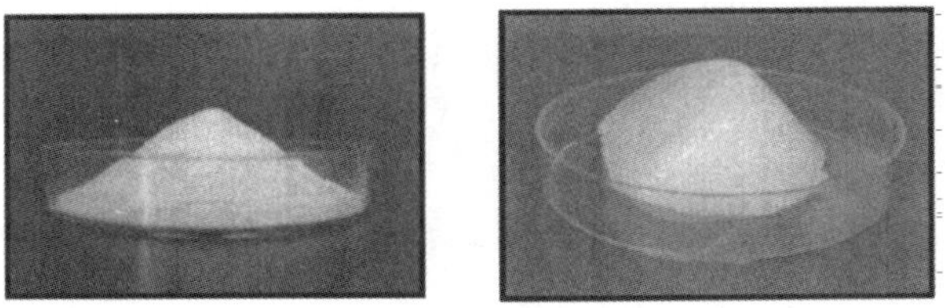

(a) (b)

Figure 1 *CMC powder (a) turned into a gel (b) when CMC (degree of substitution : 1.34) was mixed with 1M phosphoric acid in the ratio of CMC/phosphoric acid = 20/80 (wt/wt).*

or sheet with a thickness of 1 mm. The samples were kept at 30°C avoiding weight loss during storage. The cylinders and sheet were subjected to compression and elongation tests, respectively. The samples were immersed in water to remove soluble materials. Water was exchanged several times until the conductivity of the washing water became nearly equal to that of the pure water. Gel was defined as an insoluble fraction in water and remained on 180 μm stainless steel mesh when the wet sample was passed through the mesh. The remaining fraction on the mesh was weighed and dried at 50°C. The gel fraction and water absorption by the gel were calculated according to the following equations.

$$\text{Gel fraction (\%)} = 100 \times W_{dg} / W_i$$

$$\text{Water absorption by gel (g-water/g-gel)} = (W_{wg} - W_{dg}) / W_{dg}$$

where, W_{dg} , W_i and W_{wg} are dry weight of the remaining fraction on the mesh, initial weight of CMC and wet weight of the remaining fraction on the mesh, respectively.

3 RESULTS AND DISCUSSION

3.1. Gel Fraction and Water Absorption

In Figure 2, the gel fraction is plotted against initial pH value of acids mixed with CMC. The mixture was stored at 30°C for 7 days. CMC formed a gel when mixed with any acid, if the pH value of the acid solution was less than 2 for organic acids and less than 1 for inorganic acids. The difference between organic acids and inorganic acids is due to the difference of the dissociation of the acids. Longer storage time, higher storage temperature up to 50°C and higher concentration of CMC up to 50% allow gelation at higher pH than those mentioned above.

CMCs with various viscosities and DS were used to form gels with 3 M phosphoric acid, however, gel fractions depended on the viscosities of CMC as shown in Figure 3, rather than DS. Gel fraction increased with the increase of viscosities of CMC due to increased molar mass[6].

Figure 4 shows the relation between gel fraction and water absorption by the gel. Regardless of acid species, the gel with lower gel fraction absorbed more water. The gel with higher gel fraction absorbed only a small amount of water. The tendency is also observed in gels obtained by γ-irradiation[1,2]. The progress of gelation reaction induces crosslinking between molecules. Water absorbency relates to the crosslinkng density.

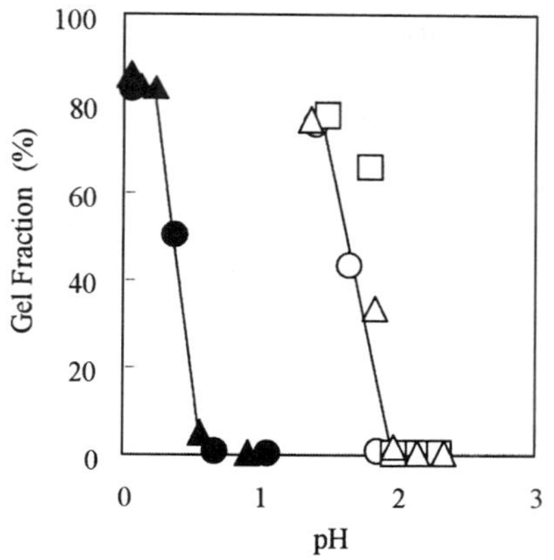

Figure 2 *Relation between initial pH of acid solution and gel fraction. Symbols: citric acid (○), lactic acid(□), formic acid (△), hydrochloric acid (●) and sulfuric acid (▲).*

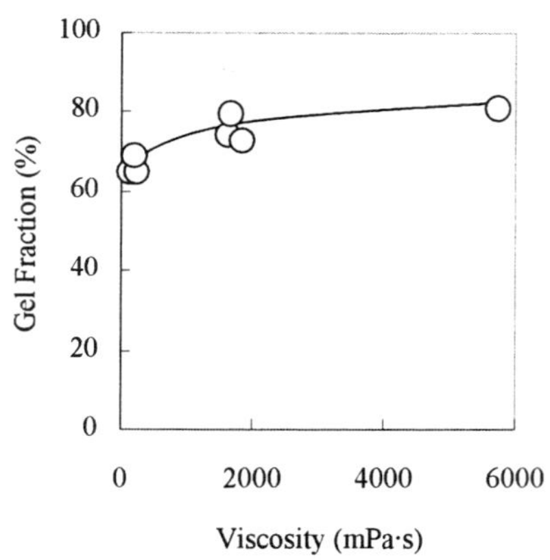

Figure 3 *Relation between viscosity of CMC and gel fraction.*

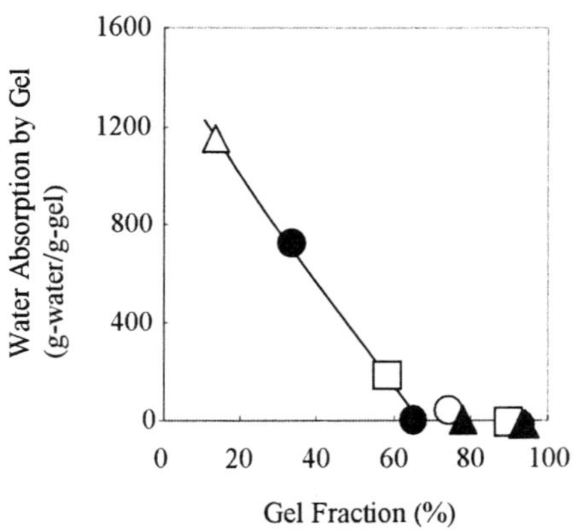

Figure 4 *Relation between gel fraction and water absorption by gel. Notations of the symbols are the same with those shown in Figure 2.*

In radiation induced gels, gels can absorb water even after drying, however, gels made by mixing CMC and acid can not absorb much water once they were dried. It infers a different mechanism of gel formation with that of radiation induced gels.

3.2. Mechanism of Gel Formation

Gels were immersed in water to remove water soluble materials including uncrosslinked CMC and acid in the same way mentioned in the section of preparation of gels.

FT-IR spectra of the dried gels were measured by Magna 560 + Continuμ MII (Nicolet, Thermo Fisher Scientific K. K.).

As shown in Figure 5, FT-IR spectra of dried CMC gel were the same in gels with higher gel fraction regardless of the acid species. Original CMC has a strong and broad absorption around 1600 cm^{-1}, whereas, CMC-acid gels have a strong absorption around 1730cm^{-1}. The absorption at 1730 cm^{-1} is especially strong in gels made from inorganic acids, however, that in gels made from organic acid is not very strong. The absorbance at 1600 cm^{-1} still remains in gels made from organic acids. The absorptions at 1600 and 1730 cm^{-1} were related to associated carboxyl groups and free carboxyl groups, respectively; sodium as a counter ion of carboxyl group was replaced with hydrogen.

The amount of sodium contained in the gel was evaluated by inductivity coupled plasma atomic emission spectrometry (ICP, Optima 4300DV, Perkin Elmer) after degrading the gel with nitric acid (d=1.38, for analysis of poisonous metal, Wako Pure Chemical Industries, Ltd., Japan) using a microwave oven. The amount of sodium contained in CMC was 11%, whereas those in gels made by mixing CMC with 1M hydrochloric acid and 1M citric acid were 1.6% and 2.3%, respectively. Sodium content in gels was much less than that in CMC.

Elements contained in acids such as phosphorous in phosphoric acid and sulfur in sulfuric acid were analyzed for gels formed using such acids, however, they were not detected in the gels. Any gel did not show any trace of such elements.

Radiation-induced CMC gel absorbs much water after drying, however, once the CMC-acid gel is dried, it absorbs a very small amount of water. CMC-acid gel is transparent when it is wet, however, once it is dried, it becomes turbid. These phenomena suggest that the gel is not formed by covalent bond like radiation-induced gels.

The results mentioned above suggest the mechanism of gel formation as shown in Figure 6. First, sodium ions existing in CMC as a counter ion of carboxyl group are replaced by hydrogen, by mixing CMC with acid. As a result, solubility of CMC in the form of an acid is reduced compared to that of CMC in sodium salt. CMC molecules locate closer, aggregate, and hydrogen bonds are formed between CMC molecules. Thus CMC gel is formed. Therefore, acid species do not give any effect on gel formation.

3.3. Characterization of CMC-Acid Gel

TG-DTA measurements of dried gels were carried out for the gels using EXSTRA6000 Series TG/DTA6200 (Seiko Instrument Inc., Japan) from 30 to 380°C. The gels were prepared with CMC and 0.5 M sulfuric acid and kept at 30°C for 1, 2, and 7 days. As shown in Figure 6, CMC showed a sharp exothermic peak at 289°C, whereas, gels showed broad exothermic peaks with onset temperature around 250°C. The peak temperature became higher with storage time; in gel stored for 7 days it was 310°C. The results infer that the gel became more rigid with time, in other words, hydrogen bond formation proceeds with time.

X-ray diffraction diagrams of the dried gels were taken with an X-ray diffractometer (RINT 2200VF/PC Series, Rigaku, Japan) from $2\theta = 5 - 45^{\circ}$. Crystallinity increased by

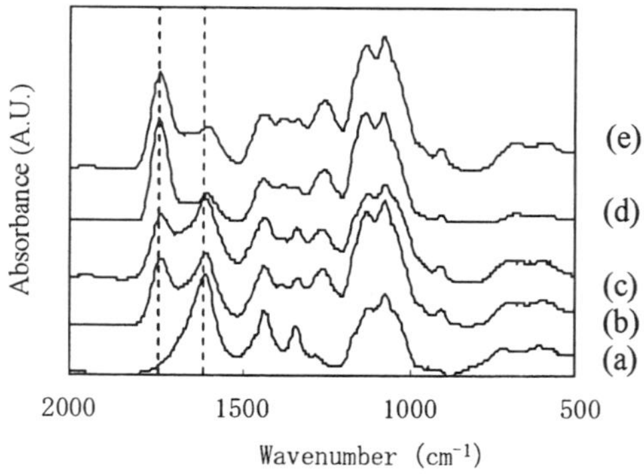

Figure 5 *FT-IR spectra of CMC and CMC-acid gels. CMC (a), CMC + 0.5M citric acid (b), CMC + 0.5M formic acid (c), CMC + 0.5M sulfuric acid (d), and CMC + 0.5M hydrochloric acid (e).*

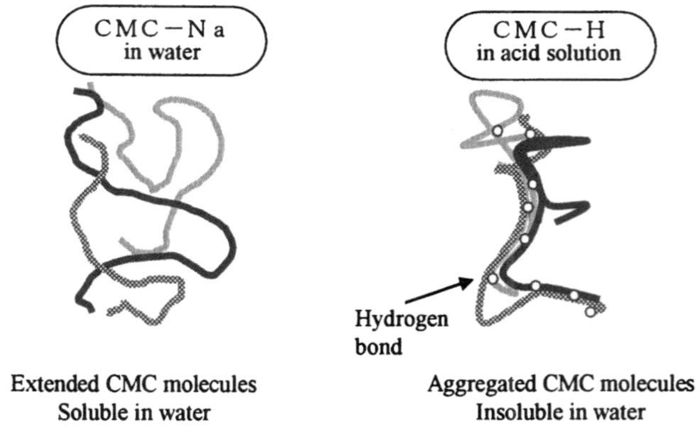

Figure 6 *Schematic model of CMC-acid gel formation.*

gelation depending on the strength of acid as shown in Figure 7. When the same kind of acid with different concentrations was used for gelation, crystallinity of the gel increased with acid concentration. The stronger the acidity, the higher the crystallinity.

For compression tests, a cylinder with a diameter of 20 mm and thickness of 10 mm was made with a mixture of CMC and sulfuric acid and kept at 30°C for 7 days. The cylinder was compressed with a compact table-top universal tester (EZTest-L, Shimadzu Corp., Japan) and compression modulus at 1 N to 2 N was measured. With the increase of acid concentration, it increased as shown in Figure 8. The cylinders did not break even if they were compressed to half thickness and recovered the initial thickness after removing load. Some of the cylinders were very strong and recovered the initial thickness after being compressed to 2 mm as shown in Figure 9. For elongation tests, a sheet with a thickness of 1 mm was prepared. Some of the sheet elongated to 4 - 5 times of the initial length.

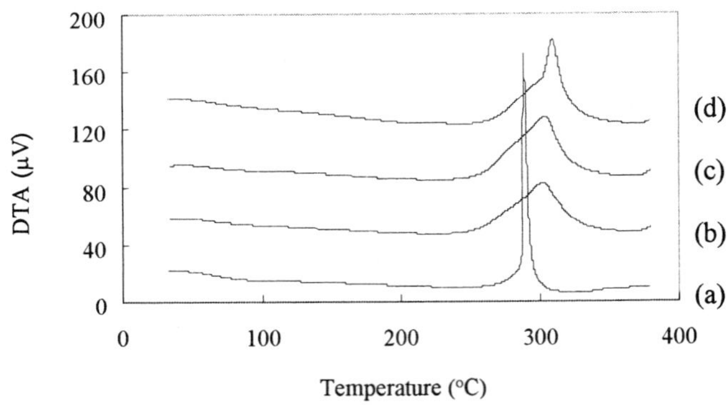

Figure 6 *DTA of CMC and gels. CMC (a) and gels made from CMC and 0.5M sulfuric acid. They were kept at 30°C for 1 day (b), 2 days (c) and 7 days(d).*

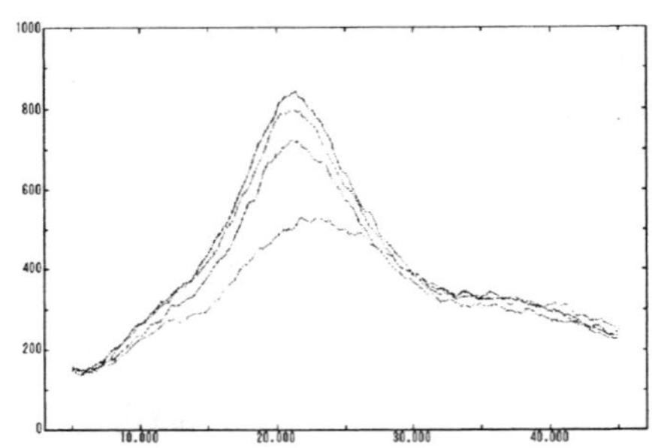

Figure 7 *X-ray diffraction diagrams of CMC and CMC-acid gels. From the bottom: CMC, gel made from CMC and 1M formic acid, gel made from CMC and 1M citric acid, gel made from CMC and1M hydrochloric acid.*

Mixing with other compounds such as pectin, microcrystalline cellulose, alginic acid, chitosan, polylactic acid, titanium oxide or aluminum oxide, CMC-acid gels can be formed. The gels are sometimes stronger than the CMC-acid gel without another component. For example, the breaking strength of CMC-acid gel containing poly(lactic acid) was more than twice of that of CMC-acid gel without containing another component. The additives play the role of fillers in composite materials.

CMC-acid gels with various properties can be prepared by changing the following factors.

(a) CMC: Degree of substitution, concentration and viscosity.

(b) Acid: Species and concentration.

(c) Gelation: Temperature and storage time.

(d) Removing acid after gelation.

(e) Adding other compounds.

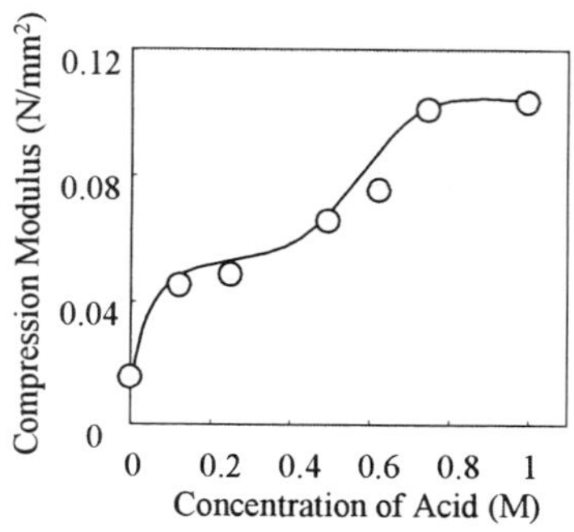

Figure 8 *Compression modulus of CMC-citric acid gel.*

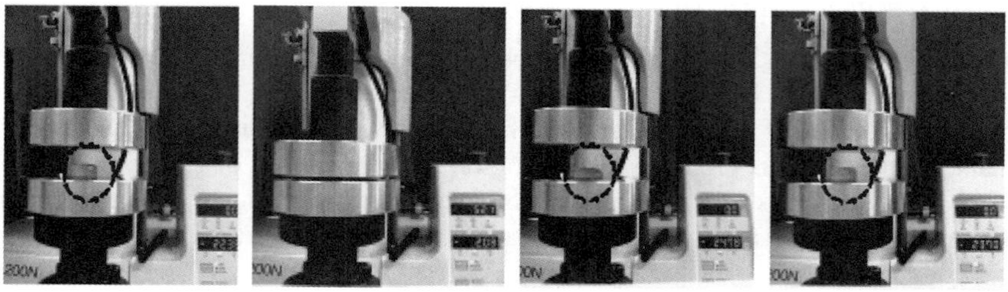

Figure 9 *Compression test of gel. Before compression (a), compressed to 20% of the initial thickness (b), immediately after removing load (c), and after 3 hours (d).*

3.4. Application of CMC-Acid Gel

Utilizing the properties of CMC-acid gel, various attempts have been made to develop new application areas, such as carriers of fragrance, deodorant agents, and anti-fungal agents. Radiation induced CMC gel does not adhere to human skin, however, CMC-acid gel after removing acid from the gel can adhere to human skin. So that wound dressings and cosmetic face masks may be an area of application. In leisure activities, lure and binder of feed in fishing might be promising. As CMC forms gel with edible acid, CMC-acid gels do not contain anything toxic and can be used in food.

As CMC-acid gel is soft and has similar touch with human body, it is suitable for human-body model. As an application, a gel was developed to improve clearness of magnetic resonance imaging (MRI). MRI sometimes gives artifacts caused by motion of the patient or unevenness of magnetic field. The artifact can be reduced putting CMC-acid gel around the neck as shown in Figure 10. In T1 weighted image, signal owing to artifact (bright part in circle) was not suppressed completely without using CMC-acid gel (left image). Once CMC-acid gel was placed around the neck, the signal owing to the artifact was reduced (right image) and clear MRI was provided.

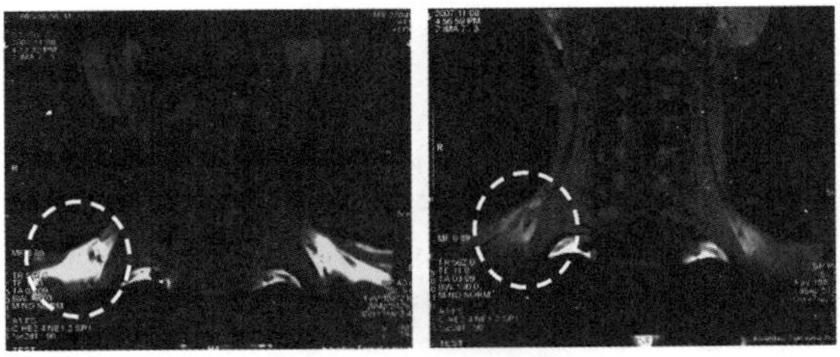

(a) (b)

Figure 10 *Magnetic resonance imaging of human. Without using CMC-acid gel (a) and using the gel (b).*

4 CONCLUSION

CMC forms gels when mixed with acid solution. The mechanism of gel formation was elucidated by hydrogen bond formation between CMC molecules, which is caused by the decrease in solubility of CMC in water by the replacement of sodium with hydrogen. The resultant gels had various properties depending on the combination of CMC and acid. Both soft and hard gels, weak and strong gels can be obtained by changing the combination of CMC and acid. Utilizing the properties, applications in many fields are in progress.

Acknowledgement

The authors are grateful for the sponsorship of Japan Science and Technology Agency to achieve Gunma Prefecture Collaboration of Regional Entities for the Advancement of Technological Excellence.

References

1 B. Fei, R.A. Wach, H. Mitomo, F. Yoshii and T. Kume, J. Appl. Polym. Sci., 2000, **78**, 278.
2 R.A. Wach, H. Mitomo, F. Yoshii and T. Kume, J. Appl. Polym. Sci., 2001, **81**, 3030.
3 A. Sannino, S. Pappda, M. Mafaghiele, A. Maffezzoli, L. Ambrosio, and L. Nicolais, Polym., 2005, **46**, 11206.
4 A. Sannino, M. Madaghiele, F. Conversano, G Mele, A. Maffezzoli, P.A. Netti, L. Ambrosio, and L. Nicolais, Biomacromol., **5**, 2004, 92.
5 M. Takigami, H, Amada, N. Nagasawa, T. Yagi, T. Kasahara, S. Takigami, and M. Tamada, Trans. Mater. Res. Soc. Jap., 2007, **32**, 713.
6 M. Takigami, A. Hiroki, N. Nagasawa, T. Kasahara, S. Takigami, and M. Tamada, Mat. Res. Soc. Jpn., 2009, **34**, accepted for publication.

Subject Index